CW00592053

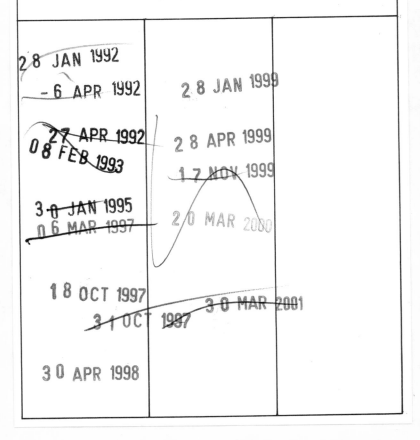

INTRODUCTION TO SEMICONDUCTOR TECHNOLOGY

INTRODUCTION TO SEMICONDUCTOR TECHNOLOGY:

GaAs and Related Compounds 1990

Edited by

Cheng T. Wang

Device Research Institute

WILEY

A Wiley-Interscience Publication

JOHN WILEY & SONS

New York • Chichester • Brisbane • Toronto • Singapore

Library of Congress Cataloging in Publication Data:

Introduction to semiconductor technology: GaAs and related compounds
/ edited by Cheng T. Wang.
    p.   cm.
"A Wiley-Interscience publication."
Bibliography: p.
Includes index.
    1. Gallium arsenide semiconductors. I. Wang, Cheng T.
TK7871.15.G3I58 1989
621.3815'2--dc19
ISBN 0-471-63119-1   89-30939
                                       CIP

Printed in the United States of America

10 9 8 7 6 5 4 3 2 1

*To Nancy
and
To the memory of
My Mother Liu Chi*

CONTENTS

PREFACE

Most undergraduates who are interested in solid-state electronics probably know gallium arsenide (GaAs) only as a material that has a direct bandgap suitable for optical applications and that has a negative differential resistance property resulting from intervalley transfer at high field. Considering the current trend in undergraduate education, which is to do away with tedious derivation of band structure and introduce students only to the concepts of silicon devices, it is very likely that in the near future new graduates will not even have these basic concepts, let alone a knowledge of what gallium arsenide is really used for. While GaAs technology is gradually being eradicated from undergraduate teaching, federal government funding on research on the other hand has been shifted almost entirely from silicon to GaAs and related compounds for their potential payoffs in defense as well as other high-speed digital and high-performance analog applications. As a result, many academics who previously worked in the area of silicon are being forced to reorient their priorities and are trying very hard to jump on the bandwagon. Unfortunately, what they find out is that there is no single book that can adequately describe this state-of-the-art technology to the extent that it can be used for self-education or for reeducating their graduate students. This state of affairs also exists in industry; one expert complained that graduates he hired did not know much about the basics of device operation especially in the area of heterostructure devices, and he had to create internal short courses in order to educate those people for their jobs.

These phenomena come as no surprise to me, for as a theoretician and an educator, I, along with many others, had long recognized the potentials of this technology and the need for a good textbook for our students. However, it did surprise me to learn after receiving manuscripts from industrial authors, just how fast the technology has come of age in terms of applications, especially of heterostructure devices. The theory or modeling efforts, on the other hand, seemed to be still in an unsettled stage. As Professor Michael Shur has stated, a better understanding of device physics is needed to help the industry mature. And the purpose of this book was and still is to help educate professionals, graduate students, and possibly advanced senior undergraduates, to motivate them to enter this exciting field: in essence, to help the industry mature.

The idea for this book came to me during the winter of 1982, while I was at the University of Miami, Florida, finishing a proposal for a National

Science Foundation grant. The proposal later became a paper published in *Solid-State Electronics*.[†] At the time, I was trying to generalize the formulas more realistically to GaAs devices, yet found no book that could give me clear ideas of GaAs devices and applications, especially about heterostructure devices, so I could apply the simulation. A chance to realize a book that furnishes academia as well as industry with a healthy dose of know-how and theoretical background did not actually materialize until the spring of 1987, when Wiley-Interscience agreed to publish this book, which at first glance might look like an encyclopedia on GaAs technology. Because of the diversity and complexity of this technology, I was forced to use an approach in which industrial experts contributed practical know-how and design experiences while academics wrote on theory and modeling. It took four months to gather the authors in the winter of 1986, and then nearly three years to complete the book. As the editor, I thank all the contributors who found precious time in their busy schedules and, through much admirable efforts, made this book possible.

The book is all about GaAs and related compounds technology. It contains adequate material for a one- or two-semester graduate course on compound semiconductor technology. Professors can freely choose subjects of interest in organizing the course contents. The adequate background for the graduate courses will be knowledge of solid-state electronics at a senior level and knowledge of quantum mechanics at a first-year graduate level. Some knowledge of microwave engineering in terms of impedance matching and transformation, as well as waveguides, will be helpful in reviewing some of the chapters in this book.

The book starts with an introduction to the physics of GaAs and related compounds by Dr. Harold Grubin of Scientific Research Associates, with particular emphasis placed on transport theory and its consequences for device applications. Professor Michael Shur of the University of Virginia follows with device models for GaAs MESFETs and heterostructure field effect transistors (HFETs), focusing in particular on useful models for circuit simulation. Drs. S. S. Pei and Nitin J. Shah of AT&T Bell Laboratories then provide a comprehensive overview of the development and design considerations of HFETs; the chapter presents one of the central themes of the book, which is heterostructure devices and applications. The choice of device parameters and its practical effects on device performance are succinctly described here and should be of considerable help to students and professionals alike. Drs. Peter M. Asbeck, M. F. Chang, K. C. Wang, and D. L. Miller of Rockwell International present a self-contained introduction to heterojunction bipolar transistors (HBTs), starting from models, through fabrication, to practical design and circuit applications. It provides a great deal of insights and the reading is a must for those interested in the

[†]C. T. Wang, "A New Set of Semiconductor Equations for Computer Simulation of Submicron Devices," *Solid-State Electron.* **28**, 783–788 (Aug. 1985).

differences between silicon and GaAs devices, as well as differences between HFETs and HBTs. Like silicon technology, field effect and bipolar technologies have their respective advantages and disadvantages, and Dr. Asbeck and his colleagues were able to give some of their critical assessments—even though the debate will continue just the same as in silicon technology. The utilization of quantum effects in device design, which otherwise perhaps absent in silicon technology at the present time, is covered by Drs. Susanta Sen and Federico Capasso of AT&T Bell Laboratories in their chapter on resonant tunneling diodes and transistors. Quantum-based transistors have long been anticipated by theoreticians. These authors and their colleagues at AT&T Bell Laboratories have been the forerunners in this field—pushing the ideas into reality. Resonant tunneling bipolar transistors (RTBTs), which constitute a very new technology, were not demonstrated until late 1986. The technology and its many derivatives are discussed in the chapter, along with potential applications in multiple valued logic. And perhaps one of the most interesting subjects discussed is the quantum wire transistor, where size quantization effects down to only one dimension may further improve the device performance.[†] For a conventional approach, Mr. William Geideman of McDonnell Douglas presents a comprehensive review of junction field effect transistors (JFETs) and their applications. Practical design issues, such as yield, temperature dependence of device parameters, and radiation hardness, are covered here, along with an introduction to several chip designs in McDonnell Douglas's quest for a 200 MHz 32-bit GaAs microprocessor. Not to be outdone, Dr. Christopher T. M. Chang of Texas Instruments presents a brief description of TI's efforts in digital IC design and provides an introduction to different GaAs logic families in terms of MESFET technology. This chapter and part of Dr. Asbeck's chapter on digital IC design considerations should be reviewed together for a better insight into practical design issues. For analog or microwave amplifier design, Mr. Thomas Apel of Teledyne Monolithic Microwave presents step-by-step circuit design procedure in a framework of impedance matching and transformation and gives an example for a manual design of a two-stage power amplifier. Passive elements realization in microwave monolithic ICs (MMICs) is also discussed. It is a very useful guide and provides considerable insights into manual design procedure. To increase the bandwidth or the throughput of the circuits further, optoelectronic ICs (OEICs) will probably have to be used. Professor Paul Yu of the University of California at San Diego and Dr. P. C. Chen of Ortel provide an excellent and comprehensive review of the practical design issues of optoelectronic devices such as lasers, LEDs, and detectors used in optical fiber communications and OEICs. Professor

---

[†]M. Okada, T. Ohshima, M. Matsuda, N. Yokoyama, and A. Shibatomi, "Quasi-one-dimensional Channel GaAs/AlGaAs Modulation Doped FET using Corrugated Gate Structure," *Extended Abstracts of the 20th* (1988 *International*) *Conference on Solid State Devices and Materials*, Tokyo, Japan Society of Applied Physics, 1988, pp. 503–506.

William Chang then gives his critical assessment of the possibility of using heterostructure and/or quantum well structure for guided wave devices. Even though many problems still exist, the field of OEICs, in particular the area of optical computing, is expected to continue to grow in order for us to meet certain challenges, especially those existing in defense applications. Last but not least, Dr. Michael Stroscio of Duke University offers a concise review of quantum transport theory. This field is not that old, and students are encouraged to work on related research topics, such as quantum noise theory, power limitation, and circuit representation of quantum-based transistors, so we can better devise and characterize future quantum-based devices and circuits.

CHENG T. WANG

Torrance, California
November 1989

ACKNOWLEDGMENTS

I would like to express my sincere thanks to the contributing authors whom I persuaded to write on one simple theme: that their admirable efforts will help motivate students into III–V compound device research and will help professionals understand more about the potential of this technology. I believe they have tried their best to achieve that goal. I would also like to thank the many reviewers who provided useful insights and suggestions on the presentation of the materials: Professors Karl Hess, University of Illinois, and R. Main, University of Michigan; Drs. D. A. B. Miller, AT&T Bell Labs; Michael Kim, TRW; P. C. Chao, General Electric; Yetzen Liu, Fermionics; Bruce Paine, Hughes Aircraft; Emilio Mendez, IBM; and Naoki Yokoyama, Fujitsu Labs. Next I would like to thank George Telecki of Wiley-Interscience and Professor S. K. Ghandhi of Rensselaer Polytechnic Institute, both of whom realized the importance of the technology and helped bring about this book. I am also very much indebted to Rose Ann Campise of Wiley for her admirable assistance in the production stage of this book. Finally, I would like to thank my family, my brother Kems-Gwor Wang and my father Yu-chu Wang, who made this task possible in the past three years.

LIST OF ABBREVIATIONS
AND ACRONYMS

APD	Avalanche PhotoDiodes
BH	Buried-Heterostructure
BFL	Buffered FET Logic
CBE	Chemical Beam Epitaxy
CDFL	Capacitor-Diode FET Logic
CML	Current Mode Logic
CMOS	Complementary MOS
CPU	Central Processing Unit
COD	Catastrophic Optical Damage
CSP	Channel Substrate Planar
DBQW	Double-Barrier Quantum Well
DCFL	Direct-Coupled FET Logic
DDE	Drift Diffusion Equation
DH-FET	Double Heterostructure FET
D-HFET	Depletion-mode Heterostructure FET
DFB	Distributed FeedBack
DBR	Distributed Bragg Reflection
DMT	Doped-channel, MIS-like FET
EA	Electro-Absorption
EBIC	Electron Beam Induced Current
EE	Edge-Emitting
E-JFET	Enhancement-mode JFET
E-HFET	Enhancement-mode HFET
E-MESFET	Enhancement-mode MESFET
ER	Electro-Refraction
ECL	Emitter-Coupled Logic
FET	Field Effect Transistors
FWHM	Full Width at Half Maximum
GRIN	GRaded-INdex
GRIN-SCH	GRIN- waveguide Separate Confinement Heterostructure
HBT	Heterojunction Bipolar Transistor
HEMT	High Electron Mobility Tansistor
HFET	Heterostructure Field Effect Transistor
HIGFET	Heterostructure Insulated-Gate FET
IGFET	Insulated-Gate Field Effect Transistor

IHEMT	Inverted-structure HEMT
I^2L	Integrated Injection Logic
JFET	Junction Field Effect Transistor
LEC	Liquid-Encaspulated Czochrolski
LED	Light Emitting Diode
LHP	Left Half Plane
LET	Linear Energy Transfer
LOC	Large Optical Cavity
LP	Low-Pass
LPE	Liquid Phase Epitaxy
LSI	Large Scale Integration
MBE	Molecular Beam Epitaxy
MBTE	Moment Boltzmann Transport Equation
MAG	Maximum Available power Gain
MESFET	MEtal-Semiconductor FET
MIMIC	MIllimeter and Microwave Monolithic Integrated Circuits
MISFET	Metal-Insulator-Semiconductor FET
MMIC	Microwave Monolithic Integrated Circuits
MOCVD	Metal-Organic Chemical Vapor Deposition
MODFET	Modulation-Doped heterostructure FET
MOS	Metal-Oxide-Semiconductor
MOSFET	Metal-Oxide-Semiconductor FET
MOMBE	Organo-Metallic MBE (Metal-Organic MBE)
MSG	Maximum Stable power Gain
MSI	Medium Scale Integration
MQW	Multiple Quantum Well
NDR	Negative Differential Resistance
NMOS	N-channel MOS
OEIC	Opto-Electronic Integrated Circuits
OMVPE	Organo-Metallic Vapor Phase Epitaxy
OPFET	OPtical FET
PBT	Permeable Base Transistor
PIN	p-i-n structure
QCSE	Quantum Confined Stark Effect
QHP	Quasi High-Pass
QLP	Quasi-Low-Pass
QW	Quantum Well
RAM	Random Access Memory
RHEED	Reflection High-Energy Electron Diffraction
RHET	Resonant Tunneling Hot Electron Transistor
RHP	Right Half Plane
RIE	Reactive Ion Etching
RIN	Relative Intensity Noise
RISC	Reduced Instruction SET Computer
RT	Resonant Tunneling

RTBT	Resonant Tunneling Bipolar Transistor
RT-FET	Resonant Tunneling FET
SAINT	Self-Aligned Implantation for N^+ Technology
SDFL	Schottky-Diode FET Logic
SDHT	Selectively-Doped Heterojunction Transistor
SEED	Self-Electro-optic-Effect Device
SE-LED	Surface Emitting LED
SEU	Single Event Upset
SLD	Super Luminescent Diode
SISFET	Semiconductor-Insulator-Semiconductor FET
SPICE	Simulation Program with Integrated Circuits Emphasis
SQW	Single Quantum Well
SRAM	Static Random Access Memory
SWAT	SideWall Assisted Transistor
SSI	Small Scale Integration
TE	Transverse Electric
TEGFET	Two dimensional Electron GAS FET
2-DEG	Two-Dimensional Electron Gas
TEM	Transmission Electron Micrograph (Microscope)
TJS	Transverse Junction Stripe
TM	Transverse Magnetic
VLSI	Very Large Scale Integration

CONTRIBUTORS

THOMAS R. APEL* AVANTEK, Inc., Santa Clara, California

PETER M. ASBECK Rockwell International Science Center, Thousand Oaks, California

FABIO BELTRAM AT&T Bell Laboratories, Murray Hill, New Jersey

FEDERICO CAPASSO AT&T Bell Laboratories, Murray Hill, New Jersey

CHRISTOPHER T. M. CHANG Texas Instruments Incorporated, Dallas, Texas

MAU-CHUNG FRANK CHANG Rockwell International Science Center, Thousand Oaks, California

WILLIAM S. C. CHANG Department of Electrical and Computer Engineering, University of California, San Diego, La Jolla, California

PEI-CHUANG CHEN Ortel Corp., Alhambra, California

WILLIAM A. GEIDEMAN McDonnell Douglas Electronic Systems Company, Huntington Beach, California

HAROLD L. GRUBIN Scientific Research Associates, Inc., Glastonbury, Connecticut

DAVID L. MILLER Pennsylvania State University, State College, Pennsylvania

SHIN-SHEM PEI AT&T Bell Laboratories, Murray Hill, New Jersey

SUSANTA SEN† AT&T Bell Laboratories, Murray Hill, New Jersey

NITIN J. SHAH AT&T Bell Laboratories, Murray Hill, New Jersey

MICHAEL SHUR‡ Department of Electrical Engineering, University of Minnesota, Minneapolis, Minnesota

*Present address: Teledyne Monolithic Microwave, Mountainview, California.
†Present address: Institute of Radio Physics and Electronics, University of Calcutta, Calcutta, India.
‡Present address: Department of Electrical Engineering, University of Virginia, Charlottesville, Virginia.

MICHAEL A. STROSCIO[§]   Electronics Division, U.S. Army Research Office, Research Triangle Park, North Carolina

KEH-CHUNG WANG   Rockwell International Science Center, Thousand Oaks, California

PAUL KIT-LAI YU   Department of Electrical Engineering, University of California at San Diego, La Jolla, California

[§]Also affiliated with: Department of Electrical Engineering and Physics, Duke University, Durham, North Carolina; and Department of Electrical and Computer Engineering, North Carolina State University, Raleigh, North Carolina.

1 Introduction to the Physics of Gallium Arsenide Devices

HAROLD L. GRUBIN

Scientific Research Associates, Inc., Glastonbury, Connecticut

1.1 INTRODUCTION

It is arguable that the history of gallium arsenide semiconductor devices, from the early 1960s to the present time, falls into three groups. First, there was the experimental work of Gunn [1], demonstrating the generation of sustained oscillations upon application of a sufficiently large dc bias. This work opened up the possibility of fabricating bulk microwave and millimeter-wave devices, and hastened additional and intense studies of the properties of compound semiconductor devices. Second, there was the study of Ruch [2], whose results suggested that the transient, or nonsteady-state, aspects of semiconductor transport would improve the speed of devices by almost an order of magnitude. This, of course, is the argument behind much of the move toward submicron and ultrasubmicron structures. The third era, the one we are presently in, involves the incorporation of gallium arsenide into material-engineered highly complex structures, some of which have provided remarkable millimeter wave characteristics, such as the pseudomorphic HEMT [3]. Much of this book is concerned with this third era, and thus this chapter will only briefly touch upon it. Rather, this section will present a road map of the consequences of using compound semiconductors for device applications, using gallium arsenide as the paradigm example.

The band structure of gallium arsenide is familiar to most and is displayed in Fig. 1. [4]. It is a direct bandgap material. The minimum in the conduction band is at Γ with relevant subsidiary conduction band minima at L and X. The curvature at Γ is such that the effective mass of the Γ-valley is lower than that of the next two adjacent subsidiary L- and X-valleys. For the valence band, the two valleys of significance are those associated with the light and heavy holes. We will concentrate on transport contributions from these five valleys.

In equilibrium, the relative population of electrons in the valleys is dependent on the density of available states and the energy separation, for

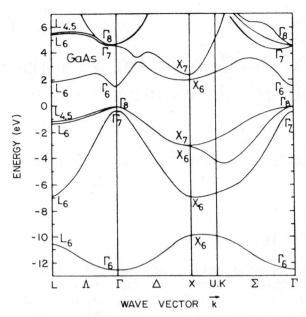

Figure 1.1 Band structure of the semiconductor gallium arsenide [4]. (Reprinted, with permission, from *Physical Review* [*Section*] *B*: *Solid State*.)

example:

$$n_\Gamma^0 = n_L^0 \left(\frac{m_\Gamma}{m_L} \right)^{3/2} \frac{\exp \Delta}{kT} \tag{1}$$

where n_Γ and n_L denote the equilibrium density of the Γ- and L-valley carriers, respectively, m_Γ and m_L are their effective masses, and Δ is the $\Gamma - L$ energy separation. Thus, in equilibrium virtually all of the electrons of interest are in the Γ-valley. For the holes, the valleys are degenerate.

Gallium arsenide is a compound semiconductor. At low values of electric field, apart from carrier–carrier scattering, there are three important scattering mechanisms: polar optical phonon scattering, acoustic phonon scattering, and impurity scattering. For Γ-valley electrons, the contribution to the momentum scattering rate from polar optical phonons is approximately two orders of magnitude larger than that of the acoustic phonon. Since, with regard to mobility, scattering rates are additive, the polar optical phonon is the dominant scatterer. Ideal room-temperature electron mobilities are in the range of 8000 to 9000 cm^2/V·s. For the subsidiary valleys, the effective masses of the carriers are much larger than that of the Γ-valley, and the relative contribution of the acoustic phonon increases. Nevertheless, the polar phonon dominates the transport. For holes, the situation is mixed, with the dominant scattering being polar and nonpolar deformation potential coupling. For momentum scattering, the nonpolar deformation potential scattering dominates.

At high values of electric field and for electrons, nonpolar phonons enter the picture, intervalley transfer from Γ to L takes place, and the situation becomes complex. For example, the spatially uniform, field-dependent velocity characteristics of gallium arsenide, ignoring electron-hole interaction, display a region of negative differential mobility, as shown in Fig. 1.2 [5], where at values of field in excess of 3 kV/cm the mean carrier velocity begins to decrease with increasing electric field. This is an unusual situation and it is perhaps important to recognize that the mean electron velocity of a given species of carrier, assuming a parabolic band, is not decreasing with increasing electric field. Rather, the numbers of high-mobility electrons are decreasing, due to transfer to the subsidiary larger effective mass valleys.

The situation with holes is different. Here, the dominant transport is through the heavy hole. Interband hole scattering is always present even at very low fields, however the relative population is fixed through the ratio of the effective masses, and the existence of a dc negative conductance for holes, on the basis of available data, is ruled out. The field dependence of the mean hole velocity, ignoring interaction with the electrons, is displayed in Fig. 1.3 [6], and there are two important features of note. First, there is the extremely low mobility of the holes at low-field values. Second, there is the saturated drift velocity, which is expected to be higher than that of electrons at high fields. We note there is no hard data on the high-field carrier velocity of holes in gallium arsenide.

Calculations of the type displayed in Figs. 1.2 and 1.3 have been described by many workers and are routinely incorporated into simulation codes. Of more recent interest, because of mixed conduction heterostructure devices, are the modifications that may be expected when electron-hole

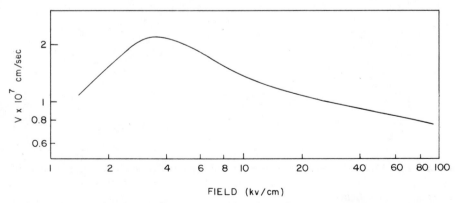

Figure 1.2 Field-dependent electron mean velocity for gallium arsenide [5].

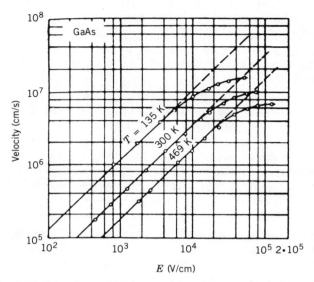

Figure 1.3 Field-dependent hole mean velocity for gallium arsenide [6]. (Reprinted, with permission, from *Journal of Applied Physics*.)

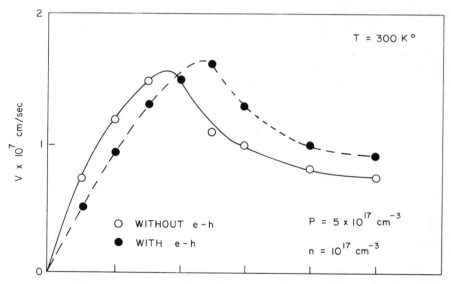

Figure 1.4 Field-dependent electron mean velocity for gallium indium arsenide assuming an interaction with heavy holes [8]. (Reprinted, with permission, from *Applied Physics Letters*.)

scattering occurs. However, because of the limited number of studies with GaAs and because of similarities with other compound semiconductors, results of InGaAs studies are presented. Additionally, because of experimental work in InGaAs [7], the role of carrier–carrier scattering has been most extensively studied for that material. Monte Carlo calculations incorporating electron-hole scattering are displayed in Fig. 1.4 [8]. The results require some detailed discussion and are considered later. Here, we simply note that the presence of holes leads to reduction in the low-field mobility but an increased peak carrier velocity. These intriguing results are also anticipated for GaAs.

The question of interest is how may we expect the role of the complicated compound semiconductor band structure to affect the performance of devices. This is considered next.

1.2 THE ROLE OF BAND STRUCTURE ON THE OPERATION OF ELECTRON DEVICES

In examining the role of band structure on the operation of electron devices, there are several items of immediate interest: the effective mass, the low-field mobility, and the direct bandgap energy of the binary III–V materials (see Table 1.1 [9]). Additionally, the energy separation of the conduction band minima to the subsidiary valleys is listed in Table 1.2 [10]. Note that of the seven binary materials listed, five are direct bandgap materials, and two, GaP and AlAs, are indirect materials. The indirect

TABLE 1.1   Critical Parameters of Select Compound Semiconductors

Compound	Effective Mass[a]	Electron Low-field Mobility $(cm^2/V \cdot s)$	Direct Energy Bandgap (eV)
GaAs	0.063^b	9,200	1.424
GaP[c]	$0.25^t/0.91^l$	160	2.78
GaSb	0.042	3,750	0.75
InAs	0.0219	33,000	0.354
InP	0.079	5,370	1.344
InSb	0.0136	77,000	0.230
AlAs[d]	0.71^b	300	2.98

Source: Ref. 9. Reprinted with permission of Springer-Verlag.
[a]Multiples of free electron mass at the conduction band minima.
[b]DOS.
[c]The minima in the conduction band are at Δ-axis near zone boundary.
[d]The minimum in the conduction band is at X.

TABLE 1.2   Intervalley Energy Separation

Compound	Γ-L (eV)	Γ-X (eV)	L-X (eV)
GaAs	0.34	0.48	0.14
GaP	−0.27	−0.39	−0.37
GaSb	0.08	0.37	0.23
InAs	1.27	1.60	0.33
InP	0.63	0.73	0.10
InSb	0.41	0.97	0.56
AlAs	−0.15	−0.79	−0.64

Source: Ref. 10. Reprinted with permission from Oxford University Press.

bandgap materials have the highest effective masses of the group and also the lowest mobility. Of these materials, GaAs, InAs, InP, and InSb possess regions of negative differential mobility. GaSb, GaP, and AlAs do not. It is perhaps not surprising that the first four mentioned materials possess a region of negative differential mobility, nor that the last two materials do not. In the latter case, the minima in energy is associated with a large effective mass, high density-of-states energy level. The situation with GaSb is peculiar. But here, while the effective mass of the Γ-valley is the smallest of the three, its closeness in energy to that of the subsidiary L-valley is such that at low values of field conduction, contributions arise from both the Γ- and L-valley, effectively suppressing the contributions of intervalley transfer to negative differential conductivity.

The presence of a region of bulk negative differential mobility has, as a major consequence, the possibility of electrical instabilities. These instabilities manifest themselves either as large-signal dipole-dominated oscillations, often referred to as the Gunn effect, or as circuit-controlled oscillations, where the semiconductor behaves electrically as a van der Pol oscillator. The binary semiconductors GaAs, InP, and InAs have exhibited electrical instabilities associated with bulk negative differential mobility. While InSb has also sustained electrical instabilities, the interpretation of the instability is complicated by the small direct bandgap and the possibility of avalanching at low bias levels.

An additional feature of importance is the intrinsic carrier concentrations of some of these materials, as shown in Table 1.3 [9]. It is clear that the intrinsic concentration of InAs and InSb make them unsuitable for a unipolar source. Indeed, all transport calculations using these latter materials must necessarily include multispecies transport.

In choosing materials for electron devices, particularly as power sources, a figure of merit has been the peak to saturated drift velocity ratio. From this point of view, indium phosphide is an attractive candidate, but this must be weighed with the fact that the low-field mobility of InP is less than that of

TABLE 1.3   Intrinsic Concentration

Compound	$n \, (/\mathrm{cm}^2)$
GaAs	2.1×10^6
InAs	1.3×10^{15}
InSb	2.0×10^{16}
InP	1.2×10^8

Source: Ref. 9. Reprinted with permission of Springer-Verlag.

gallium arsenide. A recent study comparing these features suggests that the Γ-valley mobility is the dominant material parameter of submicron structures, whereas the high-field saturated drift velocity is the dominant material parameter of micron-length structures [11].[†] Additionally, if a choice for two terminal sources is to be made between, for example, InP and GaAs, other issues emerge. For instance, the scattering rates in InP indicate a shorter energy relaxation time than that of GaAs. The consequence of this are higher-frequency operation for InP. Thus, at least with respect to these materials, the peak to valley ratio of the materials is only one factor in the design of an electrical source.

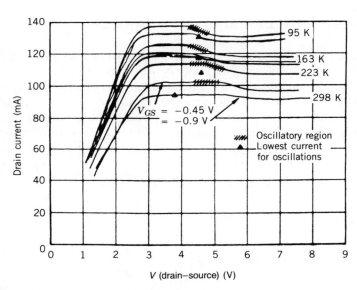

Figure 1.5 Temperature-dependent pulsed data for a GaAs FET, with a 3.0 μm gate length, a source to drain separation of 8.5 μm, and an epitaxial thickness of 3000 ± 500 A. Nominal background doping is $10^{17}/\mathrm{cm}^3$ [12].

[†](Editor's Note) See, however, Chap. 2, page 82, last paragraph and reference [12] cited within for an additional view on this issue.

There is less to say about the effects of negative differential mobility on the operation of avalanche diodes. Here, the effects of negative differential mobility conductivity are present but are overshadowed by the effects of avalanching. For example, recent simulation studies show the presence of domains in IMPATTs, whose presence is a direct consequence of negative differential conductivity. These domains can complicate the actual transit time of dipole layers associated with the avalanche generation, but the negative differential mobility is a marginal issue. Such is not the case with three-terminal devices.

For three-terminal device observations of bias-dependent white light in GaAs FETs, as from either the drain side of the gate contact and the gate side of the drain contact, are consistent with numerical calculations showing the presence of local high-field dipole layers near the gate and drain contacts. In addition, for a range of bias, some devices display a current dropback consistent with bias-dependent formation of high-field domains and concurrent current oscillations. This last result is shown in Fig. 1.5 [12]. Remaining questions of interest focus on the manner in which transport in these devices is examined. We begin with the equilibrium description of transport.

1.3 EQUILIBRIUM DESCRIPTION OF TRANSPORT

The steady-state equilibrium description of transport has traditionally provided most details of device behavior. Nevertheless, the description ignores acceleration. It assumes that the carrier velocity is determined by the local electric field and that the total current is governed by a balance of a drift component and a diffusive component. Typically, the continuity equation is solved simultaneously with the current equation, which for electrons is of the form

$$\mathbf{J}_n = -e\left(n\mathbf{v}_n - D_n \frac{\partial n}{\partial \mathbf{x}}\right) \tag{2}$$

and for holes:

$$\mathbf{J}_p = +e\left(p\mathbf{v}_p - D_p \frac{\partial p}{\partial \mathbf{x}}\right) \tag{3}$$

Here, n and p denote electron and hole concentration, respectively, $\mathbf{v}$ velocity, and $\mathbf{D}$ diffusivity. The usual derivations of Eqs. (2) and (3) proceed from a linearization of the Boltzmann transport equation. The assumption is then made that the equation is valid for high-field nonlinear transport.

Typically, the field-dependent velocity assumed in these equations is of the type displayed in Fig. 1.2.

While the use of the field-dependent velocity in these equations is universal, the type of diffusivity coefficient used in these studies is almost as numerous as the numbers of workers involved in numerical studies. However, a number of important issues are at stake in the description of the diffusivity. For example, if the Einstein relation

$$D = \frac{\mu kT}{e} \tag{4}$$

is used, then, under equilibrium and/or zero current conditions, the dependence of carrier density on conduction and valence band energy is given by either the equilibrium Boltzmann or Fermi distribution. However, under nonequilibrium conditions (and near-zero current conditions), the Einstein relation inadequately describes diffusive transport [13]. To correct for the latter deficiency, the field-dependent diffusivity often used in calculations is of a form similar to that shown below [14]:

$$D = \frac{\mu kT}{e} + \tau v_{\text{sat}}^2 \tag{5}$$

where at high values of electric field, the diffusivity only gradually decreases. While the diffusivity coefficient of Eq. (5) more adequately represents high-field phenomena, because its field dependence is conceptually consistent only with the assumption of nonequilibrium conditions it is conceptually inconsistent with equilibrium conditions, and will lead to incorrect built-in potentials [15].

While the drift and diffusion equations (DDE) clearly offer conceptual difficulties with respect to consistency of physics, they nevertheless offer considerable insight into the physics of device operation and are useful providing their limitations are kept in mind. For example, instabilities in long GaAs structures are known to depend critically on conditions at the contacts. A study in 1969 [16] demonstrated that by experimentally creating different conditions at the boundaries to the active region of GaAs, a wide range of different electrical instabilities could be obtained. Corresponding numerical studies were performed through solutions to the above drift and diffusion equations, in which a value for the electric field was specified at the cathode (and anode) boundary. It was found that the boundary-dependent electrical behavior could be broken into three categories, as summarized in Fig. 1.6 [16]. The key conclusion of the study was that the electrical behavior of compound semiconductors devices was dependent in a detailed way on contact conditions. This same critical result has reappeared numerous times in a variety of different types of structures.

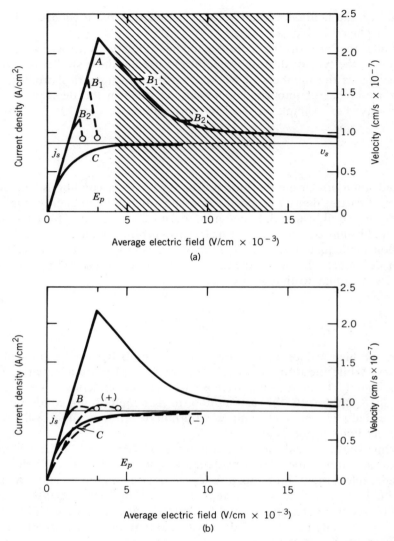

Figure 1.6 (a) The $v(E)$ curve and the computer simulated current density j as a function of average electric field $\langle E \rangle$ (A through C) for various cathode boundary fields. The boundary field is zero for curve A, 24 kv/cm for curve C, and is indicated by the arrow for curves B_1 and B_2. The sample is 10^{-2} cm long and has $n = 10^{15}$ cm^{-3} and $\mu = 6860$ cm^2/V·s. The right- and left-hand ordinates are related by $j = nev$, $v_2 = 0.86 \times 10^7$ cm/s. (b) Experimental $j - \langle E \rangle$ curves (+) and (−) (dashed) and theoretical curves B and C (solid). The only significance in the fact that the low-field slopes differ is that the theoretical curve is for a mobility of 6860 cm^2/V·s, whereas the experimental curve is for a mobility of 4000 cm^2/V·s [16].

1.4 NONEQUILIBRIUM DESCRIPTIONS OF TRANSPORT

The situation of most interest lies in nonequilibrium transport. The most critical area of interest is the incorporation of acceleration into the governing equations.

In examining nonequilibrium transport, several approaches have been used. One is the Monte Carlo method, where the trajectory of a particle is followed through its acceleration and subsequent scattering events. In the discussion below, results of Monte Carlo calculations will be presented, but we first concentrate on nonequilibrium phenomena as described by the moments of the Boltzmann transport equation. These equations, in their simplest form for parabolic bands, a position-dependent conduction band, and a position-dependent effective mass, take the form shown below [17]:

Carrier balance:

$$\frac{\partial n}{\partial t} + \nabla_{\mathbf{r}} \cdot n \frac{\hbar \mathbf{k}_d}{m} = \frac{2}{(2\pi)^3} \int \frac{\partial f}{\partial t}\bigg|_{\text{coll}} d^3\mathbf{k} \tag{6}$$

Momentum balance:

$$\frac{\partial}{\partial t} n\hbar\mathbf{k}_d + \nabla_{\mathbf{r}} \cdot \frac{(n\hbar^2\mathbf{k}_d\mathbf{k}_d)}{m} = -n\nabla_{\mathbf{r}}E_c + q n\mathbf{v} \times \mathbf{B} - \nabla_{\mathbf{r}}nkT$$

$$+ \left(n \frac{\hbar^2\mathbf{k}_d \cdot \mathbf{k}_d}{2m} + \frac{3}{2} nkT \right) \frac{\nabla_{\mathbf{r}}m}{m} + \frac{2}{(2\pi)^3} \int \frac{\partial f}{\partial t}\bigg|_{\text{coll}} \hbar\mathbf{k}\, d^3\mathbf{k} \tag{7}$$

Energy balance:

$$\frac{\partial}{\partial t}\left[n\left(\frac{\hbar^2\mathbf{k}_d \cdot \mathbf{k}_d}{2m} + \frac{3}{2}\, kT \right) \right] + \nabla_{\mathbf{r}} \cdot n v\left(\frac{\hbar^2\mathbf{k}_d \cdot \mathbf{k}_d}{2m} + \frac{5}{2}\, kT \right)$$

$$= -nv \cdot \nabla_{\mathbf{r}}E_c + \frac{2}{(2\pi)^3} \int \frac{\partial f}{\partial t}\bigg|_{\text{coll}} \frac{\hbar^2\mathbf{k} \cdot \mathbf{k}}{2m}\, d^3\mathbf{k} \tag{8}$$

In the above, $\hbar\mathbf{k}_d$ is the mean momentum of the carriers, T is the carrier temperature, and, for electrons, E_c is the position-dependent conduction band energy. $\mathbf{B}$ is an applied magnetic field. The terms on the right side represent scattering and/or electron-hole interaction, as through avalanching. For example, the right side of Eq. (6) represents intervalley scattering. If avalanching occurs, generation is expressed through an energy-dependent ionization coefficient [18]. If a carrier temperature model is assumed, then carrier generation is given by

$$n\alpha(T) \qquad\qquad (9)$$

where $\alpha(T)$ is the ionization coefficient. In the absence of a first-principle determination of $\alpha(T)$, the following relation can be assumed as a starting point:

$$\alpha(T) = \alpha^*(F)v(F) \qquad\qquad (10)$$

where $\alpha^*(F)$ and $v(F)$ are the equilibrium ionization rates and field-dependent velocities, respectively and the relation between T and F is determined from the equilibrium solution. While the Eq. (10) relation is uncertain, it has the conceptual advantage of relating ionization to energy, rather than field.

But, perhaps the most significant feature of these equations is the presence of acceleration, both spatial and temporal in the momentum balance equation. These acceleration terms are absent from the drift and diffusion equations. Additionally, under equilibrium conditions, and hence, zero current (i.e., $n\hbar\mathbf{k}_d/m = 0$), the electron temperature model teaches that for any spatially nonuniform structure, such as a p-n junction, the electron temperature is everywhere constant and equal to the ambient. Thus, conceptual problems arising from the form of the diffusion contribution to the drift and diffusion equations do not enter here. Note that a generalized drift and diffusion current term is obtained when the left side of Eq. (7) is set to zero.

Equations of the type shown above provide a considerable amount of information with regard to transport. For example, with a Γ–L–X orientation in GaAs the distribution of carriers as a function of field is shown in Fig. 1.7 [5]. Here the relative distribution of carriers in each of the valleys is determined by the distribution of temperature in each of the valleys, which in turn is driven by the electric field, as shown in Fig. 1.8 [5]. Note that for fields below 4 kV/cm, the carriers reside in the Γ-valley. At fields above 6 kV/cm, the L-valley population exceeds that of the Γ-valley. It should be emphasized, however, that because of the very low subsidiary valley mobility, most of the current, for fields up to 50 kV/cm, is carried by the Γ-valley carriers.

The interest in nonequilibrium equations lies not in the steady-state uniform field distribution, but in transients and nonuniform fields. The transient distribution of carrier density and velocity for electrons subject to a sudden change in electric field is shown in Fig. 1.9 [5], where the high peak velocity can be noted.

The high peak velocity in Fig. 1.9a is primarily associated with Γ-valley transport. Indeed, the Γ-valley velocity at 2.8 ps is near 4×10^7 cm/s. The

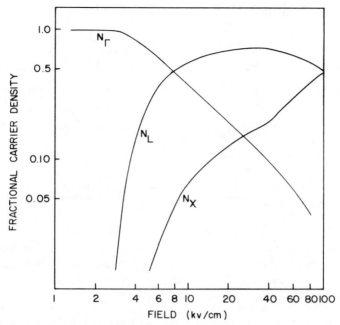

Figure 1.7 Steady-state uniform field carrier distribution for $\Gamma-L-X$ orientation in GaAs [5].

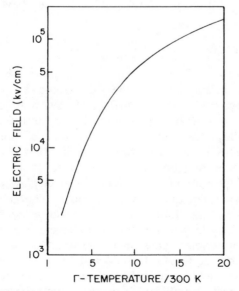

Figure 1.8 Temperature dependence of electric field for a $\Gamma-L-X$ orientation in GaAs. Electric field is the independent variable [5].

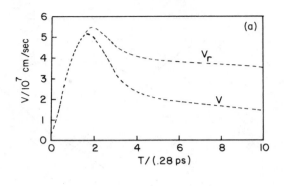

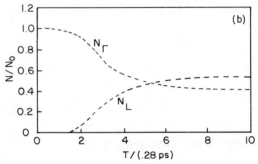

Figure 1.9 Transient overshoot for a field of 9.6 kV/cm [5]. $\Gamma-L-X$ orientation, and an applied field of (a) Γ and mean velocity, (b) Γ- and L-valley carrier density.

mean velocity

$$v = (n_\Gamma v_\Gamma + n_\Gamma v_L + n_X v_X)/N_0 \qquad (11)$$

where n denotes net population of the Γ-, L-, and X-levels and N_0 is the total carrier density, is also shown in Fig. 1.9. We note that the significant drop in mean velocity is a consequence of electron transfer from the central to the satellite valleys (see Fig. 1.9b). Also note a transient decrease in Γ-valley velocity. This is a consequence of the difference between the energy (longer) and momentum (shorter) relaxation times in GaAs. The time-independent spatially nonuniform situation also displays overshoot effects.

The situation when, in one dimension, space charge effects are introduced is displayed in Figs. 1.10 and 1.11 [5], where for a gallium arsenide device of different lengths we show the field and space charge distribution of the Γ-valley electrons and the current–voltage characteristics. The feature to note is that as the device length decreases the current drive increases. Note that in all cases the field is nonuniform and increases toward the anode. Electron transfer exists for all four structures, with the greatest amount of transfer occurring for the longest device. Additionally, since high-field

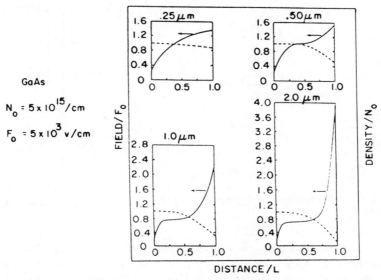

Figure 1.10 Electric field and Γ-valley carrier distribution for GaAs (with a two-level model) two-terminal devices of different lengths and a mean field of 5 kV/cm [5].

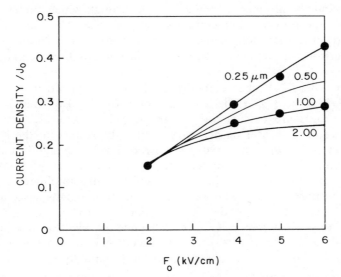

Figure 1.11 Current-voltage characteristics, as a function of lengths, for the structures of Fig. 1.10 [5].

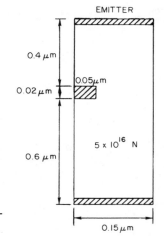

Figure 1.12 Dimensions and doping level of the simulated PBT [20].

values are synonymous with carrier accumulation, we see that electron transfer here is accompanied by an accumulation of L-valley carriers. Saturation in the current density occurs at high bias, and even for the shortest device there is electron transfer at the anode side of the device.

The role of nonequilibrium transport on two-dimensional simulations is discussed for the vertical three-terminal GaAs permeable base transistor [19], one cell of which is displayed in Fig. 1.12 [20]. One important advantage of this structure is the parallel placement of the source and drain contacts* and the absence of any substrate through which current can flow and reduce the transfer characteristics of the device. The simulations were performed for a 1 μm long source to drain region and a 200 Å gate. Also, for comparison, results of the drift and diffusion equation simulations were included.

The computed I–V characteristics of the device shown in Fig. 1.12 are presented in Fig. 1.13 [20]. The moments of the Boltzmann transport equation results are extrapolated to the origin, as indicated by the long broken lines. The shorter dashed curves show the results for the DDE. The comparison shows that the predicted current levels are significantly higher for the moment equation solutions, a result consistent with FET calculations performed by Cook and Frey [21], who used a highly simplified momentum-energy transport model. The present calculation results also indicate a region of negative differential forward conductivity at $V_{BE} = 0.6$ V. The origin of this phenomena is a consequence of electron transfer.[†] The presence of a dc negative forward conductance is also a feature of PBT measurements but is clearly absent from DDE simulations [20].

*(Editor's Note) They are also called emitter and collector in PBT terminology.

[†](Editor's Note) See also Chap. 2, Section 2.2.4 for additional views on the issue of negative differential conductance.

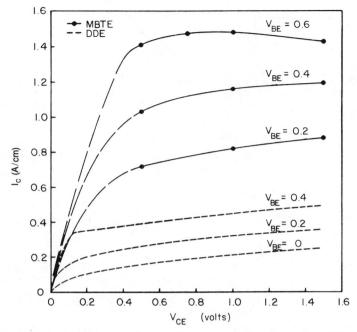

Figure 1.13 Collector current versus collector–emitter voltage for different values of base–emitter voltage [20].

A comparison between the total carrier density distribution along the center of the channel for drift and diffusion and MBTE solutions is shown in Fig. 1.14 [20] for $V_{CE} = 1.0$ V and $V_{BE} = 0.4$ V. The moment equation prediction for the Γ-valley carrier density is also shown.

As seen in Fig. 1.14 for the DDE simulations, the carrier density reaches a maximum between base contacts and displays a significant dipole layer. Here, with the velocity in saturation and the cross-sectional area at a minimum, the carrier density must increase to maintain current continuity. In the moment equation simulation, the constraints of current continuity are more complex. First, a decrease in the cross-sectional areas is, as in the DDE, accompanied by an increase in field along the channel. The field increase under both equilibrium and nonequilibrium conditions is qualitatively similar, as may be observed from Fig. 1.15 [20], which shows the potential distribution along the center of the PBT channel. However, consequent changes in electron temperature, both increasing and decreasing, lag behind the equilibrium state. This leads to velocity overshoot (Fig. 1.16 [20]) and a delay in electron transfer. As a result, for nearly the first half of the device, transport is almost exclusively Γ-valley transport. The implication is that if the Γ-valley carrier velocity increases with increasing field, then the product of density and cross-sectional area normal to current

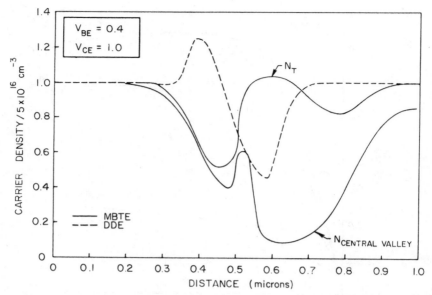

Figure 1.14 Carrier density versus distance along center of channel for the PBT [20].

flow must decrease to maintain current continuity. Since the velocity increases faster than the area decreases, the carrier density decreases.

At moderate bias levels, typical FET calculations show a decreasing field as the gate region is passed. This also occurs in the PBT. Now, as the cross-sectional area increases, the Γ-valley carriers exhibit a decrease in velocity. It must be noted, however, that for the parameters of the calcula-

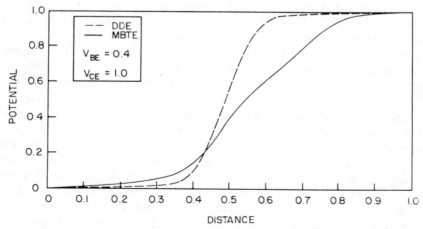

Figure 1.15 Potential versus distance along center of channel for the PBT [20].

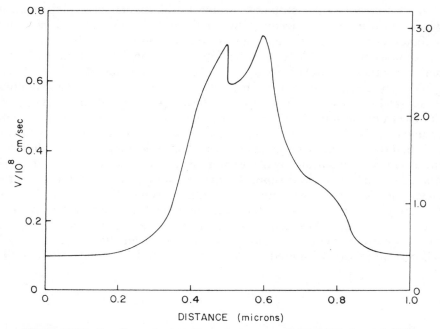

Figure 1.16 Γ-valley velocity along the center of the PBT channel [20].

tions the L-valley carriers make a negligible contribution to current. Thus, a decrease in carrier velocity results in a net increase in carrier concentration. However, initially, the decrease in field is not accompanied by a corresponding temperature decrease (as experienced in the uniform field calculations). Thus, the high Γ-valley temperature results in transfer to the L-valley, giving rise to the second minimum in the Γ-valley carrier density shown in Fig. 1.14. Further toward the drain, the field decreases. However, relaxation is incomplete and the field at the collector is not equal to the field at the emitter. Also note that the moment equation potential distribution gives rise to a slightly higher field upstream of the base, and a lower field, over a longer distance, downstream of the base compared with the drift and diffusion result. More significantly, the electron temperature at the collector exceeds that at the emitter. It is noted that as the field relaxes, the electrons transfer back to the central valley.

1.5   NONEQUILIBRIUM ELECTRON-HOLE TRANSPORT

Additional nonequilibrium studies were mentioned at the beginning of this chapter. This concerns nonequilibrium electron-hole transport, which for specificity was discussed for InGaAs. The details are considered below.

In this recent study, ensemble Monte Carlo studies were performed in which electrons were injected into p-type InGaAs [8]. In one case, the acceptor doping was 10^{17} cm^{-3}, and in the second case 5×10^{17} cm^{-3}. The calculations were at 300 K and the ratio of the injected electrons to the majority holes was taken to be 1:5. (Note: The ensemble Monte Carlo avoids any assumptions on the magnitudes of the energy and momentum exchange in an electron-hole [e-h] scattering process and the evolution of the electron and hole distribution functions.) The electrons and holes were assumed to be in equilibrium with the lattice when the electric field was switched on, and the band model consisted of three nonparabolic valleys for the conduction band and two parabolic light- and heavy-hole bands. The role of the light holes was suppressed in this study. The model includes the elastic acoustic phonon, impurity scattering using Ridley's model, alloy scattering, deformation potential, and intervalley and intravalley phonon scattering process. The e-h and screened carrier-phonon scattering are calculated from the expressions given in Osman and Ferry [22], using a self-consistent screening model. There is only one LO phonon mode in this

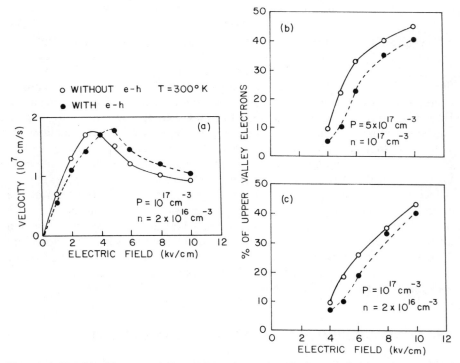

Figure 1.17 (a) Electron drift velocity in p-type InGaAs, (b–c) percentage of upper-valley electrons in p-type InGaAs [8]. (Reprinted, with permission, from *Applied Physics Letters*.)

calculation. The electron transport parameters for $In_{0.53}Ga_{0.47}$ As are the same as those reported in Ahmed et al. [23]. However, for hole effective masses and deformation potential constants, appropriate to GaAs were used. The interaction between the L-valley electrons and the heavy holes was ignored.

The drift velocities of the electrons injected into the p-type InGaAs as a function of the applied electric fields are plotted in Figs. 1.4 and 1.17a [8], for doping levels of 5×10^{17} and 10^{17} cm^{-3}, respectively. The curves connecting the open circles in these figures correspond to situations in which the interaction with the mobile holes is ignored and only the interaction with the ionized acceptor impurities is taken into account.

When the interaction between the minority electrons and the mobile majority holes is taken into account, the electron velocities are lower for fields below 4 kV/cm compared with majority electrons, as can be seen from these figures. At these low fields, the electron transfer to the upper valleys is negligible, as shown in Fig. 1.17b [8], and the energy loss through e-h interaction is not significant because the rate at which the electrons gain energy from the electric field is small, as can be seen in Fig. 1.18 [8]. However, the e-h scattering, which has the same angular distribution as the impurity scattering, has the same effect on the electron mobility as doubling the doping level. Consequently, the mobility of the electrons is reduced, leading to lower velocities. As the electric field is increased, the fraction of electrons with enough energy to transfer to the upper valleys increases.

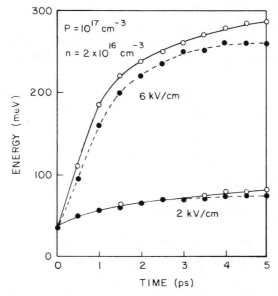

Figure 1.18 Electron energy in p-type InGaAs as a function of field [8].

However, as the energy of the electrons increases, the rate at which they lose energy through e-h scattering is increased [8]. This results in retaining more electrons in the central valley when scattering of the electron by the majority holes is taken into account, as can be seen in Fig. 1.17, and results in higher electron velocities above 4 kV/cm in the present situation. An additional feature of the velocity field curve of the minority electrons is that it converges to that of the majority electrons at higher electric fields. This reflects the fact that as these high fields the rate at which electrons gain energy from the electric field exceeds the rate at which electrons lose energy to the heavy holes. Consequently, the population of electrons in the upper valleys increases.

It is worthwhile to note that the experimental measurement [8] of minority carrier velocity does not exhibit any negative differential resistance. The origins of this are unclear. We do point out that the measurements are performed for two-terminal systems, and the highly nonuniform field distribution may prevent the appearance of NDR.

1.6 TRANSPORT IN ULTRASUBMICRON DEVICES

The entire discussion of transport has been predicated on a semiclassical picture. Certainly, quantum effects [24] take place on short-time scale where the Fermi–Golden rule breaks down and where spatial feature sizes are the order of tens of angstroms. As a rule of thumb, it is thought that quantum mechanical effects become prominent when the feature size is of the order of a thermal de Broglie wavelength or shorter, as shown in Fig. 1.19 [25].

The quantum transport formulation for devices is extremely rich and new approaches are necessary.[†] For example, it appears necessary to resort to solutions, for example, the density matrix or some equivalent form as the Wigner distribution function [26]. Moment equations are also applicable.

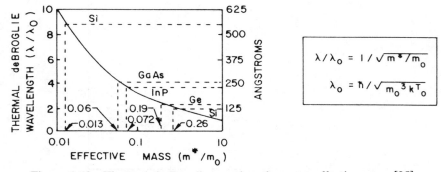

Figure 1.19 Thermal de Broglie wavelength versus effective mass [25].

[†](Editor's Note) See also Chap. 11 for a detailed review of the modeling of quantum transport effects.

For instance, from the equation of motion of the density matrix, for a system including mobile carriers and scattering centers, the first three moment equations have the following form [27]:

$$\langle\!\langle \dot{P}^{(0)} \rangle\!\rangle + \frac{1}{m} \frac{\partial}{\partial x} \langle\!\langle P^{(1)} \rangle\!\rangle = \frac{i}{\hbar} \langle\!\langle [H_s, P^{(0)}] \rangle\!\rangle \tag{12}$$

$$\langle\!\langle \dot{P}^{(1)} \rangle\!\rangle + \frac{1}{m} \frac{\partial}{\partial x} \langle\!\langle P^{(2)} \rangle\!\rangle = -\left(\frac{\partial V}{\partial x}\right)\langle\!\langle P^{(0)} \rangle\!\rangle + \frac{i}{\hbar} \langle\!\langle [H_s, P^{(0)}] \rangle\!\rangle \tag{13}$$

$$\langle\!\langle \dot{P}^{(2)} \rangle\!\rangle + \frac{1}{m} \frac{\partial}{\partial x} \langle\!\langle P^{(3)} \rangle\!\rangle = -2\left(\frac{\partial V}{\partial x}\right)\langle\!\langle P^{(1)} \rangle\!\rangle + \frac{i}{\hbar} \langle\!\langle [H_s, P^{(2)}] \rangle\!\rangle \tag{14}$$

where the $\langle\!\langle\ \rangle\!\rangle$ denote the quantum ensemble averages, and using Dirac notation, the operators of interest are of the form

$$P^{(0)} = |x_0><x_0| \tag{15}$$

$$P^{(1)} = (\tfrac{1}{2})(P|x_0><x_0| + |x_0><x_0|P) \tag{16}$$

$$P^{(2)} = (\tfrac{1}{2})^2(P|x_0><x_0| + 2P|x_0><x_0|P + |x_0><x_0|P^2) \tag{17}$$

$$P^{(3)} = (\tfrac{1}{2})^3(P^3|x_0><x_0| + 3P|x_0><x_0|P^2 + 3P^2|x_0><x_0|P$$
$$+ |x_0><x_0|P^3) \tag{18}$$

where P is the momentum operator. Note that the terms involving H_s incorporate dissipation. In a diagonal representation, the ensemble average of the first three operators breaks down into the following form:

$$\langle\!\langle P^{(0)} \rangle\!\rangle = \sum \rho_{ii}n_i(x_0) = n(x_0) \tag{19}$$

$$\langle\!\langle P^{(1)} \rangle\!\rangle = \sum \rho_{ii}n_i mv_i \equiv n(x_0)mv_d \tag{20}$$

$$\langle\!\langle P^{(2)} \rangle\!\rangle = \sum \rho_{ii}n_i m^2[(v_i - v_d) + v_d]^2 - \frac{\hbar^2}{4} \sum \rho_{ii}n_i \frac{\partial^2}{\partial x^2} \ln n_i$$

$$= \Omega_{xx} + m^2nv_d^2 - \frac{\hbar^2}{4} \sum \rho_{ii}n_i \frac{\partial^2}{\partial x^2} \ln n_i \tag{21}$$

where ρ_{ij} is the diagonal element of the density matrix, and

$$\Omega_{xx} = \sum \rho_{ii}n_i m(v_i - v_d)^2 \tag{22}$$

It is clear that with the exception of the third operator, which contains a term involving Planck's constant, the equations are of a classical form. Thus, in the simplest approximation, there appears to be a close similarity between the classical moment equations and that obtained quantum mechanically. The difficulty is, of course, in solving these equations.

There is, however, an interesting situation to consider: that in which $\rho_{ii} = 1/N$ the system. In this case, the first two moment equations, including dissipation in momentum, reduce to

$$\frac{\partial n}{\partial t} + div(nv) = 0 \qquad (23)$$

$$\frac{\partial nv}{\partial t} + \frac{\partial}{\partial x} nv_d^2 + \frac{nv}{\tau} = -\left(\frac{\partial V}{\partial x} + \frac{\partial Q}{\partial x}\right) \frac{n}{m} \qquad (24)$$

where

$$Q = -\frac{\hbar^2}{2m} \frac{1}{\sqrt{n}} \frac{\partial^2 \sqrt{n}}{\partial x^2} \qquad (25)$$

The quantity V represents an imposed barrier and the self-consistent energy associated with Poisson's equation. The potential Q [24] is density-dependent and tends to become significant near strong barriers, where the curvature of $\sqrt{n}$ will either enhance or diminish the imposed barrier. Tunneling and resonance arise from Q.

In multiple dimensions, these equations are subject to the constraint

$$\oint mv \cdot dx = nh \qquad (26)$$

or in a gauge that includes a vector potential, the constraint

$$\oint \left(mv + \frac{e}{c} \mathbf{A}\right) \cdot dx = nh \qquad (27)$$

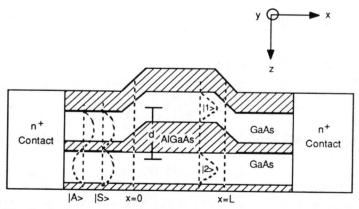

Figure 1.20 Configuration suitable for the Aharonov–Bohm constraint [29]. (Reprinted, with permission, from *Applied Physics Letters*.)

Presently, device systems are being constructed which are influenced by the constraint of Eq. (27), often called the Aharonov–Bohm condition [28]. In particular, structures are being constructed in which the path of an incident beam of electrons is split and then recombined. The split path lengths of the original beam are different, and under coherent reconstruction in which Eq. (27) is satisfied, conduction oscillations are anticipated. A structure originally proposed to deal with this is displayed in Fig. 1.20 [29].

ACKNOWLEDGMENTS

The author acknowledges numerous conversations with G. I. Iafrate, J. P. Kreskovsky, M. A. Osman, M. Meyyappan, and D. K. Ferry. Portions of this work were supported by ONR, AFOSR, and ARO.

REFERENCES

1. J. B. Gunn, *Solid State Commun.* **1**, 88 (1963).
2. J. G. Ruch, *IEEE Trans. Electron Devices* **ED-19**, 652 (1972).
3. A. A. Ketterson, W. T. Masselink, J. G. Gedymin, J. Klem, C.-K. Peng, W. F. Kopp, H. Morkqc, and K. R. Gleason, *IEEE Trans. Electron Devices* **ED-33**, 564 (1986).
4. J. R. Chelikowsky and M. L. Cohen, *Phys. Rev. B: Solid State* [14] 556 (1976).
5. H. L. Grubin, *Lecture Notes in* "The Physics of Sub-Micron Semiconductor Devices," (H. L. Grubin, D. K. Ferry, and C. Jacoboni, eds.). Plenum Press, 1988.
6. V. L. Dalal, A. B. Dreeben, and A. Triano, *J. Appl. Phys.* **42**, 2864 (1971).
7. R. J. Degani, R. F. Leheny, R. E. Nahory, and J. P. Heritage, *Appl. Phys. Lett.* **39**, 569 (1981).
8. M. A. Osman and H. L. Grubin, *Appl. Phys. Lett.* **51**, 1812 (1987)
9. O. Madelung and M. Schulz, eds., "Numerical Data and Functional Relationships in Science and Technology/Landolt-Bornstein," Vo.. 22. Springer-Verlag, Berlin, 1987.
10. B. K. Ridley, "Quantum Processes in Semiconductors." Oxford Univ. Press (Clarendon), London and New York, 1982.
11. H. L. Grubin and J. P. Kreskovsky, *Physica* **134B + C**, 67 (1985).
12. H. L. Grubin, D. K. Ferry, and K. R. Gleason, *Solid-State Electron.* **23** 157 (1980).
13. P. N. Butcher, *Rep. Prog. Phys.* **30**, 97 (1967).
14. K. Yamaguchi, S. Asai, and H. Kodera, *IEEE Trans Electron Devices* ED-23, 1283 (1976).
15. M. Meyyappan, J. P. Kreskovsky, and H. L. Grubin, to be published.
16. M. P. Shaw, P. R. Solomon, and H. L. Grubin, *IBM J. Res. Dev.* **13**, 587 (1969).

17. E. M. Azoff, *Solid State Electron.* **30**, 913 (1987).

18. R. K. Froelich, *Avion. Lab. Rep.* **AFWAL-TR-82-1107,** 1982.

19. C. O. Bozler and G. D. Alley, *IEEE Trans. Electron Devices* **ED-27,** 6 (1980).

20. J. P. Kreskovsky, M. Meyyappan, and H. L. Grubin, *in* "Proceeding of NUMOS I" (J. J. H. Miller, ed.). Boole Press, 1987.

21. R. F. Cook and J. Frey, *COMPEL* **1**, 2 (1982).

22. M. A. Osman and D. K. Ferry, *J. Appl. Phys.* **61**, 5330 (1987).

23. S. R. Ahmed, B. R. Nag, and M. D. Roy, *Solid-State Electron.* **28**, 1193 (1985).

24. C. Philippidis, D. Bohm, and R. D. Kaye, *Nuovo Cimento Soc. Ital. Fis. B* [11] **71B**, 75 (1982).

25. G. J. Iafrate, H. L. Grubin, and D. K. Ferry, *J. Phys. (Orsay, Fr.)* **42**, C7-307 (1981).

26. E. Wigner, *Phys. Rev.* **40**, 749 (1932).

27. H. L. Grubin, to be published.

28. Y. Aharonov and D. Bohm, *Phys. Rev.* **115**, 485 (1959).

29. S. Datta, M. R. Mellcoh, S. Bandyopadyay, and M. S. Lundstrom, *Appl. Phys. Lett.* **48**, 487 (1986).

2 Modeling of GaAs and AlGaAs/GaAs Field Effect Transistors

MICHAEL SHUR*

Department of Electrical Engineering
University of Minnesota, Minneapolis, Minnesota

2.1 INTRODUCTION

Compound semiconductor field effect transistors occupy an important niche in the electronics industry. GaAs FET amplifiers, oscillators, mixers, switches, attenuators, modulators, and current limiters are widely used, and high-speed integrated circuits based on GaAs FETs have been developed. The basic advantages of GaAs devices include a higher electron velocity and mobility which lead to smaller transit time and faster response, and semi-insulating GaAs substrates that reduce parasitic capacitances and simplify the fabrication process. Other material systems such as AlGaAs/InGaAs, InGaAs/InP, etc., have also exhibited superior device properties for applications in very high speed circuits.

The poor quality of oxide on GaAs and a correspondingly high density of surface states at the GaAs–insulator interface make it difficult to fabricate GaAs MOSFETs or MISFETs. Only very recently, the new approach of oxidizing a thin silicon layer grown by molecular beam epitaxy (MBE) on the GaAs surface offered hope for the development of a viable GaAs MOSFET technology [87]. Schottky barrier metal semiconductor field effect transistors (MESFETs), junction field effect transistors (JFETs), and heterostructure AlGaAs/GaAs transistors are the most commonly used GaAs devices. In many cases, GaAs MESFETs and JFETs are fabricated by direct ion implantation into a GaAs semi-insulating substrate.

GaAs technology has a significant advantage over silicon in terms of speed, power dissipation, and radiation hardness. However, compound

*Present address: Department of Electrical Engineering, University of Virginia, Charlottesville, Virginia.

semiconductor technology is much less developed. In an elemental semiconductor, such as silicon, the device quality depends on the material purity. In a compound semiconductor, such as GaAs, the material composition is of utmost importance. For example, defects that degrade device performance may be caused by a deficiency of arsenic atoms.

Another serious problem associated with GaAs MESFETs, JFETs, and heterostructure field effect transistors (HFETs) is the gate leakage current. This current limits the allowed gate-voltage swing, thus reducing the noise margin in circuits. Finally, the fabrication processes for compound semiconductor devices are not as well developed or understood as for silicon. Fabrication is expensive and the fabrication cycle usually takes several weeks.

All this makes the accurate simulation of device fabrication and realistic device and circuit modeling especially important for compound semiconductor technology, even more so than for its silicon counterpart.

Different people have a different image of GaAs technology. It may be compared with a racehorse that has four legs: device and fabrication process models, circuit design, device characterization, and fabrication technology (Fig 2.1). All these "four legs", have to be fully and proportionately developed if we want this horse to win the race for speed.

In this chapter we will describe models for GaAs MESFETs and HFETs. We will briefly discuss short-channel devices and so-called nonideal effects which interfere with ideal device behavior. The considered models relate the device $I–V$ and $C–V$ characteristics to the device parameters such as doping density, doping profile, and active layer thickness. They have been incorporated into a GaAs IC simulator that we called UM-SPICE and used for the design and optimization of GaAs devices and integrated circuits.

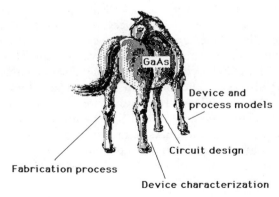

Figure 2.1 Four "legs" of GaAs technology.

2.2.   MODELS FOR GaAs METAL SEMICONDUCTOR FIELD EFFECT TRANSISTORS

2.2.1   Shockley Model

In metal semiconductor field effect transistors (MESFETs), the Schottky barrier gate contact is used in order to modulate channel conductivity (Fig. 2.2). This allows one to bypass the problems related to traps in the gate insulator in metal oxide semiconductor field effect transistors (MOSFETs), such as the hot-electron trapping, threshold-voltage shift due to charge trapped in the gate insulator, etc. Silicon has a very stable natural oxide that can be grown with a very low density of traps, and the problems related to the gate insulators for silicon MOSFETs have been minimized. However, compound semiconductors, such as GaAs, do not have a stable oxide, so most of compound semiconductor field effect transistors use a Schottky gate.

The drawback of MESFET technology is a limitation related to the gate-voltage swing, limited by the turn-on voltage of the Schottky gate. However, this limitation is less important in low-power circuits operating with a low supply voltage.

In n-channel MESFETs, n^+ drain and source regions are connected by an n-type channel (see Fig. 2.2). This channel is partially depleted by voltage applied to the gate. In normally-off (enhancement mode) MESFETs, the channel is totally depleted by the gate built-in potential, even at zero gate voltage. The threshold voltage of enhancement mode devices is positive. In normally-on (depletion mode) MESFETs, the conducting channel has a finite cross section at zero gate voltage (Fig. 2.3).

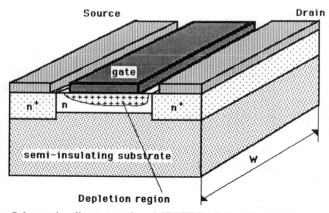

Figure 2.2 Schematic diagram of a MESFET. Schottky barrier gate creates a depletion region modulating the conductivity of the n-type channel between ohmic source and drain contacts.

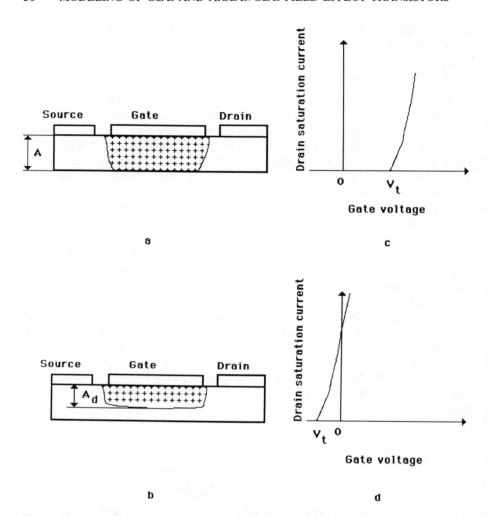

Figure 2.3 Depletion regions in (a) normally-off (enhancement mode) and (b) normally-on (depletion mode) MESFETs at zero gate-to-source and drain-to-source voltages. Also shown are schematic transfer characteristics (drain-to-source saturation current versus gate voltage for (c) normally-off and (d) normally-on MESFETs.

The depletion region under the gate of a MESFET for a finite drain-to-source voltage is schematically shown in Fig. 2.2. The depletion region is wider closer to the drain because the positive drain voltage provides an additional reverse bias across channel-to-gate junction. The analytical expressions describing the shape of the depletion region and the device current–voltage characteristics may be found using the so-called gradual channel approximation. According to the gradual channel approximation, the thickness of the depletion region at every point under the gate may be found using conventional one-dimensional equations for the Schottky barrier (in case of MESFETs) or a p-n junction (in case of JFETs) by assuming the potential drop $V_G - V(x)$ where V_G is the gate potential, $V(x)$ is the channel potential, and x is the longitudinal coordinate (i.e. coordinate along the channel). This approximation is valid when the derivative of the electric field with respect to the coordinate along the channel is much smaller than the derivative of the electric field in the depletion region with respect to the coordinate perpendicular to the channel.

For simplicity, we consider here a uniform, n-type doping profile in the channel. We will first consider small drain-to-source voltages. Then the longitudinal electric field everywhere in the channel is smaller than the velocity saturation field, F_s, so the electron drift velocity is proportional to the longitudinal electric field F (Fig. 2.4):

$$v_n = \mu_n F \qquad (1)$$

We also assume that the conducting channel is neutral and the space charge region between the gate and the channel is totally depleted. The boundary between the neutral conducting channel and the depleted region is assumed to be sharp. (In fact, the transition region between the conducting channel and the space charge region is on the order of three Debye radii.)

According to the gradual channel approximation, the potential across the channel varies slowly with distance x, so at each point the width of the depleted region can be found from the solution of the Poisson equation valid for a one-dimensional junction. Under such assumptions, we find the incremental change of channel potential

$$dV = I_{ds}dR = \frac{I_{ds}dx}{q\mu_n N_D W[A - A_d(x)]} \qquad (2)$$

where q is the electronic charge, I_{ds} the channel current, dR the incremental channel resistance, A the thickness of the active layer, $A_d(x)$ the thickness of the depletion layer (see Fig. 2.3), and W the gate width.

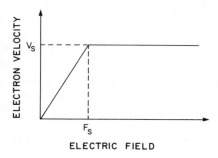

Figure 2.4 Two piece-wise linear approximation for electron velocity.

The depletion region thickness at a distance x can be found using the abrupt-junction depletion approximation:

$$A_d(x) = \left\{ \frac{2\epsilon[V(x) + V_{bi} - V_G]}{qN_D} \right\}^{1/2} \tag{3}$$

Here, V_{bi} is the built-in voltage of the Schottky barrier, ϵ the dielectric permittivity of GaAs, V_G the gate potential, N_D the effective donor concentration in the conducting channel, and $V(x)$ the potential of the neutral channel (here we neglect the variation of the potential in the conducting channel in the direction y perpendicular to x).

Substituting Eq. (3) into Eq. (2) and integrating with respect to x from $x = 0$ (the source side of the gate) to $x = L$ (the drain side of the gate), we derive the equation called the fundamental equation of MESFETs:

$$I_{ds} = g_0 \left\{ V_D - \frac{2[(V_D + V_{bi} - V_G)^{3/2} - (V_{bi} - V_G)^{3/2}]}{3V_{po}^{1/2}} \right\} \tag{4}$$

where

$$g_0 = \frac{q\mu_n N_D WA}{L} \tag{5}$$

is the conductance of the undepleted doped channel, L is the gate length,

$$V_{po} = \frac{qN_D A^2}{2\epsilon} \tag{6}$$

is the channel pinch-off voltage, and V_D is the voltage drop in the channel under the gate.

Equation (4) is applicable only for such values of the gate voltage V_G, and drain-to-source voltage V_D that the neutral channel still exists even in the most narrow spot at the drain side, that is

$$A_d(L) = \left[\frac{2\epsilon(V_D + V_{bi} - V_G)}{qN_D} \right]^{1/2} < A \tag{7}$$

If we neglect the effects of velocity saturation in the channel, then we may assume that when the pinch-off condition, that is

$$A_d(L) = A \tag{8}$$

is reached the drain-to-source current saturates. That is why the saturation voltage V_{Dsat} predicted by the constant mobility model (also called the Shockley model) is given by

$$V_{Dsat} = V_{po} - V_{bi} + V_G \tag{9}$$

Substitution of Eq. (9) into Eq. (4) leads to the following expression for the drain-to-source saturation current:

$$(I_{ds})_{sat} = g_0 \left[\frac{V_{po}}{3} + \frac{2(V_{bi} - V_G)^{3/2}}{3V_{po}^{1/2}} - V_{bi} + V_G \right] \tag{10}$$

From Eq. (4) we find that in the linear region the device transconductance

$$g_m = \frac{\partial I_{ds}}{\partial V_G | V_D} = \text{const} \tag{11}$$

is given by

$$g_m = g_0 \frac{(V_D + V_{bi} - V_G)^{1/2} - (V_{bi} - V_G)^{1/2}}{V_{po}^{1/2}} \tag{12}$$

For small drain-to-source voltages

$$V_D \ll V_{bi} - V_G \tag{13}$$

Equations (4) and (12) can be simplified:

$$I_{ds} = g_0 \left[1 - \left(\frac{V_{bi} - V_G}{V_{p0}} \right)^{1/2} \right] V_D \tag{14}$$

$$g_m = \frac{g_0 V_D}{2[V_{po}^{1/2}(V_{bi} - V_G)^{1/2}]} \tag{15}$$

From Eq. (10), we find the transconductance in the saturation region:

$$(g_m)_s = g_0 \left(1 - \left(\frac{V_{bi} - V_G}{V_{po}} \right)^{1/2} \right] \tag{16}$$

The current–voltage characteristics predicted by this model (that coincides with the Shockley model for JFETs) may be presented in the universal dimensionless form using the following dimensionless variables (Fig. 2.5):

$$i_d = \frac{I_{ds}}{g_0 V_{po}} \tag{17}$$

$$i_s = \frac{(I_{ds})_{sat}}{g_0 V_{po}} \tag{18}$$

$$u_G = \frac{V_{bi} - V_G}{V_{po}} \tag{19}$$

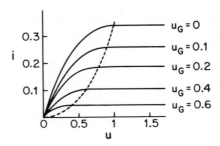

Figure 2.5 Dimensionless current–voltage characteristic of a MESFET predicted by the Shockley model. Dashed line shows the dependence of the dimensionless drain saturation current on drain saturation voltage. (After Shur [78, p. 306].)

and

$$u_i = \frac{V_D}{V_{po}}$$ (20)

2.2.2 Effects of Velocity Saturation

According to the Shockley model, current saturation occurs when a conducting channel is pinched off at the drain side of the gate. At this point, the cross section of the conducting channel, predicted by the Shockley model, is zero and hence the electron velocity has to be infinitely high to maintain a finite drain-to-source current. In reality, the electron velocity saturates in a high electric field and this velocity saturation causes the saturation of the current. The importance of the field dependence of the electron mobility for the understanding of the current saturation in field effect transistors was first mentioned by Dasey and Ross [23]. This concept was later developed in many theoretical models used to describe FET characteristics and interpret experimental results. We first consider a very simple two piece-wise linear approximation for the field dependence of electron velocity (see Fig. 2.4).

Velocity saturation is first reached at the drain side of the gate where the electric field is the highest. It occurs when

$$F(L) = F_s$$ (21)

where $F(L)$ is the magnitude of the electric field in the conducting channel at the drain side of the gate. Using the dimensionless variables

$$u = \frac{V(x)}{V_{po}}$$ (22)

and

$$z = \frac{x}{L}$$ (23)

Eq. (21) may be written as

$$\left|\frac{du}{dz}\right|_{z=1} = \alpha \qquad (24)$$

where

$$\alpha = \frac{F_s L}{V_{po}} \qquad (25)$$

At drain-to-source voltages smaller than the saturation voltage, the electric field in the channel may be found from equations of the Shockley model given in Section 2.2.1:

$$\left|\frac{du}{dz}\right| = \frac{u - (2/3)(u + u_G)^{3/2} + (2/3)u_G^{3/2}}{1 - (u + u_G)^{1/2}} \qquad (26)$$

The dimensionless saturation voltage u_s is then determined from Eq. (24), which may be rewritten as

$$\alpha = \frac{u_s - (2/3)(u_s + u_G)^{3/2} + (2/3)u_G^{3/2}}{1 - (u_s + u_G)^{1/2}} \qquad (27)$$

For large $\alpha \gg 1$ (i.e., $V_{po} \ll F_s L$) the solution of Eq. (27) approaches

$$u_s + u_G = 1 \qquad (28)$$

or

$$V_{Dsat} = V_{po} - V_{bi} + V_G$$

which is identical to the corresponding equation for the Shockley model (see Eq. (9)). The opposite limiting case $\alpha \ll 1$ corresponds to the velocity saturation model:

$$u_s = \alpha \qquad (29)$$

or

$$V_{Dsat} = F_s L$$

For this solution to be valid it is also necessary to have

$$\alpha \ll 2(1 - u_G^{1/2})u_G^{1/2} \qquad (30)$$

The numerical solution of Eq. (27) can be interpolated using a simple interpolation formula [77]:

$$u_s = \frac{\alpha(1 - u_G)}{\alpha + 1 - u_G} \qquad (31)$$

or

$$V_{Dsat} = \frac{F_s L(V_G - V_T)}{F_s L + (V_G - V_T)}$$

where $V_T = V_{bi} - V_{po}$. The saturation current is given by

$$(I_{ds})_{sat} = q N_d v_s W[A - A_d(L)] \qquad (32)$$

where the depletion width $A_d(L)$ at the drain side of the gate is found as

$$A_d(L) = A(u_G + u_s)^{1/2} \qquad (33)$$

In dimensionless units, Eq. (32) may be rewritten as

$$i_s = \alpha[1 - (u_s + u_G)^{1/2}] \qquad (34)$$

In the limiting case $\alpha \to \infty$ (this corresponds to a long gate device with a small pinch-off voltage Eq. (31) can be reduced to the corresponding equation of the Shockley model (see Eq. (28)) and $(I_{ds})_{sat}$ is given by Eq. (10). In the opposite case $\alpha \ll 1$ (short gate and/or large pinch-off voltage):

$$i_s = \alpha(1 - u_G^{1/2}) \qquad (35)$$

This expression corresponds to a simple analytical model of GaAs MES-FETs assuming the complete velocity saturation in the channel [76, 91].

A simple interpolation formula approximates the result of the numerical solution for all values of α (Fig. 2.6):

$$i_s = \frac{\alpha}{1 + 3\alpha}(1 - u_G)^2 \qquad (36)$$

This equation may be rewritten as

$$(I_{ds})_{sat} = \beta(V_G - V_T)^2 \qquad (37)$$

where

$$\beta = \frac{2\epsilon\mu v_s W}{A(\mu V_{po} + 3 v_s L)} \qquad (38)$$

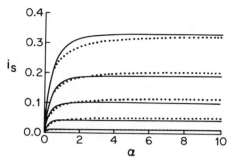

Figure 2.6 Dimensionless saturation current i_s versus α for different values of dimensionless gate voltage u_G [77]: solid lines, numerical calculation; dashed lines, interpolation formula (Eq. (36)). Top curve for $u_G = 0$, step 0.2. (From Shur [77]. Reprinted with permission, copyright 1987, IEE.)

and

$$V_T = V_{bi} - V_{po} \tag{39}$$

In practice, the "square law" (i.e., Eq. (37)) is fairly accurate for devices with relatively low pinch-off voltages, $(V_{po} \leq 1.5 \sim 2\,V)$ (Fig. 2.7). For devices with higher pinch-off voltages, an empirical expression proposed by

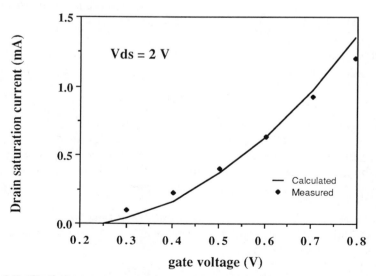

Figure 2.7 Drain-to-source saturation current versus gate voltage for a low pinch-off voltage GaAs MESFET. Solid Line is calculated using Eq. (37). (From Peczalski et al. [63]. Reprinted with permission, copyright 1986, IEEE.)

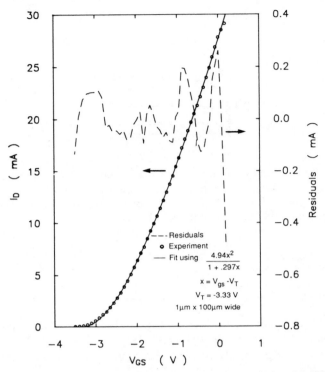

Figure 2.8 Drain-to-source saturation current versus voltage for high pinch-off voltage GaAs MESFET. (After Statz et al. [84]). Reprinted with permission, copyright 1987, IEEE.)

Statz et al. [84]

$$(I_{ds})_{sat} = \frac{\beta(V_G - V_T)^2}{1 + b(V_G - V_T)} \tag{40}$$

provides an excellent fit to the experimental data (Fig. 2.8).

Equation (38) can be used to determine the dependence of β on channel doping, gate length, electron mobility, and saturation velocity. The variation of β with pinch-off voltage is shown in Fig. 2.9. The most interesting feature of these curves is the dramatic increase in transconductance at low pinch-off voltages for devices with submicron gates.

As can be seen from Eq. (38), the values of β (and hence the values of the transconductance for a given voltage swing) increase with the decrease of active layer thickness and with the increase of doping. This increase of β is accompanied by a similar increase in gate capacitance C_g:

$$C_g = \frac{\epsilon L W}{A} \tag{41}$$

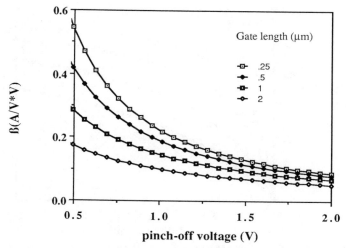

Figure 2.9 Dependence of MESFET transconductance parameter on pinch-off voltage. Parameters used in the calculation: electron mobility $\mu = 0.45 \, \text{m}^2/\text{V} \cdot \text{s}$, electron saturation velocity $v_s = 1.3 \times 10^5 \, \text{m/s}$, dielectric permittivity $\epsilon = 1.14 \times 10^{-10} \, \text{F/m}$, device width 1 mm, doping density $N_d = 10^{17} \, \text{cm}^{-3}$; pinch-off voltage V_{po} is changed by varying the thickness of the active channel A (see Eq. (6)).

However, parasitic capacitances do not increase with an increase in doping or with a decrease in device thickness, therefore thin and highly doped active layers should lead to a higher speed of operation.

Another important advantage of highly doped devices is the reduction of active layer thickness for a given value of pinch-off voltage. This allows to minimize short-channel effects, which become quite noticeable when $L/A < 3$ [20].

Equation (38) shows that low-field mobility becomes increasingly important in low pinch-off voltage devices (Fig. 2.10). As can be seen from Fig. 2.10 for enhancement mode devices with low pinch-off voltages, the high values of low-field mobility in GaAs (up to $4500 \, \text{cm}^2/\text{V} \cdot \text{s}$ in highly doped active layers) lead to a substantial improvement in performance even for short-channel devices where velocity saturation effects are very important. Increase in mobility to $8000 \, \text{cm}^2/\text{V} \cdot \text{s}$ (such values may be achieved in modulation doped field effect transistors (MODFETs) at room temperature [see Section 2.4]) leads to even better performance. However, a further increase of the low field-mobility, say, to $20,000 \, \text{cm}^2/\text{V} \cdot \text{s}$ (such as in modulation doped structures at liquid nitrogen temperature) should not lead to much further improvement. The improvement in MODFET performance at 77 K is probably due to the higher saturation velocity rather than to higher values of low-field mobility. The higher values of low-field mobility may still help reduce the series source resistance, especially in devices with nonself-aligned gates.

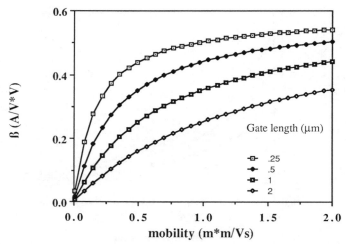

Figure 2.10 Dependence of MESFET transconductance parameter on low-field mobility. Parameters used in the calculation: electron saturation velocity $v_s = 1.3 \times 10^5$ m/s, dielectric permittivity $\epsilon = 1.14 \times 10^{-10}$ F/m, device width 1 mm, doping density $N_d = 10^{17}$ cm^{-3}, pinch-off voltage $V_{po} = 0.6$ V.

The effect of the saturation velocity on the transconductance at the onset of the velocity saturation is shown in Fig. 2.11, which clearly demonstrates the importance of high values of the electron velocity in short-channel devices.

The analysis given above applies only to the values of drain-to-source current and transconductance at the onset of the current saturation when the electron velocity becomes equal to the electron saturation velocity at the drain side of the gate. At higher drain-to-source voltages, the electric field in the conducting channel increases, leading to velocity saturation in a larger fraction of the channel (so-called gate length modulation effect). The implications of gate length modulation were considered, for example, by Grebene and Ghandi [31] and Pucel et al. [68].

Source and drain resistances, R_s and R_d, respectively, may play an important role in determining the current–voltage characteristics of GaAs MESFETs. These resistances can be taken into account as follows. The gate-to-source voltage V_{GS} is given by

$$V_{GS} = V_G + I_{ds}R_s \tag{42}$$

Substituting $V_G = V_{GS} - (I_{ds})_{sat}R_s$ into Eq. (37) and solving for $(I_{ds})_{sat}$ we obtain

$$(I_{ds})_{sat} = \frac{1 + 2\beta R_s(V_{GS} - V_T) - [1 + 4\beta R_s(V_{GS} - V_T)]^{1/2}}{2\beta R_s^2} \tag{43}$$

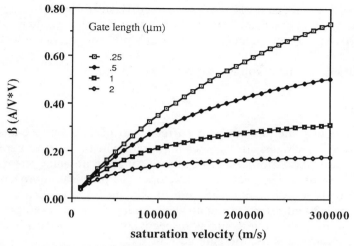

Figure 2.11 Dependence of MESFET transconductance parameter on electron saturation velocity. Parameters used in the calculation: electron mobility $\mu = 0.45 \, \text{m}^2/\text{V} \cdot \text{s}$, dielectric permittivity $\epsilon = 1.14 \times 10^{-10} \, \text{F/m}$, device width 1 mm, doping density $N_d = 10^{17} \, \text{cm}^{-3}$, pinch-off voltage $V_{po} = 0.6 \, \text{V}$.

For the device modeling suitable for computer-aided design (CAD) one has to model the current–voltage characteristics in the entire range of the drain-to-source voltages, not only in the saturation regime. Curtice [19a] proposed the use of a hyperbolic tangent function for the interpolation of MESFET current–voltage characteristics. Based on this idea, Shur [77a] used the following expression:

$$I_{ds} = (I_{ds})_{sat}(1 + \lambda V_{DS}) \tanh \left[\frac{g_{ch}V_{DS}}{(I_{ds})_{sat}} \right] \tag{44}$$

where $(I_{ds})_{sat}$ is given by Eq. (43),

$$g_{ch} = \frac{g_{chi}}{1 + g_{chi}(R_s + R_d)} \tag{45}$$

is the channel conductance at low drain-to-source voltages, and

$$g_{chi} = g_0 \left[1 - \left(\frac{V_{bi} - V_G}{V_{po}} \right)^{1/2} \right] \tag{46}$$

is the intrinsic channel conductance at low drain-to-source voltages predicted by the Shockley model.

Constant λ in Eq. (44) is an empirical constant that accounts for the output conductance. This output conductance may be related to the short-channel effects [68] and also to parasitic currents in the substrate such as space charge limited current. Hence, the output conductance may be greatly reduced using a heterojunction buffer which prevents carrier injection into the substrate [26].

The drain-to-source voltage is given by

$$V_{DS} = V_D + I_{ds}(R_s + R_d) \tag{47}$$

As previously mentioned, gate current also plays an important role in GaAs MESFET circuits. It becomes important when a positive gate voltage forward-biases the Schottky gate junction. An accurate analytical model for the gate current has not yet been developed. A practical approach used in order to fit the experimental data and to account for gate current in circuit simulations is to introduce two equivalent diodes connecting the gate contact with the source and drain contacts respectively. This approach leads to the following expressions for the gate current I_g:

$$I_g = I_{gs} + I_{gd} \tag{48}$$

where

$$I_{gs} = I_{g0} \exp\left[\frac{V_{GS} - I_g R_g - (I_{ds} + I_{gs})R_s}{n_g k_B T}\right] \tag{49}$$

$$I_{gd} = I_{g0} \exp\left[\frac{V_{GD} - I_g R_g - (I_{ds} + I_{gd})R_d}{n_g k_B T}\right] \tag{50}$$

Equations (43–50) form a complete set of equations of the analytical MESFET model used in the GaAs circuit simulator UM-SPICE [36]. The parameters of the model are related to the device geometry, doping and material parameters such as saturation velocity and low-field mobility.

Figure 2.12 compares the I_{DS} and V_{DS} characteristics calculated using this model with the experimental data. MESFET device parameters used in these calculations were determined from the measured $I-V$ characteristics using the following procedure. First, a suitable λ is chosen from the I_{DS} versus V_{DS} curve for one value of V_{GS}. Then, saturation currents are extrapolated to $V_{DS} = 0$ for all values of V_{GS}, keeping λ constant. To obtain the source resistance R_s, one plots $\sqrt{I_{DS}}$ versus $V_{GS} - I_{DS}R_s$ for different values of R_s until a best least square fit is obtained. The slope and intercept of this line give one β and V_T, respectively. The value of built-in voltage V_{bi} is determined from the gate $I-V$ characteristic. Channel thickness and doping are determined using Eq. (6) and the implant dose (i.e., $N_D A$) data. Assuming $R_s = R_d$, one obtains the intrinsic channel resistance R_i from the slope of the I_{DS} versus V_{DS} characteristic in the linear region at large V_{GS}:

$$R_i = R_{DS} - R_s - R_d \tag{51}$$

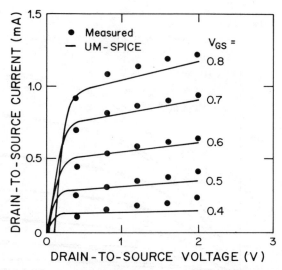

Figure 2.12 I_{DS} versus V_{DS} characteristics of a GaAs MESFET: solid lines, calculated curves; dark circles, experimental data. (From Hyun et al. [36]. Reprinted with permission, copyright 1986, IEEE.)

R_i is related to the gate voltage as follows:

$$R_i = \frac{L}{qA\mu N_d W\{1 - [V_{bi} - V_{GS})/V_{po}]^{1/2}\}} \qquad (52)$$

This equation is used to determine μ. Once μ is known, the saturation velocity is calculated using Eq. (38).

For more accurate device characterization, more sophisticated characterization techniques may be required, such as "end" resistance measurements [50, 95], channel resistance measurements [34] gated transmission line model (GTLM) measurements [77b], and geometric magnetoresistance measurements [38].

The ability to approximate experimental data and the simple procedure for parameter determination make the simple "square law" model attractive for use in computer simulations of GaAs MESFET circuits with relatively low pinch-off voltage FETs. However, this model does not work very well for some ion-implanted devices where the nonuniformity of the doping profile may be important. The transfer characteristics (i.e., $(I_{ds})_{sat}$ versus V_{GS} curves) for transistors with low pinch-off voltages are relatively insensitive to the doping profiles. However, the channel conductance at low drain-to-source voltages is very much dependent on the doping profile. Peczalski et al. [65] proposed an extension of the analytical model described in this section that takes into account the effects of the nonuniform doping profile on the channel conductance. This improved model is in good

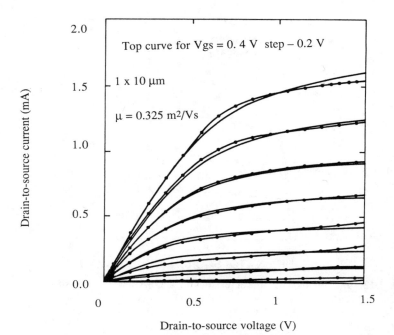

Figure 2.13 I_{DS} versus V_{DS} characteristic of a GaAs MESFET calculated taking into account the nonuniformity of channel doping profile: solid lines, calculated curves; dark circles, experimental data. (From Peczalski et al. [65].)

agreement with experimental data for a large variety of different devices (see, e.g., Fig. 2.13).

For high pinch-off voltage devices, the model can be improved by using an additional empirical parameter b, as was proposed by Statz et al. [84] (see Eq. (40)). Statz et al. [84] also proposed an interpolation formula for drain-to-source current–voltage characteristics that is more computationally efficient compared with the hyperbolic tangent formula (see Eq. (44)):

$$I_{ds} = \frac{\beta(V_G - V_T)^2}{1 + b(V_G - V_T)} \left[1 - \left(\frac{1 - \alpha V_{ds}}{3} \right)^3 \right] (1 + \lambda V_{ds}) \qquad (53)$$

for $0 < V_{ds} < 3/\alpha$ where $\alpha = g_{ch}/(I_{ds})_{sat}$ and

$$I_{ds} = \frac{\beta(V_G - V_T)^2}{1 + b(V_G - V_T)} (1 + \lambda V_{ds}) \qquad (54)$$

for $V_{ds} \geq 3/\alpha$. As can be seen from Fig. 2.14, this model is in excellent agreement with the experimental data. The values of b depend on the device pinch-off voltage and on the channel doping profile, with a more gradual doping profile giving a lower value of b. (The values of b for devices

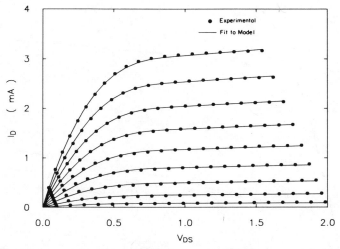

Figure 2.14 I_{DS} versus V_{DS} characteristics of a GaAs MESFET. (From Statz et al. [84]. Reprinted with permission, copyright 1987, IEEE.)

modeled by Statz et al. [84] varied from $0.45\,(V)^{-1}$ to $2.6\,(V)^{-1}$.) Recently, this model was incorporated into the circuit simulator SPICE 3.3 and into the latest version of UM-SPICE.

The analytical models discussed above are suitable for computer-aided design of GaAs MESFETs and GaAs MESFET circuits. However, these models do not explicitly take into account many important and complicated effects: deviations from the gradual channel approximation (which may be especially important at the drain side of the channel [68]), possible formation of a high-field region (i.e., a dipole layer) at the drain side of the channel [27, 28, 80]; inclusion of diffusion and incomplete depletion at the boundary between the depletion region and conducting channel [94]; ballistic or overshoot effects [11, 55, 70, 75, 90] effects of donor diffusion from the n^+ contact regions into the channel [15]; effects of the passivating silicon nitride layers [3, 15]; and effects of traps [14]. These effects may be still included indirectly by adjusting the model parameters, such as μ, v_s, N_d etc. In a rigorous way they can be only treated using numerical solutions. Such solutions provide an insight into the device physics. However, for a practical circuit simulator used in the circuit design, analytical or very simple numerical models remain a must. Beside considerations related to computer time involved in the simulation of hundreds transistors in a circuit, simple models make device characterization easier because of a relatively small number of parameters. Hence, these parameters can be readily measured and compared for different wafers, fabrication processes, etc.

In a short-channel GaAs MESFET, effects related to the overshoot transport become especially important. These effects occur because the

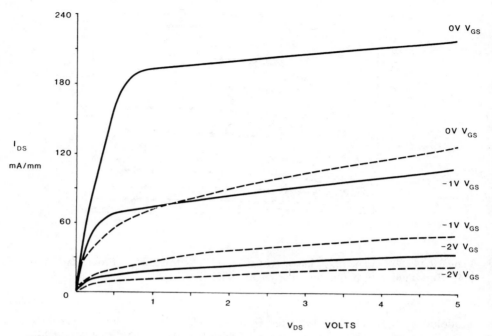

Figure 2.15   Computed current–voltage characteristics of a GaAs MESFET with a 0.3 μm gate length: dashed lines, diffusion-drift model (neglecting overshoot effects); solid lines, energy transport model (taking overshoot effects into account). (From Snowden and Loret [83]. Reprinted with permission, copyright 1987, IEEE.)

electron transit time under the gate becomes comparable to the energy relaxation time. Their importance is illustrated by Fig. 2.15 [83], where we show the current–voltage characteristics of a GaAs MESFET, with a 0.3 μm gate length, calculated with and without taking into account the overshoot effects. As can be seen from the figure, considerably larger currents are predicted by a more accurate model.

2.2.3   GaAs MESFET Capacitances and Small-signal Equivalent Circuit

Internal device capacitances play an important role in determining the speed of GaAs MESFETs and GaAs MESFET circuits. The simplest approach is to model two FET capacitances, gate-to-source capacitance C_{gs} and gate-to-drain capacitance C_{ds}, as capacitances of equivalent Schottky barrier diodes connected between the gate and source and drain, respectively:

$$C_{gs} = \frac{C_{g0}}{(1 - V_{gs}/V_{bi})^{1/2}} \tag{55}$$

$$C_{gd} = \frac{C_{g0}}{(1 - V_{gd}/V_{bi})^{1/2}} \tag{56}$$

Here

$$C_{g0} = \frac{\epsilon WL}{2A_0} = \frac{WL}{2} \left(\frac{q\epsilon N_d}{2V_{bi}} \right)^{1/2} \tag{57}$$

$$A_0 = \left(\frac{2\epsilon V_{bi}}{qN_d} \right)^{1/2} \tag{58}$$

This capacitance model was applied, for example, to junction field effect transistors (JFETs) in a circuit simulator, SPICE. However, this model is completely inadequate for gate voltages smaller than the threshold voltage. Indeed, below the threshold the channel under the gate is totally depleted and the variation of the charge under the gate is only related to the fringing (sidewall) capacitance (Fig. 2.16).

A better capacitance model was proposed by Takada et al. [86]. This model assumes that at gate voltages above the threshold (at gate voltages V_g higher than $V_{t2} > V_T$) the device capacitances are equal to the Schottky gate capacitances and the fringing sidewall capacitances:

$$C_{gs} = \frac{C_{g0}}{(1 - V_{gs}/V_{bi})^{1/2}} + \frac{\pi \epsilon W}{2} \tag{59}$$

$$C_{gd} = \frac{C_{g0}}{(1 - V_{gd}/V_{bi})^{1/2}} + \frac{\pi \epsilon W}{2} \tag{60}$$

(The fringing sidewall capacitances are represented by second terms in the right-hand side of Eqs. (59) and (60).) Well below the threshold ($V_g < V_{t1} <$

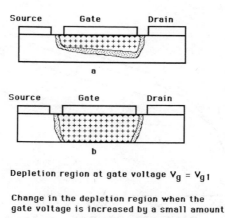

| Depletion region at gate voltage $V_g = V_{g1}$ |
| Change in the depletion region when the gate voltage is increased by a small amount |

Figure 2.16 Changes in the depletion region shape in response to a small variation of gate voltage: (a) gate voltage above threshold, (b) gate voltage below threshold. Below the threshold, the channel under the gate is totally depleted and the variaiton of the charge under the gate is only related to the fringing (sidewall) capacitance.

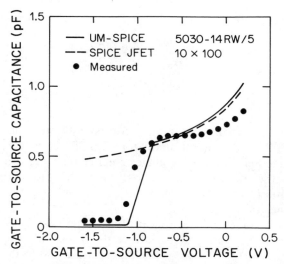

Figure 2.17   Gate-to-source capacitance of a long-channel GaAs MESFET at zero drain-to-source voltage. The curve marked UM-SPICE is calculated using the model proposed by Takada et al. [86]. (From Hyun et al. [36]. Reprinted with permission, copyright 1986, IEEE.)

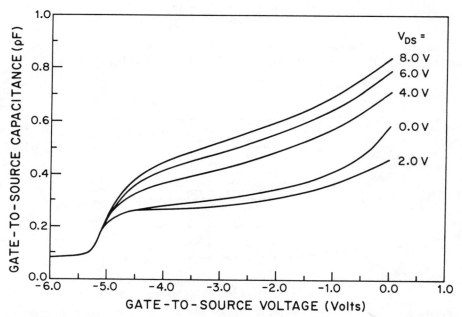

Figure 2.18   Gate-to-source capacitance of a high pinch-off voltage, ion-implanted MESFET ($V_{po} = 5.8$ V) versus gate-to-source voltage. (From Chen and Shur [13]. Reprinted with permission, copyright 1985, IEEE.)

V_T), the internal capacitances are equal to the fringing sidewall capacitances:

$$C_{gs} = \frac{\pi \epsilon W}{2} \tag{61}$$

$$C_{gd} = \frac{\pi \epsilon W}{2} \tag{62}$$

For the gate-voltage range $V_{t1} < V_g < V_{t2}$, capacitances are found by using an interpolation formula. In Fig. 2.17, we compare the predictions of this model with experimental data for a GaAs MESFET gate capacitance [36] and with the simple SPICE model given by Eqs. [55] and [56]).

A more sophisticated model for GaAs MESFET capacitances that takes into account nonuniformity of the doping profile, the effect of traps in a semi-insulating substrate, and a possible formation of a high-field domain (a dipole layer) at the drain side of the gate was developed by Chen and Shur [13]. This model predicts the complicated dependence of the gate-to-source capacitance and gate-to-drain capacitance on the gate and drain voltages (Figs. 2-18–2.23). As can be seen from Figs. 2.20 and 2.23, the results of the calculation qualitatively agree with experimental data.

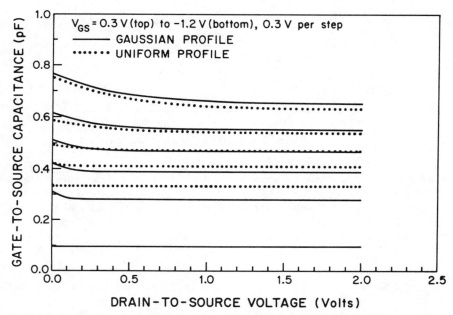

Figure 2.19 Gate-to-source capacitance of a low pinch-off voltage, ion-implanted MESFET ($V_{po} = 1.89$ V) versus drain-to-source voltage. Results of the calculation assuming a uniform doping profile are shown for comparison. (From Chen and Shur [13]. Reprinted with permission, copyright 1985, IEEE.)

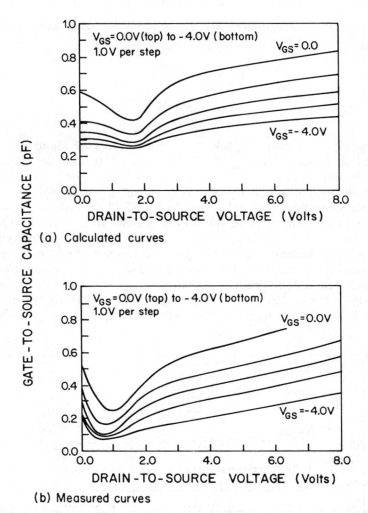

Figure 2.20   Gate-to-source capacitance of a high pinch-off voltage, ion-implanted MESFET ($V_{po} = 5.8$ V) versus drain-to-source voltage: (a) calculated (from Chen and Shur [13] reprinted with permission, copyright 1985, IEEE.), (b) measured (from Willing and de Santis [92] and Willing et al. [93]; reprinted with permission, copyrights 1977, 1978, IEEE).

Using a simplified equivalent circuit of a GaAs MESFET with equivalent gate-to-source and gate-to-drain capacitances, as shown in Fig. 2.24, the circuit model of a GaAs MESFET was implemented in a GaAs circuit simulator, UM-SPICE [36]. (The capacitance model in UM-SPICE is based on the model developed by Takada et al. [86].) This circuit simulator is quite adequate for the design of digital circuits.

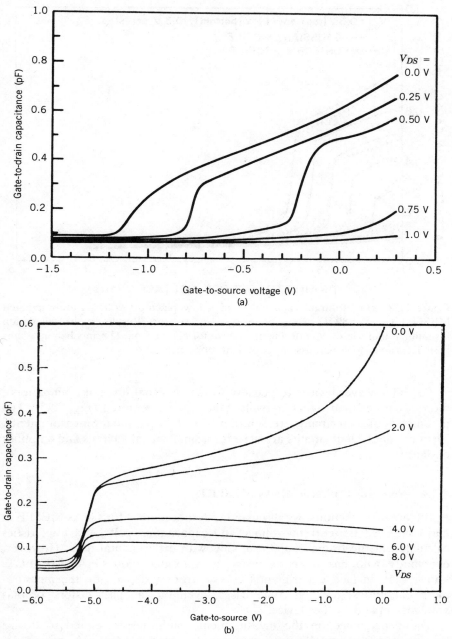

Figure 2.21 Gate-to-drain capacitance versus gate-to-source voltage: (a) low pinch-off voltage, ion-implanted MESFET ($V_{po} = 1.89$ V); (b) high pinch-off voltage, ion-implanted MESFET ($V_{po} = 5.8$ V). (From Chen and Shur [13]. Reprinted with permission, copyright 1985, IEEE.)

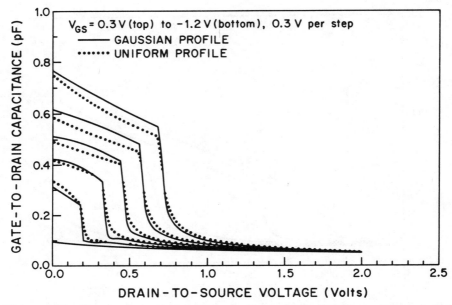

Figure 2.22 Gate-to-drain capacitance of a low pinch-off voltage, ion-implanted MESFET ($V_{po} = 1.89$ V) versus drain-to-source voltage. Results of the calculation assuming a uniform doping profile are shown for comparison. (From Chen and Shur [13]. Reprinted with permission, copyright 1985, IEEE.)

For microwave devices (especially for small-signal low-noise amplifiers), more accurate small-signal equivalent circuits are required (Fig. 2.25). The parameters of such circuits are determined from *S*-parameter measurements. Then the equivalent circuits are used for a small-signal analysis and amplifier design.

2.2.4 Nonideal Effects in GaAs MESFETs

In the previous sections, we discussed different models for GaAs MESFETs and compared the current–voltage and capacitance–voltage characteristics calculated in the frame of these models with experimental data. However, everybody who has done extensive measurements on GaAs MESFETs knows that, in fact, experimental curves strongly depend on the measurement conditions, and, in some cases, even on the past history of bias voltages applied to the device.

The dependence on the measurement conditions is related to device heating and is especially important for depletion mode, high-current devices. In such devices, the *I–V* characteristics measured at dc frequently show regions of negative differential resistance (see, e.g., Fig. 2.26). The current–voltage characteristics of the same device measured using short

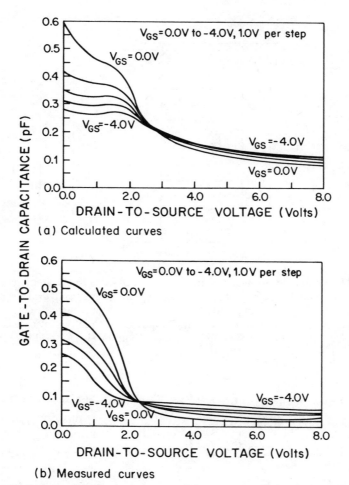

Figure 2.23 Gate-to-drain capacitance of a high pinch-off voltage, ion-implanted MESFET ($V_{po} = 5.8$ V) versus drain-to-source voltage: (a) calculated (from Chen and Shur [13] reprinted with permission, copyright 1985, IEEE.), (b) measured (from Willing and de Santis [92] and Willing et al. [93]; reprinted with permission, copyrights 1977, 1978, IEEE).

pulses does not show any negative resistance. A very clear example of such behavior was recently presented by Pucel [67]. He also pointed out that the conditions of device breakdown are quite different at dc and ac, as avalanche breakdown takes time to develop.

Deep traps are another important factor. Both shallow impurities and deep traps exist in GaAs substrates [39, 56]. The density of so-called EL2 traps typically varies between 1 and 6×10^{16} cm^{-3} [42]. The electron emission time from the trap is an exponential function of temperature and trap

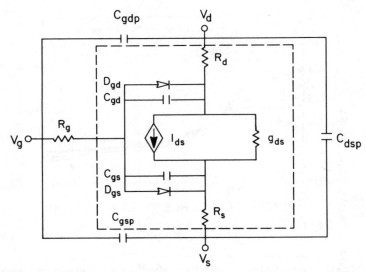

Figure 2.24 Simplified equivalent circuit of a GaAs MESFET implemented in a GaAs circuit simulator, UM-SPICE: R_s and R_d are source and drain series resistances, respectively; diodes D_{gd} and D_{gs} account for the gate current; C_{gdp}, C_{dsp}, and C_{gsp} are parasitic capacitances. The same equivalent circuit is also used for the modeling of MODFETs (see Section 2.4.3, and Hyun et al. [36] and [37]; reprinted with permission, copyrights 1986, IEEE.)

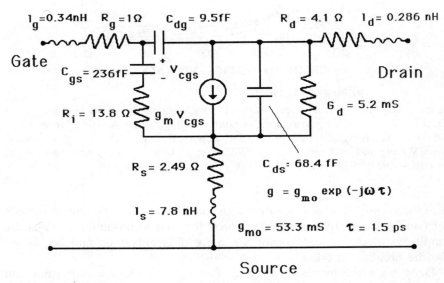

Figure 2.25 Small-signal equivalent circuit of GaAs MESFET fabricated on silicon substrate: $L = 0.25\ \mu\text{m}$, $W = 100\ \mu\text{m}$, $V_{ds} = 3.1\ \text{V}$, $V_{gs} = 0.53\ \text{V}$, $I_{ds} = 32\ \text{mA}$. (From Aksun et al. [1]. Reprinted with permission, copyright 1986, IEEE.)

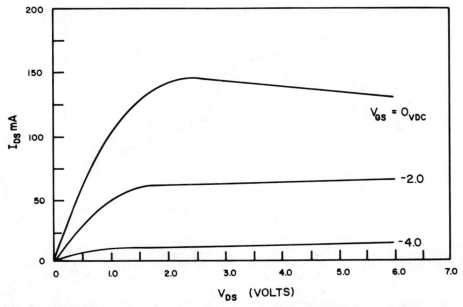

Figure 2.26 Current–voltage characteristics of GaAs MESFETs measured at dc. Notice the region of negative differential resistance (related to thermal effects) at high drain currents. (From Willing and de Santis [92] and Willing et al. [93]; reprinted with permission, copyrights 1977, 1978, IEEE.)

energy level. This leads to the temperature and bias dependence of the effective carrier concentration in the channel and makes the device response dependent on previous bias conditions and the signal frequency at relatively low frequencies. For example, let us consider the temperature dependence of the threshold voltage [14]. At high temperaturs, most EL2 levels located above the Fermi level are empty. At low temperatures, even the traps located above the Fermi level may be fully occupied. Because of the exponential temperature dependence of the emission rate, the transition between these two regimes is rather sharp. The temperature variation of the GaAs MESFET threshold voltage shown in Fig. 2.27 is stongly affected by traps (see the abrupt change in threshold voltage at $T \approx 0°C$).

Another important nonideal effect is the dependence of the parameters of GaAs FETs on the gate orientation on the wafer with respect to the crystallographic directions. Such an orientation effect was reported by Lee et al. [45] and by Yokoyama et al. [96]. Asbeck et al. [3] suggested that this orientation dependence could be related to a piezoelectric stress effect. The dielectric passivation layers (typically silicon nitride) in the gate–source and gate–drain regions cause compressive or tensile stress, leading to piezoelectric charges induced in the channel and in the substrate (Fig. 2.28). The stress depends upon gate orientation on the wafer, thickness of the passivat-

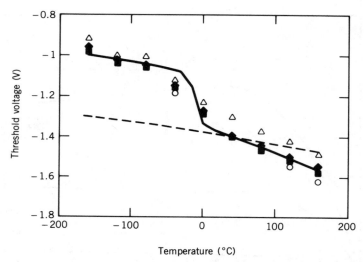

Figure 2.27 The temperature variation of GaAs MESFET threshold voltage. Different symbols correspond to different techniques of measuring the threshold voltage: solid line, calculation taking into account traps; dotted line, calculation neglecting the effect of traps on the threshold voltage. The implanted dose is 2.4×10^{12} cm^{-2} and the gate length is 10 μm. (From Chen et al. [14]. Reprinted with permission, copyright 1986, IEEE.)

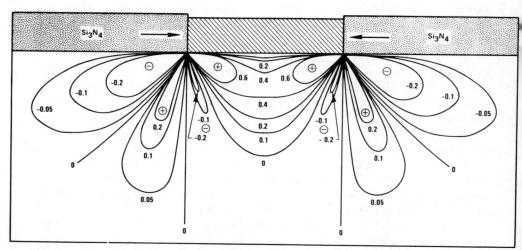

Figure 2.28 Calculated normalized piezoelectric charge density induced in the channel and in the substrate of a GaAs MESFET. (From Asbeck et al. [3]. Reprinted with permission, copyright 1984, IEEE.)

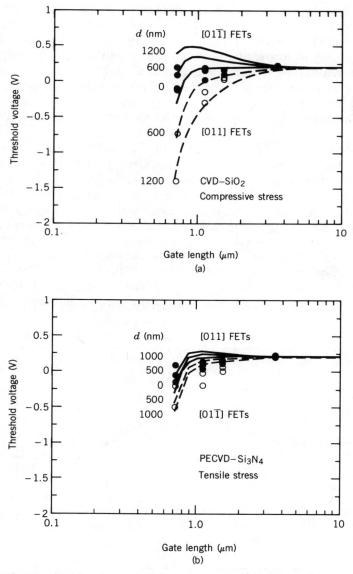

Figure 2.29 Threshold voltage versus gate length with the dielectric overlayer thickness as the parameter: (a) silicon dioxide overlayer, (b) silicon nitride overlayer. Calculated curves from Chen et al. [15], (reprinted with permission, copyright 1987, IEEE.); measured data from Ohnishi et al. [61].

ing dielectric film, and the gate length. Hence, the piezoelectric effect leads to dependence of the device parameters, such as the threshold voltage and transconductance parameter, on gate orientation on the wafer, thickness of the passivating dielectric film, and the gate length (Figs. 2.29 and 2.30).

In the self-aligned device, the gate acts as an implantation mask for the n^+ source and drain contacts. During implant annealing, the donors diffuse from the n^+ contacts into the channel, changing the carrier concentration there. The schematic effective doping profile versus depth is shown in Fig. 2.31.

Matsumoto et al. [57] fabricated GaAs MESFETs with gate lengths of 0.8 μm, 0.6 μm, and 0.4 μm. Se ions were implanted to form ohmic contacts for the drain and source with no implantation for the n-channel. The device with a 0.8 μm, gate acted as a normal FET. The output conductance and current level of the device with a 0.6 μm gate at the same drain and gate voltages were larger than those of the 0.8 μm device. The MESFET with a 0.4 μm gate exhibited a much larger pinch-off voltage and less controllability of the drain current by the gate bias, suggesting that the two n^+ regions were effectively in contact with each other. These results clearly showed that n^+ dopants diffused heavily into the channel. For a realistic description of the gate length dependence of the threshold voltage

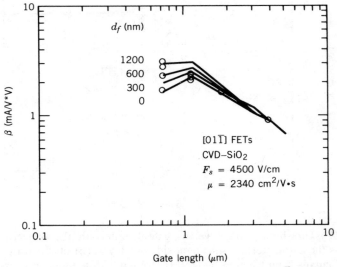

Figure 2.30 Transconductance parameter β versus gate length with the dielectric overlayer thickness as the parameter. Device transconductance $g_m \approx \beta(V_{gs} - V_T)$ where V_T is the threshold voltage (this equation is valid only at relatively small gate-to-source voltages when the gate current may be neglected). Calculated curves from Chen et al. [15], (reprinted with permission, copyright 1987, IEEE.); measured data from Onodera et al. [62].

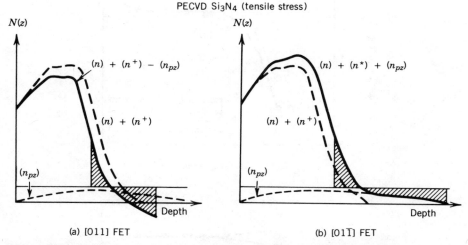

Figure 2.31 Effective doping profiles for GaAs ion-implanted FETs. The term (n^+) represents the donors diffused from the n^+ ohmic contact regions into the channel, the term (n_{pz}) represents the density of induced piezoelectric charges. Shaded areas correspond to the depletion region at the channel–substrate interface. (After Chen et al. [15]. Reprinted with permission, copyright 1987, IEEE.)

of self-aligned MESFETs, the lateral spread of the n^+ layer should be taken into account [15].

More recently, Ueto et al. [88] demonstrated that, by using a 0.2 μm thickness of SiO_2 sidewall (Fig. 2.32), the threshold voltages of submicron devices could be made to be about the same as those of the long-gate devices (≥ 3 μm). Kato et al. [41] achieved a similar result by using a T-gate structure (Fig. 2.33).

So-called "backgating" and "sidegating" effects play an important role in GaAs circuits. These effects associated with traps in the substrate [10, 29, 44, 53, 69] are illustrated by Fig. 2.34, which shows how a change of the potential of a side contact affects the drain-to-source saturation current of a GaAs MESFET at zero gate voltage.

The nonideal effects discussed above cause so-called technological problems in fabrication of GaAs devices and circuits, such as a large spread in device parameters from run to run and across the wafer. In addition, factors such as a large temperature variation of device parameters, strong dependence of these parameters on the properties and thickness of a passivating dielectric layer, and large subthreshold currents and output conductance complicate the circuit design and reduce design flexibility, making large-scale commercialization of GaAs technology a challenge. An adequate understanding of the nature of many nonideal effects in GaAs devices is a necessary prerequisite for meeting this challenge.

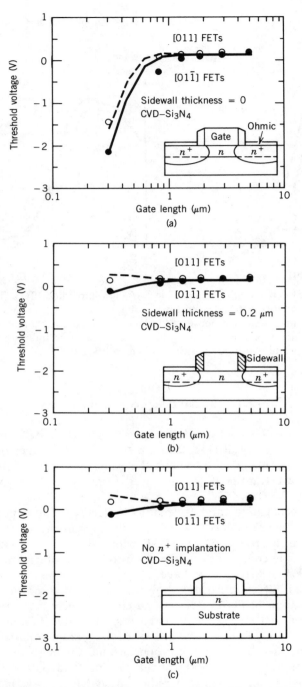

Figure 2.32 Effect of sidewall on threshold voltage of ion-implanted GaAs MES-FETs. (a) Schematic drawing of the doping implants made without the sidewall (the sidewall was made after n⁺ implantation) and the dependence of the threshold voltage on the gate length for two different gate orientations of this structure. (b) Schematic drawing of the doping implants with sidewall structure (sidewall is shown

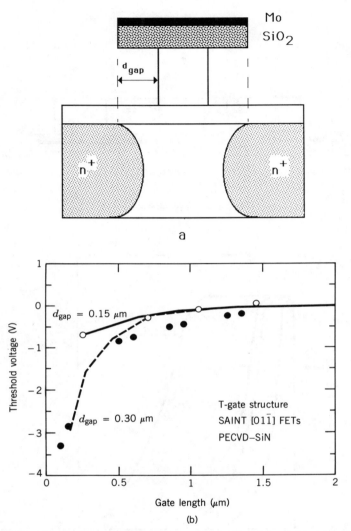

Figure 2.33 Effect of T-gate structure on threshold voltage of ion-implanted GaAs MESFETs: (a) schematic drawing of the T-gate structure, (b) threshold voltage versus gate length for two different T-gate structures. Calculated curves from Chen et al. [15]; measured data from Kato et al. [41]. (Reprinted with permission, copyrights 1987 and 1983, IEEE.)

by shaded region) and the dependence of the threshold voltage on the gate length for two different gate orientations of this structure. Sidewall thickness is 0.2 μm. (c) Schematic drawing of the doping implant without the n[+] implants (the sidewall was added for the process convenience) and the dependence of the threshold voltage on the gate length for two different gate orientations for this structure. Calculated curves from Chen et al. [15]. (Reprinted with permission, copyright 1987, IEEE.); measured data from Ueto et al. [88].

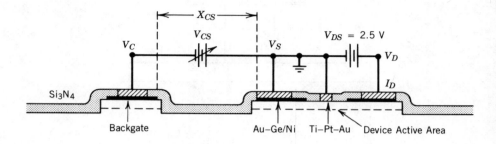

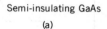

(a)

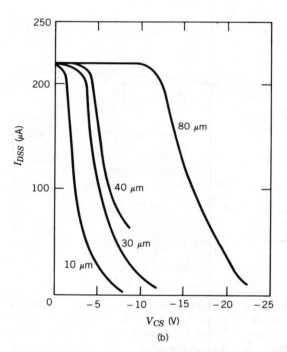

(b)

Figure 2.34 (a) Cross section of a GaAs MESFET with a backgating contact, (b) drain-to-source saturation current of a GaAs MESFET at zero gate voltage versus voltage applied to backgating contact for different distances from the backgating contact to the source. (From Birittella et al. [10]. Reprinted with permission, copyright 1982, IEEE.)

2.3   UNGATED GaAs FETs

Saturated resistors (or ungated FETs) [18, 46, 51, 52] are frequently used as load elements in GaAs logic gates. Current saturation in these structures is a consequence of electron velocity saturation.

The free surface of GaAs is depleted and the surface Fermi level is pinned by the high density of the surface states. The shape and the thickness of this depletion layer determine the low-field resistance and saturation current of ungated loads. Hariu et al. [33] proposed a model for GaAs MESFETs which took into account the change of the surface potential due to voltages applied to the gate and drain. The surface potential was assumed to vary linearly along the surface, provided that the source-to-gate separation is much larger than the depletion depth, and the electric field is almost perpendicular to the surface. Under these conditions, the resistivity of the channel is modulated not only by the channel potential but also by the surface potential. Baek et al. [7] employed this model to describe ungated GaAs FETs with uniform and nonuniform (ion-implanted) doping profiles and to deduce the electron saturation velocity as a function of device length.

Figure 2.35 shows a schematic cross section of an ungated FET with applied drain voltage. Using the gradual channel approximation, the depth of the surface depletion layer $h(x)$ may be found as

$$\frac{qN_d}{2\epsilon}\, h^2(x) = V_{sbi} + V(x) - \phi_s(x) \tag{63}$$

where N_d is the uniform doping density, V_{sbi} is the surface built-in voltage, $V(x)$ the channel potential, and $\phi_s(x)$ the surface potential. This is equivalent to an FET with a distributed gate whose potential is $\phi_s(x)$. The surface potential is assumed to vary linearly along the surface according to

$$f_s(x) = \frac{V_D}{L}\, x \tag{64}$$

where V_D is the applied drain voltage and L the source-to-drain separation.

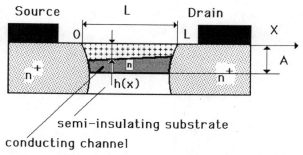

Figure 2.35   Schematic cross section of ungated FET with applied drain voltage. (After Baek et al. [7]. Reprinted with permission, copyright 1985, IEEE.)

The surface potential at the source end is equal to zero, whereas the channel potential at the source side is equal to $+I_D R_c$, where I_D is the device current and R_c is the contact resistance. The surface potential at the drain end is equal to V_D whereas the channel potential at the drain side is equal to $V_D - I_D R_c$. Hence, the source side of channel is more reverse-biased with respect to the surface than the drain side. As a consequence, the conducting channel is more narrow at the source end (see Fig. 2.35), which is quite different from what is expected in gated MESFETs (compare with Fig. 2.2).

An ungated FET model may be developed using a two-piece linear approximation for the dependence of electron velocity on electric field shown in Fig. 2.4. First, we model the ungated FET in the linear region of velocity versus electric field curve, assuming constant electron mobility μ, and then in the saturation region, assuming constant saturation velocity v_s.

In the linear region, the drain current I_D is given by

$$I_D = q\mu N_d W [A - h(x)] \frac{dV(x)}{dx} \tag{65}$$

The boundary conditions for $V(x)$ are

$$V(0) = R_c I_D \tag{66}$$

$$V(L) = V_D - R_c I_D \tag{67}$$

where R_c is the ohmic contact resistance under the source and the drain. Integrating Eq. (65) from $x = 0$ to L leads to an implicit expression for I_D [33]:

$$KL = \frac{1}{2} [h^2(0) - h^2(L)] - d[h(0) - h(L)] + d(d - A) \ln \left[\frac{h(0) + d - A}{h(L) + d - A} \right] \tag{68}$$

where

$$h(0) = \left[\left(\frac{2\epsilon}{qN_d} \right) (V_{sbi} + R_c I_D) \right]^{1/2} \tag{69}$$

$$h(L) = \left[\left(\frac{2\epsilon}{qN_d} \right) (V_{sbi} - R_c I_D) \right]^{1/2} \tag{70}$$

$$K = \frac{\epsilon V_D}{qN_d L} \tag{71}$$

$$d = \frac{L I_D}{G V_D} \tag{72}$$

$$G = Wq\mu N_d \tag{73}$$

The solution of Eq. (68) yields I_D in the linear region for a given V_D. In the limiting case of $R_c = 0$, the thickness of the depletion layer becomes uniform throughout the channel, and under this condition, the drain current is given by

$$I_D = GA_c \frac{V_D}{L} \tag{74}$$

where A_c is the active channel thickness:

$$A_c = A\left[1 - \left(\frac{V_{sbi}}{V_{po}}\right)^{1/2}\right] \tag{75}$$

Velocity saturation starts when the electric field reaches the critical value, $F_s = v_s/\mu$, at the source end of the channel. The corresponding saturation current I_{Dsat} is given by

$$I_{Dsat} = WqN_d\left\{A - \left[\frac{2\epsilon}{qN_d}\left(V_{sbi} + R_cI_{Dsat}\right)\right]^{1/2}\right\}v_s \tag{76}$$

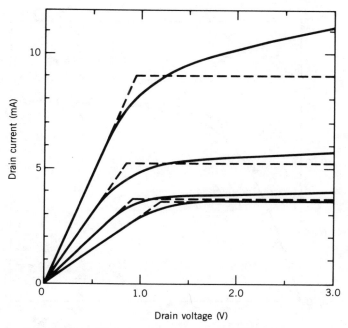

Figure 2.36 Calculated and measured I_D–V_D characteristics of an ungated FET. Parameters used in the calculation: $V_{sbi} = 0.46\,\text{V}$, $v_s = 1.2 \times 10^5\,\text{m/s}$, $R_c = 0.88\,\Omega$-mm. (After Baek et al. [7]. Reprinted with permission, copyright 1985, IEEE.)

Solving Eq. (76) for I_{Dsat}, we find

$$I_{Dsat} = I_{FC}\left[1 + \frac{\eta}{2} - \left(\frac{\eta^2}{4} + \eta + \zeta\right)^{1/2}\right] \qquad (77)$$

where $I_{FC} = WAqN_d v_s$, $\eta = R_c I_{FC}/V_{po}$, $\zeta = V_{sbi}/V_{po}$, and $V_{po} = qN_d A^2/(2\epsilon)$. In the limiting case of $R_c = 0$, I_{Dsat} is given by

$$I_{Dsat} = WqN_d A_c v_s \qquad (78)$$

The saturation voltage V_{Dsat} can be found from Eq. (68) by substituting I_{Dsat} for I_D and solving for V_D. The results obtained above are valid for a uniform doping profile. Similar expressions for a nonuniform, ion-implanted profile were obtained by Baek et al. [7].

Figure 2.36 compares calculated and measured I_D–V_D characteristics of an ungated FET. As can be seen from the figure, below saturation the ungated FET behaves like a resistor:

$$R_{off} = \left(\frac{V_D}{I_D}\right)\bigg|_{I_D \to 0} = 2R_c + R_{ch} \qquad (79)$$

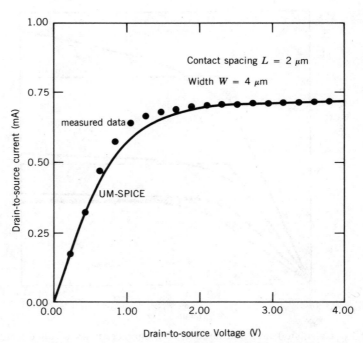

Figure 2.37 Comparison of measured I_D–V_D characteristics of an ungated FET with an empirical equation used in UM-SPICE. (After Hyun [35]. Reprinted with permission, copyright 1985, IEEE.)

where $R_{ch} = L/Wq\mu N_d A_c$. Even close to the saturation, the deviation of the slope of the current–voltage characteristics from R_{off} is less than 3%. This may be attributed partly to the assumption that the low-field mobility is constant up to the saturation, and partly to the fact that the depletion thickness at the source end increases with the current, while at the drain end it decreases so that the overall change in the shape of the depletion region is not enough to cause a large variation of the resistivity of the active channel.

Measured current–voltage characteristics do not exhibit an abrupt saturation. A simple empirical model of an ungated FET used in the circuit simulator UM-SPICE [36] is similar to the model used for GaAs MESFETs:

$$I = I_{Dsat} \tanh\left(\frac{V_D}{V_{lss}}\right)(1 + \lambda V_D) \tag{80}$$

where

$$V_{lss} = I_{Dsat} R_{off} \tag{81}$$

As can be seen from Fig. 2.37, this equation is in reasonable agreement with experimental data.

2.4 MODELS FOR AlGaAs/GaAs HETEROSTRUCTURE FIELD EFFECT TRANSISTORS

2.4.1 Heterostructure Field Effect Transistors

Heterostructure AlGaAs/GaAs and AlGaAs/InGaAs/GaAs field effect transistors (HFETs), also called modulation doped field effect transistors (MODFETs), high electron mobility transistors (HEMTs), and selectively doped heterojunction transistors (SDHTs), offer excellent ultrahigh-speed performance. Propagation delays as low as 5.8 ps at 77 K [74] and 6 ps at 300 K [60] have been obtained in HFET ring oscillator circuits.

Most HFETs are n-channel devices. A schematic cross section of an n-channel HFET and the band diagram of the structure are shown in Fig. 2.38. As can be seen from the figures, a two-dimensional (2D) electron gas is formed in the unintentionally doped GaAs buffer layer at the heterointerface. More recently, complementary n- and p-channel heterostructure insulated-gate field effect transistors (HIGFETs) were developed (Fig 2.39), offering potential for high-speed low-power operation [19, 21, 22, 43, 72].

A different type of an HFET utilizes a plane of dopants in the AlGaAs layer. A band diagram and a schematic cross section of such a device (that, to our knowledge, was first proposed by Eastman [25]) are shown in Fig. 2.40. This device has two distinct advantages. First of all, doping of AlGaAs in conventional HFETs leads to many undesirable effects related to traps associated with dopants (see, e.g., Shur [78, p. 583]). In planar doped

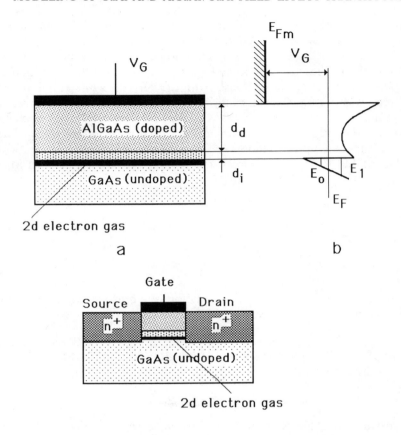

Figure 2.38 (a) Schematic diagram of n-channel modulation doped structure, (b) band diagram of modulation doped structure, (c) cross section of a self-aligned n-channel MODFET: d_d is the thickness of doped AlGaAs layer, d_i is the thickness of undoped AlGaAs spacer layer, E_0 and E_1 are energies of two lowest subbands, E_F is the electron quasi-Fermi level, and V_G is the applied gate voltage.

devices (sometimes called delta-doped devices), the effect of traps is diminished. Also, as will be discussed below, planar doped devices can operate with larger gate-voltage swings.

The same goal—trying to diminish the effects of traps in the AlGaAs layer—led to the development of superlattice HFETs [5, 74, 82]. In this device, dopants are incorporated into several thin GaAs quantum wells that form a superlattice structure in the AlGaAs layer.

Since the lattice constants of GaAs and AlGaAs are nearly equal, the above mentioned AlGaAs/GaAs heterostructures are "lattice-matched." However, the lattice constants of InGaAs and GaAs (or AlGaAs) differ more appreciably. Nevertheless, AlGaAs/InGaAs/GaAs structures of very

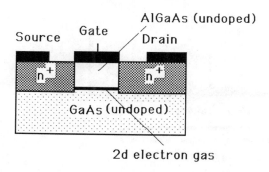

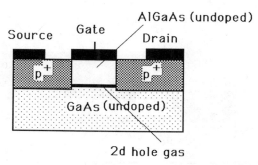

Figure 2.39 Complementary n- and p-channel heterostructure insulated-gate field effect transistors (HIGFETs).

high quality can be grown by molecular beam epitaxy (MBE) if the thickness of the InGaAs layer is kept fairly small (typically on the order of 150 Å). The advantage of such structures is higher electron velocity and electron mobility in InGaAs. The AlGaAs/InGaAs/GaAs device structures are called "pseudomorphic" or "strained-layer". A schematic cross section of a recently developed n-channel AlGaAs/InGaAs/GaAs HFET and the band diagram of the structure are shown in Fig. 2.41 [12].

All these different HFETs operate in a similar way. A two-dimensional gas of carriers is induced at the heterointerface. The density of this gas and, hence, the device current are modulated by the gate voltage. The composition and doping profile of the AlGaAs layer (sometimes called the charge control layer) determine the device threshold voltage and gate-voltage swing.

2.4.2 Density of Two-dimensional Gas

The charge density of the two-dimensional (2D) electron gas in an n-channel device, n_s, and the charge density of the 2D hole gas, p_s, versus the gate voltage may be calculated numerically by solving Poisson's equation and the

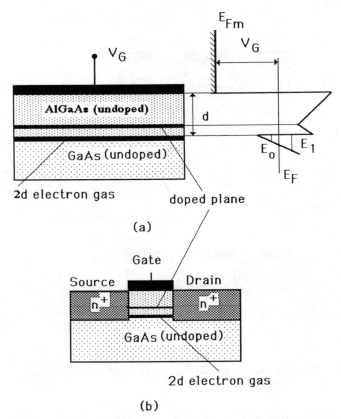

Figure 2.40 Band diagram and schematic cross section of planar doped HFET.

Schrödinger equation in a self-consistent manner [9, 47, 85]. To a first order, the concentration of 2D electron gas induced into the device channel at zero drain-to-source voltage is given by [24]:

$$n_s \approx \frac{C_0(V_g - V_t)}{q} \qquad (82)$$

Here

$$C_0 \approx \frac{\epsilon_1}{(d + \Delta d)} \qquad (83)$$

is the gate capacitance per unit area, V_t is the threshold voltage:

$$V_t = \phi_b - \Delta E_c - \frac{q}{\epsilon_1} \int_0^d N_d(x)x\,dx \qquad (84)$$

where ϕ_b is the metal barrier height, N_d is the doping density in the $Al_xGa_{1-x}As$ layer, q is the electronic charge, d is the distance between the

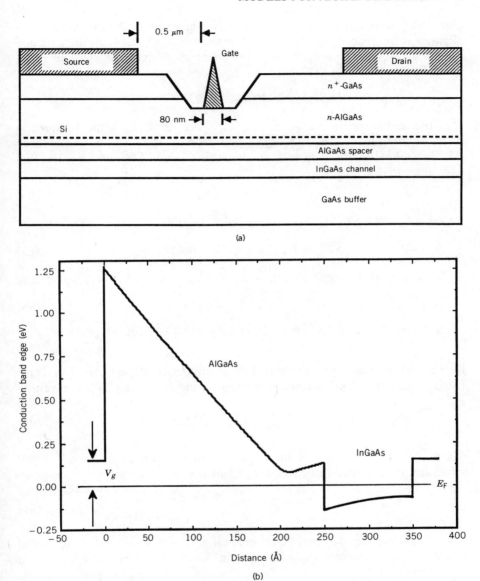

(a)

(b)

Figure 2.41 (a) The cross-sectional view of the 80 nm gate length planar doped pseudomorphic HFET. (b) computed energy band diagram of planar doped Al-GaAs/InGaAs/GaAs quantum well structure. The applied gate voltage V_g is negative. Notice a large conduction band discontinuity at the heterointerface between the AlGaAs layer and the InGaAs quantum well. Also notice a conduction band discontinuity between the InGaAs quantum well and GaAs buffer layer that limits the electron spillover into the buffer, resulting in a reduction in the output conductance. The planar doping leads to a shallow potential minimum in the charge-control layer, limiting the electron transfer into the AlGaAs layer. (From Chao et al. [12]. Reprinted with permission, copyright 1989, IEEE.)

gate and heterointerface, Δd is the effective thickness of the two-dimensional gas, x is the space coordinate ($x = 0$ at the boundary between the gate and AlGaAs layer), $\epsilon_1 = \kappa_1 \epsilon_0$ is the dielectric permittivity of $Al_xGa_{1-x}As$, $\epsilon_0 = 8.854 \times 10^{-14}$ F/cm,

$$\kappa_1 \approx 13.18 - 3.12X \tag{85}$$

X is the molar fraction of Al in $Al_xGa_{1-x}As$, and ΔE_c is the conduction band discontinuity. For a general discussion of ΔE_c let us consider an $Al_xGa_{1-x}As/In_yGa_{1-y}As$ system.[†] For this system, ΔE_c can be calculated as follows [54]:

$$\Delta E_c = \Delta E_g - \Delta E_v \tag{86}$$

where $\Delta E_v = 0.4 \Delta E_{gg}$ is the valence band discontinuity,

$$\Delta E_{gg} = 1.247X + 1.5Y - 0.4Y^2 (eV) \tag{87}$$

is the difference between Γ-valleys in $Al_xGa_{1-x}As$ and $In_yGa_{1-y}As$, and

$$\Delta E_g = \Delta E_{gg} \quad \text{for } X < 0.45 \text{ (eV)} \tag{88}$$

$$\Delta E_g = 0.476 + 0.125X + 0.143X^2 + 1.5Y - 0.4Y^2 \quad \text{for } X \geq 0.45 \text{ (eV)} \tag{89}$$

is the energy gap discontinuity. For a uniformly doped AlGaAs layer ($N_d(x) = $ constant), the threshold voltage, found from Eq. (84), is given by

$$V_t = \phi_b - \Delta E_c - \frac{qN_dd_d^2}{2\epsilon_1} \tag{90}$$

where d_d is the thickness of the doped AlGaAs layer. For a planar doped structure, the evaluation of the integral in Eq. (84) yields

$$V_t = \phi_b - \Delta E_c - \frac{qn_dd_d}{\epsilon_1} \tag{91}$$

where d_d is the distance between the metal gate and the doped plane and n_d is the surface concentration of donors in the doped plane. The calculated dependence of V_t on d_d for uniformly doped and planar doped devices is compared in Fig. 2.42.

Equation (83) does not take into account several effects that may be quite important, such as real-space transfer of electrons into the AlGaAs layer, which may play a crucial role in modulation doped devices, nonlinear dependence of the Fermi level on the carrier concentration of the 2D

[†](Editor's Note) Discussions of band discontinuity for systems like $Al_xGa_{1-x}As$/GaAs can be found in Chap. 1, Section 1.2; Chap 3, Section 3.2.1.5; Chap. 4, Section 4.2.1; and references cited within. One may note that if $Y = 0$, Eq. (88) will be the same as Eq. (3) in Chap. 3.

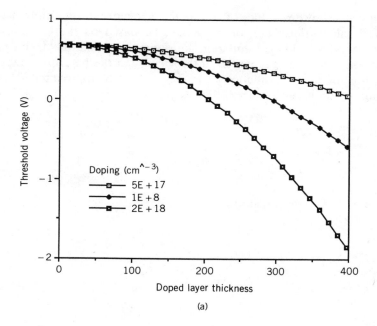

(a)

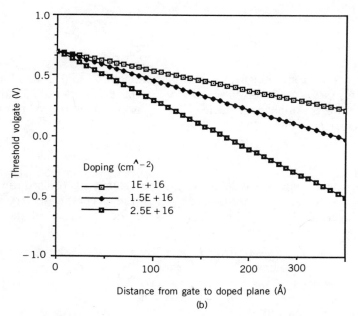

(b)

Figure 2.42 Threshold voltage of (a) conventional and (b) planar doped HFETs. (From Chao et al. [12]. Reprinted with permission, copyright 1989, IEEE.)

electron gas, n_s, at low values of n_s gate leakage current, etc. More accurate numerical calculations of n_s on V_g should be based on the self consistent solution of the Schrödinger equation and Poisson's equation which describe the electron density and potential in the vicinity of the heterointerface [85]. The results of such calculations can be very well approximated by the following interpolation formulas [9]:[†]

$$n_s = \frac{C_0[V_g - V_{cn} - K_n(V_g - V_{rn})^{1/2}]}{q} \tag{92}$$

$$p_s = \frac{C_0[-V_g - V_{cp} - K_p(-V_g - V_{rp})^{1/2}]}{q} \tag{93}$$

where V_g is the intrinsic gate-to-source voltage,

$$V_{cn} = V_{bin} - V_{on} - \frac{C_0 A_n^2}{2q} \tag{94}$$

$$V_{rn} = V_{bin} - V_{on} - \frac{C_0 A_n^2}{4q} \tag{95}$$

$$V_{cp} = V_{bip} - V_{op} - \frac{C_0 A_p^2}{2q} \tag{96}$$

$$V_{rp} = V_{bip} - V_{op} - \frac{C_0 A_p^2}{4q} \tag{97}$$

$$V_{bin} = \phi_b - \Delta E_c - \frac{q N_d d_d^2}{2\epsilon_1} \tag{98}$$

ϕ_b is the metal barrier height, ΔE_c is the conduction band discontinuity at the (AlGA)As/GaAs heterointerface, N_d is the donor density in the Al-GaAs layer of the n-channel device, and d_d is the thickness of the doped AlGaAs layer (see Fig. 2.38).

$$V_{bip} = E_{gA} - \phi_b - \Delta E_v + \frac{q N_a d_d^2}{2\epsilon_1} \tag{99}$$

E_{gA} is the energy gap of AlGaAs, ΔE_v is the valence band discontinuity at the (AlGa)As/GaAs heterointerface, N_a is the acceptor density in the AlGaAs layer of the p-channel device, d_d is the thickness of the doped AlGaAs layer,

$$K_n = \left(\frac{C_0 A_n^2}{q}\right)^{1/2} \tag{100}$$

$$K_p = \left(\frac{C_0 A_p^2}{q}\right)^{1/2} \tag{101}$$

[†](Editor's Note) Ref. [9] discusses a simple relation between n_s and V_g in a (AlGa)As/GaAs system. The approach was later expanded and applied to a (AlGa)As/InGaAs system in [12], to be discussed in the latter part of this section.

and A_n, A_p, V_{on}, and V_{op} are functions of temperature T:

$$A_n = 5.22 \times 10^{-16}T^2 + 1.74 \times 10^{-12}T + 1.88 \times 10^{-9}(\text{V/m}) \quad (102)$$

$$V_{on} = 1.66 \times 10^{-7}T^2 + 2.36 \times 10^{-4}T + 3.43 \times 10^{-2} \quad (\text{V}) \quad (103)$$

$$A_p = 2.31 \times 10^{-12}T + 8.06 \times 10^{-10} \quad (\text{V/m}) \quad (104)$$

$$V_{op} = 1.41 \times 10^{-7}T^2 + 4.12 \times 10^{-4}T + 4.48 \times 10^{-3} \quad (\text{V}) \quad (105)$$

The computed dependence of n_s and p_s on V_g is shown in Figs. 2.43 and 2.44, where they are compared with the simple linear approximation given by Eq. (82). However, these calculations do not account for possible electron transfer into the AlGaAs layer. Let us consider how we can compute surface electron carrier concentrations in the 2-D electron gas and in the charge-control AlGaAs layer, n_s and n_t, and the potential distribution in the modulation doped structure for different gate biases under zero drain bias (e.g., see Chao et al. [12], and Fig. 2.41). We choose the position of the Fermi level in the quantum well as an energy reference point ($E_f = 0$) and assume a given value of n_s and, hence, a given value of surface electric field, F_i, in the AlGaAs charge-control layer at the heterointerface:

$$F_i = \frac{q n_s}{\epsilon_1} \quad (106)$$

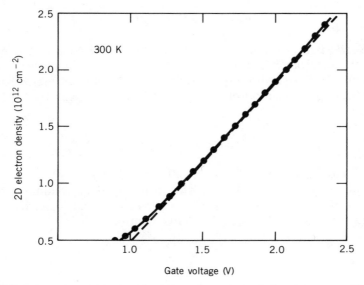

Figure 2.43 Density of 2D electron gas versus gate voltage: dots, exact calculation; solid line, analytical model. (After Baek et al. [9]. Reprinted with permission, copyright 1987, IEEE.)

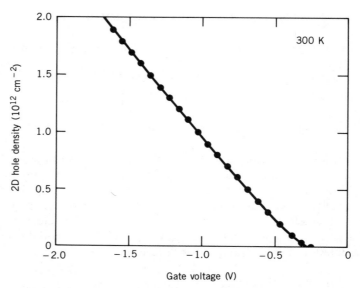

Figure 2.44 Density of 2D hole gas versus gate voltage: dots, exact calculation; solid line, analytical model. (After Baek et al. [9]. Reprinted with permission, copyright 1987, IEEE.)

The position of the conduction band edge in the AlGaAs charge control layer at the heterointerface, $E_c^-(d)$, is given by

$$E_c^-(d) = \Delta E_c + E_c^+(d) \tag{107}$$

where $E_c^+(d)$ is the position of the conduction band edge in the InGaAs layer at the heterointerface and ΔE_c is the conduction band discontinuity. The position of $E_c^+(d)$ with respect to the Fermi level is calculated based on an analytical approximation of numerical calculations given by Baek et al. [9]:

$$E_c^+(d) = -A_n\sqrt{n_s} + V_{on} \tag{108}$$

where A_n and V_{on} are given by Eqs. (102) and (103).

The values of F_i and $E_c^-(d)$ provide two boundary conditions for the integration of Poisson's equation, which allows us to calculate the conduction band variation in the AlGaAs charge-control layer:

$$\frac{dF}{dx} = \frac{\rho}{\epsilon_1} \tag{109}$$

$$\frac{dE_c}{dx} = F \tag{110}$$

Here, F is the electric field, E_c is in eV,

$$\rho = q(N_d - n) \tag{111}$$

is the space charge density, N_d is the density of the ionized donors,

$$n = N_c F_{1/2}(E_F) \tag{112}$$

is the electron concentration in the AlGaAs layer,

$$N_c = 4.7 \times 10^{17} \left(\frac{T}{300}\right)^{3/2} (1 + 1.24X)^{3/2} \quad (\text{cm}^{-3}) \tag{113}$$

is the density of states in AlGaAs, $F_{1/2}(E_F)$ is the Fermi integral that we calculate using an approximation proposed by Shur [79]. The integration yields the value of the bottom of the conduction band in the AlGaAs charge-control layer at the metal–AlGaAs interface, $E_c^+(0)$. The position of the Fermi level in the metal is given by

$$E_{Fm} = E_c^+(0) - \phi_b \tag{114}$$

Finally, the gate voltage is found as

$$V_g = E_F - E_{Fm} \tag{115}$$

The total surface electron concentration in AlGaAs is calculated as

$$n_t = \int_0^d n \, dx \tag{116}$$

In a similar way, we can calculate the number of electrons in donor and/or trap levels in the AlGaAs layer.

In Fig. 2.45, we show the computed dependences of n_s and n_t on the gate voltage. As can be seen from the figure, at high gate voltages the concentration of electrons in the AlGaAs layer increases sharply and the slope of the n_s versus V_g dependence decreases. This result can be understood by analyzing the energy band diagram of the modulation doped structure. As an example, we plot simplified band diagrams for planar doped HEMTs for different values of the gate voltage V_g in Fig. 2.46 [12]. As can be seen from this illustrations, at large gate-voltage swings the electron quasi-Fermi level in the AlGaAs structure reaches the bottom of the conduction band so that the carriers are induced into the conduction band minimum in the AlGaAs layer.

As can be seen from Fig. 2.45, the dependence of n_s on V_g is quite different from linear at large gate-voltage swings, since the electrons are induced into the potential minimum in the AlGaAs layer. At larger gate

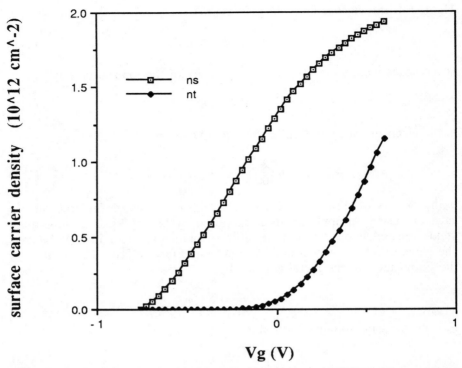

Figure 2.45   Computed surface carrier densities of electrons in 2D electron gas, n_s and in AlGaAs layer, n_t versus gate voltage. At gate voltages close to zero, the real-space transfer of electrons into the AlGaAs layer becomes important, leaving fewer electrons induced into the quantum well and limiting the gate-voltage swing. (After Chao et al. [12]. Reprinted with permission, copyright 1989, IEEE.)

voltages, the electron quasi-Fermi level in the AlGaAs layer is split from the electron quasi-Fermi level in GaAs because of the large gate current. (This effect was first studied by Ponse et al. [66] and then by Ruden et al. [71].) As a consequence, the gate capacitance versus V_g curve may actually have two peaks, as shown in Fig. 2.47 [71].

2.4.3   Charge-control Model

Analytical models play an important role in the development and design of semiconductor devices. They are used for device characterization, design, and circuit simulation. They are also helpful in understanding device physics. In this section, we describe an approximate analytical model for the description of devices based on the equations developed by Grinberg and Shur [31a]. We start from the equation for drain current, valid at low drain-to-source voltages when the electric field in the channel is small and

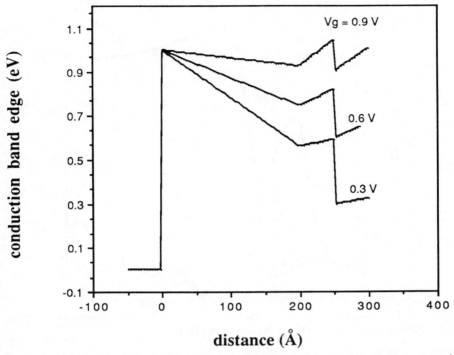

Figure 2.46 Band diagrams for planar doped HEMTs at different values of gate voltage. At small gate voltages, the energy separation between the electron quasi-Fermi level and the conduction band minimum in the AlGaAs charge-control layer is large and the electron concentration in the AlGaAs layer is negligible. At large gate voltages, the electron quasi-Fermi level nearly touches the conduction band minimum in the AlGaAs charge-control layer, and the transfer of electrons into this layer becomes dominant, sharply reducing the device transconductance. (After Chao et al. [12]. Reprinted with permission, copyright 1989, IEEE.)

the electron velocity, v is proportional to the electric field F:

$$I_D \approx q\mu n_{xs} F \qquad (117)$$

where $n_{xs}(x)$ is the surface carrier density at position x of the channel (from the source), μ the electron mobility, q the electronic charge, and

$$F = \frac{dV}{dx} \qquad (118)$$

the absolute value of the electric field. In the framework of the gradual channel approximation,

$$n_{xs} \approx n_s - \frac{\epsilon_1 V(x)}{q(d + \Delta d_1)} \qquad (119)$$

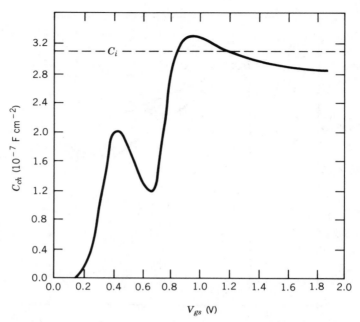

Figure 2.47   Computed gate capacitance of HFET under the conditions when the gate current is important. (From Ruden et al. [71].)

where ϵ_1 is the dielectric permittivity of the AlGaAs layer, d is the distance between the gate and the channel, and $n_s = n_s(V_g)$ is the surface carrier density at the source (independent of V_D), Δd_1 is the effective width of the 2D gas in the channel [97].[†] Substituting Eqs. (118) and (119) into Eq. (117), we obtain the following expression for the current–voltage characteristic at drain voltages smaller than the saturation voltage V_D:

$$I_D \approx q\mu \frac{W}{L} \left(n_s V_D - \frac{\epsilon_1 V_D^2}{2qd + \Delta d_1} \right) \tag{120}$$

If we now use the approximate relation between n_s and V_g proposed by Drummond et al. [24] (see Eq. [82]) we obtain

$$I_D \approx \mu\epsilon_1 \frac{W}{L} \left[\frac{(V_g - V_t)V_D}{(d + \Delta d)} - \frac{V_D^2}{2(d + \Delta d_1)} \right] \tag{121}$$

This equation is somewhat different from a similar expression derived by Drummond et al. [24] using the conventional charge-control model:

$$I_D \approx \left[\frac{\mu\epsilon_1 (W/L)}{d + \Delta d} \right] \left[(V_g - V_t)V_D - \frac{V_D^2}{2} \right] \tag{122}$$

[†](Editor's Note) The width of the 2-D electron gas is normally not constant along the channel and will be affected by the drain voltage. The author is believed to try to distinguish it from Δd in his latest analysis.

This difference may be essential, especially in devices with thin AlGaAs layers (d comparable to the effective thickness of the 2D gas, $\Delta d \approx 80$ Å at room temperature). More importantly, however, Eq. (120) yields a simple analytical expression for the current–voltage characteristics with an arbitrary dependence of n_s on V_g. As was done in Section 2.2 for GaAs MESFETs, we will assume that the current–voltage characteristics in short-channel devices saturate when the electric field F at the drain side of the channel reaches the electric field of velocity saturation $F_s = v_s/\mu$, where v_s is electron saturation velocity. The electric field at the drain F can be found as

$$F = \frac{I_D}{q\mu W n_d} \tag{123}$$

or, using Eq. (119),

$$F = \frac{I_D}{q\mu W[n_s - \epsilon_1 V_D/(qd_0)]} \tag{124}$$

where $d_0 = d + \Delta d_1$. Substituting $F = F_s$ into Eq. (124) and using Eq. (120), we find the drain saturation voltage V_D:

$$V_{Dsat} = \frac{qd_0 n_s}{\epsilon_1}[1 + a - (1 + a^2)^{1/2}] \tag{125}$$

where

$$a = \frac{\epsilon_1 F_s L}{qn_s d_0} \tag{126}$$

and n_s is a function of V_g (see Section 2.4.2). From Eqs. (120) and (125), we find the drain saturation current [31a]:

$$I_{Dsat} = qn_s \mu F_s W[(1 + a^2)^{1/2} - a] \tag{127}$$

This model uses a piece-wise linear approximation for electron velocity:

$$v = \begin{cases} \mu F & F \leq F_s \\ v_s & F > F_s \end{cases} \tag{128}$$

where μ is the low-field mobility, F is the electric field, v_s is the effective saturation velocity, and $F_s = v_s/\mu$. Based on the results of Monte Carlo simulation [11] and transient calculations [75], we know that v_s in short-channel devices is very much different from the electron saturation velocity in a long bulk semiconductor sample. It is considerably higher and dependent on the effective gate length L_{eff}. One-dimensional simulations for GaAs n-i-n structures led to the following interpolation formula for v_s [81]:

$$v_s \approx \frac{(0.22 + 1.39 L_{eff}) \times 10^7}{L_{eff}} \quad \text{(cm/s)} \tag{129}$$

where the effective gate length L_{eff} is in μm. Also, numerical simulations clearly show that electron velocity in short structures does not saturate in high electric fields. As a matter of fact, the velocity depends on both electric field and potential. Hence, this simple model relies on an effective value of the electron velocity in high electric fields that gives a general idea of the electron velocities in a device channel, but does not give any information about the exact shape of the velocity profile in the channel. For devices with recessed gates, the effective gate length can be estimated as

$$L_{\text{eff}} \approx L + \beta(d + \Delta d) \tag{130}$$

where Δd is the effective thickness of the 2D electron gas and β is a constant. The length $\beta(d + \Delta d)$ in the equation represents the total lateral depletion width and is a function of gate recess width in an HFET. Chao et al. [12] estimated that $\beta \approx 2$.

The source and drain series resistances, R_s and R_d, respectively, may play an important role in determining the current–voltage characteristics of HFETs. These resistances can be taken into account as follows. The extrinsic (applied) gate-to-source voltage V_{GS} is given by

$$V_{\text{GS}} = V_{\text{G}} + (I_{\text{ds}} + I_{\text{g}})R_s \tag{131}$$

where I_{g} is the gate current. For depletion mode HFETs, the gate current does not usually play an important role because electron transfer into the AlGaAs layer (see Fig. 2.46) severely limits device transconductance at lower voltages than the turn-on voltage for gate leakage current. On the other hand, for enhancement mode HFETs, the gate current can play a dominant role and, as discussed below, may even affect the value of the "intrinsic" drain current I_{ds}. Equation (131) has to be solved together with Eq. (127) in order to find the drain saturation current as a function of the gate voltage.

Figure. 2.48 (from Chao et al. [12]) shows a comparison of this model with experimental data for device transconductance and saturation current. Because of the ultra-short gate length in our devices, the maximum device transconductance is primarily determined by the effective saturation velocity and only slightly affected by the low-field mobility. This is clearly illustrated Fig. 2.49a and b, where we show computed transconductance for different values of μ and v_s. As a matter of fact, the maximum device transconductance is fairly close to

$$g_{\text{m}} = \frac{g_{\text{m0}}}{1 + g_{\text{m0}}R_s} \tag{132}$$

where

$$g_{\text{m0}} = \frac{\epsilon_1 v_s}{d + \Delta d} \tag{133}$$

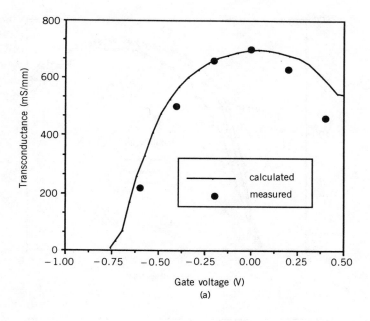

(a)

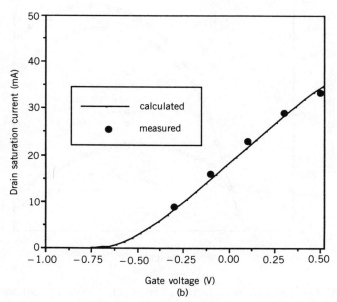

(b)

Figure 2.48 (a) Calculated and measured device transconductance (b) drain saturation current for planar doped AlGaAs/InGaAs/GaAs. HEMT. (After Chao et al. [12]. Reprinted with permission, copyright 1989, IEEE.)

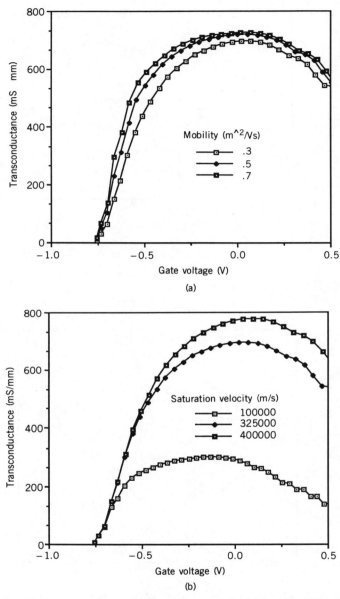

Figure 2.49   Computed transconductance versus gate voltage for different values of (a) low-field mobility and (b) saturation velocity. Notice that a variation of the field effect mobility by more than a factor of two does not change the computed curves very much, whereas the change of the effective saturation velocity affects the computed curves very significantly. This is a consequence of very high electric fields in an ultrashort gate device. (After Chao et al. [12]. Reprinted with permission, copyright 1989, IEEE.)

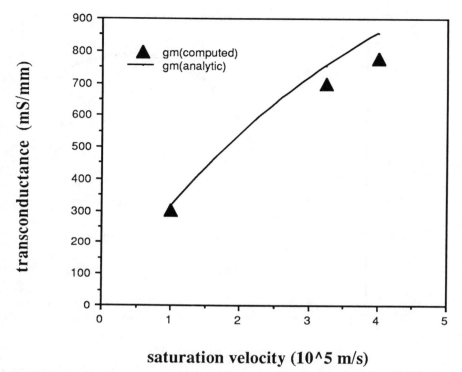

saturation velocity (10^5 m/s)

Figure 2.50 Comparison of computed maximum values of device transconductance with the device transconductance given by Eqs. (132) and (133). This comparison shows that the maximum device transconductance is close to the maximum possible value for a given value of the effective electron saturation velocity. (After Chao et al. [12]. Reprinted with permission, copyright 1989, IEEE.)

is the maximum value of device transconductance predicted by the velocity saturation model, and R_s is the source resistance (Fig. 2.50).

Figure 2.51 [12] presents a comparison between a planar doped barrier and conventional modulation doped structure having the same pinch-off voltage. As can be seen from the figure, in a conventional HEMT the electrons are induced into the AlGaAs layer at smaller gate voltages. This leads to a decrease in both maximum transconductance and gate-voltage swing. The effect of the gate-to-channel spacing is shown in Fig. 2.52. With a decrease in d, the maximum transconductance increases if the source series resistance remains constant (as was assumed in the calculations for Fig. 2.52). In fact, the source series resistance is expected to rise with the increase in the threshold voltage. This resistance plays a very important role in limiting the device transconductance. Experimentally, a gate–channel spacing of 200–250 Å yields maximum device transconductance. This depends, however, on the doping profile and on the shape of the gate recess.

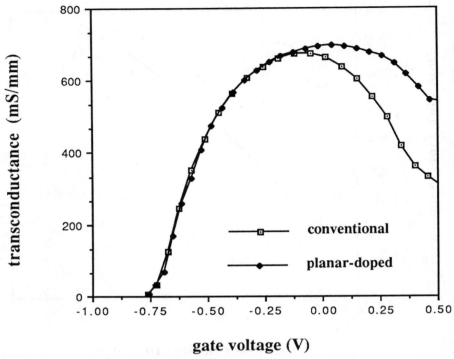

gate voltage (V)

Figure 2.51   Comparison of transconductance between planar doped and conventional modulation doped structures with the same threshold voltage. A broader high-transconductance region is observed for the planar doped structure. (After Chao et al. [12]. Reprinted with permission, copyright 1989, IEEE.)

One other important device parameter is the separation, $d_1 = d - d_d$, between the doped plane and the channel (Fig. 2.53). A smaller distance leads to both a more negative threshold voltage, and, hence, to an expected decrease in a source resistance, as well as higher gate-voltage swing and, as a consequence, higher transconductance.

The effect of gate length on device transconductance is primarily related to the velocity enhancement due to ballistic and overshoot effects in short device structures. Figure 2.54 shows the computed dependence of the device transconductance on the gate voltage, assuming that the effective saturation velocity is given by Eq. (129). As can be seen from the figure, the transconductance nearly doubles when the device length is scaled down from 1 μm to 80 nm.

Simpler output current–voltage characteristics of HFETs can be calculated using the same approach as for GaAs MESFETs (see Section 2.2): [37]

$$I_{ds} = I_{Dsat}(1 + \lambda V_{DS}) \tanh\left[\frac{g_{ch}V_{DS}}{I_{Dsat}}\right] \tag{134}$$

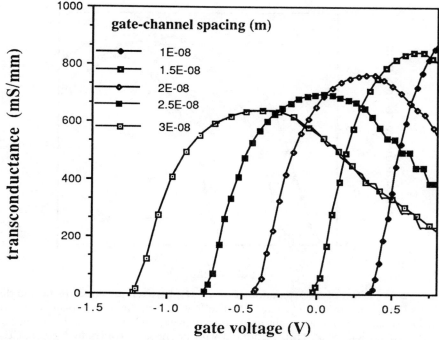

Figure 2.52 Effect of the gate-to-channel spacing on device transconductance. This calculation does not take into account the dependence of the source series resistance on the gate recess depth. Such a dependence may decrease the device transconductance for devices with deep recess (small gate-to-channel spacing). Also, notice the decrease in the gate-voltage swing for devices with small gate-to-channel spacing. The gate–voltage swing in these devices may be maintained by increasing the doping level proportionally to the decrease in the gate-to-channel spacing. (After Chao et al. [12]. Reprinted with permission, copyright 1989, IEEE.)

where I_{Dsat} can be given by Eq. (43)[†] with a slight change to the definition for threshold voltage in order to take into consideration the band discontinuity [37], and

$$g_{\mathrm{ch}} = \frac{g_{\mathrm{chi}}}{1 + g_{\mathrm{chi}}(R_{\mathrm{s}} + R_{\mathrm{d}})} \qquad (135)$$

is the channel conductance at low drain-to-source voltages,

$$g_{\mathrm{chi}} = \frac{q\mu n_{\mathrm{s}}W}{L} \qquad (136)$$

is the intrinsic channel conductance at low drain-to-source voltages. The constant λ in eq. (134) is an empirical constant which accounts for output

[†](Editor's Note) As mentioned by the author, one may otherwise use Eqs. (127) and (131) for a more accurate estimation of the drain saturation current as a function of the gate-to-source voltage. See also [9] for models related to HIGFETs, which utilized Eqs. (92) and (93).

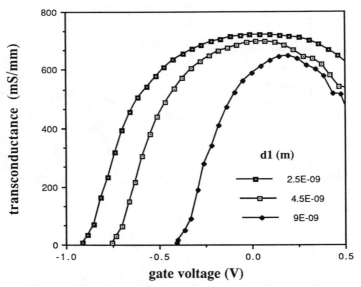

Figure 2.53   Effect of the separation, $d_1 = d - d_d$, between the doped plane and the channel on device transconductance. Notice how moving the doped plane closer to the heterointerface increases the gate-voltage swing and the device transconductance. (After Chao et al. [12]. Reprinted with permission, copyright 1989, IEEE.)

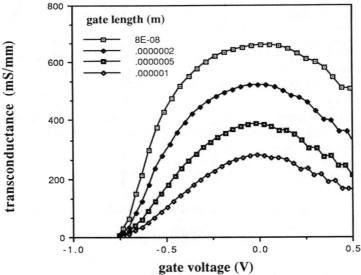

Figure 2.54   Effect of gate length on device transconductance. The device transconductance increases with the decrease in gate length due to the increase of the effective electron velocity caused by the ballistic and overshoot effects. (After Chao et al. [12]. Reprinted with permission, copyright 1989, IEEE.)

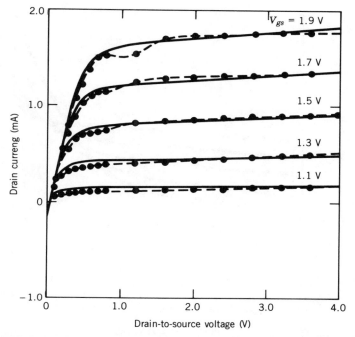

Figure 2.55 Current–voltage characteristics for n-channel HIGFETs: dots, measured data; solid line calculated curves. (After Baek et al. [9]. Reprinted with permission, copyright 1987, IEEE.)

conductance. This output conductance may be related to short-channel effects as well as effects related to the space charge injection into the buffer layer [32].

As can be seen from Figs. 2.55 and 2.56, this model agrees quite well with experimental data for both n- and p-channel devices. This model has been implemented in the GaAs circuit simulator UM-SPICE [36]. The equivalent circuits of an HFETs used in UM-SPICE are shown in Fig. 2.57. Figure 2.57a shows a conventional equivalent circuit similar to that used for other field effect transistors. This circuit was used in the old version of UM-SPICE [37]. In this circuit, the gate current can be modeled by equivalent Schottky diodes connected from the gate to source and to the drain, respectively. Using the well-known diode equation, we find for the total gate current

$$I_g = J_s WL \left[\exp\left(\frac{V_{gs}}{nV_{th}}\right) + \exp\left(\frac{V_{gd}}{nV_{th}}\right) - 2 \right] \tag{137}$$

where V_{th} is the thermal voltage, n is the diode ideality factor, and J_s is the reverse saturation current density. More accurate models describing gate current in MODFETs were developed by Ponse et al. [66], Ruden et al. [71], Chen et al. [16], and Baek and Shur [6]. As shown in Fig. 2.58, the

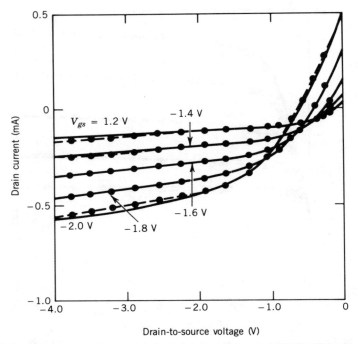

Figure 2.56 Current–voltage characteristics for p-channel HIGFETs: dots, measured data; solid line, calculated curves. (After Baek et al. [9]. Reprinted with permission, copyright 1987, IEEE.)

gate current mechanism changes at $V_g = \phi_b - \Delta E_c$ where ϕ_b is the Schottky barrier height at the metal–semiconductor interface and ΔE_c is the conduction band discontinuity. At smaller gate voltages, the gate current is primarily determined by the Schottky barrier at the metal–semiconductor interface. At larger gate voltages, the gate current is limited by conduction band discontinuity, or more precisely, by the effective barrier height equal to the difference between the bottom of the conduction band in the AlGaAs layer at the heterointerface and the electron quasi-Fermi level in the 2D gas. This barrier height changes very little with the gate voltage (and only as a consequence of the dependence of the quasi-Fermi level on electron concentration in the 2D electron gas and, hence, on the gate voltage). Thus, the interpolation of the experimental dependence of the gate current on the gate voltage in this regime, given by the diode equation, has a very large ideality factor (usually from 5 to 20). Another interesting effect that is seen in the measured current–voltage characteristics at high gate voltages is a negative differential resistance (see curve corresponding to $V_{gs} = 1.9\,\text{V}$ in Fig. 2.55). This effect may be much more pronounced [61a] and is related to the real-space transfer mechanism, first proposed by Hess et al. [33a], that is, to the transfer of hot electrons from the channel over the barrier created by the

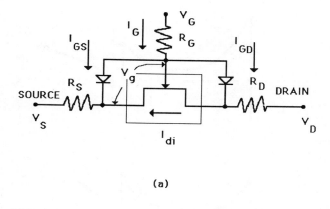

(a)

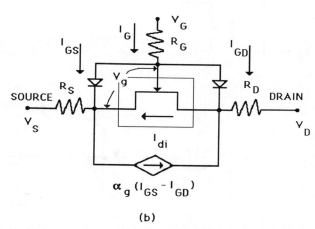

(b)

Figure 2.57 HFET equivalent circuits: (a) conventional equivalent circuit similar to that used for other field effect transistors (the complete equivalent is given in Fig. 2.24), (b) equivalent circuit that takes into account the effect of the gate current on the channel current. (After Ruden et al. [73]. Reprinted with permission, copyright 1989, IEEE.)

conduction band discontinuity. A similar effect was observed in double heterojunction MODFETs by Chen et al. [17].

The equivalent circuit shown in Fig. 2.57b takes into account the effect of gate current on channel current. Indeed, the gate current is distributed along the channel, with the largest gate current density near the source side of the channel. This leads to the redistribution of electric field along the channel, with an increase in electric field near the source side of the device and a decrease in channel current. A numerical solution of coupled differential equations describing the gate and channel current distributions along the channel is given by Baek and Shur [6]. This calculation provides a justifica-

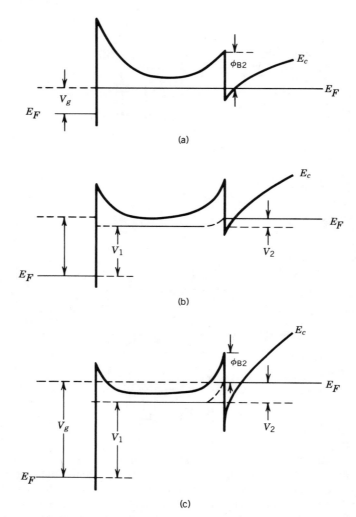

Figure 2.58 MODFET band diagram at different gate biases: (a) $V_g < \phi_b - \Delta E_c$, (b) $V_g \approx \phi_b - \Delta E_c$, (c) $V_g > \phi_b - \Delta E_c$. Gate current mechanism changes at $V_g = \phi_b - \Delta E_c$. ϕ_{b2} is the effective barrier height for high voltages and V_1 and V_2 are gate-voltage drops across the Schottky barrier and the conduction band discontinuity, respectively. (After Chen et al. [16]. Reprinted with permission, copyright 1988, IEEE.)

tion for the new equivalent circuit shown in Fig. 2.57b. As illustrated by Fig. 2.59, this new equivalent circuit allows us to obtain much better agreement with the experimental data.

The nonlinear capacitances $C_{gs}(V_{gs}, V_{ds})$ and $C_{gd}(V_{gd}, V_{ds})$ shown in Fig. 2.24 represent channel charge storage effects. These capacitances can be

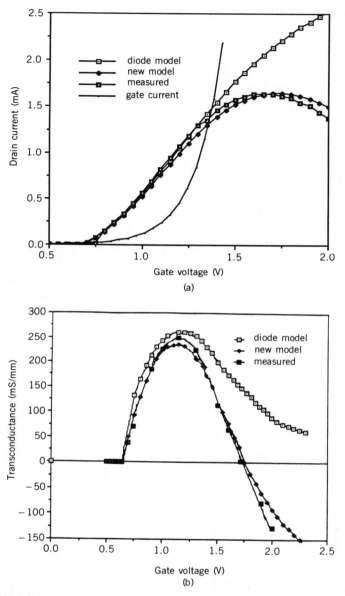

Figure 2.59 Measured and calculated (a) drain current and (b) transconductance for a self-aligned AlGaAs/GaAs MODFET. The parameters used in the calculation: threshold voltage $V_t = .0582$ V, thickness of doped AlGaAs layer $d_d = 281$ Å, undoped AlGaAs spacer layer thickness $d_i = 75$ Å, mole fraction of Al in $Al_xGa_{1-x}As$, $x = 0.3$, electron saturation velocity $v_s = 1.6 \times 10^5$ m/s, electron mobility $\mu = 0.35$ m^2/V·s, gate saturation current $I_{gs} = 1.14 \times 10^{-7}$ A, effective ideality factor $m = 5.61$, drain and source series resistances $R_d = R_s = 49$ Ω, gate series resistance $R_g = 45$ Ω, gate length $L = 1$ μm, gate width $W = 10$ μm, output conductance parameter $\lambda = 0.07$, $a_g = 0.26$, $V_{DS} = 3$ V. (After Ruden et al. [72]. Reprinted with permission, copyright 1988, IEEE.)

estimated using a modified Meyer's model [58]:

$$C_{gs} = \frac{2C_g}{3}\left[1 - \frac{(V_{Dsat} - V_{ds})^2}{(2V_{Dsat} - V_{ds})^2}\right] \tag{138}$$

$$C_{gd} = \frac{2C_g}{3}\left[1 - \frac{V_{Dsat}^2}{(2V_{Dsat} - V_{ds})^2}\right] \tag{139}$$

(where V_{Dsat} the drain saturation voltage, is given by Eq. (125)) in the linear region. In the saturation region, we have $C_{gs} = 2C_g/3$ and $C_{gd} = 0$. The capacitance C_g is the total gate-to-channel capacitance at $V_{ds} = 0$. An approximate analytical expression for C_g

$$C_g = C_0\left\{1 - \frac{(1 + \sqrt{2})^2 V_{C_0}^2}{[-(2 + \sqrt{2})V_{C_0} + V_{gst} + \Delta V_T]^2}\right\} \tag{140}$$

where V_{gst} is the gate-to-source voltage less the threshold voltage, was used to fit the numerically calculated C_g versus V_{gs} curve, which includes the Fermi level dependence on n_s (Fig. 2.60 [37]). Here, $\Delta V_T = 0.03$ V and $V_{C_0} = -3V_{th}$. As can be seen from the figure, there is good agreement between numerical calculation and Eq. (140).

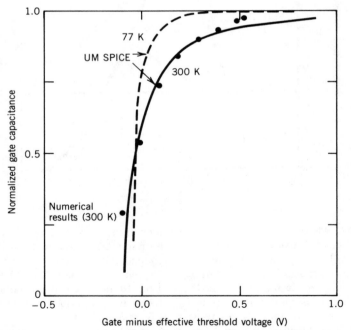

Figure 2.60   HFET gate capacitance versus gate voltage. (After Hyun et al. [37]. Reprinted with permission, copyright 1986, IEEE.)

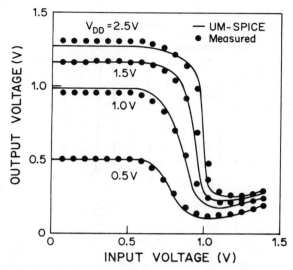

Figure 2.61 Inverter transfer characteristics for different power supply voltages. Gate length and width of the MODFET drivers are 1 μm and 20 μm, respectively. The width of the ungated FET load is 4 μm and the separation between the contacts is 2 μm. (After Hyun et al. [37]. Reprinted with permission, copyright 1986, IEEE.)

Figure 2.61 shows HFET inverter transfer curves simulated using the UM-SPICE. Figure 2.62 gives an example of the simulated output waveform of an 11-stage HFET ring oscillator.

2.4.4   Nonideal Effects.

The discussion regarding nonideal effects in GaAs MESFETs, given in Section 2.2 is also relevant for HFETs. They exhibit similar orientation and backgating effects and dependence of the threshold voltage on temperature and gate and drain bias conditions [40, 89]. The dependence of the threshold voltage on temperature and bias is primarily caused by deep traps in the doped AlGaAs layer (so-called DX centers). In addition, the parallel conduction of electrons induced into the charge control AlGaAs layer at high gate biases may cause them to interact with traps. Effects related to the filling and emptying of traps also lead to persistent photoconductivity (see Ref. 78 for a brief review of these phenomena). Eliminating doping in heterostructure insulated gate FETs results in much smaller variation of the threshold voltage temperature (Fig. 2.63). As was mentioned in Section 2.4.1, the effect of traps may be also diminished by using planar doped or superlattice HFETs. To the first order, the effect of traps can be modeled by accounting for their temperature-dependent contribution into the device threshold voltage and gate capacitance [89]. A more detailed theory will require more detailed information about the nature and location of traps.

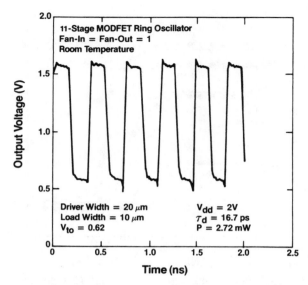

Figure 2.62   UM-SPICE simulation of MODFET ring oscillators. (After Hyun et al. [37]. Reprinted with permission, copyright 1986, IEEE.)

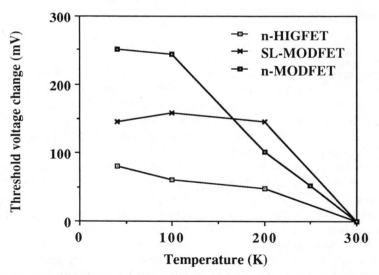

Figure 2.63   Variation of MODFET threshold voltage with temperature for different HFETs. Conventional MODFET and HIGFET data are from Cirillo et al. [19]; superlattice FET data are from Arch et al. [2]. (Reprinted with permission, copyrights 1985 and 1987, IEEE.).

2.5. CONCLUSION

We described analytical models for GaAs and AlGaAs/GaAs devices that provide some insight into the complicated device physics. They have been used as design and characterization tools. A better understanding of device physics and further development of both analytical and computer models are required to help compound semiconductor technology to mature and fully realize its apparent potential.

ACKNOWLEDGMENTS

I would like to thank Jun-ho Baek, Chung-Hsu Chen, Tzu-hung Chen, Choong Hyun, Kwyro Lee, Kang Lee, P.C. Chao, Paul Ruden, Andy Peczalski, Nick Cirillo, Jr., Robert Daniels, David Arch, Obert Tufte, John Abrokwah, Phil Jenkins, Tim Drummond, Hadis Morkoç, Steve Baier, and Ho-Kyoon Chung, all of whom contributed to the development of GaAs and AlGaAs/GaAs device models. I am also grateful to Jun-ho Baek, Michael Norman, Young Byun, Phil Jenkins, and Byung Moon for the critical reading of this manuscript.

REFERENCES

1. M. I. Aksun, H. Morkoç, L. F. Lester, K. H. G. Duh, P. M. Smith, P. C. Chao, M. Longerbone, and L. P. Erickson, *Tech. Dig.—Int. Electron. Devices Meet.* pp. 752–754 (1986).

2. D. K. Arch, M. Shur, J. K. Abrokwah, and R. R. Daniels, *J. Appl. Phys.* **61**, 1503–1509 (1987).

3. P. M. Asbeck, C. P. Lee, and F. M. Chang, *IEEE Trans. Electron Devices* **ED-31**, 1377 (1984).

4. Y. Awano, M. Kasugi, T. Mimura, and M. Abe, *IEEE Electron Device Lett.* **EDL-8**, 451–453 (1987).

5. T. Baba, T. Muzutani, M. Ogawa, and K. Ohata, *Jpn. J. Appl. Phys.* **23** L654 (1983).

6. J. H. Baek and M. Shur, unpublished (1989).

7. J. H. Baek, M. Shur, K. Lee, and T. Vu, *IEEE Trans. Electron Devices* **ED-32**, 2426–2430 (1985).

8. J. H. Baek, M. Shur, R. R. Daniels, D. K. Arch, and J. K. Abrokwah, *IEEE Electron Device Lett.* **EDL-7** 519–521 (1986).

9. J. Baek, M. Shur, R. R. Daniels, D. K. Arch, J. K. Abrokwah, and O. N. Tufte, *IEEE Trans. Electron Devices* **ED-34** 1650–1657 (1987).

10. M. S. Birittella, W. C. Seelbach, and H. Goronkin, *IEEE Trans. Electron Devices* **ED-29** 1135–1142 (1982).

11. A. Cappy, B. Carnes, R. Fauquembergues, G. Salmer, and E. Constant, *IEEE Trans. Electron Devices* **ED-27** 2158–2168 (1980).

12. P. C. Chao, M. Shur, R. C. Tiberio, K. H. G. Duh, P. M. Smith, J. M. Ballingall, P. Ho, and A. A. Jabra, *IEEE Trans. Electron Devices* **ED-36** 461–473 (1989).

13. C. H. Chen and M. Shur, *IEEE Trans. Electron Devices* **ED-32** 883–891 (1985).

14. C. H. Chen, M. Shur, and A. Peczalski, *IEEE Trans. Electron Devices* **ED-33,** 792–798 (1986).

15. C. H. Chen, A. Peczalski, M. Shur, and H. K. Chung, *IEEE Trans. Electron Devices* **ED-34** (7) 1470–1481 (1987).

16. C. H. Chen, S. Baier, D. Arch, and M. Shur, *IEEE Trans. Electron Devices* **ED-35** 570–577 (1988).

17. Y. K. Chen, D. C. Radulescu, G. W. Wang, A. N. Lepore, P. J. Tasker, L. F. Eastman, and E. Strid, in *Proc. Microwave Theory Tech. IEEE Symp.* p. 871 (1987).

18. N. C. Cirillo, Jr., J. K. Abrokwah, and M. Shur, *IEEE Trans. Electron Devices* **ED-31** No 12, p. 1963 (1984).

19. N. C. Cirillo, Jr., M. Shur, P. J. Vold, J. K. Abrokwah, R. R. Daniels, and O. N. Tufte, *IEEE Electron Device Lett.* **EDL-6,** 645–647 (1985).

19a. W. R. Curtice, A. MESFET Model for Use in The Design of GaAs Integrated Circuits, *IEEE Trans. Microwave Theory Tech.* **MTT-28** No. 5, pp. 488–456 (1980).

20. H. Dambkes, W. Brokerhoff, and K. Heime, *Tech. Dig.—Int. Electron Devices Meet.* pp. 621–624 (1983).

21. R. R. Daniels, R. Mactaggart, J. K. Abrokwah, O. N. Tufte, M. Shur, J. Baek and P. Jenkins, *IEDM Tech. Dig.* pp. 448–449, (1986).

22. R. R. Daniels, P. P. Ruden, M. Shur, D. E. Grider, T. Nohava, and D. Arch, *IEEE Trans. Electron Devices* **EDL-9** pp. 355–357 (1988).

23. G. C. Dasey and I. M. Ross, *Bell Syst. Tech. J.* **34** 1149–1189 (1955).

24. T. J. Drummond, H. Morkoç, K. Lee, and M. Shur, *IEEE Electron Device Lett.* **EDL-3,** 338–341 (1982).

25. L. F. Eastman, private communication (1983).

26. L. F. Eastman and M. Shur, *IEEE Trans. Electron Devices* **ED-26,** 1359–1361 (1979).

27. R. W. H. Engelman and C. A. Liehti, *IEEE Trans. Electron Devices* **ED-24,** 1288–1296 (1977).

28. T. A. Fjeldly, *IEEE Trans. Electron Devices* **ED-33,** 874–880 (1986).

29. S. T. Fu and M. B. Das, *IEEE Trans. Electron Devices* **ED-34,** 1245–1252 (1987).

30. S. Furukawa, H. Matsumura, and H. Ishiwara, *Jpn. Appl. Phys.* **11** 134 (1972).

31. A. B. Grebene and S. K. Ghandi, *Solid-State Electron.* **12,** 573–589 (1969).

31a. A. A. Grinberg and M. Shur, *J. Appl. Phys.*, vol. **65** No. 5, p. 2116 (1989).

32. C. J. Han, P. P. Ruden, D. Grider, A. Fraasch, K. Nestrom, P. Joslyn, and M. Shur, *IEDM Tech. Dig.—Int. Electron Devices Meet.*, pp. 696–699, (1988).

33. T. Hariu, K. Takahashi, and Y. Shibata, *IEEE Trans. Electron Devices* **ED-30,** 1743–1749 (1983).

33a. K. Hess, H. Morkoç, H. Shichijo, and B. G. Streetman, *Appl. Phys. Lett.*, **35** 459 (1979).

34. P. Hower and G. Bechtel, *IEEE Trans. Electron Devices* **ED-20,** 213–220 (1973).

35. C. H. Hyun, Ph. D. Thesis, University of Minnesota, Minneapolis (1985).

36. C. H. Hyun, M. Shur, and A. Peczalski, *IEEE Trans. Electron Devices* **ED-33,** 1421–1426 (1986).

37. C. H. Hyun, M. Shur, and N. C. Cirillo, Jr., *IEEE Trans. Comput.-Aided Des.* **CAD-5,** 284–292 (1986).

38. P. R. Jay and R. H. Wallis, *IEEE Electron Device Lett.* **EDL-2,** 265–267 (1981).

39. E. J. Johnson, J. A. Kafalas, and R. W. Davies, *J. Appl. Phys.* **51** 2840 (1983).

40. A. A. Kastalski and R. A. Kiehl, *IEEE Trans. Electron Devices* **ED-33** No. 3, pp. 419–423 (1986).

41. N. Kato, Y. Matsuoka, K. Ohwada, and S. Moriya, *IEEE Electron Device Lett.* **EDL-4,** 417 (1983).

42. U. Kaufmann, J. Windschief, M. Baeumlër, J. Schnieder, and F. Kohl, *in* "Semi-insulating III–V Materials" (D. C. Look and J. S. Blakemore, eds.), Natwich, Chesire, England: Shiva Pub., p. 247 1984.

43. R. K. Kiehl, D. A. Frank, S. L. Wright, and J. H. Magerlein, *Tech. Dig—Int. Electron Devices Meet. pp.* 70–73 (1987).

44. *C. Kocot and C. A. Stolte, IEEE Trans. Electron Devices* **ED-29,** 1059–1064 (1982).

45. C. P. Lee, R. Zucca, and B. M. Welch, *Appl. Phys. Lett.* **37**(3), 311 (1980).

46. C. P. Lee, B. M. Welch, and R. Zucca, *IEEE Trans. Electron Devices* **ED-29** 1103–1109 (1982).

47. K. Lee, M. Shur, T. J. Drummond, and H. Morkoç, *J. Appl. Phys.* **54** 2093–2096 (1983).

48. K. Lee, M. Shur, T. J. Drummond, and H. Morkoç, *IEEE Trans. Electron Devices* **ED-30,** 207–212 (1983).

49. K. Lee, M. Shur, T. J. Drummond, and H. Morkoç, *IEEE Trans. Electron Devices* **ED-31,** 29–35 (1984).

50. K. Lee, M. Shur, T. Vu, P. Roberts, and M. Helix, *IEEE Trans. Electron Devices* **ED-32,** 987–992 (1985).

51. K. Lehovec and R. Zuleeg, *IEEE Trans. Electron Devices* **ED-27,** 1074–1091 (1980).

52. M. S. Levinstein and M. Shur, *Sov. Phys. Semicond.* (Engl. Transl.) **3** No. 7 (1969).

53. P. F. Lindquist and W. M. Ford, *in* "Gallium Arsenide FET Principles and Technology" (J. V. DiLorenzo and D. D. Khandewal, eds.). Artech House, Dedham, Massachusetts, p.1, 1982.

54. O. Madelung, ed., "Landolt-Börnstein, Numerical Data and Functional Relationships in Science and Technology", Vol. 22, New Ser., Group III, Springer-Verlag, Berlin and New York.

55. T. J. Maloney and J. Frey, *IEEE Trans. Electron Devices* **ED-22,** 357–358 (1975).

56. G. M. Martin, A. Mitoneau, D. Pons, A. Mircea, and D. W. Woodard, *J. Chem. Phys.* 3855 (1980).

57. K. Matsumoto, N. Hashizume, N. Atoda, K. Tomizuwa, T. Kurosu, and M. Ioda, *Conf. Ser.—Inst. Phys.* **65**, 317 (1983).

58. J. E. Meyer, *RCA Rev.* **32**, 42–63, March (1971).

59. V. Milunovic (Ed.), *Computer*, Spec. Issue, **19** (10), (1986).

60. U. K. Mishra, J. F. Jensen, A. S. Brown, M. A. Thompson, L. M. Jelloian, and R. S. Beaubien, *IEEE Electron Device Lett.* **9** (9) 482–484 (1988).

61. T. Ohnishi, Y. Yamaguchi, T. Onodera, N. Yokoyama, and H. Nishi, *Ext. Abst. Conf. Solid State Devices Mater.* **16**, 391 (1984).

61a. M. S. Shur, D. K. Arch, R. R. Daniels, and J. K. Abrokwah, "New Negative Resistance Regime of Heterostructure Insulated Gate Transistor (HIGFET) Operation," *IEEE Electron Device Lett.* **EDL-7** No. 2, pp. 78–80 (1986).

62. T. Onodera, T. Ohnishi, N. Yokoyama, and H. Nishi, *IEEE Trans. Electron Devices* **ED-32**, 2314 (1985).

63. A. Peczalski, M. Shur, C. H. Hyun, and T. T. Vu, *IEEE Trans. Comput.-Aided Des.* **CAD-5**, 266–273 (1986).

64. A. Peczalski, C. H. Chen, M. Shur, and S. M. Baier, *IEEE Trans. Electron Devices* **ED-34**, 726–732 (1987).

65. A. Peczalski, M. Shur, and C. H. Chen, *in* "The Physics of Semiconductor Materials and Applications" (C. Lee and W. Paul, eds.), pp. 237–254, Korea Advanced Institute of Science and Technology, Seoul, 1987.

66. F. Ponse, W. T. Masselink, and H. Morkoç, *IEEE Trans. Electron Devices* **ED-32**, 1017 (1985).

67. R. A. Pucel, *Non-linear CAD Workshop*, IEEE MTT Symp. 1987, (oral presentation).

68. R. A. Pucel, H. Haus, and H. Statz, *Adv. Electron. Electron Phys.* **38** 195–205 (1975).

69. J. W. Roach, H. H. Wieder, and R. Zuleeg, *IEEE Trans. Electron Devices* **ED-34**, 181–184 (1987).

70. J. G. Ruch, *IEEE Trans. Electron Devices* **ED-19**, 652–654 (1972).

71. P. P. Ruden, C. J. Han, and M. Shur, *J. Appl. Phys.* **64** (3), 1541–1546 (1988).

72. R. R. Daniels, P. P. Ruden, M. Shur, D. E. Grider, T. Nohava, and D. Arch, "Quantum Well p-channel AlGaAs/InGaAs/GaAs Heterostructure Insulated Gate Field Effect Transistors with Very High Transconductance, "*IEEE Electron Device Lett.* **EDL-9** 355–357 (1988).

73. P. P. Ruden, M. Shur, A. I. Akinwande, and P. Jenkins, *IEEE Trans. Electron Devices* **ED-36** No. 2, pp. 453–456 (1989).

74. N. J. Shah, S. S. Pei, and C. W. Tu, *IEEE Trans. Electron Devices* **ED-33**, 543 (1986).

75. M. Shur, *Electron. Lett.* **12** 615–616 (1976).

76. M. Shur, *IEEE Trans. Electron Devices* **ED-25**, 612–618 (1978).

77. M. Shur, *Electron. Lett.* **18** (21) 909–911 (1982).

77a. M. S. Shur, *IEEE Trans. Electron Devices* **ED-32** No. 1, pp. 70–72 (1985).

77b. S. M. Baier, M. S. Shur, K. Lee, N. C. Cirillo, and S. A. Hanka, *IEEE Trans. Electron Devices* **ED-32** No. 12, pp. 2824–2829 (1985).

78. M. Shur, GaAs Devices and Circuits. Plenum, New York, 1987.

79. M. Shur, "Physics of Semiconductor Devices." Prentice-Hall, New York, 1989.

80. M. Shur and L. F. Eastman, *IEEE Trans. Electron Devices* **ED-25**, 605–617 (1978).

81. M. Shur and D. Long, *IEEE Electron Device Lett* **EDL-3**, 124 (1982).

82. M. Shur, J. K. Abrokwah, R. R. Daniels, and D. K. Arch, *J. Appl. Phys.* **61** (4) 1643–1645 (1987).

83. C. M. Snowden and D. Loret, *IEEE Trans. Electron Devices* **ED-34**, 212–233 (1987).

84. H. Statz, P. Newman, I. W. Smith, R. A. Pucel, and H. A. Haus, *IEEE Trans. Electron Devices* **ED-34**, 160–169 (1987).

85. F. Stern, *CRC Crit. Rev. Solid State Sci.* p. 499 (1974).

86. T. Takada, K. Yokoyama, M. Ida, and T. Sudo, *IEEE Trans. Microwave Theory Tech.* **MTT-30**, 719–723 (1982).

87. S. Tiwari, S. L. Wright, and J. Batey, *IEEE Electron Device Lett.* **9** (9) 488–489 (1988).

88. K. Ueto, T. Furutsuka, H. Toyoshima, M. Kanamori, and A. Higashisaka, *Tech. Dig.—Electron Devices Meet.* p. 82 (1985).

89. A. J. Valois, R. Robinson, K. Lee, and M. Shur, *J. Vac. Sci. Technol.* **B1** [2], 190–195 (1983).

90. R. A. Warriner, IEE J. Solid-State Electron Devices **1** 105 (1977).

91. R. E. Williams and D. W. Shaw, *IEEE Trans. Electron Devices* **ED-25**, 600–605 (1978).

92. H. A. Willing and P. de Santis, *Electron. Lett.* **13**, No. 18, pp. 537–537 (1977).

93. H. A. Willing, C. Rausher, and P. de Santis, *IEEE Trans. Microwave Theory Tech.* **MTT-26**, No. 12, pp. 1017–1023 (1978).

94. K. Yamaguchi and H. Kodera, *IEEE Trans. Electron Devices* **ED-23**, 545–553 (1976).

95. L. Yang and S. T. Long, *IEEE Electron Device Lett.* **EDL-7**, 75–77 (1986).

96. N. Yokoyama, H. Onodera, T. Ohnishi, and A. Shibatomi, *Appl. Phys. Lett.* **42** (3), 270 (1983).

97. Y. Byun, K. Lee, M. Shur, to be published.

3 Heterostructure Field Effect Transistors

SHIM-SHEM PEI and NITIN J. SHAH

AT&T Bell Laboratories, Murray Hill, New Jersey

3.1 INTRODUCTION

The III–V compound semiconductor family consists of many binary, ternary, and quaternary compound materials. By tailoring the composition of these compounds, it is possible to grow lattice-matched epitaxial layer structures with different bandgaps and distinct electronic and optical properties. The most widely used are the lattice-matched $GaAs/Al_xGa_{1-x}As$ and $InP/In_{0.53}Ga_{0.47}As/In_{0.52}Al_{0.48}As$ material systems. Heterostructure photonic devices such as the double heterojunction laser as well as electronic devices such as heterstructure FETs (HFETs) have been successfully developed on these material systems.

Most of the heterostructure FETs reported in the literature have a two-dimensional electron gas (2DEG) channel formed at a heterojunction interface or in a quantum well. The presence of an inversion or accumulation electron layer located at the interface of certain heterojunctions was first predicted by Anderson [1] in 1960. In 1978, Dingle et al. [2] showed that in a modulation doped AlGaAs/GaAs superlattice, that is, alternate AlGaAs/GaAs layers where only the center region of the wider bandgap material (AlGaAs) was doped, ionized impurities and the free electrons forming the 2DEG could be spatially separated to reduce the Coulomb scattering and, therefore, to enhance the electron mobility. It was demonstrated later by Stormer et al. [3] that mobility enhancement also exists in selectively doped single heterojunction structures. Since the realization of the first heterostructure FET (also known as SDHT for selectively doped heterostructure transistor, MODFET for modulation doped FET, TEGFET for two-dimensional electron gas FET, HEMT for high electron mobility transistor, etc.), discrete devices have been engineered to achieve excellent high-frequency characteristics and integrated to make ultrahigh-speed SSI and MSI integrated circuits. Low-noise HFET amplifiers have demonstrated a 0.5–1 dB lower noise figure than the MESFET amplifiers with comparable gate length. A ring oscillator propagation delay as low as 9.2 ps at 300 K and

5.8 ps at 77 K has been achieved. Frequency dividers have operated at input frequencies up to 13 GHz, and an access time of 0.5 ns has been demonstrated on a 4 kbit static RAM. The n^+-AlGaAs/GaAs HFETs have already made a dramatic impact on high-frequency device and high-speed circuit technology.

Advanced HFET structures with superlattice or δ-doped donor layers and with quantum well or doped channels have also been explored for improved charge control in the FET channel. Another variation is the insulated-gate HFET, which utilizes an undoped wide-bandgap material such as AlGaAs or InAlAs as the dielectric material separating the channel from the gate. This class of device overcomes many of the limitations of the conventional HFET.

An enhancement in hole mobility has also been observed in the selectively doped heterostructures, which open the possibility of high-performance p-channel devices for complementary circuits. Recently, HFETs in materials systems other than AlGaAs/GaAs have been demonstrated. Devices based on lattice-matched InAlAs/InGaAs and slightly mismatched strained-layer (also called pseudomorphic) material systems have been fabricated. The history and development of the HFET technology, from the early device demonstration to the state-of-the-art performance of microwave devices and digital integrated circuits, are discussed here.

3.2 HETEROSTRUCTURE FIELD EFFECT TRANSISTORS

The design and the device characteristics of the conventional n^+-AlGaAs/ GaAs HFET are reviewed in this section. The charge control due to the field effect of the gate on the conduction channel is first discussed in terms of the device structure. Then, the transport of carriers in the device, which results in the measurable extrinsic characteristics such as the drain I–V dependence, is explained. The effect of Al concentration on the device behavior as well as trapping effects associated with doping of AlGaAs is covered. The Schottky gate characteristics and the backgating in HFETs are two other important factors in the performance of the HFET that limit its capabilities for digital circuits. The scaling of device properties with gate length, which is a key to understanding the possibilities for very high speed performance, and the microwave performance of HFETs are reviewed.

3.2.1 Device Design of n^+-AlGaAs/GaAs HFET

3.2.1.1 Two-dimensional Electron Gas. When two semiconductor materials of different bandgap are joined together to form a heterojunction, discontinuities in the conduction and valence band edges occur at the heterojunction interface. In an n^+-AlGaAs/GaAs heterojunction, a triangular potential well is formed on the GaAs side of the interface, as illustrated in Fig.

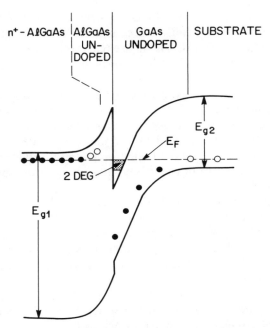

Figure 3.1   Band diagram of a n^+-AlGaAs/GaAs heterojunction.

3.1. Due to the difference in the electron affinity of the two materials, electrons accumulate in this potential well and form a sheet of electron gas similar to the inversion layer in an SiO_2/Si metal–oxide–Semiconductor (MOS) structure. The width of the well is approximately 10 nm, which is more than an order of magnitude smaller than the de Broglie wavelength of the electron, and therefore the electron wave function in the direction perpendicular to the interface is quantized to form a two-dimensional system. The existence of a two-dimensional electron gas (2DEG) is verified by observing the Shubnikov–de Haas oscillations of the conduction electrons and their dependence on the angle between the magnetic field and the normal of the interface [4].

The spatial separation between the 2DEG and the ionized donors in the wide-bandgap material (AlGaAs) reduces the Coulombic electron–donor interaction and enhances the low-field mobility of the electrons. At room temperature, the electron mobility in GaAs is less than 9000 cm^2/V, being limited by the electron–phonon interactions. As the temperature is lowered, the optical phonon scattering is reduced. Ionized impurity scattering becomes dominant in the bulk GaAs at low temperature and limits the mobility to low 10^3 cm^2/V·s for a doping level of 10^{17} cm^{-3}.

In selectively doped heterostructures, the background doping level in the GaAs is typically~10^{14} cm^{-3}. Furthermore, the residual Coulomb scattering is electrostatically screened by the electrons in the 2DEG, which has an extremely high electron density ($\sim 10^{18}$ cm^{-3}, corresponding to a sheet electron density of 10^{12} cm^{-2}, confined to within ~ 0.01 μm of the interface). Hence, the mobility in a selectively doped heterostructure is limited only by the piezoelectric and acoustic phonon scattering. The electron mobility in the 2DEG can be further enhanced by inserting an undoped AlGaAs spacer layer between the n^{+}-AlGaAs donor layer and the 2DEG [5], although this is accompanied by a reduction in the electron density in the 2DEG. A mobility of 5×10^{6} cm^{2}/V·s at 2 K has been demonstrated with a 75 nm spacer layer [6].

Although the low-field mobility is an important parameter in determining the parasitic resistance and, therefore, the performance of HFETs, it should be recognized that in practice there is a trade-off between the carrier mobility and other device parameters such as the 2DEG charge density and the transconductance, to achieve the optimum FET performance.

The conduction in the 2DEG can be modulated and controlled by placing a Schottky gate on top of the doped AlGaAs layer to form an FET structure, as shown in Fig. 3.2a. In a typical HFET structure, the AlGaAs layer, which is 25–60 nm thick doped at $\sim 1 \times 10^{18}$ cm^{-3}, is fully depleted. The application of a bias voltage to the gate modulates the charge in the 2DEG and thus the channel current, thereby giving rise to the field effect action. Figure 3.2b shows a typical conduction band edge diagram of an HFET under forward gate bias. The short gate–channel spacing, the high mobility and effective velocity of the 2DEG, and the good confinement of the 2DEG in the quantum well provide the HFET with a significant performance edge over the MESFET for high-speed and high-frequency applications.

Although the basic operation principles of the n^{+}-AlGaAs/GaAs resemble those of the Si MOSFET in many aspects, the details of material parameters and device physics are different. In particular, the use of more exact 2DEG formulism is needed for the HFET due to the much smaller effective mass of electrons in GaAs [7]. However, the final results are similar to those of MOSFETs. The sheet charge–voltage relationship is given by [8]

$$n_{\mathrm{s}} = \frac{\epsilon}{q(d_{\mathrm{d}} + d_{\mathrm{i}} + \Delta d)} \, (V_{\mathrm{g}} - V_{\mathrm{th}}) \tag{1}$$

and the threshold voltage

$$V_{\mathrm{th}} = \frac{\phi_{\mathrm{b}}}{q} - \frac{\Delta E_{\mathrm{c}}}{q} - \frac{N_{\mathrm{d}} d_{\mathrm{d}}^{2}}{2\epsilon} + \frac{\Delta E_{\mathrm{F0}}}{q} \tag{2}$$

where $\Delta d \sim 8$ nm and $\Delta E_{\mathrm{F0}} \sim 0$ and 25 meV at 300 K and 77 K, respectively. Δd and ΔE_{F0} are fitting parameters used in the linear approximation of Fermi potential–2DEG charge density relationship.

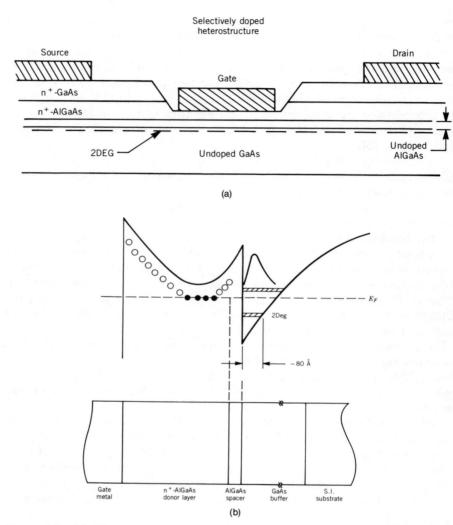

Figure 3.2 (a) The cross section and (b) the conduction band edge of an n⁺-AlGaAs/GaAs under positive gate bias.

The charge control due to the field effect of the gate on the conducting channel is determined by the doping, composition, and thickness of the layers between the gate and the channel, as well as the background doping in the channel itself. An approximate solution of the charge-control problem can be obtained using the Poisson equation invoking Fermi–Dirac statistics. However, an accurate solution is afforded by the one-dimensional quantum mechanical solution of the coupled Poisson and Schrödinger equations in a self-consistent manner. This yields the structure of the conduction and valence band edge as well as the charge distribution throughout the structure as a function of the gate bias. Foisy et al. [9] recently introduced the concept of "modulation efficiency" of the charge in a FET channel. Their modeling of the channel charge and the parasitic charge within the device demonstrated the importance of charge confinement within the channel for good device performance.

The charge-control model, Eqs. (1) and (2), shows the dependence of the charge density of 2DEG and the threshold voltage on the thickness d_d and doping level N_d of the n^+-AlGaAs donor layer, the thickness d_i of the undoped AlGaAs spacer layer, and the aluminium mole fraction of the AlGaAs layers. The aluminium mole fraction determines the conduction band edge discontinuity ΔE_c at the heterojunction interface as well as the barrier height at the Schottky interface. Since the threshold voltage is a strong function of these parameters, it is important to know the degree of control needed to produce the required threshold-voltage uniformity and reproducibility for integrated circuits.

3.2.1.2. Donor Layer Thickness and Doping Level.

The donor layer is depleted at the interface by the heterojunction and at the surface by the Schottky barrier. It is desirable that the two depletion regions overlap each other over the whole range of gate voltages. If the AlGaAs layers are not fully depleted, a conduction channel develops between the gate electrode and the 2DEG. This channel shields the 2DEG from the potential changes in the gate, interfering with the field effect action, and degrading the performance of the HFET.

Figure 3.3 shows the depletion of the donor layer and 2DEG as functions of donor layer thickness and doping level [10]. When the donor layer is too thick or the doping level too high, the donor layer is not depleted. Poor or no field effect action is expected in this regime. Depletion mode HFETs (D-HFET or normally-on HFETs) can be obtained in the shaded area, where the donor layer is fully depleted, but the built-in voltage of the Schottky barrier is not sufficient to deplete the 2DEG completely. Enhancement mode HFETs (E-HFET or normally-off HFETs) can be obtained in the region just below the shaded area, where the gate built-in voltage overcomes the heterojunction built-in voltage and fully depletes the 2DEG such that a positive gate voltage is needed to induce a 2DEG in the channel. One major advantage of HFETs is that the small separation between the gate and channel contributes to the high transconductance of the device. Hence it is desirable to dope the donor layer as highly as possible. The doping level using Si as the donor species during MBE growth is limited to $\sim 2 \times 10^{18}$ cm^{-3}. For an E-HFET with the donor layer doped at $\sim 1 \times 10^{18}$ cm^{-3}, the donor layer thickness is less than 30 nm, in order to be fully depleted at zero applied bias.

The threshold-voltage sensitivity can be modeled by solving the coupled Poisson's and Schrödinger's equation self-consistently. At 1×10^{18} cm^{-3}

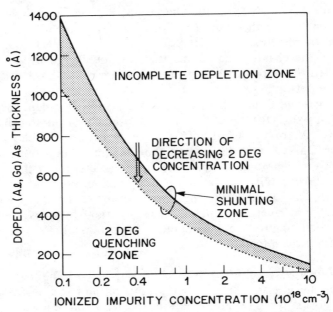

Figure 3.3 Dependence of the depletion of donor layer and 2DEG on the doping level and thickness of donor layer.

donor layer doping and with a 2 nm spacer layer, the threshold-voltage sensitivities are found to be $-55\,\text{mV}/10^{17}\,\text{cm}^{-3}$ and $-47\,\text{mV/nm}$. Other experimental and theoretical analyses also produced similar results [9, 11]. Figure 3.4 shows the dependence of three contributions of charge in the conventional HFET as a function of gate voltage for a depletion mode device. The charge in the 2DEG rises and then saturates, and this is the dominant charge which contributes to conduction within the FET channel. At a large enough V_g, significant charge is created at bound donor sites in the n^+-AlGaAs. This charge does not contribute to the conduction in the HFET but appears as a parasitic capacitance. The third component of the charge is free electrons in the conduction band of the AlGaAs layer, which act as a "parallel MESFET" in the low-mobility AlGaAs layer.

These three components are also reflected in the capacitance–voltage (C–V) characteristics of the HFET. The gate capacitance of a conventional HFET was measured by Moloney et al. [12] and is illustrated in Fig. 3.5. The schematic plot of gate capacitance versus gate voltage is divided into three regions. At V_g below the threshold voltage, the capacitance is due to the doping of the GaAs buffer layer; above threshold, it is due to the charge in the 2DEG. Finally, as the gate bias rises and saturates the 2DEG, the AlGaAs layer rapidly contributes to the capacitance. This capacitance arises due to bound donors and free electrons in the n^+-AlGaAs donor layer. All three components of capacitance were accounted for in this model. The rise

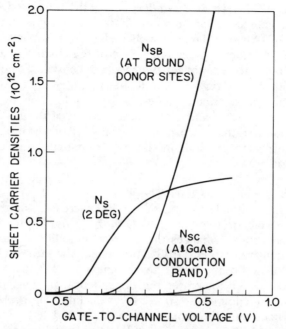

Figure 3.4   Charge control in the n^+-AlGaAs/GaAs HFET.

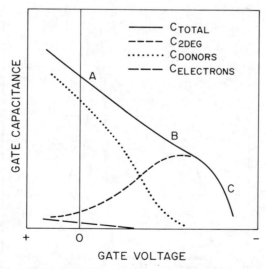

Figure 3.5   Schematic behavior of capacitance as a function of V_{gs} for a conventional HFET. (From Moloney et al. [12]. Reprinted with permission, copyright 1985, IEEE.)

in capacitance coincides with a rapid decrease in the transconductance of the HFET when the gate is forward-biased. These effects are both due to charge storage in the n^+-AlGaAs layer, which adds to the capacitance, and yet makes little contribution to the conductance of the channel. Several models have been published to account for the capacitance–voltage behavior of the HFET, since this parameter is a key to the potential speed of the device, and its prediction in a circuit model; for example, Lee et al. [13] had characterized both the I–V and C–V characteristics of the HFET. A model has been formulated that incorporates donor neutralization in the n^+-AlGaAs, in contrast to previous models which assume complete ionization of donor impurities [14].

3.2.1.3   Spacer Layer Thickness. The insertion of an undoped AlGaAs layer between the donor layer and the 2DEG increases the spatial separation between the electrons in the 2DEG and the doping impurities in the donor layer and further enhances the electron mobility of the 2DEG [15]. However, the density of the 2DEG and, hence, the current level and the transconductance are reduced as the thickness of the spacer layer is increased [16]. Figure 3.6 illustrates the dependence of the maximum transconductance as a function of the spacer layer thickness. For a HFET with no spacer layer, a maximum transconductance of 250 and 400 mS/mm has been achieved at 300 and 77 K, respectively. But the improvement of the trans-

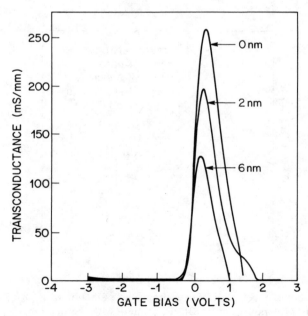

Figure 3.6   Transconductance of HFET with 0, 2, and 6 nm thick spacer layers.

conductance is offset by the reduction in mobility when the spacer layer is thinner than 2 nm. A reasonable compromise is ≥ 2 nm and is widely used for both low-noise devices and digital integrated circuits.

The trade-off between the transconductance and the low-field mobility illustrates that, although the high mobility at low fields is an important characteristic of the HFET, it is not the most important parameter determining the performance of the HFET. Schubert and Ploog [17] explained that although electrons in the lowest subband in the 2DEG had a velocity of 2×10^7 cm/s, the average velocity of all electrons was 1.5×10^7 cm/s at 77 K, with a field of 500 V/cm. At higher fields, the electron mobility of the 2DEG is quenched, due to scattering of electrons into lower-mobility subbands.

3.2.1.4 Scattering Mechanisms in HFET.
The scattering mechanisms affecting the 2DEG in HFETs are Coulomb scattering, due to ionized donors and deep levels primarily in the doped AlGaAs region, electron–phonon scattering, due to the polar optical phonons in the channel, and the interface scattering. In addition, there is momentum-space transfer of electrons from the Γ- to the L-valley in the presence of acceleration in high electric fields. The scattering of electrons not only limits the low-field mobility but also causes the removal of carriers from the lowest subband of the 2DEG. The mechanism of the removal of carriers is illustrated in Fig. 3.7 [17]. There is real-space transfer of electrons [18] into the AlGaAs layer by thermionic emission or thermionic field emission (Fig. 3.7a). Electrons may be trapped in immobile states on the heterointerface or in the AlGaAs near the interface (Fig. 3.7b). They may also be scattered into the GaAs buffer layer, away from the gate (Fig. 3.7c), and lose their two-dimensional properties, or into other subbands within the well (Fig. 3.7d).

The reduction of mobility, carrier density, and effective velocity in the 2DEG at high gate bias results in a rapid drop in the plot of g_m versus V_g at high V_g. The effect is seen in the I–V curves of the D-HFET in Fig. 3.8. The peak transconductance occurs at $V_g = -0.15$ V, and then drops rapidly due to the population of the donor sites and low-mobility conduction band in the AlGaAs layer. This should be contrasted with other structures, such as the superlattice donor or insulated-gate FETs discussed in the following sections. In these structures, the carriers are better confined in the GaAs channel. Therefore, they are able to sustain high transconductance over a much wider range of V_g. The turn-on of the Schottky gate, which contributes to the degradation of E-HFET characteristics at high V_g will be discussed later.

3.2.1.5 Aluminum Mole Fraction.
The ratios of the band edge discontinuities (ΔE_c and ΔE_v) to the total energy bandgap difference (ΔE_g) are among the most important parameters in determining the electrical properties of a heterojunction. For $Al_x Ga_{1-x} As$ with an aluminum mole fraction

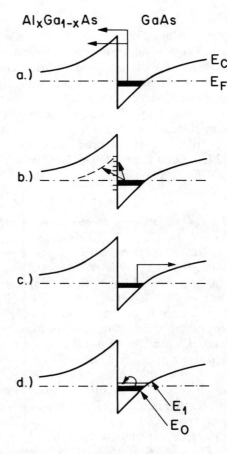

Figure 3.7 Schematic illustration of mechanisms of the removal of electrons from the lowest subband of the 2DEG. (From Schubert and Ploog [17]. Reprinted, with permission, from *Applied Physics*)

less than approximately 0.45, the alloy has a direct bandgap and the energy bandgap difference of a n^+-AlGaAs/GaAs heterojunction at 300 K is given by [19]:

$$\Delta E_g \quad (\text{eV}) = 1.247x \qquad x \leq 0.45 \tag{3}$$

The ratio of the valence band edge discontinuity ΔE_v to the direct Γ bandgap difference has been studied extensively. The majority of the measured ratios lie in the range 0.33 to 0.41 and are independent of the aluminum mole fraction. An empirical relation for the valence band edge discontinuity for $x \geq 0.45$ is given by [20]:

$$\Delta E_v \quad (\text{eV}) = 0.45x \tag{4}$$

and, therefore, the conduction band edge discontinuity

$$\Delta E_c \quad (\text{eV}) = 0.8x \tag{5}$$

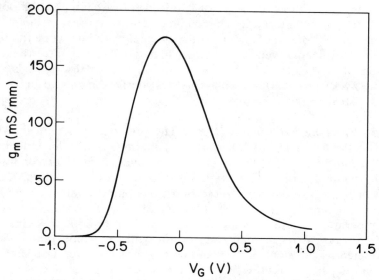

Figure 3.8 Transconductance curve of a D-HFET.

For aluminum mole fractions larger than 0.45, the X-valleys drop below the Γ-valley, and the conduction band edge discontinuity decreases with increasing aluminum mole fraction. The dependence of valence and conduction band edge discontinuities on aluminum mole fraction is shown in Fig. 3.9.

The significance of the conduction band edge discontinuity is that this

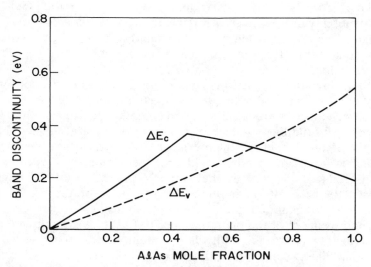

Figure 3.9 Dependence of valence and conduction band edge discontinuities on aluminum mole fraction.

parameter determines the extent of confinement of charge in the potential well at the heterointerface. The confinement is directly related to the efficiency of the control of charge in the channel by the field due to the gate. In addition, a deeper well (and a larger bandgap discontinuity) reduces the emission of hot electrons into the donor layer. As discussed later, the strained-layer n^+-AlGaAs/InGaAs HFET derives many of its superior characteristics due to the larger ΔE_c.

3.2.1.6 Trapping Behavior in HFET. The presence of silicon in AlGaAs results in the formation of DX centers in the material, independent of the growth conditions [21]. It was proven that over a range of conditions of growth, the doping of AlGaAs with Si led to the presence of the DX centers [22], which are responsible for the persistent photoconductivity and drain I–V collapse effects in HFETs. Most studies have concentrated on the low-temperature characteristics of the traps in Si-doped AlGaAs [23]. However, even at room temperature and with or without illumination, a nonexponential transient in the drain current was observed in response to a pulse on the gate of an HFET [24]. The transient response was attributed to trapping of emission of electrons at deep donors in the n^+-AlGaAs layer.

3.2.1.7 Device Characteristics at 77 K. The operation of semiconductor devices at low temperature has been reviewed in Kirschman [25], where the topics cover Si technologies, as well as HFETs. The reduction of the temperature of a 2DEG system results in enhancement of the carrier mobility in both n- and p-channel devices. Typical values for the electron mobility are 8,000 $cm^2V \cdot s$ at 300 K, and 100,000 at 77 K. Secondly, the saturated electron velocity rises from 1.1×10^7 to 2.3×10^7 [26, 27]. The low-field mobility of HFETs at 77 K depends on the precise device structure, as explained above; however, the fields found in real devices degrade the mobility to an extent that the often-quoted values of low-field mobility have little correlation to device operation. [28].

Pei et al. investigated the low-temperature behavior of the conventional HFET [29]. In discrete devices, these effects result in the increase of the peak transconductance from 195 to 305 mS/mm for the conventional HFET, and lowering of the knee voltage of the drain I–V characteristics. Furthermore, the improvement of the device characteristics also improves the performance of circuit speed. In addition, the subthreshold current of the HFET is reduced, as well as the output conductance of the device. In summary, all these effects improve the suitability of the FET for both ultrahigh-speed digital and microwave applications.

The improvement of the operating speed of MSI-level integrated circuits was described by Arch et al. [30], where a multiplication speed of 1.8 ns was measured at 300 K, compared with 1.08 ns at 77 K. This 50% improvement in speed is typical for most SSI/MSI circuits that have been reported. Although the low-temperature performance of HFETs is attractive, there are some practical problems that limit the conventional HFET. The problem with the low-temperature characteristics of n[+]-AlGaAs/GaAs heterostructures is that Si-doped AlGaAs contains DX centers [31], which are deep donor traps [32]. These traps result in persistent photoconductivity, a shift in the threshold voltage of devices, and "collapse" of the drain I–V characteristics when the device operates at 77 K without illumination [33].

Baba et al. [34] solved the low-temperature degradation problem by replacement of the Si-doped AlGaAs layer with an AlAs/n[+]-GaAs short period superlattice. In this structure, the AlAs is not doped, and yet macroscopically the effect of the superlattice is similar to that of the n[+]-AlGaAs in conventional HFETs. The threshold-voltage dependence on temperature is illustrated in Fig. 3.10. This showed that the superlattice structure had a much smaller shift in threshold voltage with temperature than the conventional HFET. Subsequently, Tu et al. [35] implemented a similar structure, and demonstrated that the "superlattice donor' HFET had higher g_m than the conventional HFET over a wider range of V_g. It also exhibited little (<50 mV) shift in V_{th} from 300 K to 77 K, compared with the conventional HFET of 150 mV. Finally, this device also showed no collapse of the I–V characteristics at 77 K.

The use of selenium as an n-type dopant rather than silicon results in elimination of collapse [36]. The DX centers associated with Se have a different energy to the Si traps, and so the trapping and emission constants are different. Another method to reduce collapse is to use $Al_xGa_{1-x}As$ with $x < 0.22$, to lower the conduction band edge below energy state of the DX center as described by Huang et al. [37]., where they used an $Al_{0.2}Ga_{0.8}As$ donor, and an undoped $Al_{0.45}Ga_{0.55}As$ spacer layer. Finally, it should be noted that p-doped HFETs do not show collapse [38], consistent with the fact that holes are not trapped as effectively by the deep donors in the

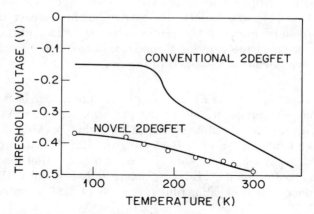

Figure 3.10 Threshold voltage of the superlattice donor HFET and conventional HFET as a function of temperature. (From Baba et al. [34].)

AlGaAs. Channel devices were fabricated and tested at 77 K. They had a peak transconductance of 35 mS/mm and showed little persistent photo-conductivity effects, unlike their n-channel counterparts. A comprehensive study of the low-temperature degradation of n^+-AlGaAs/GaAs HFETs was performed by Kastalsky and Kiehl [39]. Collapse of the drain I–V charac-teristics at 77 K in the absence of illumination was not due to hot-electron injection of electrons from the 2DEG into the donor layer, but an intrinsic property of n^+-AlGaAs when under a voltage bias, which then leads to the trapping behavior. The field dependence and location of trapping of elec-trons at 77 K was examined in the "inverted" structure HFET by Kinoshita et al. [40].

3.2.1.8   Gate Barrier on HFETs. The forward I–V characteristics of the gate contact on the HFET determines the operating characteristics of the E-HFET. This is a limitation of the DCFL circuit technology[†], which can benefit by the increased noise margins if the gate current in forward gate bias can be reduced. The study of the gate current behavior across the Schottky barrier in HFETs have indicated methods to improve the gate turn-on of the device [41]. The mechanism of gate current was identified as being controlled by two barriers: one was the metal–semiconductor junction and the second the AlGaAs–GaAs junction. The former is the controlling mechanism at low gate bias, but at high forward bias the latter determined the gate current. Therefore, incorporation of an AlAs spacer layer at the heterojunction between the n^+-AlGaAs layer and the GaAs improved the gate characteristics at forward bias. The typical gate turn-on voltage of gates for HFETs is measured to be ~0.1 V higher than on GaAs MESFETs, that is, 0.8 V. The enhancement of gate barrier height has been studied for HFETs [42] where a thin (8 nm) p^+-GaAs layer was grown on the surface of the wafer. Using E-mode drivers and resistive loads, a 1.1 V logic swing was measured, with a propagation delay of 15 ps. This method was also studied by Priddy et al. [43] who observed a barrier height increase of 0.8 V for epitaxially grown p^+ layers on the HFET. The importance of this technique is that the turn-on of the gate for E-mode devices is suppressed, such that larger logic swings can be sustained in the devices.

3.2.1.9   Backgating in HFETs. The phenomenon of backgating and sidegating has an important role in the performance of GaAs digital ICs [44], especially as the technology moves to higher packing densities and larger scales of integration [45]. Backgating in HFET structures has been characterized by Arnold et al. [46]. They demonstrated that the insertion of a superlattice buffer in the HFET structure between the 2DEG and the substrate reduced the backgating effect. A study of the backgating threshold

[†]For description of a typical logic family, see Chapter 7.

voltage for HFET structures [47] showed that the threshold voltage for backgating in HFETs was very low, implying excessive leakage current in the GaAs buffer layer. Yokoyama et al. [48] investigated backgating on five different devices, including HFETs, MBE-grown MESFETs, and ion-implanted MESFETs. They found that, independent of the substrate source, the use of MBE resulted in a large backgating effect on both HFETs and MESFETs (open symbols in Fig. 3.11). If the device had no MBE growth (i.e., ion-implanted MESFETs) or if a thermal etching procedure was used prior to MBE growth of the epitaxial structure, the backgating characteristics were greatly improved (closed symbols in Fig. 3.11). Thus, the conclusion was that there exists a conducting layer at the substrate–epitaxial layer interface which contributes to the large backgating effect, but this can be removed by thermal etching of the wafer surface in situ, prior to growth.

3.2.1.10   Gate Length Dependence of HFET Characteristics. The scaling of dc characteristics of self-aligned ion-implanted HFETs was reported by Cirillo et al. [49]. The early reports of the device behavior included the gate length dependence of transconductance and output conductance of the devices (Fig. 3.12). There is evidence of a short-channel effect due to lateral straggling of the silicon ions, which reduced the actual channel length and the impurity distribution under the gate. Engineering of these early device structures, implant and anneal conditions, and the fabrication of sidewall spacer structures or a lightly doped drain implant promises to aid in the reduction of these short-channel effects.

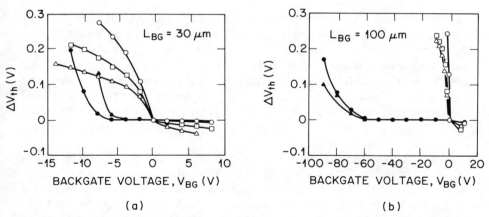

Figure 3.11   Plot of change in threshold voltage as a function of backgate voltage for 30 μm and 100 μm spacing backgate electrodes. (From Yokoyama et al. [48]. Reprinted with permission, copyright 1987, IEEE.)

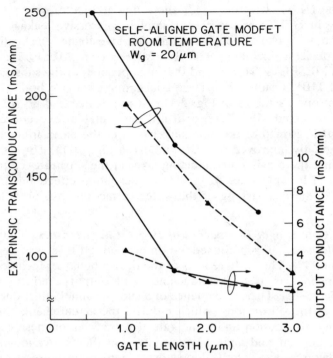

Figure 3.12 Plot of extrinsic transconductance and output conductance versus gate length for self-aligned refractory gate HFETs. (From Cirillo et al. [49]. Reprinted with permission, copyright 1984, IEEE).

With the wet chemical recess-etch technology, the output resistance of 0.8 μm and 0.35 μm gate length HFETs dropped from 250 to 120 Ω-mm, due to the short-channel effect, but the threshold voltage and transconductance remained the same [50]. Awano et al. [51] recently compared the HFET and GaAs MESFET with gates recessed by reactive ion etching and their threshold voltage as a function of gate length. A shift in threshold voltage of only 30 mV was observed for the HFET gate length change from 1.4 to 0.28 μm, in contrast to the MESFET, which had almost an order-of-magnitude higher change.

None of these empirical studies examined the fundamental scaling rules for HFETs, to include scaling of the rest of the device. The systematic analysis of scaling of gate length and the importance of the complete device design for power HFETs was performed by Das [52]. The aspect ratio (gate length/depletion depth under the gate) must be kept high enough(~10) to ensure adequate voltage gain g_m/g_d. A lower limit of 0.25 μm for the gate length of AlGaAs/GaAs HFETs was calculated, in order to obtain satisfactory power gain of the device. In addition, the effect of leading parasitics such as the gate–drain capacitance C_{gd} and the source resistance were accounted for. One observation was that the doping of the channel must be raised to maintain the aspect ratio and the performance. These predictions have been borne out in practice, which has led to exploration of alternative materials systems for HFETs (discussed later). A two-dimensional Monte Carlo study of the HFET over the gate length 2 to 0.5 μm was done to predict the switch-on and switch-off characteristics of the device [53]. The current and transconductance of the device were predicted to increase, with a corresponding decrease in gate capacitance, so the switching time for a 0.5 μm gate length HFET was six times faster than the 2 μm device.[†]

3.2.2   Microwave Heterostructure FETs

The most significant application in which HFETs have had an immediate impact due to their superior performance is for microwave low-noise and power devices. The figures of merit for performance of such devices are reviewed below and compared with MESFETs. Microwave HFET devices have been commercially produced by several companies for the past few years, using MBE- or MOCVD-grown wafers, and gate lengths as short as 0.25 μm, fabricated with direct-write electron beam lithography. Laboratory demonstrations have reached gate lengths below 0.1 μm, with state-of-the-art performance of speed and noise.

3.2.2.1   *Low-noise HFETs.*   The low-noise properties of HFETs were recognized in the early stages of development of the HFET technology [54]. The progress in improvement of the device characteristics is charted by the figures of merit in Table 3.1 [55–64], where improvements in device design and fabrication technology are reflected. The values of minimum noise figure and associated gain for the devices are presented. Additional references may be found in Berentz et al. [65].

A few simple equations from the small-signal analysis of the FET indicate the features of the HFET which make it suitable for this application. The

---

[†](Editor's Note) Recently, DelaHoussaye et al. [53a] compared experimentally the DC performance of conventional GaAs channel HFETs and pseudomorphic or strained-layer InGaAs channel HFETs (see Section 3.5.2) with gate length reducing from 2 μm down to 550 Å. The transconductance and effective saturation velocity of conventional HFETs were found to increase first, but then decrease below about 0.15 μm gate length. Similar yet much less pronounced behavior was also observed for pseudomorphic InGaAs channel devices and a peak in the saturation velocity (observed by the Editor) exists at the gate length about 0.07 μm. The decrease of velocity below certain gate length was believed to be the effect of "finite-length acceleration", that is, it requires a certain distance to accelerate carriers to nominal saturation velocity [53b].

cutoff frequency

$$f_t = \frac{g_m}{2\pi C_{gs}} \tag{6}$$

$$= \frac{v_s}{2\pi L_g} \tag{7}$$

where g_m is the transconductance, C_{gs} the gate–source capacitance, v_s the effective carrier velocity, and L_g the gate length. Figure 3.13 is a plot of the transit time of the carriers for the FET versus gate length, for GaAs MESFETs and AlGaAs/GaAs HFETs, from published data. There is a clear indication of the shorter transit time for the heterostructure devices, which is attributed to a higher effective velocity in the confined channels of HFETs.

The noise figure

$$NF = 1 + K \frac{f}{f_t} \sqrt{g_{m0}(R_s + R_g)} \tag{8}$$

where f is frequency of operation, g_{m0} is the intrinsic transconductance, and R_s and R_g are the source and gate resistance, respectively, from Fukui's analysis of GaAs MESFETs [66]. The high transconductance and effective velocity of carriers in the HFET contribute to the noise and speed performance. The noise figures of microwave HFETs have been compared with GaAs MESFETs [67, 68]. Figure 3.14 from Duh et al. is a plot of minimum

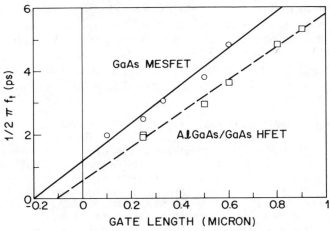

Figure 3.13 Transit time for carriers under channel versus gate length for GaAs FETs and n$^+$-AlGaAs/GaAs HFETs.

TABLE 3.1 Noise Figures and Associated Gains of Various HFETs

Technology	Size (μm)	g_m (mS/mm)	f (GHz)	NF_{min} (dB)	G_a (dB)	Year/Company	Reference
n+-AlGaAs/GaAs HFET	0.5		10	1.3		1984/Th-CSF	[55]
			18	0.9			
	0.5		35	2	5	1986/Rockwell	[56]
	0.25	320	12	0.66	11.8	1985/Toshiba	[57]
			18	0.85	9.4		
	0.25	430	18	1.2	11.6	1985/GE	[58]
			30	1.8	9.7		
			40	2.1	7.0		
	0.25	480	30	1.5	10	1986/GE	[59]
			40	1.8	7.5		
			62	2.7	3.8		
	0.25		60	2.5	4.4	1987/GE	[60]
	0.2	575	12	0.61	12.58	1987/TRW	[61]
AlGaAs/GaAs (MISFET)	1	280	10	2.1	6.9	1987/ITT	[62]
n+-AlGaAs/InGaAs (strained layer)	0.25	495	18	0.9	10.4	1986/III-GE	[63]
			62	2.4	4.4		
	0.1	540	60	2.4	5.8	1987/GE	[64]†

†(Editor Note) For recent developments, see relevant papers in *IEDM Technical Digest*, 1988. A minimum noise figure of 1.9 dB with 9.2 dB associated gain at 59.3 GHz was reported for a 80 nm strained-layer (pseudomorphic) HEMT by GE group (see Chao et al. [12] cited in Chapt. 2.)

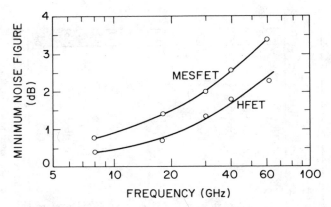

Figure 3.14 Minimum noise figure versus frequency for 0.25 μm gate length GaAs MESFETs and n⁺-AlGaAs/GaAs HFETs. (From Duh et al. [68]. Reprinted with permission, copyright 1988, IEEE.)

noise figure against frequency over the range 8 to 60 GHz, measured at room temperature, for 0.25 μm gate length HFETs and MESFETs. The noise figure for the HFET is consistently lower over the frequency range by 0.5–1 dB. In addition, the HFET had better output match and larger gain bandwidth product than the MESFET. In this work, the room-temperature noise figure at 8 GHz was 0.4 dB, with 15 dB gain. The low-temperature noise characteristic of HFETs is another feature of interest, due to the enhanced mobility of the carriers at cryogenic temperatures. By cooling to 12.5 K, the minimum noise temperature was 5.3 K at 8.5 GHz achieving state-of-art low-noise cryogenic operation. The device was optimized to have minimal distortion of the drain I–V characteristics at low temperature.

The measurements of minimum noise figure for a 0.35 μm gate length distributed amplifier were made by Bandy et al. [69]. They correlated the high transconductance and low output conductance of the HFET with better noise figures than in the MESFET; the gain was 12 dB for the HFET and 9.5 dB for the MESFET. The need to have short gate length and low gate resistance has led to developments of novel gate structures. T-shaped gates were developed [70], using a trilevel electron beam resist to achieve gates with a short footprint and yet large conductance, to improve the noise figure.

It was claimed that in a broadband amplifier built with GaAs FETs, the performance of a 0.5 μm HFET was better than a 0.25 μm MESFET [71]. A discussion of the trade-offs of device geometry in the context of the equations that control device performance was made by Jay et al. [72]. The importance of reducing the source–gate spacing and, of necessity, improving the ohmic contact technology was indicated. A number of improvements to the conventional HFET structure have been suggested for microwave applications, including an undoped surface AlGaAs layer, to reduce input capacitance and improve gate leakage and breakdown voltage [73]. However, the most significant improvement in microwave performance has been observed in the HFETs with a strained-layer InGaAs channel, which will be discussed later.

3.2.2.2 Microwave Power HFETs. The HFET for power devices has received less attention than for low-noise applications. However, as the design of HFETs has evolved from the simple single heterojunction device to double and multiple heterojunction channel structures, the possibility of obtaining large gain and power at high frequencies is now being realized. This is perhaps the major growing application for very high performance HFETs, as the device engineering challenges are met. The microwave performance of HFETs as power devices has been studied by a number of authors [74]. The microwave power FET is a large-signal device, in contrast to the low-noise FET. The requirement for high power has led to the use of multiple heterojunction devices, to increase the current density in the FET [75]. The gate-drain breakdown voltage was improved by lowering the surface doping adjacent to the gate. Saunier et al. characterized the multiple channel HFET as a power device over the range 10 to 60 GHz [76] and achieved a power of 1 W/mm from a multiple channel HFET device, at 21 GHz, with 3.9 dB gain, in a 0.4 μm gate length technology [77]. These are the state-of-the-art figures in the power FET technology.[†]

3.3 PROCESS AND MATERIALS TECHNOLOGY FOR HFETs

HFET processing technologies have largely been derived from GaAs MESFET technology [78]. Because the device structure and geometry are similar to those of the GaAs MESFET, often the early development of the HFET technology occurred alongside a MESFET fabrication program at many companies.

In this section, the techniques that have been employed for HFET fabrication are reviewed. The methods of interconnection, isolation, lithography, and metallization for HFETs are very similar to MESFET processes. The structure of the gate, and its alignment to the source and drain, are the critical parts of the HFET processing technology, as these determine the device characteristics. The technologies for the transistor gate formation encompass the recessed-gate process, self-aligned ion implantation using a refractory gate, self-aligned ion implantation with a "dummy" gate, a buried

[†](Editor's Note) For recent developments, see, for example, Smith et al. [77a] where maximum output power of 100 mW was obtained at 60 Ghz for a 0.25 μm gate technology. At 35 Ghz, saturated power density of 0.9 W/mm with 5.0 dB power gain was measured. The HEMTs exhibit peak extrinsic dc transconductance of 600 mS/mm, full channel current of 600 mA/mm, and a gate-drain reverse breakdown voltage of 9 V.

gate, and a sidewall regrowth method. The overall aims of the gate technology are to minimize the variation in threshold voltage of the FETs, reduce the leading parasitics (in particular the source–gate resistance), and achieve submicron gate geometry for the maximum device speed.

3.3.1 Gate Technology

The gate technology has been divided into five distinct methods which have been reported for the formation of the gates of HFETs. They are the recessed-gate technology, the self-aligned ion-implantation method, the dummy-gate, sidewall-assisted regrowth, and buried-gate technologies. Each technique is aimed at obtaining the best gate length and threshold-voltage control, particularly for submicron gate features to achieve ultrahigh-speed performance.

3.3.1.1 Recessed-gate Technology. For HFETs with a recessed-gate structure, the cap layer of n^+-GaAs must be removed before deposition of the Schottky metallization on the AlGaAs donor layer. The conventional process which has been used is a selective or nonselective etch, which etches the GaAs and AlGaAs. The threshold voltage of the devices is determined by the remaining doping and thickness of the AlGaAs layer after etching. This thickness is normally reached by using an iterative sequence of etching, measurement of I_{ds}, and re-etching of the gate recess until a predetermined value of I_{ds} corresponding to the required V_{th} is reached. In practice, the control in such a technique is poor, especially when both enhancement and depletion mode devices have to be fabricated to match each other's characteristics, (e.g., in a DCFL circuit), since two etch steps to two different depths are required.

An etch consisting of a pH-adjusted mixture of hydrogen peroxide and ammonium hydroxide in aqueous solution is used by a number of laboratories for the recess etching of heterostructure devices. This solution has a selectivity of 5:1 for etching of GaAs over $Al_{0.3}Ga_{0.7}As$. Another etchant which has less selectivity is the phosphoric acid and hydrogen peroxide etch mixture. Recently, the use of citric acid was also reported [79]. In this study, a comparison of selective and nonselective etches was conducted, with the conclusion that best results are obtained for citric acid which is a nonselective etch. The results indicated that the precise shape of the etch trough in the recessed-gate HFET has a large influence on the effective source resistance of the device. This study has implications on the etching of recesses not only for wet etching but also dry etching of HFETs. They measured a peak transconductance of 520 mS/mm at 300 K and a gate–drain breakdown voltage of over 10 V, with an optimized etch solution. The high transconductance is due to reduction of the source resistance of the device as a result of control of the etch trough.

The selective etching of GaAs over AlGaAs has proven to be a significant factor in the development of the recessed-gate process for heterostructure devices. The chemical difference between different III–V compounds is used to aid in the processing and process control of the device structure. The reactive ion etching (RIE) of GaAs and AlGaAs in CCl_2F_2 has been explored for processing of HFETs. The principal requirement of the recess etching is that the etchant must have high selectivity, to ensure that the as-grown layer thicknesses are preserved by a "dead stop" at the AlGaAs layer. In addition, the residual state of the semiconductor, that is, the damage in the near-surface region and the quality of the surface (e.g., residual deposits), must give good Schottky behavior from the gate metallization.

The first demonstrations of such selective etching was by Hikosaka et al. [80], where a selectivity of 200:1 for GaAs to $Al_{0.3}Ga_{0.7}As$ was reported. Selectivity as high as 1000:1 has also been demonstrated [81]. A threshold-voltage uniformity of 16 mV for EFETs and 24 mV for DFETs on a full 2 in. MBE-grown wafer were reported by Abe et al. [82]. The short-range uniformity achieved was a standard deviation over a $10 mm^2$ area of 6 mV for E-mode and 11 mV for D-mode devices. The etch rate of the AlGaAs layer is 2 nm/min, compared with 520 nm/min for GaAs [83]. The reason for the high selectivity is that the AlGaAs does not react in the CCl_2F_2 plasma environment, due to the formation of a passivating layer of aluminum fluoride on the surface. In this work, the MBE-grown layer structure is not the simple n^+-AlGaAs/GaAs structure employed in wet etching, but incorporates thin etch-stop layers of AlGaAs (Fig. 3.15). The process sequence illustrated shows that the photoresist is first opened to etch the E-mode device to a first etch stop (Fig. 3.15b). Then the photoresist is re-exposed, to reveal the E-mode gates, and a second RIE step (Fig. 3.15c) gives an E-mode and D-mode gate which can be metallized in one step (Fig. 3.15d). The threshold voltages are therefore determined largely by the thickness and doping of the as-grown material, rather than the processing. Lin et al. [84] conducted an extensive study of the machine-dependent and growth-dependent V_{th} control of reactive-ion etched HFETs. He observed that a load–lock system with parallel plates operated in the RIE mode was best for minimal damage and that rotation of the wafers during Schottky metal deposition was responsible for attaining better V_{th} control.

Application of this technique to E–D circuits has been reported for some of the largest circuits made in HFETs, the complexity of which is comparable to those made by a self-aligned ion-implantation process. Sheng et al. [85] reported a 1 kbit SRAM fabricated with a RIE process which had a threshold-voltage standard deviation of 4 mV for 100 devices in a $1 mm^2$ area. Recently, Notomi et al. reported a 0.5 µm gate length 4 kbit SRAM in this technology for HFETs has aided in the evolution of the circuits from small to medium and large scales of integration.

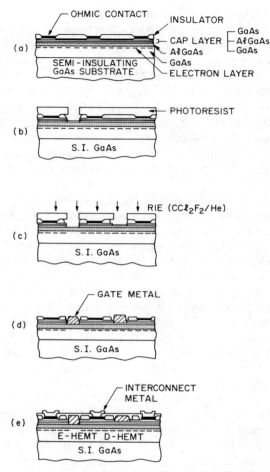

Figure 3.15   Process sequence for the E- and D-mode HFET. (From Abe et al. [83]. Reprinted with permission, copyright 1986, IEEE.)

The use of recessed gates for the inverted structure and subsequently the insulated gate inverted-structure HFETs eliminates the need for the use of reactive etching. An argon ion-milling technique, with a low-energy beam is used to fabricate the recess [87], and yet minimal damage is observed in the underlying semiconductor [88]. Device results of these structures are reviewed in the survey of HFET devices.

3.3.1.2 Self-aligned Refractory Gate Technology. The use of W–Si as a refractory metal for the formation of a Schottky barrier on GaAs, its thermal stability after the activation of the implanted dopants, and its suitability for LSI MESFET circuits have been demonstrated by a number of groups [89]. The process is planar, using subtractive patterning of a refractory gate metallization, followed by a self-aligned n^+ ion implantation, to obtain low ohmic contact and source resistance. The refractory metallization for GaAs is usually W–Si, although other compounds have been successfully implemented [90]. The application to HFET structures has been pursued by Cirillo et al. [91, 92] for both n^+-AlGaAs/GaAs HFETs and superlattice donor HFETs (Fig. 3.16). The typical ion-implantation dose was $\sim 10^{13}$ cm^{-2} Si ions, and a rapid thermal anneal was used for activation of the implanted ions. With this process, the peak transconductance was 200 mS/mm and rose to over 300 mS/mm at 77 K [93]. It was observed that the stability of the 2DEG and the abruptness of the heterointerface in the device structure is a concern, unlike the conventional GaAs MESFET [94]. The degradation of a heterostructure has been observed after rapid thermal annealing, resulting in a reduction in the 2DEG mobility [95]. However, the effects can be reduced such that the device characteristics are not compromised.

The Schottky barrier height of the W–Si gates in the devices fabricated by Cirillo et al. was 1.0 to 1.2 eV, confirming the thermal stability of the W–Si/AlGaAs interface. The self-aligning scheme reduced the source resistance to 0.77 Ω-mm, and threshold-voltage uniformity was better than 12 mV standard deviation over a 0.6 mm^2 area, with a mean of 0.44 V and 30 to 40 mV across a 3 in. wafer. The uniformity and yield of this process have served to make integrated circuits with high levels of integration [96].

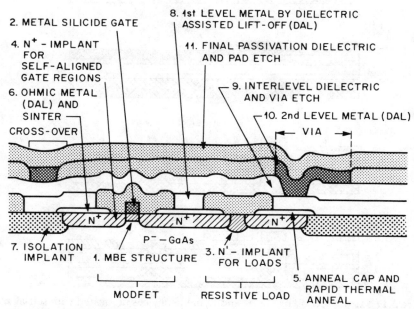

Figure 3.16 Cross section of the planar self-aligned ion-implanted HFET. (From Cirillo et al. [91, 92]. Reprinted with permission, copyright 1984, IEEE.)

This technology is planar and eliminates the threshold-voltage nonuniformity which may be caused by the variations in the gate-to-channel spacing as a result of the gate recess-etching process. Thus, the threshold voltage is largely determined by the uniformity, thickness, and doping of the materials growth, rather than the processing. A notable limitation of this technology is that it allows only one threshold voltage for the HFETs. For circuits needing load devices, selective are ion-implanted saturated resistors are used. It is highly desirable to have a technology which allows HFETs with different threshold voltages to be fabricated on the same chip while maintaining some advantages of the fully planar technology.

Pei [97] proposed a process that takes advantage of the high selectivity between the GaAs and AlGaAs layers to fabricate self-aligned ion-implanted HFETs with different threshold voltages. Figure 3.17 shows the cross-sectional view of two devices at various states of the fabrication process. A shallow mesa is created with two sequential selective etches. The photoresist stencil is then removed and a blanket selective etch is applied to remove the

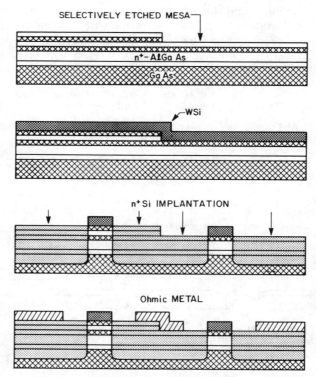

Figure 3.17   Cross section of the E- and D-mode devices fabricated with self-aligned ion-implantation technology. (From Pei [97]. Reprinted with permission, copyright 1988, IEEE.)

AlGaAs layer, which serves as a protection of the underlying GaAs surface, from the whole wafer just prior to the deposition of the refractory metal for gates. The gate metal is then patterned and etched, followed by ion implantation as in the previous process. By creating shallow mesas, HFETs with different threshold voltages can be fabricated with one refractory metal deposition, one gate metal patterning step, and one ion-implantation step.

3.3.2   Other Self-aligned HFET Technologies

Two other techniques for creating a self-aligned HFET structure without using ion implantation and high temperature annealing have been reported. Sheng et al. [98] used a close drain–source technique, using a "dummy-gate" process illustrated in Fig. 3.18. This method does not require the high-temperature anneal associated with the ion-implanted self-aligned technology. A selective area regrowth technique employing the sidewall-assisted transistor (SWAT) process was used [99]; in this process, after the gate is patterned, a sidewall is formed and then the n^+-GaAs contact layer is grown. The attractiveness of this method is not only the self-aligned pattern, but also the cap layer, which is doped with Se at a concentration of 5×10^{18} cm^{-3}, leading to an ohmic contact resistance to the cap layer of <0.1 Ω-mm. However, the additional step of regrowth and the capacitive coupling between the gate and contact regions make the process unattractive. Neither of these techniques has been pursued extensively for integrated circuits.

3.3.3   Ohmic Contacts to HFETs

The technology for fabrication of ohmic contacts to HFETs is derived from the alloyed ohmic contacts used for GaAs [100]. However, the fabrication of ohmic contacts to HFETs is different from that of contacts to GaAs MESFETs for two major reasons. One is that the semiconductor contains AlGaAs, which has a higher bandgap than GaAs, and the second is that the conducting channel, the 2DEG, lies below this barrier. Feuer [101] analyzed the recessed-gate HFET in terms of a two-layer model, where conduction occurs in the n^+-GaAs cap layer and in the 2DEG, and the two layers are separated by the high bandgap AlGaAs donor layer. He concluded that the presence of the n^+-GaAs cap layer is important in reducing the source resistance of the device, since at room temperature the source resistance is dominated by the barrier resistance of the AlGaAs layer.

Because of the presence of the AlGaAs, a number of authors have observed that conventional alloying treatments suited to GaAs processing must be modified to penetrate the AlGaAs layer, for minimum contact resistance. In particular, Ketterson et al. [102] demonstrated that a high-temperature alloy up to 800°C reduced contact resistance to 0.035 Ω-mm. Lower temperatures (500–600°C) gave contact resistance below 0.1 Ω-mm.

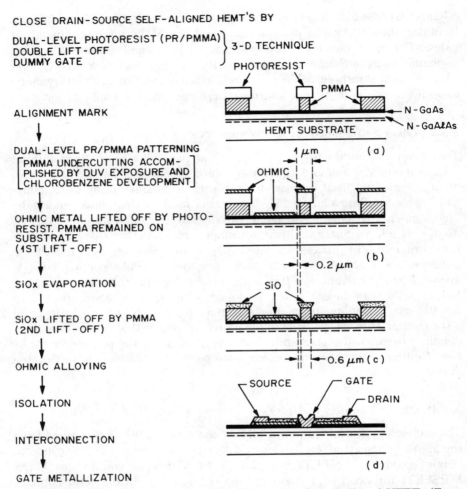

CLOSE DRAIN-SOURCE SELF-ALIGNED HEMT'S BY

DUAL-LEVEL PHOTORESIST (PR/PMMA)
DOUBLE LIFT-OFF   } 3-D TECHNIQUE
DUMMY GATE

PHOTORESIST

PMMA

ALIGNMENT MARK

N-GaAs
HEMT SUBSTRATE   N-GaAℓAs

DUAL-LEVEL PR/PMMA PATTERNING
[PMMA UNDERCUTTING ACCOM-
PLISHED BY DUV EXPOSURE AND
CHLOROBENZENE DEVELOPMENT]

1 μm   (a)
OHMIC

OHMIC METAL LIFTED OFF BY PHOTO-
RESIST. PMMA REMAINED ON
SUBSTRATE
(1ST LIFT-OFF)

0.2 μm   (b)

SiOx EVAPORATION

SiO

SiOx LIFTED OFF BY PMMA
(2ND LIFT-OFF)

0.6 μm (c)

OHMIC ALLOYING

SOURCE   GATE
DRAIN

ISOLATION

INTERCONNECTION

(d)

GATE METALLIZATION

Figure 3.18 Process scheme for the close source–drain self-aligned HFET. (From Sheng et al. [98]. Reprinted with permission, copyright 1984, IEEE.)

Rapid annealing has been used in achieving low contact resistance to heterostructures by several authors. Hong et al. [103] compared the behavior of Ni/Ge/Au/Ni/Au and Ni/Ge/Au/Ni/Ti/Au evaporated layers, and measured a minimum contact resistance of $0.12\,\Omega$-mm, or specific contact resistivity of $4 \times 10^{-7}\,\Omega$-cm^{-2}, using a two-step halogen lamp heating system. Jones and Eastman [104] showed that a minimum contact resistance of $0.1\,\Omega$-mm was obtained with Ni/Au–Ge/Ag/Au layers, after sequential alloying at increasing temperatures.

The morphology and precise location of the alloyed metallization have not been widely reported. However, by examining ohmic contacts to typical HFET structures by cross-sectional transmission microscopy [105], a lateral diffusion of the metal contact of 0.12 μm was observed under normal alloy conditions. This has implications in the understanding of current transport at the ohmic contacts, which is geometry-dependent. In addition, there may be limitations to the fabrication of small-geometry devices due to this effect, which, however, may be overcome by the use of n^+-InGaAs cap layers and refractory, nonalloyed metal contacts [106].

3.3.4 Other Processing Techniques

Much of the lithography for the fabrication of HFET devices and circuits has been performed by optical techniques. The application of advanced electron beam techniques to GaAs devices has a long history, based on the advantages in performance, especially for low-noise microwave devices. In addition, there is considerable interest in reduction of device dimensions in silicon MOS technology to enhance the speed of circuits using both advanced optical lithographic techniques, and electron beam lithography.

There has been a corresponding trend in the fabrication of submicron gates for GaAs and HFET devices, to achieve improved performance. The principal application has been for microwave devices, with the development of multilevel resists [107] to make short gates with a "T" shape to optimize performance. Shah et al. [108] employed direct-write electron beam lithography with single-layer PMMA to explore device and small-scale circuit performance down to gate lengths of 0.35 μm, with a recessed-gate process. Recently, Resnick et al. [109] described a novel trilevel resist scheme to obtain controlled undercut profiles in an electron beam resist, for a recessed-gate lift-off process for HFETs.

Most of the process schemes that have been reported use a mesa technique for isolation between devices, since the resistivity in the MBE-grown undoped GaAs buffer layer is sufficiently large for circuit operation. For the fabrication of MSI and LSI circuits, oxygen or boron ion-implantation isolation is commonly used [110, 111]. The questions of interconnect technology, the use of dielectric isolation, air-bridge technology, etc., are common to other high-speed technologies. The realization of ultrahigh-speed performance from HFET circuits will have to rely not only on the intrinsically high speed of the device, but on an integration technology compatible with that speed.

3.3.5 Material Technology

The epitaxial growth of materials for HFETs has been performed by molecular beam epitaxy (MBE) and metal-organic chemical vapor deposition (MOCVD). The requirements of the device govern the choice of growth technique. The abruptness of the heterojunction between the narrow and wide bandgap semiconductors must be sufficient to ensure that a 2DEG of the optimal charge density and mobility is created at the heterointerface. Secondly, the ability to control the composition of the compounds and their doping, as well as the background doping levels in the absence of doping, has an influence on the choice of growth method. Some important criteria for a high-throughput growth technique include multiple wafer loading, computer control, reliability, and reproducibility. These factors will determine the future of growth technologies for the commercial realization of HFET circuits.

3.3.5.1 Molecular Beam Epitaxy (MBE).

MBE has been used for most of the early studies on HFETs. MBE has been extensively reviewed and characterized by Cho and Arthur [112, 113]. The essential components of an MBE system are a wafer-loading chamber, sample preparation chamber, and then one or more growth chambers, where each section is independently pumped to a pressure $\leq 10^{-10}$ torr. Wafers are mechanically transported from one chamber to the next. Most commercial MBE systems handle one 3 in. diameter wafer at a time during growth, with noncontact mounting and radiant heating. In experimental and older systems, wafers are often mounted with indium onto molybdenum heater blocks to control the substrate temperature during growth. This method is undesirable, since indium reacts with GaAs at the typical growth temperatures, leading to pitting of the backface. Thus, noncontact radiant heating has prevailed [114].

The growth of epitaxial layers is achieved by directing a molecular beam of the elemental sources (e.g., As, Ga, Al), which are heated in effusion cells, onto a heated GaAs substrate. Dopants (e.g., Si, Be) are also heated in such cells, and the choice of composition and doping is determined by opening and closing of shutters, the temperature of the cells (which determines beam flux rate), and the temperature of the substrate. The proparation of the wafer before growth determines several features of the grown layers which are critical to the circuit manufacturer. One is the uniformity of the layers, which determines the yield of circuits (e.g., for DCFL circuits, V_{th} value must be within 30 mV of the design value [115], which translates into only a few angstroms of thickness of the layers across a circuit). Therefore, short-range and global variations are important. Second is the electrical quality of the buffer layer; for example, the backgating effect in a HFET was reduced by thermal etching of the wafer prior to growth, which eliminated a conduction path at the substrate–epi-layer interface [48]. Another phenomenon is oval defects. These are hillocks oriented in the $\langle 110 \rangle$ direction on a (001) surface, ranging in size from 1 to 100 μm, which are characteristic of MBE-grown layers. Much work has been done to reduce their density [116]. The oval defect has two effects. One is the effect on lithography of small features, as it gives rise to anomalous topography. The second is that the defect can cause electrical degradation of the I–V characteristics of the FET [117] such that the device cannot be pinched off. Therefore, the uniformity, oval defect density, and absolute control of layers and their doping will determine applicability of any of the growth methods for LSI. Note that most detailed uniformity studies have been done only on conventional HFETs, not on other devices.

3.3.5.2 Metal Organic Chemical Vapor Deposition (MOCVD).

In recent years, the technique of MOCVD, where gaseous sources of the organic compounds of the epitaxial layers (e.g., tri-methyl gallium [TMG] and tri-methyl arsenic [TMA]) have been used for the growth of the HFETs. Previous results indicated inferior properties compared with MBE, but 2DEG mobilities and device characteristics now rival those made from MBE-grown layers [118]. The use of MOCVD for HFETs was reviewed by Kawai et al. [119]. There are two features of MOCVD that appear to be particularly attractive. Two groups have reported that the problem of the deep donor traps in Si-doped AlGaAs layers grown by MBE can be reduced in MOCVD materials [120, 121]. The other advantage is that MOCVD is suited to multiwafer growth [122]. Bhat et al. [120] describe a three-wafer-at-a-time growth system, where thickness and doping uniformity of 2% and 1.5% were measured for AlGaAs, respectively, on 2 in. wafers. They also realized E- and D-HFETs with V_{th} sigma of 23 and 35 mV. Also, the first application of MOCVD to low-noise HFETs by Takakuwa et al. [123] indicated the potential for the production of these devices by MOCVD. Subsequently, a number of companies now offer commercial low-noise microwave HFET devices based on the MOCVD growth technique.

3.3.5.3 Other Growth Techniques.

A novel technique for the growth of high-quality epitaxial layers of III–V compound semiconductors is called metal organic MBE (MOMBE). The technique was developed as chemical beam epitaxy (CBE) by Tsang [124], and gas source MBE by Panish et al. [125]. This method is being developed to obtain high-throughput heterostructure wafers, with many of the advantages of both MBE and MOCVD.

3.3.6 Summary of Processing Techniques

Two techniques have emerged as important for the fabrication of MSI and LSI digital circuits for HFETs. One is the selective RIE of the multi-layered AlGaAs/GaAs structure, relying on the selectivity of etching between AlGaAs and GaAs for threshold-voltage control. The second is the self-aligned ion-implantation technology, which is well established for GaAs MESFETs, and has been successfully applied to HFETs. These two methods take advantage of the uniformity and reproducibility of MBE and MOCVD epitaxial growth methods to achieve the control of the device characteristics. Only in the case of discrete devices and small-scale circuits, where the requirements are less stringent, is wet chemical etching continuing to be used.[†]

[†](Editor's Note) The above three sections were contributed by Dr. N. J. Shah. Dr. S. S. Pei on the other hand contributed the following three sections.

3.4 SURVEY OF HETEROSTRUCTURE FETs

The simple single heterojunction n^+-AlGaAs/GaAs FET, known as the SDHT, MODFET, HEMT, or TEGFET, has been described in the previous sections. This device is the basis of most of the pioneering work on device characterization and circuit demonstrations. However, as the HFET technology has evolved, the motivation to improve on this structure comes from the potential applications. The needs for a high-current-drive FET for digital ICs [126], a high-current device for power FETs, and low-noise microwave and millimeter-wave FETs have created the three most important applications for this new class of FETs. In all cases, the maximum charge density in the conventional HFET (1×10^{12} cm^{-2}) limits the current density of the device. For digital circuits, an increase in the transconductance over a large range of gate bias is desirable, as well as for the power FETs. In the case of microwave devices, the peak transconductance is important, and in all cases the minimum gate–channel capacitance is desired for effective modulation of the charge in the channel. The confinement of carriers to a well-defined channel eliminates two unwanted effects. One is the real-space transfer of carriers from the channel to the higher-bandgap confining layers, leading to the formation of a parallel MESFET in the AlGaAs. The second is the space charge injection of carriers into the substrate, leading to an increase in the drain output conductance, especially for short-channel devices. The progress to submicron geometry gates has been driven by the marked improvement in both digital and microwave device performance over the years, and this is a key factor in the development of the HFET technology.

With these simple guidelines, there are two basic variants of the HFET. In the first, the structure is optimized to achieve the best channel confinement by adjusting the layer thicknesses and compositions. This approach is often referred to as "bandgap engineering." In the other, the doping is distributed to allow charge to populate the device. The donor layer may be a bulk-doped layer, a δ-doped layer, or a modulation doped superlattice. The conducting channel itself may also be doped. It is also possible to create an inversion channel in a totally undoped heterostructure. The family tree of HFETs, shown in Table 3.2, conveniently categorizes the large variety of HFET structures that has been developed. Particular examples of each of these devices are now reviewed.

3.4.1 HFETs with Improved Channel Confinement

3.4.1.1 Single Quantum Well HFET. The single quantum well HFET is a structure in which the GaAs conducting channel is bounded on either side by AlGaAs confinement layers. This is also referred to as the double heterostructure (DH) FET. The effect of the confinement is to increase the effective carrier velocity in the channel, and also to improve the drain

TABLE 3.2   Family Tree of HFET Devices

Donor layer	Selective doping	Bulk doped donor layer (SDHT, MODFET, TEGFET, HEMT, etc.) δ-doped donor layer Superlattice donor layer
	Insulated gate	MISFET, HIGFET
		SISFET
Channel confinement	Quantum well channel (SQW, etc.) Inverted structure (I-HEMT, I^2-HEMT, etc.)	
Channel doping	Undoped Doped (DMT, etc.)	

conductance and short-channel properties of the FET, by improving the charge confinement in the channel. In most cases, both of the confining layers are doped to maximize the charge in the intervening channel layer. The early demonstrations of such a structure [127] showed that the double heterostructure gave a sheet conductivity in the 2DEG of 24 mS/□ at 77 K when the undoped AlGaAs spacer layer was eliminated. Although the device has poor pinch-off characteristics due to the large distance between the gate and the underlying doped AlGaAs confinement layer, the sheet electron concentration was twice that of the conventional single heterojunction device [128], which made this an attractive alternative to the simple HFET. This poor subthreshold behavior with a quantum well bounded on both sides by a doped AlGaAs layer was also observed by Sheng et al. [129]. The calculation of electron states using a self-consistent technique showed that charge is accumulated at both of the heterointerfaces. The charge control by the field from the gate electrode was modeled, and explained the empirically observed charge-control relationship in terms of occupation of different eigen-states in the quantum well [130]. The structure of a single quantum well HFET, and the associated eigen-states which are confined in the well, are shown in Fig. 3.19. These investigations point out the need for precise engineering of the quantum well structure to suppress subthreshold current and to obtain the optimal charge control from the gate of the FET. A multiple quantum well channel FET was proposed [131] to give a larger sheet charge density. A six-channel structure had a 77 K electron density of 5.3×10^{12} cm^{-2} and a current of 800 mA/mm.

The growth of a DH structure is more demanding than the conventional HFET, since a good-quality GaAs layer is required to be grown on AlGaAs. By control of the growth rate and temperature, the heterointerface is made suitable for device fabrication [132]. The diffusion of the impurities from the lower AlGaAs layer into the channel also demands control of the substrate temperature during growth [133]. Unintentional diffusion of silicon into the

(a)

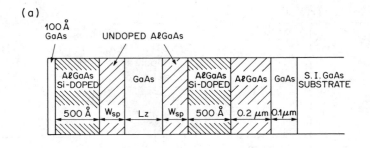

(b)

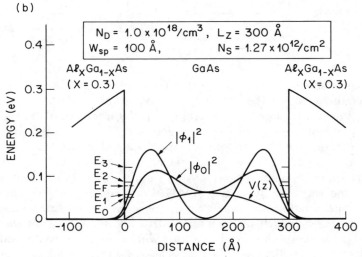

Figure 3.19 The structure of the single quantum well HFET, and the associated eigen-states which are confined in the well. (From Inoue et al. [130]. Reprinted with permission, from *Japanese Journal of Applied Physics*.)

channel region leads to a reduction in the mobility of the 2DEG in the GaAs quantum well channel.

3.4.1.2 Inverted-structure HFET. The early attempts at making inverted HFETs used a recessed-gate process [134]. More recently, Cirillo et al. [135] fabricated an inverted HFET using a self-aligned ion-implantation technique. This device had a very high peak transconductance of 1180 mS/mm at 300 K, but the cutoff frequency of a 1.75 μm gate device was only 7.5 GHz. Much work has been done on the I^2-HEMT, where the gate is placed on an undoped AlGaAs layer, under which is the GaAs channel and then the n^+-AlGaAs donor layer [136]. The significant feature of this device was that the gate leakage was small under forward bias up to a voltage of 1.4 V, compared with the Schottky turn-on voltage of 0.8 V for the conventional HFET. Comparisons between the I-HEMT and I^2-HEMT were made

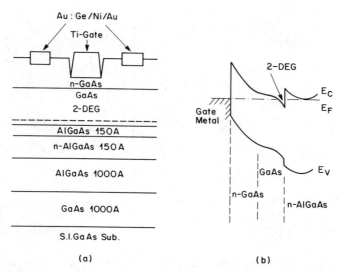

Figure 3.20 Conduction band edge diagram and device structure of the inverted HFET (I-HEMT). (From Kinoshita [137]. Reprinted with permission, copyright 1986, IEEE.)

by Kinoshita et al. [137]. The conduction band edge and device structures are illustrated in Figs. 3.20 and 3.21. The I-HEMT has no doping between the gate and the channel, so the threshold-voltage sensitivity to the recess depth in the I-HEMT is less than that in the conventional HFET. Also, it does not suffer the problems of the parallel MESFET due to population of

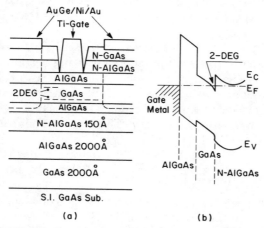

Figure 3.21 Conduction band edge diagram and device structure of the insulated-gate inverted HFET (I^2-HEMT). (From Kinoshita [137]. Reprinted with permission, copyright 1986, IEEE.)

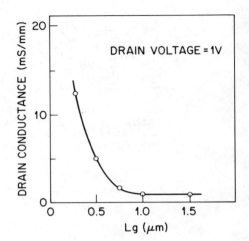

Figure 3.22 Drain conductance versus gate length for an inverted-structure HFET. (From Saito [138].)

the n^+-AlGaAs donor layer in the conventional structure. Whereas the gate of the I-HEMT is placed on undoped GaAs, the I^2-HEMT has undoped AlGaAs as the barrier under the gate. The improved barrier height at the metal-semiconductor interface and the better confinement of carriers were seen as advantages of this device over the I-HEMT. The I^2-HEMT has remarkable short-channel behavior, with drain conductance of only 2 mS/mm at 300 K and a peak transconductance of 230 mS/mm. The shift in threshold voltage from a gate length of 1.2 to 0.7 μm was 50 mV, compared with 250 mV for the inverted structure without the top insulating layer. The devices had a peak current of 300 mA/mm, and DCFL inverters were made with a 1.2 V logic swing. It should be noted that the fabrication scheme of these devices relies on a recessed-gate process, but the recessed was achieved by a low-energy ion-milling technique, rather than the conventional wet or dry etch.

Saito et al. reported an inverted-structure HFET fabricated using a recessed-gate and lift-off structure and employing angle evaporation to achieve the short gates. The gate lengths were as short as 0.28 μm in these devices [138]. The drain conductance was 13 mS/mm at this gate length, and ring oscillators were fabricated with a propagation delay of 11.9 ps/gate with 0.56 mW/gate, for a 0.5 μm gate length. The plot of drain conductance as a function of gate length is illustrated in Fig. 3.22.

3.4.2 HFETs with Different Donor Layers

3.4.2.1 Superlattice Donor Layer. The advantages of a superlattice donor layer for low-temperature operation of HFETs were reviewed in Section 3.2.1.7. The charge control in a superlattice donor structure was discussed by Baba et al. [139] and Tu et al. [140]. In a superlattice donor layer under forward bias, due to the higher bandgap of AlGaAs, there is no parallel

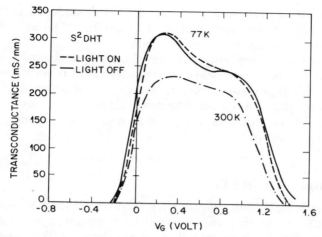

Figure 3.23 Transconductance versus gate voltage of a HFET with superlattice donor layer.

conduction in the low mobility and low velocity AlGaAs layers. This is important for the E-mode devices used in digital circuits. These features are borne out in the experimental characteristics as reported by Tu et al. and are shown in Fig. 3.23, where a large transconductance is sustained over a greater range of gate bias than in the conventional HFETs.

The superlattice donor layer has been used extensively for the fabrication of devices and circuits by the self-aligned ion-implantation process [141].

It should be noted that the role of superlattices in the AlGaAs/GaAs material system goes beyond 2DEG FET devices. A superlattice buffer for a conventional GaAs MESFET has been used to harden the devices to radiation effects [142]. The superlattice creates potential barriers to carriers generated by flash X-rays and reduces the primary photocurrent due to substrate leakage current by two orders of magnitude.

3.4.2.2 δ-Doping.

δ-Doping is also referred to planar or pulse or atomic layer or spike doping, whose characteristics and subband structure were described by Schubert et al. [143]. This doping technique consists of growth interruption, and incorporation of a doping layer of density $>1 \times 10^{13} \, \text{cm}^{-2}$. The first δ-doped HFET was described by Hueshen et al. [144]. Low-noise microwave GaAs MESFETs with an AlGaAs buffer layer and a δ-doped channel were fabricated with a gate length of 0.1 μm [145]. The device, which had a minimum noise figure of 0.8 dB at 18 GHz, reduced short-channel effects compared with conventional MESFETs. A "doped quantum well" FET was described recently, although the discrete device performance was poor. The structure consisted of a δ-doped GaAs quantum well bounded on both sides with undoped AlAs [146]. In this structure, the

2DEG is created in two wells, formed in the two undoped GaAs regions outside the well, and this resulted in a characteristic "dual-humped" plot of transconductance as a function of gate bias.

A self-aligned ion-implanted version of the single quantum well device employing a δ-doped layer below the channel has also been demonstrated [147]. The advantage of this structure was the ability to obtain a large gate turn-on voltage (0.5 V more than the conventional HFET) and, by tailoring the grading of the AlGaAs composition under the channel, to obtain improved charge control.

3.4.3   Insulated-gate HFETs

The heterostructure insulated-gate FET (IGFET) is similar to the conventional HFET except that the epitaxial layer structure is undoped. The attraction of this device is that there is no doped AlGaAs layer, which would have trapping behavior, and the threshold voltage of the device is governed by the material and its composition, not its thickness and doping, unlike the n^+-AlGaAs/GaAs HFET. Since the layers are undoped, however, there is a need for a self-aligned process rather than the simpler recessed-gate technology suited to the conventional HFET, in order to reduce parasitic source resistance. These devices have evolved in two forms, one where the gate consists of a metal (HIGFET) and one where the gate is a semiconductor (SISFET).

3.4.3.1   Semiconductor–Insulator–Semiconductor FET (SISFET). The GaAs-gate FET is analogous to the polysilicon-gate FET in Si MOS devices. The concept of the device was introduced by Solomon et al. [148] and by Matsumoto et al. [149]. Figure 3.24 shows the comparison of the band diagram of the GaAs-gate HFET and the conventional HFET, and Fig. 3.25 is a schematic cross section of the device. Only the GaAs gate is doped in the device, whereas the 2DEG is formed in an inversion channel formed at the undoped AlGaAs–undoped GaAS heterojunction. Solomon described the advantages of this device over HFETs which contain an n^+-AlGaAs donor layer. The threshold voltage of the GaAs-gate HFET is primarily determined by the work function difference between the gate material and the channel, which in this case is close to zero. This threshold voltage is also almost independent of the AlGaAs layer thickness, in contrast to the n^+-AlGaAs/GaAs HFET, where there is a strong sensitivity to doping and thickness [150]. The mean threshold voltage of GaAs-gate HFETs fabricated by Matsumoto et al. was +35 mV, with a standard deviation of 13 mV, over a $10 \times 5 \, \text{mm}^2$ area. The absence of doping in the AlGaAs layer also eliminates problems associated with deep donor traps in the device. Another consequence of using an undoped AlGaAs layer is the elimination of

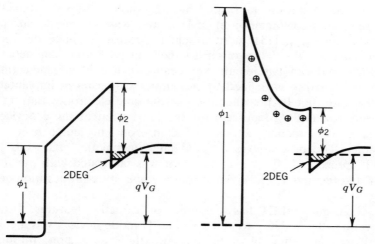

Figure 3.24 Comparison of the conduction band diagrams of the n$^+$-GaAs SISFET and the conventional HFET. (From Solomon [148]. Reprinted with permission, copyright 1984, IEEE.)

parasitic charge storage in this layer, which is a problem in the conventional n$^+$-AlGaAs/GaAs HFET at forward gate bias. This structure also has the potential for in situ formation of the gate contact in the epitaxial growth chamber.

The process employed for this device uses a molybdenum layer on top of the n$^+$-GaAs gate to act as a mask for the self-aligned ion implantation that is required to make contact to the channel. This top layer also serves to reduce gate resistance. The GaAs gate is undercut by plasma etching with a selective etch. Ion implantation is used to reduce the source and drain resistance of this device.

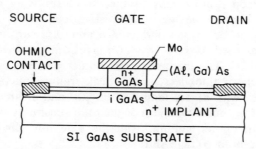

Figure 3.25 Cross section of the n$^+$-GaAs SISFET. (From Solomon [148]. Reprinted with permission, copyright 1984, IEEE.)

The devices that were fabricated had threshold voltages of 0.05–0.2 V, and a peak transconductance of 250 mS/mm was measured. Subsequent work by Chen et al. [151] used a similar process to make devices with different gate lengths. They observed that the peak transconductance increased with reduced gate length over a range of 1 to 20 μm, indicating that the channel length was defined by the lateral scattering of implanted ions under the gate, not by the dimension of the gate electrode. Baratte et al. [152] used a self-aligned process for the fabrication of these devices and observed no dependence of transconductance on gate length over 1–4 μm dimensions. The process was a "cold-gate" technique, analogous to the "SAINT" process [153]. A ring oscillator was fabricated with the E-mode SISFET and resistive loads, giving a room-temperature propagation delay of 35 ps/gate.

The GaAs-gate HFET has a number of attractive features. However, until recently, the inability to make E- and D-mode devices was a concern. Selected area implantation of Si ions into the gate region, through the refractory gate of the GaAs SISFET into the undoped AlGaAs layer, has been used to tailor the threshold voltage of the devices, giving a change in threshold voltage of up to -0.6 V [154].

Since the threshold voltage of the insulated-gate FET depends on the difference in electron affinity between the gate material and the channel, Arai et al. chose n^+-Ge as the gate material in their structure [155]. The threshold voltage is ~ 0 V for this structure, and furthermore, the Ge can be deposited in situ in the MBE system and then selectively etched to form the gate pattern. Such a SISFET employing Ge rather than GaAs as the semiconductor gate has been implemented into static frequency dividers operating at 11.3 GHz [156].

Arai et al. also demonstrated the first circuit with the n-channel SISFETs. They used an n^+-Ge gate on an undoped AlGaAs and GaAs layer, with a self-aligned ion implantation for the source and drain regions. They observed a threshold voltage of 0.25 V at 300 K which only changed by 30 mV by cooling to 77 K, unlike the n^+-AlGaAs/GaAs HFET. This was attributed to the lack of deep donor traps in the structure. Ring oscillators were realized with the E-mode devices and resistive loads.

The trade-off between transconductance and gate current behavior as a function of the undoped AlGaAs insulator layer thickness was performed by Maezawa et al. [157]. The intrinsic transconductance was inversely proportional to the insulator thickness, and a peak transconductance of 430 mS/mm was observed, with a gate length of 0.8 μm and a threshold voltage of 0.4 V. The large threshold voltage indicates the sensitivity of this technique to the in-diffusion of germanium from the gate into the FET structure, resulting in a threshold-voltage shift.

3.4.3.2 Heterojunction Insulated-gate FET (HIGFET). The use of an undoped epitaxial layer for an n-channel device was reported by Katayama et

al. [158]. The uniformity of these MIS-like HFETs for a threshold voltage of 0.25 V was a standard deviation of 11 mV for a quadrant of a 2 in. wafer. The control was attributed to the lack of dependence of doping and thickness on the threshold voltage of this device.

In contrast to the SISFETs, the W–Si refractory metal-gate HIGFET has a threshold voltage of +0.9 V, and the metallization is performed after the epitaxial growth, rather than in situ [159]. The cross section of this device is shown in Fig. 3.26. A number of studies have been made on the detailed operation of the HIGFET [160].

The epitaxial layer structure for the HIGFET is undoped, making threshold-voltage variation and susceptibility to collapse comparable to those in the SISFET. However, since the threshold voltage of the E-mode device is very positive, it is too high for E- and D-mode DCFL circuits and is more appropriate for complementary logic circuits. This is because in the same structure, both p- and n-channel devices can be fabricated by selective implantation, as described in the next section.

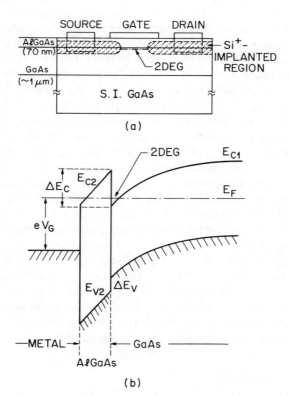

Figure 3.26 Conduction band edge of the HIGFET. (From Katayama et al. [158]. Reprinted, with permission, from *Japan Journal of Applied Physics*.)

3.4.3.3   p-HFET and Complementary HFETs. A number of authors have

investigated AlGaAs/GaAs heterostructure FETs which contain a two-
dimensional hole gas (2DHG), using p-type doping of the AlGaAs acceptor
layer. The motivation was not to implement solely p-type logic, but to
determine the suitability of the 2DHG device as the p-channel device in
complementary HFET technology. The room-temperature transconductance
of these devices was poor, but due to enhancement of hole mobility at 77 K,
a peak transconductance of 28 mS/mm was measured for a 2 μm FET [161].
It is interesting to note that the p-channel device does not exhibit "collapse"
of the drain *I–V* characteristics, in contrast to the n-channel device [162]. As
a demonstration of the performance of the device, ring oscillators with a
propagation delay of 233 ps/gate and a power-delay product of only 9.1 fJ at
77 K were reported [163]. This had spurred interest in p-channel devices in
low-power complementary logic circuits, employing AlGaAs/GaAs heterostructures.

The absence of doping in the active layers of the HIGFET and SISFET
devices allows the creation of n- and p-channel devices by choice of
implantation species for the self-aligned implant. This is in contrast to the
earlier work of Kiehl et al. [164], who fabricated complementary devices,
using p-channel HFETs and n-channel MESFETs, on a complex epitaxially
grown layer structure.

Cirillo et al. realized both n- and p-channel HFETs by a self-aligned
ion-implantation process, with W–Si as the refractory gate and selective Si
implantation for the n-channel device and Be for the p-channel device [165].
The threshold voltage was −0.3 V for the p-channel HFETs and 0.9 V for
the n-channel HFETs. Although a more symmetrical characteristic is desirable for a complementary circuit, ring oscillators and SRAM cells have been
successfully demonstrated. The n$^+$-Ge-gate n-SISFET and the metal-gate
p-HIGFET were also integrated to make complementary logic circuits. In
this case, an accurate simulation of the gate current was obtained, where a
thermionic emission model was used to describe gate current over the
AlGaAs/GaAs heterojunction for both the p- and n-channel devices [166].
The device cross sections of these two complementary HFET technologies
are shown in Fig. 3.27.

A novel complementary device which used a n$^+$GaAs gate was demonstrated by Matsumoto et al. [167]. The device required selective area
regrowth of the epitaxial structure, as the n- and p-channel devices are
fabricated with either a Si-doped n$^+$-GaAs gate or a Be-doped p$^+$-GaAs
gate. The fabrication steps were complex, compared with other approaches,
although the device characteristics were acceptable.

3.4.4   Doped-channel HFET

The doped-channel HFET is a significant departure from the conventional
HFET, in which the GaAs channel is not doped to minimize impurity

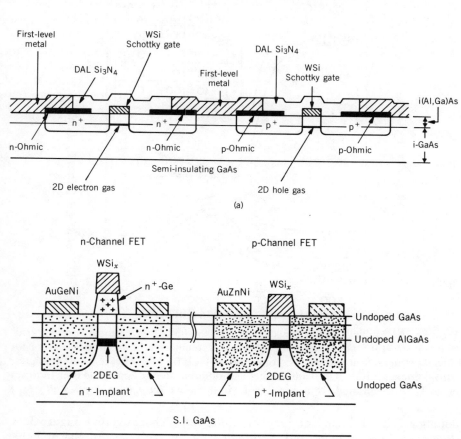

Figure 3.27 Cross sections of complementary HIGFET technologies: (a) W–Si-gate n- and p-channel devices (from Cirillo et al. [165]; reprinted with permission, copyright 1985, IEEE.), (b) n$^+$-Ge-gate n-channel and W–Si-gate p-channel devices (from Fujita and Mizutani [166]; reprinted, with permission, from *Transactions IECE Japan.*)

scattering and to maximize the mobility. The doped-channel device confirmed that the confinement of carriers at a heterointerface in a large concentration is more important in some applications than their mobility for transistor action [168]. In the case of the doped-channel device, the gate is deposited on undoped AlGaAs, which acts as an insulator, and also a confinement barrier for the underlying layer of n$^+$-GaAs, which is the channel. A self-aligned ion-implanted HFET with W–Al as the refractory metal was made with both a conventional and doped-channel structure. The

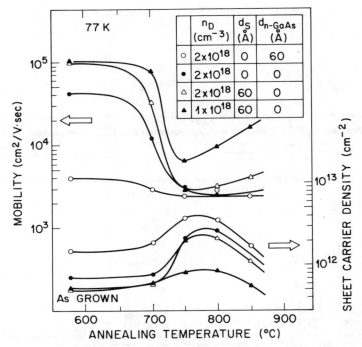

Figure 3.28   Mobility and carrier density comparison of conventional versus doped-channel structure after annealing at various temperature; open circles, doped-channel structure; closed circles, conventional HFET with no space layer; triangles, conventional HFETs with 6 nm spacers and different donor doping. (From Inomata et al. [169]. Reprinted, with permission, from *Japan Journal of Applied Physics*.)

doping was 2×10^{18} cm^{-3} in the 6 nm thick GaAs channel [169]. Figure 3.28 shows the 77 K mobility and carrier density for the conventional and the doped-channel structures for different furnace anneal temperatures. The points show that after furnace annealing, the 77 K mobility of conventional and doped-channel devices reach the same value, which is similar to that of the doped-channel device without annealing. Therefore, since this device has good FET characteristics, the degradation of channel mobility due to doping is not a concern. The doped-channel HFET had a peak transconductance of 440 mS/mm, compared with 350 mS/mm for the conventional HFET. The results show that the low-field mobility of the carriers is not the dominant parameter in the determination of the device characteristics.

Hida et al. [170] proposed a doped-channel MIS-like FET at about the same time, but without using the self-aligned ion-implanted structure. This device is commonly called the doped-channel MIS-like FET, or DMT. The gate–drain breakdown voltage was superior to the MESFET or conventional HFET, since the gate contact is on undoped AlGaAs. A high transconductance was sustained over a gate-voltage range of 2.5 V, as shown in Fig. 3.29,

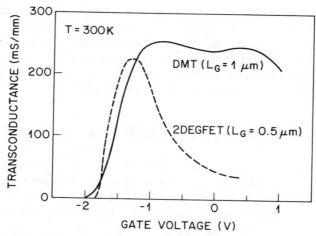

Figure 3.29   Comparison of the plot of g_m versus V_{gs} for a doped-channel HFET and conventional HFET. (From Hida et al. [170]. Reprinted with permission, copyright 1986, IEEE.)

with a peak value of 310 mS/mm and a saturation current of 650 mA/mm at 300 K for a 0.5 μm gate, which is four times that of the conventional HFET. The absence of doping in the AlGaAs resulted in no persistent photoconductivity effects, and the room-temperature electron saturation velocity was calculated as 1.5×10^7 cm/s, which rose to 2×10^7 cm/s at 77 K. A cut-off frequency of 45 GHz and other device characteristics, including noise measurements, were also reported [171]. Small-scale ICs have been fabricated in the DMT, with a buried p layer to suppress short-channel effects and improve channel confinement. The sensitivity to fan-out of the DMT was superior to that of conventional MESFETs, and a ring-oscillator propagation delay of 24 ps/gate at 1.4 mW/gate power was reported [172].

3.4.5   Summary of Heterostructure FET Structures

The simple single heterojunction n^+-AlGaAs/GaAs HFET spurred many studies and application of the high-speed capabilities of the device. A number of significant advances have been made to improve on these structures, and these advances are summarized here. The introduction of the quantum well for carrier confinement, and the insulated gate for reduction of gate current, have made significant improvements in the device behavior. The use of δ-doping, the doped channel, and the superlattice donor layer have all proven to be better than n^+-AlGaAs for the tailoring of the band structure and material properties to improve charge control in the channel. The innovations in the SISFET and HIGFET, and their applications to n-channel and complementary devices, have opened a new regime for HFET device operation, compared with MESFET technology.

The single quantum well (inverted and insulated-gate inverted) HFET, the heterojunction insulated-gate FET, and the doped-channel devices, each with their own characteristic advantages, are merging into a new class of HFETs with low drain conductance, high gate–drain breakdown voltage, low gate conduction under forward bias and a higher channel sheet charge density than the conventional n^+-AlGaAs/GaAs HFET.

3.5 HFETs IN DIFFERENT MATERIAL SYSTEMS

The small effective electron mass m^*, high mobility at low field μ_0, high electron saturation velocity v_{sat}, and large energy gap between the Γ- and L-valleys make InP and $In_{1-x}Ga_xAs$ promising materials for high-speed electronic applications. A comparison of principal physical parameters of GaAs, InP, InAs, and $In_{0.53}Ga_{0.47}As$ is summarized in Table 3.3 [173]. In particular, $In_{0.53}Ga_{0.47}As$, which is lattice-matched to InP, is a semiconductor of great interest for electronic as well as optoelectronic applications. Experimentally, mobilities as high as $13,800 \ cm^2/V \cdot s$ [174], effective saturation velocities up to $2.95 \times 10^7 \ cm/s$ [175], and an energy separation between the Γ- and L-valley as large as 0.55 eV [176] have been reported for $In_{0.53}Ga_{0.47}As$ at room temperature. The development of MESFETs on InP-based materials, however, is hindered by the low turn-on voltage and the high reverse leakage current due to the lack of a suitable Schottky barrier. For example, the Schottky barrier height is only 0.2 V for InGaAs [177] and 0.4–0.5 V for InP [178], values too low for E-mode devices. The effective Schottky barrier height can be enhanced by incorporating a thin p^+ layer between the gate and the channel. Other techniques include the use of junction FETs (JFETs) and insulated-gate FETs (MISFETs). However, the JFET suffers from excessive p-n junction capacitance, and the MISFET from a high density of surface states, which causes a drift in drain current and a degradation of transconductance. Another approach uses a heterojunction structure by growing a wide bandgap material between the channel and the gate to enhance the Schottky barrier height. A 0.63 V barrier height is obtained on an $Al/In_{0.52}Al_{0.48}As$ junction [179]. InAlAs/InGaAs HFETs,

TABLE 3.3   The Principal Physical Parameters of Selected III–V Materials

Material	m^*/m_0	μ_0 $(cm^2/V \cdot s)$	v_{sat} $(10^7 \ cm/s)$	$\Delta E_{\Gamma L}$ (eV)	E_g (eV)
GaAs	0.063	4,600	1.8	0.33	1.42
InP	0.08	2,800	2.4	0.61	1.35
$In_{0.53}Ga_{0.47}As$	0.032	7,800	2.1	0.61	0.78
InAs	0.022	16,000	3.5	0.87	0.35

TABLE 3.4   HFETs Based on Various Material Systems

Substrate	Lattice-matched HFET	Strained-layer HFET
GaAs	AlGaAs/GaAs	$Al_xGa_{1-x}As/In_yGa_{1-y}As$[a]
InP	$In_{0.52}Al_{0.48}As/In_{0.53}Ga_{0.47}As$	$In_{0.52-u}Al_{0.48+u}As/In_{0.53+u}Ga_{0.47-u}As$[b]

[a]Where $x \leq 0.15$ and $y \leq 0.25$.
[b]Where $u = 0.07$.

which take advantage of the high electron mobility of the 2DEG in InGaAs, have been demonstrated to have superior device characteristics.

Furthermore, it has also been demonstrated that rather large lattice mismatches may be accomodated by elastic deformations in thin heterostructure layers. This opens up the possibility of strained-layer (or pseudomorphic) HFETs with a lattice-mismatched InGaAs channel grown on either GaAs or InP substrates. Table 3.4 summarizes various HFET structures realized in the GaAs- and InP-based material systems.

3.5.1   Lattice-matched HFETs

Due to the larger discontinuity of the conduction band edge and the intrinsic higher electron mobility and velocity, the InGaAs HFET potentially offers performance improvements over those of AlGaAs/GaAs HFETs or other InGaAs FETs. A room-temperature mobility of $\sim 10^4 \, cm^2/V \cdot s$ has been demonstrated in a selectively doped n$^+$-InAlAs/InGaAs structure [180]. The initial attempts on n$^+$-InAlAs/InGaAs HFET failed to realize the full potential of the structure [181]. With improvements in the materials, a maximum transconductance of 440 mS/mm at 300 K and 700 mS at 77 K has been achieved with 1 μm gates. [182].

Recently, SISFETs with $In_{0.53}Ga_{0.47}As$ channels, lattice-matched $In_{0.52}Al_{0.48}As$ gate barriers, and n$^+$-$In_{0.53}Ga_{0.47}As$ gates have been successfully fabricated. For devices with a barrier thickness of 60 nm and a gate length of 1.7 μm, a maximum transconductance of 250 mS/mm at 300 K has been achieved [183]. The transconductance remains above 210 mS/mm for forward gate bias up to 1.0 V, which demonstrates the advantage of using a SISFET structure in enhancement mode InGaAs FETs.

3.5.2   Strained-layer n-HFETs

It was first predicted by van der Merwe [184] and demonstrated by Sugita and Tamura [185] that the lattice mismatch in layers thinner than the critical thickness can be totally accommodated by coherent layer strain. Figure 3.30 shows the critical thickness as a function of the percentage of lattice mismatch [186]. An $Al_xGa_{1-x}As/In_yGa_{1-y}As$ strained-layer HFET structure with $x \leq 0.15$ and $y \leq 0.25$ has been realized [187–189]. It was moti-

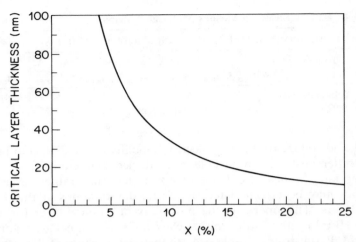

Figure 3.30   Critical thickness as a function of percentage of lattice mismatch.

vated by looking for ways to reduce or eliminate some of the problems associated with deep level traps. The conduction band edge drops below the DX centers, which reduces the possibility of trapping electrons in these deep level traps if the aluminum mole fraction is less than 0.22. Unfortunately, it also degrades the performance of the FETs due to the reduction of the conduction band edge discontinuity and, therefore, the carrier confinement. The strained-layer structure maintains sufficient conduction bandgap edge discontinuity for 2DEG confinement by using a lower-bandgap material in the channel. A mobility of $4,210 \, \text{cm}^2/\text{V} \cdot \text{s}$ with a sheet charge density of $4.5 \times 10^{12} \, \text{cm}^{-2}$ has been demonstrated at 300 K, and $18,640 \, \text{cm}^2/\text{V} \cdot \text{s}$ and $2.4 \times 10^{12} \, \text{cm}^{-2}$ at 77 K [190].

The indium mole fraction in this mismatched channel with a practical thickness ($>10 \, \text{nm}$) is limited to 0.25. It is desirable to go to higher indium mole fraction for better performance; this is achieved with device structures grown on InP substrates. A strained-layer n-$\text{In}_{0.52-u}\text{Al}_{0.48+u}\text{As}/\text{In}_{0.53+u}\text{Ga}_{0.47-u}\text{As}$ HFET structure has been demonstrated [191]. For $u = 0.07$, both the InAlAs donor layer and the InGaAs channel have a 1% lattice mismatch. The strains are equal but in opposite directions. Figure 3.31 shows the conduction band diagram of such a structure. The $\text{In}_{0.6}\text{Ga}_{0.4}\text{As}$ channel offers not only higher electron velocity and mobility, but also the largest conduction band edge discontinuity among all HFET structures reported in the literature. The higher aluminium mole fraction in the InAlAs layer also leads to larger bandgap and higher Schottky barrier height than those in the n^+-InAlAs/InGaAs HFETs [192]. Table 3.5 shows the device characteristics of various HFETs with InGaAs channel, [193–207]. The measured unity-current-gain frequencies of various GaAs- and InP-based HFETs are compared in Fig. 3.32. The data shows that the speed

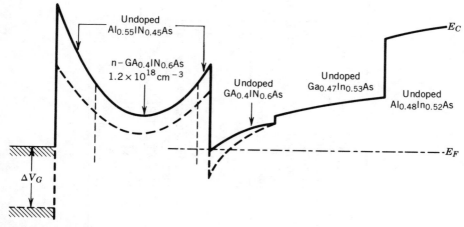

Figure 3.31 The conduction band diagram of a strained-layer n^+-InAlAs/InGaAs HFET. The dashed line illustrates the condition under positive gate bias. (From Kuo et al. [191]. Reprinted, with permission, from IEDM and copyright 1987, IEEE.)

of the HFETs with InGaAs channel are comparable to or better than those of the best AlGaAs/GaAs HFETs, as expected.

Recently, AlSb/InAs quantum well HFETs have also been realized. Room-temperature mobilities up to $11,200 \, \text{cm}^2/\text{V} \cdot \text{s}$ with sheet carrier density around $5 \times 10^{12} \, \text{cm}^{-2}$ have been demonstrated. Strained-layer HFETs with 2 μm gate length have yielded enhancement mode devices with transconductances of 75 mS/mm at room temperature and 180 mS/mm at 77 K [208].

3.5.3 Strained-layer p-HFETs

The large mismatch between the electron and hole mobilities in III–V semiconductors has limited the speed of III–V complementary logic for low-power digital applications. The previous section described attempts to improve the performance of p-channel FETs by exploiting the mobility enhancement in 2DHG system due to reduced ionized impurity scattering. Further enhancement in the mobility is possible by reducing the effective hole mass, as reported by Fritz et al. [209]. A factor of five enhancement in the 77 K mobility was observed as the compressive strain was increased from 0.5 to 1.4%. p^+-GaAs/InGaAs [210] and p^+-AlGaAs/InGaAs [211] HFETs have been successfully demonstrated with a 77 K transconductance of 11.3 and 89 mS/mm, respectively.

These devices have attempted to exploit strain-induced splitting of the light- and heavy-hole valences in the strained-layer structures. Due to the small splitting (on the order of tens of meV), the light-hole conduction is not expected to be pronounced except at very low temperatures [212].

TABLE 3.5 Characteristics Comparison of Various HFETs

Technology	Size (μm)	g_m (mS/mm)	f_t (GHz)	f_{max} (GHz)	Year/Company	Reference
n⁺-AlGaAs/InGaAs (Strained-layer)	1	310	24.5	40	1986/Ill-Cascade	[193]
	1.2	408	21.5	80	1987/Cornell	[194]
	0.25	495	80	250	1986/Ill-GE	[195]
	0.25	500	98		1987/HP	[196]
	0.1	930	97		1987/GE-Cornell	[197]
n-InAlAs/InGaAs	1	650	26.5	62	1986/Ill-Cascade	[198]
	0.3		80		1987/Hughes	[199]
InAlAs/InGaAs (SISFET) (MISFET)	1.7	251	15		1987/AT & T-BL	[200]
	1.1	220	27		1988/AT & T-BL	[201]
	1	310	32		1987/Mich-Cascade	[202]
InAlAs/n-InGaAs	2	152	12.4	50	1987/NTT	[203]
InP/n-InGaAs	1.2	330	22.2	27	1988/AT & T-BL	[204]
n-InAlAs/InGaAs (strained-layer)	1.7		15.5		1987/AT & T-BL	[205]
	1.5		14		1987/Mich	[206]
	0.12		107	125	1987/Cornell	[207]

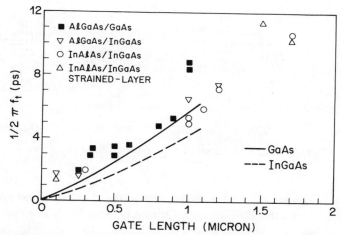

Figure 3.32 Comparison of the transit time of various GaAs- and InP-based HFETs with the calculated frequencies [172].

3.6 HFET CIRCUITS

Since the demonstration of the first HFET ring oscillator by Mimura et al. in 1981 [213], most current HFET circuits are implemented with n^+-AlGaAs/GaAs HFETs and DCFL gates. The transconductance of a conventional HFET peaks over a very narrow range of gate voltage due to the saturation of the 2DEG. To take full advantage of the conventional HFET, the choice of logic family is limited to those with a small logic swing (<1 V), such as DCFL. With the more advanced HFET structures, it is also possible to have devices with high transconductance over a much wider range, which creates the potential of using of other logic families to exploit the performance advantages of the HFETs fully.

DCFL is the simplest logic family for the FET circuits. Its advantages include high speed and low power, which makes it particularly attractive for LSI circuits. Part of the reason for its high speed is the small logic swing which also limits the noise margins of the DCFL gates. The small noise margin imposes a stringent requirement on the uniformity and reproducibility of the threshold voltage. Due to the usage of AlGaAs, which gives a higher Schottky barrier, HFET DCFL gates have slightly larger noise margins compared with the MESFET gates. The noise margins of HFET DCFL gates are still relatively small (~0.2 V). With the excellent epitaxy growth control of the MBE, HFET DCFL circuits with a complexity up to 16 kbit static RAM have been demonstrated [214].

Figure 3.33 shows various DCFL gates that have EFET drivers and DFET loads. The load device may also be a saturated resistor that is an ungated DFET. Figure 3.34 compares a DFET load with a saturated resistor load. Due to the saturation of the confined 2DEG, the current in a saturated

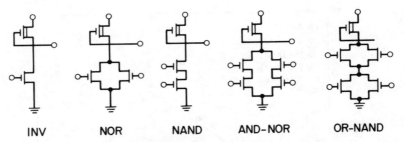

Figure 3.33 Various direct coupled FET logic gates.

load has very similar $I–V$ characteristics to a DFET. There is no appreciable difference between the DFET and saturated resistor loads at room temperature. They become almost identical to each other at 77 K. The other important feature is that the saturated current ratio of the driver and load remains essentially the same as the temperature is lowered to 77 K. It suggests that if the circuit works at room temperature it will also function at 77 K.

3.6.1 Ring Oscillators

The ring oscillator is the simplest circuit and has been widely used as a vehicle for the performance comparison of device technologies. The low knee voltage in the drain $I–V$ characteristics makes the HFET particularly suitable for circuits operating at a low bias voltage and low-power dissipation. This is illustrated in Fig. 3.35 by the ability to achieve almost the

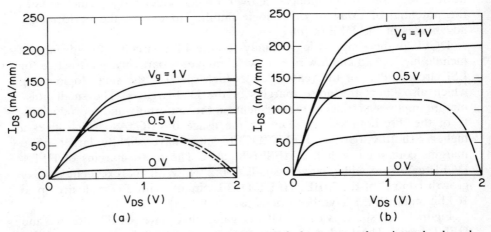

Figure 3.34 $I–V$ characteristics comparison of the saturated resistor load and D-HFET load at (a) room temperature and (b) 77 K.

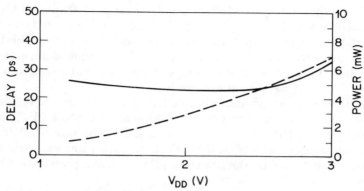

Figure 3.35 Propagation delay versus the bias voltage of a 1 μm gate HFET ring oscillator.

maximum speed at very low bias voltages. The propagation delay remains almost constant over a wide range of bias voltage, which is also partially due to the excellent saturation characteristics of the 2DEG in the load device.

Figure 3.36 shows the propagation delay comparison of HFET and MESFET ring oscillators with various gate lengths as a function of power dissipation. As the gate length is reduced to substantially below 1 μm, there is no appreciable improvement of the MESFET ring-oscillator speed due to the degradation of the output conductance at short gate lengths. For

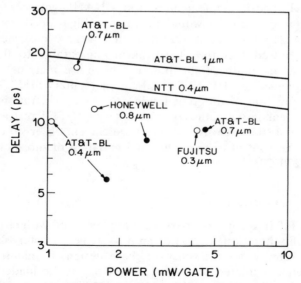

Figure 3.36 Ring oscillator delay of MESFET (solid lines) and HFET (open circles, room temperature; closed circles, 77 K).

HFETs, the short-channel effects are greatly reduced thanks to the better confinement of the 2DEG to the heterointerface, and the speed continues to scale with gate length at submicron sizes. The scalability of the HFET makes it a very attractive submicron technology. Propagation delays as low as 9.2 ps with 0.5 μm gates at room temperature [215] and 5.8 ps with 0.4 μm gates at 77 K [216] have been achieved.

Ring oscillators with more advanced HFET structures have also been reported. Saito et al. demonstrated an GaAs/n$^+$-AlGaAs HFET (inverted-structure) ring oscillator with 0.5 μm gates [217]. The ring oscillator had a propagation delay of 11.9 ps at 0.56 mW/gate. A 0.7 μm gate SISFET ring oscillator was reported by Baratte et al. [218]. It had a propagation delay of 35 ps and dissipated ~3 mW/gate.

Another interesting class of circuits uses complementary logic gates with both n- and p-HFETs. Due to the low mobility and heavy mass of the holes in III–V materials, the p-channel FET is much slower than the n-channel one. The mismatch between the p- and n-channel FET limits the performance of the III–V complementary FET circuits. Significant improvement of the p-FET characteristics has been achieved by enhancing the mobility of the 2DHG in the p-HFET, as described in previous sections. Complementary ring oscillators implemented with n-SISFETs and p-MISFETs have been successfully demonstrated with 94 ps propagation delay at 0.58 mW/gate using 1.5 μm gates [219] and 76.2 ps at ~6 mW/gate using 1 μm gates [220].

Predictions of the potential performance of this technology show that the complementary MISFET has larger logic swings and noise margins than the E/D MESFET, and its implementation in a NAND-gate configuration for large fan-in/fan-out circuits would maintain a high-speed and low-power performance [221]. Kiehl et al. [221a] simulated the complementary HFET circuits and showed that they have speed comparable to the n-channel HFET, but at one-tenth of the power. The noise margins of n-channel HFETs, complementary HIGFETs and complementary SISFETs, and their loading as a function of different fan-in/fan-out considerations, were also studied. Two weaknesses of the complementary HIGFET restrict the performance at present: the excessive gate leakage current observed on fabricated devices, and the control of the value and variation of threshold voltage of the devices [220, 221a].

3.6.2 Frequency Dividers

The early HFET frequency dividers were implemented with either a 6-NOR gate type-D flip–flop design [222] or an 8-NOR gate master–slave flip–flop design [223]. The major drawback of these designs is that the maximum toggle frequency is limited to $1/4 \, \tau_{PD}$, where τ_{PD} is the loaded gate delay. They are slower than the transmission gate [224] or the stacked logic gate design [225], which have a maximum toggle frequency of $1/2\tau_{PD}$. Using four of the DCFL AND-NOR gates shown in Fig. 3.33, master–slave flip–flop frequency dividers were demonstrated to operate at a toggle frequency of 9.1 GHz with 0.4 μm gates at room temperature [226] and 13 GHz with 0.7 μm gates at 77 K [227]. A DCFL dynamic divider with E-HFET transmission gates was also reported [228]. A toggle frequency of 6.8 GHz is achieved with 1 μm gates. Due to the small noise margins of these two circuits, it is highly desirable to have dividers implemented in more advanced HFET structures because this allows wider noise margins for these more complex logic gates.

3.6.3 Multipliers

Another popular demonstration circuit is the digital multiplier. An HFET parallel multiplier was reported by Schlier et al. [229]. This 4 × 4 multiplier used a 1 μm recessed-gate technology and had an average gate delay of 113 ps and an average power dissipation of 0.34 mW/gate. The shortest average gate delay in multiplier circuits was achieved with a 1 μm self-aligned ion-implanted HFET by Cirillo et al. [230]. In a 8 × 8 multiplier with a pipeline design, they demonstrated a 70 ps gate delay. A 73 and 43 ps gate delay were also reported for a 5 × 5 parallel multiplier at room temperature and 77 K, respectively [231].

Wantanabe et al. reported a carry-look-ahead multiplier implemented in a 1.5 kbit gate array with 1.2 μm gates [232]. The gate array had a basic gate delay of 85 ps and an additional 29 ps/fan-out and 14 ps/mm of interconnection line. The multiplier has a multiplication time of 4.9 ns at room temperature and 3.1 ns at 77 K.

The average gate delay of HFET multipliers are summarized in Fig. 3.37. The results compare favorably with ≥100 ps loaded gate delay for the MESFET multipliers.

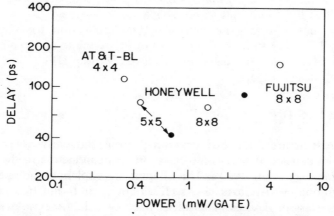

Figure 3.37 Performance comparison of HFET multipliers.

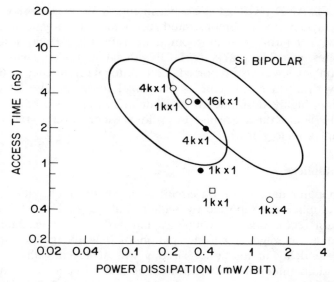

Figure 3.38 Access time versus normalized power dissipation of HFET SRAMs compared with Si bipolar and GaAs MESFET SRAMs.

3.6.4 Static RAMs (SRAMs)

A static random access memory (SRAM) implemented in HFET technology was reported by Nishiuchi et al. in 1984 [233]. A 3.4 ns access time with a power dissipation of 0.29 W was reported for a 1 kbit × 1 SRAM. Recently, a 16 kbit × 1 SRAM with an access time of 3.4 ns and dissipating 1.34 W at 77 K was demonstrated [234]. The performance of the HFET SRAM is summarized in Fig. 3.38. The shortest access times at room temperature are 0.6 ns with a total power dissipation of 0.45 W for a 1 kbit × 1 SRAM [235] and 0.5 ns with 5.7 W for a 1 kbit × 4 SRAM [236], compared with ≥1 ns for the best MESFET 1 kbit and 4 kbit SRAMs.

Memory may also be the most suitable application for GaAs complementary FET due to its low standby power. A minimum standby power dissipation of 23 μW/cell has been achieved with p-HFET loads in the memory-cell design [220].

3.7 CONCLUSION

The Heterostructure FET has developed from the early studies of the properties of carriers at hetero-interfaces in semiconductors to the practical device technology in the 1980's. This chapter has explained the operation of the conventional n^+-AlGaAs/GaAs HFETs in detail to illustrate the principles of this class of devices. The initial work on AlGaAs/GaAs HFETs in the early 1980s spurred considerable work on a wide range of HFET structures and this chapter has attempted to chart some of the major classes of devices and their potential applications. Continued engineering of better devices in the AlGaAs/GaAs and strained-layer AlGaAs/InGaAs/GaAs is still resulting in improved device performance. The use of materials grown on InP, such as the InAlAs/InGaAs and the corresponding strained layer structures, have resulted in unsurpassed discrete low noise device characteristics.

The developments in fabrication technology have accompanied the progress in device structures to enable high yield and uniform devices to be fabricated. In particular, much attention has been placed on precise control of the threshold voltage of devices, and the variation of this parameter over a wafer. Improvements in processing have also led to developments in planar processes suited to obtaining high yield on devices and circuits. A continued trend to shorter-channel devices by optical and direct-write electron-beam lithography has resulted in improvement of f_t and f_{max} of HFETs. These developments have highlighted the role of carrier confinement in HFET, and they have challenged the device engineer to design structures which suppress short-channel effects down to gate lengths of below 0.1 μm.

HFETs have inherent performance advantages over more conventional GaAs MESFETs, predominantly because of better carrier confinement and larger effective saturation velocity. There are several classes of applications which have benefited from the use of HFET structures over conventional devices. The one application which has been widely commercialized is the short-gate low noise microwave HFET and this device is under continued development. The large charge densities which can be attained in HFETs, as well as the ability to engineer the breakdown voltage, make the application to microwave power devices very attractive. Many small-scale ICs have been reported in HFET technologies, which show the potential for ultra-high speed analog and digital SSI/MSI circuits. The application of HFETs to LSI circuits is also being developed in a number of laboratories and may lead to high performance digital ICs.

REFERENCES

1. R. L. Anderson, *IBM J. Res. Dev.* **4** 283 (1960).
2. R. Dingle, H. L. Stormer, H. L. Gossard, and W. Wiegmann, *Appl. Phys. Lett.* **33** 665 (1978).
3. H. L. Stormer, R. Dingle, A. C. Gossard, W. Wiegmann, and M. D. Sturge, *J. Vac. Sci. Technol.* **16**, 1517 (1979).
4. H. Stormer, R. Dingle, A. C. Gossard, and W. Wiegmann, *Solid State Commun.* **29**, 705 (1979).
5. L. C. Witkowski, T. J. Drummond, C. M. Stanchak, and H. Morkoç, *Appl. Phys. Lett.* **37**, 1033 (1980).
6. J. H. English, A. C. Gossard, H. L. Stormer, and K. W. Baldwin, *J. Appl. Phys.* **51**, 752 (1987).

7. D. Delagebeaudeuf and N. T. Linh, *IEEE Trans. Electron Devices* **ED-29**, 955 (1982).

8. K. Lee, M. S. Shur, T. J. Drummond, and H. Morkoç, *IEEE Trans. Electron Devices* **ED-30**, 207 (1983); R. F. Pierret and M. S. Lundstrom, *ibid* **ED-31**, 383 (1984).

9. M. C. Foisy, P. J. Tasker, B. Hughes, and L. F. Eastman, *IEEE Trans. Electron Devices* **ED-35**, 871 (1988).

10. J. V. DiLorenzo, R. Dingle, M. D. Feuer, A. C. Gossard, R. H. Hendel, J. C. M. Hwang, A. Kastalsky, R. A. Kiehl, and P. O'Connor, *Tech. Dig.—Int. Electron Devices Meet.* p. 578 (1982).

11. S. Tiwari, *IEEE Trans. Electron Devices* **ED-31**, 879 (1984).

12. M. J. Moloney, F. Ponse, and H. Morkoç, *IEEE Trans. Electron Devices* **ED-32**, 1675 (1985).

13. K. Lee M. S. Shur, T. J. Drummond, and H. Morkoç, *IEEE Trans. Electron Devices* **ED-30**, 207 (1983).

14. K. Park, H. B. Kim, and K. D. Kwack, *IEEE Trans. Electron Devices* **ED-34**, 2422 (1987).

15. L. C. Witkowski, T. J. Drummond, C. M. Stanchak, and H. Morkoç, *Appl. Phys. Lett.* **37**, 1033 (1980).

16. E. F. Schubert and K. Ploog, *Appl. Phys. [Part] A* **33**, 63 (1984).

17. E. F. Schubert and K. Ploog, *Appl. Phys. [Part] A* **33**, 183 (1984).

18. K. Hess, and H. Morkoç, H. Shichijo, and B. G. Streetman, *Appl. Phys. Lett.* **35**, 469, (1979).

19. H. C. Casey, Jr., and M. B. Panish, "Heterostructure Lasers," Part A. Academic Press, New York, 1978.

20. W. I. Wang and F. Stern, *J. Vac. Sci. Technol.*, *B* [2] **3**, 1280 (1985), and the references therein.

21. T. Ishikawa, K. Kondo, S. Hiyamizu, and A. Shibatomi, *Jpn. J. Appl. Phys.* **24** (6), L408 (1985).

22. M. Mizuta, M. Tachikawa, and H. Kukimoto, *Jpn. J. Appl. Phys.* **24**, L143 (1985).

23. A. Kastalsky and J. C. M. Huang, *Solid State Commun.* **51**, 317 (1984).

24. M. I. Nathan, P. M. Mooney, P. M. Solomon, and S. L. Wright, *Appl. Phys. Lett.* **47**, 628 (1985).

25. R. Kirschman, "Low Temperature Electronics," IEEE Press, New York, 1986.

26. D. Delagebeaudeuf, P. Delescluse, M. Laviron, P. N. Tung, J. Chaplart, J. Chevrier, and N. T. Linh, *Conf. Ser.—Inst. Phys.* **65** (1982), 393.

27. T. J. Drummond, S. L. Su, W. G. Lyons, R. Fischer, W. F. Kopp, H. Morkoç, K. Lee, and M. S. Shur, *Electron. Lett.* **18**, 1057 (1982).

28. W. T. Masselink, T. S. Henderson, J. Klem, W. F. Kopp, and H. Morkoç, *IEEE Trans. Electron Devices* **ED-33**, 639 (1986).

29. S. S. Pei, N. J. Shah, R. H. Hendel, C. W. Tu, and R. Dingle, *Tech. Dig. IEEE Gallium Arsenide Integrated Circuits Symp. 1984*, p. 129 (1984).

30. D. K. Arch, B. K. Betz, P. J. Vold, J. K. Abrokwah, and N. C. Cirillo, *IEEE Trans. Electron Devices* **ED-7**, 700 (1986).

31. D. V. Lang, R. A. Logan, and M. Jaros, *Phys. Rev. B: Condens. Matter* [3] **19**, 1015 (1979).

32. M. O. Watanabe, *Jpn. J. Appl. Phys.* **23**, L103 (1984).

33. A. Kastalsky and R. Kiehl, *IEEE Trans. Electron Devices* **ED-33**, 414 (1986).

34. T. Baba, T. Mizutani, and M. Ogawa, *Jpn. J. Appl. Phys.* **22** L627 (1983).

35. C. W. Tu, W. L. Jones, R. F. Kopf, L. D. Urbanek, and S. S. Pei, *IEEE Electron Device Lett.* **EDL-7**, 552 (1986).

36. R. Bhat, W. K. Chan, A. Kastalsky, and M. A. Koza, *43rd Annu. Device. Res. Conf. Tech. Dig.* IIA-2 (1985).

37. J. C. Huang, G. W. Wicks, A. R. Calawa, and L. F. Eastman, *Electron. Lett.* **20**, 925 (1985).

38. S. Tiwari and W. I. Wang, *IEEE Electron Device Lett.* **EDL-5**, 333 (1984).

39. A. Kastalsky and R. A. Kiehl, *IEEE Trans. Electron Devices* **ED-33**, 414 (1986).

40. K. Kinoshita, M. Akiyama, T. Ishida, S. Nishi, Y. Sano, and K. Kaminishi, *IEEE Electron. Device Lett.* **EDL-6**, 473 (1985).

41. P. P. Ruden, C. J. Han, C. H. Chen, S. Baier, and D. K. Arch, *45th Annu. Device Res. Conf.*, IIA-1 (1987).

42. Y. Suzuki, H. Hida, H. Toyoshima, and K. Ohata, *Electron. Lett.* **22** 672 (1986).

43. K. L. Priddy, D. R. Kitchen, J. A. Grzyb, C. W. Litton, T. S. Henderson, C. K. Peng, W. F. Kopp, and H. Morkoç, *IEEE Trans. Electron Devices* **ED-34**, 175 (1987).

44. H. Goronkin, *Tech. Dig. IEEE Gallium Arsenide Integrated Circuits Symp.* p. 26 (1983).

45. M. S. Britella, W. C. Seelbach, and H. Goronkin, *IEEE Trans. Electron Devices* **ED-29**, 1135 (1982).

46. D. Arnold, J. Klem, T. Henderson, H. Morkoç, and L. P. Erikson, *Appl. Phys. Lett.* **45**, 764 (1984).

47. A. Ezis and D. W. Langer, *IEEE Electron. Device Lett.* **EDL-6**, 494 (1985).

48. T. Yokoyama, M. Suzuki, T. Yamamoto, J. Saito, and T. Ishikawa, *IEEE Electron. Device Lett.* **EDL-8**, 280 (1987).

49. N. C. Cirillo, J. K. Abrokwah, and S. A. Jamison, *Tech. Dig., IEEE Gallium Arsenide Integrated Circuits Symp.* p. 167 (1984).

50. N. J. Shah, S. S. Pei, C. W. Tu, and R. C. Tiberio, *IEEE Trans. Electron Devices* **ED-33**, 543 (1986).

51. Y. Awano, M. Kosugi, T. Mimura, and M. Abe, *IEEE Electron Device Lett.* **EDL-8**, 451 (1987).

52. M. B. Das, *IEEE Trans. Electron Devices* **ED-32**, 11 (1985).

53. I. C. Kizilyalli, K. Hess, J. L. Larson, and D. J. Widiger, *IEEE Trans. Electron Devices* **ED-33**, 1427 (1986).

53a. P. DelaHoussaye et al., *IEEE Electron Device Lett.* **EDL-9**, 148 (1988).

53b. C. T. Wang, submitted for publication.

54. M. Laviron, D. Delagebeaudeuf, P. Delescluse, J. Chaplart, and N. T. Linh, *Electron. Lett.* **17**, 563 (1981).

55. P. Delescluse et al. *Proc. Molecular Beam Epitaxy Int. Conf.*, *1984*.

56. E. A. Sovero, A. K. Gupta, and J. A. Higgins, *IEEE Electron Device Lett.* **EDL-7**, 179 (1986).

57. K. Kamei et al., *Conf. Ser.—Inst. Phys.* **79**, 541 (1986).

58. P. C. Chao, S. C. Palmateer, P. M. Smith, U. K. Misha, K. H. G. Duh, and J. C. M. Hwang, *Electron Device Lett.* **EDL-6**, 531 (1985).

59. K. H. G. Duh, P. C. Chao, P. M. Smith, L. F. Lester, B. R. Lee, and J. C. M. Hwang, *44th Annu. Device Res. Conf., Tech. Dig.* p. IIA-2 (1986).

60. A. W. Swanson, *Microwaves & RF* **26**, 1042 (1985).

61. W. L. Jones, S. K. Ageno, and T. Y. Sato, *Electron. Lett.* **23**, 844 (1987).

62. G. E. Menk, R. A. Sadler, M. L. Balzan, A. E. Geissberger, I. J. Bahl, and H. Lee, *Proc. IEEE/Cornell Conf. Adv. Concepts High Speed Semicond. Devices Circuits, 1987.* p.26.

63. T. Henderson, M. I. Aksun, C. K. Peng, H. Morkoç, P. C. Chao, P. M. Smith, K. H. G. Duh, and L. F. Lester, *Tech. Dig.—Int. Electron Devices Meet.* p. 464 (1986).

64. P. C. Chao, R. C. Tiberio, K.-H. G. Duh, P. M. Smith, J. M. Ballingall, L. F. Lester, B. R. Lee, A. Jarba, and G. G. Gifford, *IEEE Electron Device Lett.* **EDL-8**, 489 (1987).

65. J. J. Berentz, K. Nakano, and K. P. Weller, *IEEE Microwave Millimeter-Wave Monolithic Circuits Symp., Tech. Dig. 1984* p. 83 (1984); U. K. Mishra, S. C. Palmateer, P. C. Chao, P. M. Smith, and J. C. M. Hwang, *IEEE Electron Device Lett.* **EDL-6**, 142 (1985); P. M. Smith, P. C. Chao, K. H. G. Duh, L. F. Lester, B. R. Lee, and J. N. Ballingall, *IEEE MTT-S Int. Microwave Symp. Dig.* p. 749 (1987).

66. H. Fukui, *IEEE Trans. Electron Devices* **ED-26**, 1032 (1979).

67. H. Goronkin and V. Nair, *IEEE Electron. Device Lett.* **EDL-6**, 47 (1985).

68. K. H. G. Duh, M. W. Pospeiszalski, W. F. Kopp, P. Ho, A. A. Jabra, P. C. Chao, P. M. Smith, L. F. Lester, J. M. Ballingall, and S. Weinreb, *IEEE Trans. Electron Device Lett.* **ED-35**, 249 (1988).

69. S. G. Bandy, C. K. Nishimoto, C. Yuen, R. A. Larue, M. Day, J. Eckstein, Z. C. H. Tan, C. Webb, and G. A. Zdasiuk, *IEEE Trans. Electron Devices* **ED-34**, 2603 (1987).

70. P. C. Chao, P. M. Smith, S. C. Palmateer, and J. C. M. Hwang, *IEEE Trans. Electron Devices* **ED-32**, 1042 (1985).

71. K. Shibata, M. Abe, H. Kawasaki, S. Hori, and K. Kamei, *IEEE MTT-S Int. Microwave Symp. Dig.* p. 547 (1987).

72. P. R. Jay, H. Derewonko, D. Adam, P. Briere, D. Delagebeaudeuf, P. Delescluse, and J. F. Rochette, *IEEE Trans. Electron Devices* **ED-33**, 590 (1986).

73. H. Hida, K. Ohata, Y. Suzuki, and H. Toyoshima, *IEEE Trans. Electron Devices* **ED-33**, 601 (1986).

74. P. M. Smith, U. K. Mishra, P. C. Chao, S. C. Palmateer, and J. C. M. Hwang, *IEEE Electron Device Lett.* **EDL-6**, 86 (1985); A. K. Gupta, R. T. Chen, E. A. Sovero, and J. A. Higgins, *IEEE Microwave Millimeter-Wave Monolithic Circuits Symp. Tech. Dig. 1985* pp. 50–53 (1985); K. Hikosaka, J. Saito, T. Mimura, and M. Abe, *Proc. Nat. Conv. Inst. Electron. Commun. Eng. Jpn.* p. 286 (1983).

75. N. H. Sheng, C. P. Lee, R. T. Chen, D. L. Miller, and S. J. Lee, *IEEE Electron. Device Lett.* **EDL-6**, 307 (1985); K. Hikosaka, Y. Hirachi, T. Mimura, and M. Abe, *ibid.* pp. 341–343; E. Sovero, A. K. Gupta, J. A. Higgins, and W. A. Hill, *IEEE Trans. Electron Devices* **ED-33**, 1434–1438 (1986).

76. P. Saunier and J. W. Lee, *IEEE Electron Device Lett.* **EDL-7**, 503 (1986).

77. P. Saunier and J. W. Lee, *45th Annu. Device Res. Conf. 1987.* p II A-8.

77a. P. Smith, L. Lester, P. Chao, B. Lee, R. Smith, J. Ballingall, K. Duh, *Tech. Dig.—Int. Electron Devices Meet.* pp. 854–856 (1987).

78. N. Yokoyama, *in* "Semiconductor Technologies" (J. Nishizawa, ed.), vol. 19, pp. 195–210, North-Holland, Publ., Amsterdam, 1986.

79. H. M. Levy, H. Lee, C. J. Wu, M. Schneider, and E. Kohn, *Tech. Dig., IEEE Cornell Conf.* p. 21 (1987).

80. K. Hikosaka, T. Mimura, and K. Joshin, *Jpn. J. Appl. Phys.* **20**, L847 (1981).

81. J. Vatus, J. Chevrier, P. Delescluse, and J.-F. Rochette, *IEEE Trans. Electron Devices* **ED-33**, 934 (1986).

82. M. Abe, T. Mimura, S. Notomi, K. Odani, K. Kondo, and M. Kobayashi, *J. Vac. Sci. Technol. A* [2] **5**, 1387 (1987).

83. M. Abe, T. Mimura, K. Nishiuchi, A. Shibatomi, and M. Kobayashi, *IEEE J. Quantum Electron.* **QE-22**, 1870 (1986).

84. B. J. F. Lin, S. Kofol, C. Kocot, H. Leuchinger, J. N. Miller, D. E. Mars, B. White, and E. Littau, *Tech. Dig., IEEE Gallium Arsenide Integrated Circuits Symp.* p. 51 (1986).

85. N. H. Sheng, H. T. Wang, C. P. Lee, G. J. Sullivan, and D. L. Miller, *IEEE Trans. Electron Devices* **ED-34**, 1670 (1987).

86. S. Notomi, Y. Awano, M. Kosugi, T. Nagata, K. Kosemura, M. Ono, N. Kobayashi, H. Ishiwara, K. Odani, T. Mimura, and M. Abe, *Tech. Dig., IEEE Gallium Arsenide Integrated Circuits Symp.* p. 177 (1987).

87. H. Kinoshita, S. Nishi, M. Akiyama, and K. Kaminishi, *Jpn. J. Appl. Phys.* **24**, 1061 (1985).

88. H. Kinoshita, Y. Sano, T. Nonaka, T. Ishida, and K. Kaminishi, *Proc. Int. Ion Eng. Cong., 1983* p. 1629 (1983).

89. N. Yokoyama, T. Ohnish, K. Odani, M. Onodera, and M. Abe, *IEEE Electron. Device Lett.* **EDL-5**, 129 (1984).

90. H. Nakamura, Y. Sano, T. Nonaka, T. Ishida, and K. Kaminishi, *Tech. Dig. IEEE Gallium Arsenide Integrated Circuits Symp.* p. 134 (1983).

91. N. C. Cirillo, J. K. Abrokwah, and M. S. Shur, *IEEE Electron. Device Lett.* **EDL-5**, 129 (1984).

92. N. C. Cirillo, J. K. Abrokwah, and S. A. Jamison, *Tech. Dig. Gallium Arsenide IEEE IC Symp. 1984* p. 167 (1984).

93. N. C. Cirillo, H. K. Chung, P. J. Vold, M. K. Hibbs-Brenner, and A. M. Fraasch, *J. Vac. Sci. Technol. B* [2] **3**, 1680 (1985).

94. N. C. Cirillo, J. K. Abrokwah, and M. S. Shur, *IEEE Electron. Device Lett.* **EDL-5**, 129 (1984).

95. S. Tatsuta, T. Inata, S. Ohamura, and S. Hiyamizu, *Jpn. J. Appl. Phys.* **23**, L147 (1984).

96. D. K. Arch, B. K. Betz, P. J. Vold, J. K. Abrokwah, and N. C. Cirillo, *IEEE Electron Device Lett.* **EDL-7**, 700 (1986).

97. S. S. Pei, *Tech. Dig. IEEE Int. Symp. Circuits Syst. 1988.*

98. N. H. Sheng, M. F. Chang, C. P. Lee, D. L. Miller, and R. T. Chen, *IEEE Electron. Device Lett.* **EDL-7**(1), 11 (1986).

99. M. Miyamoto, K. Ohata, H. Toyashima, K. Suzuki, and H. Itoh, *43rd Annu. Device Res. Conf. Tech. Dig., 1985.*

100. N. Braslau, *J. Vac. Sci. Technol.* **19**, 803 (1981).

101. M. D. Feuer, *IEEE Trans. Electron Devices* **ED-32**, 7 (1985).

102. A. Ketterson, F. Ponse, T. Henderson, J. Klem, and H. Morkoç, *J. Appl. Phys.* **57**, 2305 (1985).

103. W.-P. Hong, K. S. Seo, P. K. Bhattacharya, and H. Lee, *IEEE Electron. Device Lett.* **EDL-7**, 320 (1986).

104. W. L. Jones, and L. F. Eastman, *IEEE Trans. Electron. Devices* **ED-33**, 712 (1986).

105. A. Ezis, A. K. Rai, and D. W. Langer, *Electron. Lett.* **23**, 113 (1987).

106. T. Nittono, H. Ito, O. Nakajima, and T. Ishibasi, *Jpn. J. Appl. Phys.* **25**, L865 (1986).

107. M. Hatzakis, *Jl. Vac. Sci. Technol.* **16**, 1984 (1979).

108. N. J. Shah, S. S. Pei, C. W. Tu, and R. C. Tiberio, *IEEE Trans. Electron. Devices* **ED-33**, 543 (1986).

109. D. J. Resnick, D. K. Atwood, T. Y. Kuo, N. J. Shah, and F. Ren, *Proc. SPIE—Int. Soc. Opt. Eng.* **923**, 296–303 (1988).

110. N. H. Sheng, H. T. Wang, S. J. Lee, C. P. Lee, G. J. Sullivan, and D. L. Miller, *Tech. Dig., IEEE Gallium Arsenide Integrated Circuits Symp.* p. 97 (1986).

111. N. C. Cirillo, D. K. Arch, P. J. Vold, B. K. Betz, I. R. Mactaggart, and B. L. Grung, *Tech. Dig., IEEE Gallium Arsenide Integrated Circuits Symp.* p. 257 (1987).

112. A. Y. Cho and J. R. Arthur, *Prog. Solid State Chem.* **10** 157 (1975).

113. A. Y. Cho, *Thin Solid Films* **100**, 291 (1983).

114. D. E. Mars and J. N. Miller, *J. Vac. Sci. Technol. B* [2] **4** 571 (1986).

115. R. C. Eden, *Proc. IEEE* **70**, 5 (1982).

116. K. Fujiwara, Y. Nishikawa, Y. Tokuda, and T. Nakayama, *Appl. Phys. Lett.* **48**, 701 (1986).

117. T. Mimura, M. Abe, and M. Kobayashi, *Fujitsu Sci. Tech. J.* **21**, 370 (1985).

118. G. B. Stringfellow, *in* "Semiconductors and Semimetals" (W. T. Tsang, ed.), Vol. 22, p. 209 Academic Press, Orlando, Florida, 1985.

119. H. Kawai and N. Watanabe, *Tech. Dig., IEEE Gallium Arsenide Integrated Circuits Symp.* p. 75 (1987).

120. R. Bhat, W. K. Chan, A. Kastalsky, and M. A. Koza, *43rd Annu. Device Res. Conf., Tech. Dig.* p. IIA-2 (1985).

121. A. L. Powell and P. Mistry, *Electron. Lett.* **23**, 528 (1987).

122. H. Tanaka, H. Itoh, T. O'Hori, M. Takikawa, K. Kasai, M. Takechi, M. Suzuki, and J. Komeno, *Jpn. J. Appl. Phys.* **26**, L1456 (1987).

123. H. Takakuwa, Y. Kato, S. Watanabe, and Y. Mori, *Electron. Lett.* **21**, 125 (1985).

124. W. T. Tsang, *J. Crystal Growth* **81**, 261 (1987).

125. M. B. Panish, H. Temkin, and S. Sumski, *Jl. Vac. Sci. Technol. B* [2] **3**, 657 (1985).

126. P. M. Solomon and H. Morkoç, *IEEE Trans. Electron. Devices* **ED-31**, 1015 (1984).

127. K. Inoue and H. Sakaki, *Jpn. J. Appl. Phys.* **23**, L61 (1984).

128. K. Inoue, H. Sakaki, and J. Yoshino, *Jpn. J. Appl. Phys* **23**, L767 (1984).

129. N. H. Sheng, C. P. Lee, and D. L. Miller, *Tech. Dig.—Int. Electron Devices Meet.* p. 352, (1984).

130. K. Inoue, H. Sakaki, J. Yoshino, and T. Hotta, *Japanese Appl. Phys.* **58**, 4277 (1985).

131. N. H. Sheng, C. P. Lee, R. T. Chen, D. L. Miller, and S. J. Lee, *IEEE Electron Device Lett.* **EDL-6**, 307 (1985).

132. F. Alexandre, L. Goldstein, G. Levoux, M. C. Joncour, H. Thibierge, and E. V. K. Rao, *J. Vac. Sci. Technol. B* [2] **3**, 950 (1985).

133. K. Inoue, H. Sakaki, J. Yoshino, and Y. Yoshioka, *Appl. Phys. Lett.* **46**, 973 (1985).

134. K. Lee, M. S. Shur, T. J. Drummond, and H. Morkoç, *J. Vac. Sci. Technol. B* [2] **2**, 113 (1984).

135. N. C. Cirillo, M. S. Shur, and J. K. Abrokwah, *IEEE Electron Device Lett.* **EDL-7**, 71 (1986).

136. H. Kinoshita, Y. Sano, S. Nishi, M. Akiyama, and K. Kaminishi, *Jpn. J. Appl. Phys.* **23**, L836 (1983).

137. H. Kinoshita, T. Ishida, H. Inomata, M. Akiyama, and K. Kaminishi, *IEEE Trans. Electron. Devices* **ED-33**, 608 (1986).

138. T. Saito, H. Fujishiro, S. Nishi, Y. Sano, and K. Kaminishi, *Ext. Abstr., Conf. Solid State Devices Mater.* **19**, 267 (1988).

139. T. Baba, T. Mizutani, M. Ogawa, and K. Ohata, *Jpn. J. Appl. Phys.* **23**, 654–656 (1984).

140. C. W. Tu, W. L. Jones, R. F. Kopf, L. D. Urbanek, and S. S. Pei, *IEEE Electron. Device Lett.* **EDL-7**, 552 (1986).

141. J. K. Abrokwah, N. C. Cirillo, D. Arch, R. R. Daniels, M. Hibbs-Brenner, A. Fraasch, P. Vold, and P. Joslyn, *J. Vac. Sci. Technol. B* [2] **4**, 615 (1986).

142. K. Tabatabaie-Alavi, B. W. Black, and S. E. Bernacki, *Tech. Dig., IEEE GaAs Integrated Circuits Symp.* p. 137 (1986).

143. E. F. Schubert, A. Fischer, and K. Ploog, *IEEE Trans. Electron. Devices* **ED-33**, 625 (1986).

144. M. Hueshen, N. Moll, E. Gowen, and J. Miller, *Tech. Dig., Int. Electron Devices Meet.* p. 348 (1984).

145. U. Mishra, R. Beaubien, M. Delaney, A. Brown, and L. Hackett, *Tech. Dig.—Int Electron Devices Meet.* p. 829 (1986).

146. K. Hikosaka, S. Sasa, and Y. Hirachi, *Electron. Lett.* **22**, 1240 (1986).

147. R. A. Kiehl, S. L. Wright, J. H. Margelin, and D. J. Frank, *Proc. IEEE Cornell Conf.* p. 28, (1987).

148. P. M. Solomon, C. M. Knoedler, and S. L. Wright, *IEEE Electron Device Lett.* **EDL-5**, 379 (1984).

149. K. Matsumoto, M. Ogara, T. Wada, N. Hashizume, T. Yao, and Y. Hayashi, *Electron. Lett.* **20**, 462 (1984).

150. S. Tiwari, *IEEE Trans. Electron. Devices* **ED-31**, 879 (1984).

151. M. Chen, W. J. Schaff, P. J. Tasker, and L. F. Eastman, *Electron. Lett.* **23**, 105, 800 (1987).

152. H. Baratte, D. C. LaTulipe, C. M. Knoedler, T. N. Jackson, D. J. Frank, P. M. Solomon, and S. L. Wright, *Tech. Dig.—Int. Electron Devices Meet.* p. 444 (1986).

153. K. Yamasaki, K. Arai, and K. Kuramada, *IEEE Trans. Electron. Devices* **ED-29**, 1772 (1982).

154. H. Baratte, P. M. Solomon, D. C. La Tulipe, T. N. Jackson, D. J. Frank, and S. L. Wright, *IEEE Electron Device Lett.* **EDL-8**, 486 (1988).

155. K. Arai, T. Mizutani, and F. Yanagawa, *Conf. Ser.—Int. Phys.* **79**, 631 (1985).

156. S. Fujita, M. Hirano, K. Maezawa, and T. Mizutani, *IEEE Electron Device Lett.* **EDL-8**, 226 (1987).

157. K. Maezawa, T. Mizutani, K. Arai, and F. Yanagawa, *IEEE Electron Device Lett.* **EDL-7**, 454 (1986).

158. Y. Katayama, M. Morioka, Y. Sawada, K. Ueyanagi, T. Mishima, Y. Ono, T. Usagawa, and Y. Shiroki, *Jpn. J. Appl. Phys.* **23**, L150 (1984).

159. N. C. Cirillo, M. Shur, P. I. Vold, J. K. Abrokwah, R. R. Daniels, and O. N. Tufte, *Tech. Dig.—Int. Electron Devices Meet.* p. 317 (1985).

160. M. S. Shur, D. K. Arch, R. R. Daniels, and J. K. Abrokwah, *IEEE Electron Device Lett.* **EDL-7**, 78 (1986).

161. H. L. Stormer, K. Baldwin, A. C. Gossard, and W. Weigmann, *Appl. Phys. Lett.* **44**(11), 1062 (1984).

162. S. Tiwari and W. I. Wang, *IEEE Electron Device Lett.* **EDL-5**, 333 (1984).

163. R. A. Kiehl and A. C. Gossard, *IEEE Electron Device Lett.* **EDL-5**(10), 420 (1984).

164. R. A. Kiehl and A. C. Gossard, *IEEE Electron Device Lett.* **EDL-5**, 521 (1984).

165. N. C. Cirillo, M. Shur, P. J. Vold, J. K. Abrokwah, and O. N. Tufte, *IEEE Electron Device Lett.* **EDL-6**, 645 (1985).

166. S. Fujita and T. Mizutani, *Trans. IECE Jpn.* **69**, 288 (1986).

167. K. Matsumoto, M. Ogara, T. Wada, T. Yao, Y. Hayashi, N. Hashizume, M. Kato, N. Fukuhara, N. Hirashima, and T. Miyashita, *IEEE Electron Device Lett.* **EDL-7**, 182 (1986).

168. F. Hasegawa, *43rd Annu. Device Res. Conf. Tech. Dig.* p. IIA-1 (1985).

169. H. Inomata, S. Nishi, S. Takahashi, and K. Kaminishi, *Jpn. J Appl. Phys.* **25**, L731 (1986).

170. H. Hida, A. Okamoto, H. Yoyoshimo, and K. Ohata, *IEEE Electron Device Lett.* **EDL-7**, 625 (1986).

171. H. Hida, A. Okamoto, H. Toyoshima, and K. Ohata, *IEEE Trans. Electron Devices* **ED-34**, 1448 (1987).

172. H. Hida, H. Toyoshima, and Y. Ogawa, *IEEE Electron Device Lett.* **EDL-8**, 557 (1987).

173. A. Cappy, B. Carnez, R. Fauquembergues, G. Salmer, and E. Constant, *IEEE Trans. Electron Devices* **ED-27**, 2158 (1980).

174. J. D. Oliver, Jr., and L. F. Eastman, *J. Electron. Mater.* **9**, 693 (1980).

175. S. Bandy, C. Nishimoto, S. Hyder, and C. Hooper, *Appl. Phys. Lett.* **38**, 817 (1981).

176. K. Y. Chen, A. Y. Cho, S. B. Christman, T. P. Pearsall, and J. E. Rowe, *Appl. Phys. Lett* **40**, 423 (1982).

177. K. Kajiyama, Y. Mizushima, and S. Sakata, *Appl. Phys. Lett.* **23**, 458 (1973).

178. C. A. Mead and W. G. Spitzer, *Phys. Rev.* **134**, A173 (1964).

179. K. Kamada, H. Ishikawa, M. Ikeda, Y. Mori, and C. Kojima, *Conf. Ser.—Inst. Phys.* **83**, 574 (1987).

180. K. Y. Chang, A. Y. Chao, T. J. Drummond, and H. Morkoç, *Appl. Phys. Lett.* **40**, 147 (1982); U. K. Mishra, A. S. Brown, L. M. Jelloian, L. H. Hackett, and M. J. Delaney, *45th Ann. Device Res. Conf. Tech. Dig.*, p. IIA-6 (1987).

181. C. Y. Chen, A. Y. Chao, K. Y. Cheng, T. P. Pearsall, P. O'Connor, and P. A. Garbinski, *Electron. Device Lett.* **EDL-3**, 152 (1982).

182. H. Hirose, K. Ohata, T. Mizutani, T. Itoh, and M. Ogawa, *Conf. Ser.—Inst. Phys.* **79**, 529 (1986).

183. M. D. Feuer, T. Y. Chang, S. C. Shunk, and B. Tell, *44th Annu. Device Res. Conf., Tech. Dig.* p. IIA-8 (1986).

184. J. H. van der Merwe, *J. Appl. Phys.* **34**, 123 (1963).

185. Y. Sugita and M. Tamura, *J. Appl. Phys.* **40**, 3089 (1969).

186. J. W. Matthews and A. E. Blakeslee, *J. Cryst. Growth* **27**, 118 (1974).

187. J. J. Rosenberg, M. Benlamri, P. D. Kirchner, J. M. Wodall, and G. D. Pettit, *IEEE Electron Device Lett.* **EDL-6**, 491 (1985).

188. A. Ketterson, M. Moloney, W. T. Masselink, J. Klem, R. Fisher, W. Kopp, and H. Morkoç, *IEEE Electron Device Lett.* **EDL-6**, 628 (1985).

189. T. E. Zipperian and T. J. Drummond, *Electron. Lett.* **21**, 823 (1985).

190. T. Henderson, M. I. Aksun, C. K. Peng, H. Morkoç, P. C. Chao, P. M. Smith, K. H. G. Duh, and L. F. Lester, *Tech. Dig.—Int. Electron Devices Meet.* p. 464 (1986).

191. J. M. Kuo, B. Lalevic, and T. Y. Chang, *Tech. Dig.—Int. Electron Devices Meet.* p. 460 (1986); J. M. Kuo, T. Y. Chang, and B. Lalevic, *IEEE Electron Device Lett.* **EDL-8**, 380 (1987).

192. P. Chu, C. L. Lin, and H. H. Wieder, *Electron. Lett.* **22**, 890 (1986); C. L. Lin, P. Chu, A. L. Kellner, and H. H. Wieder, *Appl. Phys. Lett.* **49**, 1593 (1986).

193. A. A. Ketterson, W. T. Masselink, J. S. Gedymin, J. Klem, C. K. Peng, W. F. Kopp, H. Morkoç, and K. R. Gleason, *IEEE Trans. Electron Devices* **ED-33**, 564 (1986).

194. Y. K. Chen, D. C. Radulescu, A. N. Lepore, M. C. Foisy, G. W. Wang, P. J. Tasker, and L. F. Eastman, *45th Annu. Device Res. Conf., Tech. Dig.* p. IIA-3 (1987).

195. T. Henderson, M. I. Aksun, C. K. Peng, H. Morkoç, P. C. Chao, P. M.

Smith, K. H. G. Duh, and L. F. Lester, *Tech. Dig—Int. Electron Devices Meet.* p. 464 (1986).

196. N. Moll, A. Fisher-Colbrie, and M. Hueschen, *45th Annu. Device Res. Conf., Tech. Dig.* p. IIA-5 (1987).

197. P. C. Chao, R. C. Tiberio, K.-H. G. Duh, P. M. Smith, J. M. Ballingall, L. F. Lester, B. R. Lee, A. Jabra, and G. G. Gifford, *IEEE Electron Device Lett.* **EDL-8**, 489 (1987); P. C. Chao, P. M. Smith, K. H. G. Duh, J. M. Ballingall, L. F. Lester, B. R. Lee, A. A. Jabra, and R. C. Tiberio, *Tech. Dig.—Int. Electron Devices Meet.* p. 410 (1987).

198. C. K. Peng, M. I. Aksun, A. A. Ketterson, H. Morkoç, and K. R. Gleason *IEEE Electron Device Lett.* **EDL-8**, 24 (1987); M. I. Aksun, C. K. Peng, A. A. Ketterson, and H. Morkoç, *Tech. Dig.—Int. Electron Devices Meet. p.* 822 (1986).

199. U. K. Mishra, A. S. Brown, L. M. Jelloian, L. H. Hackett, and M. J. Delaney, *IEEE Electron Device Lett.* **EDL-9**, 41 (1988).

200. M. D. Feuer, T. Y Chang, and S. C. Shunk, *44th Annu. Device Res. Conf., Tech. Dig.* p. IIA-8 (1986).

201. M. D. Feuer, J. M. Kuo, S. C. Shunk, R. E. Behringer, and T. Y. Chang, *IEEE Electron Device Lett.* **EDL-9**, 162 (1988).

202. K. S. Seo, P. K. Bhattacharya, and K. R. Gleason, *Electron. Lett.* **23**, 295 (1987).

203. J. A. Del Alamo and T. Mizutani, *IEEE Electron Device Lett.* **EDL-8**, 534 (1987).

204. E. F. Schubert, W. T. Tsang, M. D. Feuer, and P. M. Mankiewich, *IEEE Electron Device Lett.* **EDL-9**, 145 (1988).

205. J. M. Kuo, M. D. Feuer, and T. Y. Chang *J. Vac. Sci. Technol.* **B6**, 657–659 (1988).

206. Y. Sekiguchi, Y. J. Chan, M. Jaffe, M. Weiss, G. I. Ng, J. Singh, M. Quillec and D. Pavlidis, *Conf. Ser.—Inst. Phys.* **91**, p. 215 (1987).

207. Y. K. Chen, G. W. Wang, W. J. Schaff, P. J. Tasker, K. Kavanagh, and L. F. Eastman, *Tech. Dig—Int. Electron Devices Meet.* p. 431 (1987).

208. G. Tuttle and H. Kroemer, *45th Annu. Device Res. Conf., Tech. Dig.* p. IIA-6 (1987).

209. I. J. Fritz, B. L. Doyle, J. E. Shirber, E. D. Jones, L. R. Dawson, and T. J. Drummond, *Appl. Phys. Lett.* **49**, 581 (1986).

210. T. J. Drumond, T. E. Zipperian, I. J. Fritz, J. E. Shirber, and T. A. Pluz, *Appl. Phys. Lett.* **49**, 461 (1986).

211. C. P. Lee, H. T. Wang, G. J. Sullivan, N. H. Sheng, and D. L. Miller, *IEEE Electron Device Lett.* **EDL-8**, 85 (1987).

212. T. E. Zipperian, T. J. Drummond, and I. J. Fritz, *Proc. IEEE Cornell Conf. Adv. Concepts High Speed Semicond. Devices Circuits 11th*, p. 18 (1987).

213. T. Mimura, K. Joshin, S. Hiyamizu, K. Hikosaka, and M. Abe, *Jpn. J. Appl. Phys.* **20**, L598 (1981).

214. M. Abe, T. Mimura, S. Notomi, K. Odani, and K. Kondo, *J. Vac. Sci. Technol. A* [2] **5**, 1387 (1987).

215. Y. Awano, M. Kosugi, T. Mimura, and M. Abe, *IEEE Electron Device Lett.* **EDL-8**, 451 (1987).

216. N. J. Shah, S. S. Pei, C. W. Tu, and R. C. Tiberio, *IEEE Trans. Electron Devices* **ED-5**, 543 (1986).

217. T. Saito, H. Fujishiro, S. Nishi, Y. Sano, and K. Kaminishi, *Proc. Conf. Solid State Devices Mater.* **19**, 267 (1988).

218. H. Baratte, D. C. LaTulipe, C. M. Knoedler, T. N. Jackson, D. J. Frank, P. M. Solomon, and S. L. Wright, *Tech. Dig.—Int. Electron Devices Meet.* pp. 444–447 (1986).

219. T. Mizutani et al., *Tech. Dig., IEEE Gallium Arsenide Integrated Circuits Symp.* p. 107 (1986).

220. R. R. Daniels, R. Mactaggart, J. K. Abrokwah, O. N. Tufte, M. Shur, J. Back, and P. Jenkins, *Tech. Dig.—Int. Electron Devices Meet.* 448–451 (1986).

221. S. Fujita and T. Mizutani, *IEEE Trans. Electron Devices* **ED-34**, 1889 (1987).

221a. R. A. Kiehl, M. A. Scontras, D. J. Widiger, and W. M. Kwapien, *IEEE Trans. Electron Devices* **ED-34**, 2412 (1987).

222. R. A. Kiehl;, M. D. Feuer, R. H. Hendel, J. C. M. Hwang, V. G. Kermidas, C. L. Allyn, and R. Dingle, *IEEE Electron Device Lett.* **EDL-4**, 377 (1983).

223. M. Abe, T. Mimura, K. Nisiushi, A. Shibatomi, and M. Kobayashi, *Tech. Dig., IEEE Gallium Arsenide Integrated Circuits Symp.*, p. 158 (1983).

224. M. Rocchi and B. Gabillard, *IEEE J. Solid-State Circuits* **SC-18**, 369 (1983).

225. R. L. Van Tuyl, C. A. Liechti, R. E. Lee, and E. Gowen, *IEEE J. Solid-State Circuits* **SC-12**, 485 (1981).

226. N. J. Shah, S. S. Pei, C. W. Tu, and R. C. Tiberio, *IEEE Trans. Electron Devices* **ED-5**, 543 (1986).

227. R. H. Hendel, S. S. Pei, C. W. Tu, B. J. Roman, and N. J. Shah, *Tech. Dig.—Int. Electron Devices Meet.* p. 867 (1984).

228. S. S. Pei, *Tech. Dig., IEEE Int. Symp. Circuits Syst.*, p. 1715 (1988).

229. A. R. Schlier, S. S. Pei, C. W. Tu, and G. E. Mahoney, *Tech. Dig., IEEE Gallium Arsenide Integrated Circuits Symp.* p. 91 (1985).

230. N. C. Cirillo, D. K. Arch, P. J. Vold, B. K. Betz, I. R. Mactaggart, and B. L. Grung, *Tech. Dig., IEEE Gallium Arsenide Integrated Circuits Symp.* p. 257 (1987).

231. D. K. Arch, B. K. Betz, P. J. Vold, J. K. Vold, J. K. Abrokwah, and N. C. Cirillo, *IEEE Electron Device Lett.* **EDL-7**, 700 (1986).

232. Y. Watanabe, K. Kajii, K. Nishiuchi, M. Suzuki, I. Hanyu, M. Kosugi, and K. Odani, *Dig. Tech. Pap.—Int. Solid-State Circuits Conf.* p. 80 (1986).

233. K. Nishiuchi, N. Kobayashi, S. Kuroda, S. Notomi, T. Mimura, M. Abe, and M. Kobayashi, *Dig. Tech. Pap—Int. Solid-State Circuits Conf.* p. 48 (1984).

234. M. Abe, T. Mimura, S. Notomi, K. Odani, and K. Kondo, *J. Vac. Sci. Technol., A* [2] **5**, 1387 (1987).

235. N. H. Sheng, H. T. Wang, S. J. Lee, C. P. Lee, G. J. Sullivan, and D. L. Miller, *Tech. Dig. IEEE Gallium Arsenide Integrated Circuits Symp.* p. 97 (1986).

236. S. Notomi, Y. Awano, M. Kosugi, T. Nagata, K. Kosemura, M. Ono, N. Kobayashi, H. Ishiwari, K. Odani, T. Mimura, and M. Abe, *Tech. Dig., IEEE Gallium Arsenide Integrated Circuits Symp.*, p. 177 (1987).

4 Heterojunction Bipolar Transistor Technology

PETER M. ASBECK, MAU-CHUNG FRANK CHANG,
KEH CHUNG WANG
Rockwell International Science Center, Thousand Oaks, California

and

DAVID L. MILLER
Pennsylvania State University, State College, Pennsylvania

4.1 INTRODUCTION

4.1.1 Scope of This Chapter

This chapter describes the principles of operation, the fabrication, and the applications of the heterojunction bipolar transistor, or HBT. The HBT derives its name from its use of one or more heterojunctions, or junctions between distinct semiconductors. Although the underlying concept of the HBT dates back to the early days of the bipolar transistor, with the work of Shockley and Kroemer in the 1950s [1, 2], high-performance HBTs are a comparatively recent development [3]. High current gain and high-speed devices emerged only after the development of an advanced materials technology in III–V compounds, principally with the (Ga,Al)As/GaAs system. Realistic prospects for integraged circuit (IC) manufacture with HBTs existed only after the introduction of modern epitaxial growth techniques, molecular beam epitaxy (MBE), and metallorganic chemical vapor phase deposition (MOCVD). HBTs remain a very active area for solid-state device research. Development of new device structures, extension to new material systems, and application to new areas are being vigorously pursued. Rather than attempt to cover recent developments in detail, this chapter will concentrate on what is today the most prevalent HBT, the n-p-n (Ga,Al)As/GaAs transistor, and focus primarily on the underlying principles of its operation.

In the remainder of this section, the unique characteristics and advantages of the HBT will be described. Subsequent sections will focus on the modeling of HBTs and their fabrication (including a brief introduction to MBE and MOCVD), electrical characteristics, and areas of application developed to date.

4.1.2 Figures of Merit of Transistors, f_t and f_{max}

To understand the benefits obtained from the HBT structure, it is worthwhile to consider first what constitutes a good transistor. Two commonly used figures of merit are f_t, the cutoff frequency, and f_{max}, the maximum frequency of oscillation. Particularly in microwave (linear) applications, these figures of merit describe well the high-frequency capability of a device. For digital applications, as described below in Section 4.5.2, additional factors can be important.

The cutoff frequency f_t corresponds to the frequency for which the incremental short-circuit current gain decreases to unity. For most devices, for a wide range of frequencies below f_t (but not necessarily extending down to dc), the magnitude of the current gain G_i is given by

$$|G_i| = \frac{f_t}{f} \tag{1}$$

f_t can be simply related to the transit time τ_t of carriers (electrons) across the device within a simple charge-control model of a transistor. In this framework, the output current of a transistor i_o can be expressed as the total charge q_o of carriers (electrons) within the control region of the transistor divided by their effective transit time across the device control region, τ_t. The input current of the device supplies charge to the device equal in magnitude (by overall device quasi-neutrality) to the charge q_o of the carriers. Thus

$$i_o = \frac{q_o}{\tau_t}, \; i_{in} = -j\omega q_o, \; G_i = \frac{di_o}{di_{in}} = \frac{-1}{j\omega\tau_t} \tag{2}$$

where $\omega = 2\pi f$. From these relations, f_t is found to be equal to $1/(2\pi\tau_t)$. To increase the cutoff frequency of the transistor, the transit time of carriers across the device control region must be made as small as possible by making the distance the carriers must travel as short as possible and by causing the carriers to move at the highest attainable speed.

The maximum frequency of oscillation describes the frequency at which the power gain of the transistor falls to unity. Over a significant frequency range, the maximum available power gain of a single-stage transistor amplifier is often found to obey the relation

$$G_p \cong \frac{f^2_{max}}{f^2} \tag{3}$$

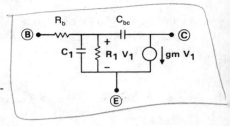

Figure 4.1 Simplified circuit model of a homo-junction or heterojunction bipolar transistor.

Under optimal input and output impedance matching conditions, the maximum available power gain G_p of the amplifier may be expressed in terms of the device current gain G_i as

$$G_p \cong \frac{|G_i|^2}{4} \frac{\mathrm{Re}(Z_{\mathrm{out}})}{\mathrm{Re}(Z_{\mathrm{in}})} \qquad (4)$$

where Z_{out} and Z_{in} are the respective output and input impedances of the device. The important factors governing these latter quantities are apparent from a simplified circuit model of a transistor, the basic hybrid π-structure shown in Fig. 4.1. At sufficiently high frequency, the real part of the input impedance is R_b, the base resistance. The output impedance is established indirectly by feedback through the capacitor C_{bc} of a current that is then amplified by the forward gain of the transistor. The corresponding $\mathrm{Re}(Z_{\mathrm{out}})$ is approximately $1/(2\pi C_{\mathrm{bc}} f_t)$. From these, an expression for f_{max} can be derived [4]:

$$f_{\mathrm{max}} \cong \sqrt{\frac{f_t}{8\pi R_b C_{\mathrm{bc}}}} \qquad (5)$$

It is apparent that to have high f_{max} it is necessary to minimize the base (access) resistance of the device and decrease the magnitude of its feedback capacitance C_{bc}.

Equipped with this knowledge of the basic factors contributing to high-speed operation of transistors, we now turn to a discussion of the HBT structure and an examination of how it achieves these objectives.

4.1.3 Principles of HBT Operation

Figure 4.2 illustrates the cross section of a representative HBT. The various epitaxial layers used in the device are listed in Fig. 4.3 along with their typical characteristics. A key feature is that the emitter consists of (Ga,Al)As, whose bandgap is wider than that of the GaAs base region, by at least 250 meV (about $10kT$ at room temperature). The corresponding band diagram of the HBT in the direction of electron travel (emitter to collector) is shown in Fig. 4.4, along with that of a conventional homojunc-

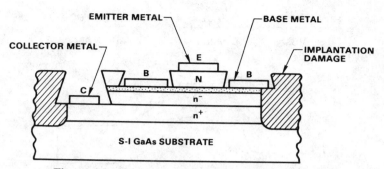

Figure 4.2   Cross section of a representative HBT.

tion bipolar device for comparison. During transistor operation, the input voltage forward-biases the base–emitter junction, allowing electrons from the n-type emitter to enter the base. The injected electrons travel across the base by diffusion or drift, and when they reach the base–collector junction, they are swept into the collector by the high fields in that region. A (usually small) fraction of the injected electrons recombines with the holes in the transistor base region, giving rise to (undesirable) base current flow. An additional component of base current arises because with the forward-biased base–emitter junction, there is a flow of holes from the base into the emitter, where they subsequently recombine. In a homojunction bipolar transistor, this latter contribution to base current is a significant design limitation. To suppress the reverse hole injection and attain high current gain, the doping of the base is made lower than that of the emitter, typically by a factor of 100 or more. Consequently, the base layer becomes resistive. In turn, the overall base resistance of the device rises, and, as seen in Eq. (5), the f_{max} value suffers. In an HBT, the presence of a wide-bandgap

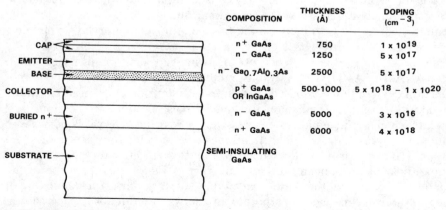

	COMPOSITION	THICKNESS (Å)	DOPING (cm^{-3})
CAP	n$^+$ GaAs	750	1×10^{19}
EMITTER	n$^-$ GaAs	1250	5×10^{17}
BASE	n$^-$ Ga$_{0.7}$Al$_{0.3}$As	2500	5×10^{17}
COLLECTOR	p$^+$ GaAs OR InGaAs	500–1000	$5 \times 10^{18} - 1 \times 10^{20}$
BURIED n$^+$	n$^-$ GaAs	5000	3×10^{16}
	n$^+$ GaAs	6000	4×10^{18}
SUBSTRATE	SEMI-INSULATING GaAs		

Figure 4.3   Schematic epitaxial layer structure of a (Ga,Al)As/GaAs HBT.

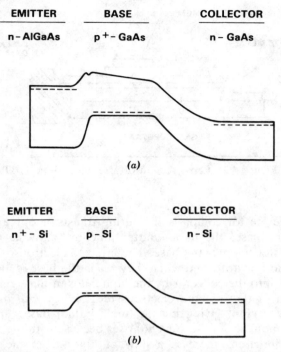

Figure 4.4 Band diagram along direction of electron travel of (a) an HBT and (b) a homojunction bipolar transistor.

emitter alleviates this problem. As seen in Fig. 4.4a, in an HBT the energy barrier for hole injection into the emitter is wider than the barrier for electron flow into the base. The base current from hole injection is automatically suppressed. The doping of the base layer can be made as large as possible from solid-state chemistry considerations, and the base resistance correspondingly reduced. At the same time, the emitter doping level can be reduced, which tends to increase the base–emitter depletion layer thickness and reduce the emitter–base capacitance.

In addition to the benefits of the wide-gap emitter just described, (Ga,Al)As/GaAs HBTs enjoy a number of other advantages over homo-junction devices (e.g., Si-based transistors). By introducing a gradient in the composition of the semiconductor that makes up the base region, and thereby changing its bandgap progressively across the layer, it is possible to build in a "quasi-electric" field that drives injected electrons across the base of an HBT by drift, rather than by the slower diffusion process. This "quasi-field" is evident in the band diagram of Fig. 4.4 as the slope of the conduction band in the base. An additional technique for reduction of electron transit time is to incorporate an energy step (an abrupt decrease in the energy of the conduction band) at the base–emitter junction, as also shown in Fig. 4.4a. If sufficiently abrupt, the electrons traversing this step will acquire considerable kinetic energy and be launched into the base at high velocity. They can retain their forward momentum over distances on the order of 300–1000 Å (depending on the composition of the base layer), again contributing to high f_t.

The high-frequency performance of GaAs MESFETs rests primarily on the fact that the low-field electron mobility in GaAs is a factor of seven higher than that of Si of comparable doping, and that due to the large bandgap of GaAs and resultant low intrinsic carrier concentration, semi-insulating substrates can be easily produced. Both of these advantages also extend to GaAs-based bipolar transistors and circuits, leading to reduced electron transit times, reduced series resistance at comparable doping, and reduced capacitance of interconnects on GaAs substrates.

Additional advantages of GaAs-based HBTs stem from the improved radiation hardness expected from the use of semi-insulating GaAs substrates together with superlattice buffer layers, and from the fact that the GaAs transistors are directly compatible with a variety of optoelectronic devices (lasers, LEDs, and detectors), which will enable their eventual integration.

4.1.4 Comparison of HBTs with FETs

There are a number of significant structural differences between HBTs and FETs that lead to corresponding differences in characteristics. The HBT is a vertical device in which the electrons travel from emitter to collector across the epitaxially grown layers. With modern crystal growth techniques, the overall distance can be reproducibly made very small, at most a few thousand angstroms. This is conducive to short transit times. In the FET, electrons traverse the device laterally, with an overall distance of travel that is determined by the lithographically defined gate length. In FET fabrication, it is a major challenge to fabricate the gates with submicron dimensions reproducibly, as needed to attain high f_t.

In the HBT, the entire emitter area contributes to current flow, rather than a thin surface channel layer, as in the case of the FET. This leads to significantly higher current-handling capability per unit chip area for the HBT and contributes to the higher output drive capability of the bipolar circuits compared with FET circuits.

For bipolar transistors, the output current varies exponentially with the input voltage. This variation is substantially more rapid than that for FETs (typically linear or quadratic). This derives from the fact that in FETs, much of the applied voltage is dropped over a spacer region between the gate and the channel, and does not directly modulate the density of carriers within the channel, as occurs in the HBT. As a result, the transconductance of bipolars exceeds that of FETs by a factor of 10 to 100, depending on the magnitude of the output current used.

An additional advantage of bipolar transistors derives from the fact that the threshold voltage for turning on the output current is, in the HBT case, established by the built-in potential of a p-n junction. This quantity is primarily dependent on the semiconductor composition and is easy to reproduce. In the case of FETs, the threshold voltage is governed by the doping and thickness of the channel layer and is relatively hard to control to tight tolerances, as needed in low-power digital circuits.

Additionally, because the intrinsic HBT is comparatively well shielded from the wafer surface and substrate, HBTs do not experience trap-induced frequency dispersion of device characteristics, hysteresis, and backgating effects ordinarily encountered in GaAs MESFETs.

The structure of the two device types leads also to some natural advantages of the FET over the bipolar transistor. One is that the FET, when appropriately biased, does not have any appreciable dc input current (the dc current gain is effectively infinite). Thus, the concerns over recombination currents in bipolar design are absent. Another advantage stems from the adjustability of their threshold voltage for current flow. The fact that in FETs the channel doping and thickness can be changed to vary V_{th} is a problem from the standpoint of uniformity and reproducibility requirements, but allows one to vary V_{th} easily, and fabricate both enhancement and depletion mode devices, for example. FETs also do not store extra charge within the device when operated in saturation mode, as do bipolars. HBTs, as described below, potentially overcome some of the limitations of homojunction bipolar transistors in this regard, but the highest circuit speeds are still limited to those designs that avoid transistor saturation. In the microwave regime, the rf noise characteristics obtained to date in FETs are better than those demonstrated in HBTs.

The distinct strengths and weaknesses of HBTs and FETs suggest that there are distinct areas where one device has a natural advantage over the other. As discussed more fully in Section 4.5, the HBT is particularly strong in application to very high speed small-scale IC, digital gate arrays, analog-to-digital converter circuits, wideband feedback amplifiers, and microwave and millimeter-wave oscillators and power amplifiers. However, the performance limits of both types of devices are still being explored in these and other application areas.

To understand HBT operation and their advantages on a more quantitative basis, we turn now to consideration of the physical relations and corresponding equations governing their performance, beginning with an introduction to heterostructures.

4.2   HETEROSTRUCTURE DEVICE PRINCIPLES

4.2.1   Band Diagrams

Band diagrams display the energy of the conduction band minimum E_c and the valence band maximum E_v as a function of position within a device.

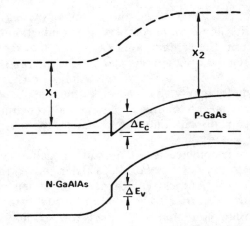

Figure 4.5 Band diagram of an abrupt n-GaAlAs/p-GaAs heterojunction in equilibrium. X_1 and X_2 denote electron affinities of the two materials.

Band diagrams for heterostructures, which contain more than one semiconductor, are somewhat more complex than those of homostructures. At an abrupt heterojunction, the conduction band energy typically has a discontinuity ΔE_c and the valence band energy has a discontinuity ΔE_v, as shown in Fig. 4.5, for an n-(Ga,Al)As/p-GaAs heterojunction. These discontinuities reflect the different binding energies of the compounds for the carriers, stemming from the different structure and bond energies of the materials. In general, ΔE_c and ΔE_v may reflect in addition the presence of an interfacial dipole that forms at the junction of the two materials (which may be dependent on the method of heterojunction preparation). The contributions ΔE_c and ΔE_v must add up to the overall difference of bandgap between the materials. No well-established theory allows the calculation of the quantities ΔE_c and ΔE_v at present. In the simplest approximation, however, ΔE_c is given by the difference between the electron affinities of the two materials, where electron affinity is the difference between the energy of an electron at the conduction band minimum in the semiconductor and one at rest in vacuum outside the semiconductor. Electron affinities are measurable in photoemission measurements and in other work-function-related experiments. This approximation to ΔE_c is known as the Anderson model [5]. It leads to quite good agreement for ΔE_c in many important cases, despite the fact that it ignores the dipole layer formation at the interface between the semiconductors, as well as that between the semiconductors and vacuum.

In the experimentally important (Ga,Al)As/GaAs system, the energy gap in $Al_xGa_{1-x}As$ has been measured to be [6]

$$E_g = 1.424 + 1.247x \tag{6}$$

for x less than 0.47. This range covers important device cases. For higher values of x, the relationship is more complex; the conduction band minimum of AlGaAs moves from the gamma point ($k = 0$) to the X point (Brillouin zone edge). In the regime of common interest, the energy gap variation is thus about 12.5 meV per mol % Al in the AlGaAs. The corresponding conduction band discontinuity has been estimated to correspond to 0.65 times the bandgap discontinuity [7, 8]; the remaining energy of 0.35-times-the-bandgap discontinuity corresponds to the valence band step.

When dissimilar semiconductors are joined, in general, electrons and holes flow from one to the other because of carrier concentration gradients or because of band discontinuities. The carrier transfer sets up electrostatic fields that inhibit further carrier transfer after steady-state conditions are reached. The equilibrium condition corresponds to a spatially constant value of the electrochemical potential (Fermi level). Figure 4.5 shows the equilibrium situation for the n-AlGaAs/p-GaAs abrupt junction, which is representative of the MODFET channel region as well as the HBT emitter–base junction. A potential barrier has formed on the AlGaAs side of the junction in the depletion layer. On the GaAs side, an accumulation of electrons is obtained. These characteristics are central to the operation of the MOD-FET, but can be detrimental to current gain in an HBT; the potential spike tends to retard electron flow, and the accumulation of carriers increases the amount of recombination. The potential spike can be reduced or eliminated by grading the Al composition of the AlGaAs in the vicinity of the junction, as shown in Fig. 4.6.

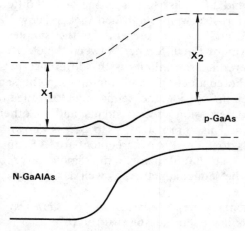

Figure 4.6   Band diagram of a graded n-GaAlAs/p-GaAs heterojunction.

4.2.2 Basic Transport Equations

To describe the local concentration of electrons $n(x)$ in quasi-equilibrium within a material of fixed composition, the equation

$$n(x) = N_c \exp\left[- \frac{E_{co} - qV(x) - E_f(x)}{kT} \right] \tag{7}$$

may be used, where N_c is the effective density of states in the conduction band, E_{co} is the conduction band energy (excluding electrostatic energy), V is the electrostatic potential and E_f is the electron quasi-Fermi level. If the material composition is variable as in a heterostructure, the same equation may be used, only in this case E_{co} and N_c are functions of position due to the changing material parameters. Similar equations apply in the case of holes.

When a semiconductor is perturbed from equilibrium, carrier flows are set up. If the perturbations are suitably small, the current density can be expressed as

$$J_n = n\mu_n \frac{dE_f}{dx} \tag{8}$$

(for the case of electrons), where μ_n is the electron mobility. This expression holds for both spatially constant and spatially varying composition. By using Eq. (7), this expression may be recast as

$$J_n = n\mu_n\left[-q \frac{dV}{dx} + \frac{dE_{co}(x)}{dx} \right] + \mu_n kT\left[\frac{dn}{dx} - \frac{n}{N_c(x)} \frac{dN_c(x)}{dx} \right] \tag{9}$$

Various components of current flow can be identified in this expression. Electron flow by drift due to electrostatic fields are represented by the dV/dx term, as in homostructures. Similarly, diffusion currents are represented by the dn/dx term (particularly transparent after recognizing that the standard diffusion coefficient D_n is $\mu_n kT/q$). A new driftlike term has arisen, where the driving force is $dE_{co}(x)/dx$, a quasi-field produced by composition variation. A new diffusionlike term has also appeared, where the driving force is $dN_c(x)/dx$, representative of the varying entropy per carrier in material of varying composition.

Equations similar to the above may be written for holes in material of spatially varying composition. The resulting equations may be combined with Poisson's equation and the electron and hole current continuity equation to form the basic set of equations for semiconductor device analysis [4, 9]. It must be remembered when using these equations that their domain of applicability is limited to small departures from equilibrium.

To complete the description of carrier behavior, it is necessary to allow for electron-hole recombination and provide a better description of carrier dynamics. In GaAs-based devices, radiative recombination, deep-level nonradiative recombination (Sah–Noyce–Shockley–Reed–Hall recombination) and Auger recombination are potentially of importance. The corresponding analytical descriptions for the net recombination rate U are given below:

$$U = B(pn - n_i^2) + \frac{pn - n_i^2}{(n + n_t)\tau_p + (p + p_t)\tau_n} + (A_1 p + A_2 n)(pn - n_i^2)$$
(10)

where B is the radiative recombination coefficient of about 10^{-10} cm^3/s^{-1} in GaAs [6], n_i is the intrinsic carrier concentration (2×10^6 cm^{-3} at 300 K), τ_n and τ_p are effective lifetimes for capture of electrons and holes, n_t and p_t are carrier densities when the Fermi level is fixed at the trap energy (usually negligible in GaAs), and A_1 and A_2 are Auger coefficient (A_1 is on the order of 10^{-30} cm^6/s in GaAs). The contribution due to traps is highly dependent on the method of preparation and history of individual samples. A wide variety of traps are known in GaAs and GaAlAs; their carrier capture cross sections are also well established in a number of cases [10]. The recombination lifetime in GaAs is much smaller than in Si of comparable doping, typically 5 ns or less.

4.2.3   Electron Dynamics

To describe GaAs devices adequately, the above equations must be supplemented with an improved description of the velocity of electrons in GaAs at various fields. For steady-state conditions, electron velocity versus applied electric field follows the behavior shown in Fig. 4.7 [11]. This relationship is characterized by high low-field mobility, followed by a region of differential

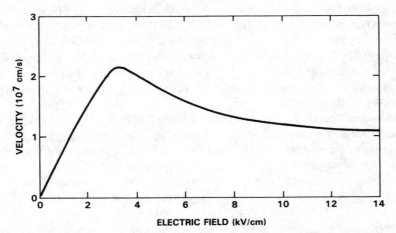

Figure 4.7   Electron velocity in GaAs versus applied electric field under steady-state conditions.

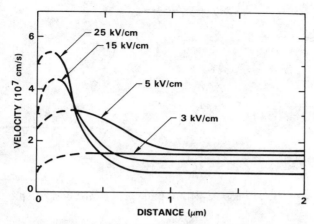

Figure 4.8 Ensemble average velocity of electrons in GaAs entering a region of high electric field versus distance from the beginning of the region. (After Maloney and Frey [12].)

negative mobility for fields above 3–4 kV/cm, and saturation of the velocity at 10^7 cm/s for fields well above 10 kV/cm. The same relationship is assumed to apply to electrons driven by composition-induced quasi-electric fields over large distances and large voltages (although this seldom, if ever, occurs). It is well known, however, that over short distances or short times, there are significant departures from the predictions of Fig. 4.7, describable as velocity-overshoot effects. Figure 4.8 shows, for example, the ensemble average velocity of electrons as a function of position after entering a region of high field, as calculated by Maloney and Frey using Monte Carlo techniques [12]. Exceptionally high velocity, up to 5×10^8 cm/s, is obtained over short distances, on the order of 1000 Å. Elsewhere in this text are described in detail the phenomena involved and the present methods of calculating the effects. Velocity overshoot of this type is of considerable significance to HBTs because of the short distances for electron travel within the devices. Unfortunately, it is difficult to handle velocity overshoot within the framework of drift–diffusion calculations, as represented in Eqs. (8) and (9). Typically, these effects are ignored in transistor analyses. Thus, while the simple drift–diffusion formulation can provide useful descriptions of many features of HBTs, they are fundamentally incapable of accurate estimation of very high frequency performance. An improved description results from the Monte Carlo technique, which has been used for the HBT structure by several groups [13, 14].

4.2.4 HBT Modeling

As an example of the application of the transport equations, let us consider motion of electrons across an HBT base region, assuming the base composi-

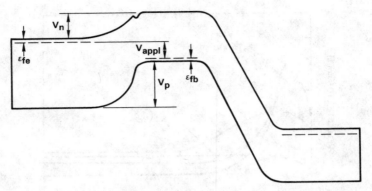

Figure 4.9 Band diagram of HBT under forward bias (simple case).

tion is linearly graded from one side to another, as illustrated in Fig. 4.9. Holes in the base of HBTs are vastly more numerous than the injected electrons, and will redistribute themselves to cancel out any effective field. Thus

$$J_p = \mu_p p \left(-q \frac{dV}{dx} + \frac{dE_v}{dx} \right) = 0$$

$$q \frac{dV}{dx} = \frac{dE_v}{dx}$$

(11)

where the small diffusionlike effects have been neglected. It is seen that an electric field is set up in the base to cancel the quasi-electric field driving holes. The electron current is

$$J_n = n\mu_n \left(\frac{dE_c}{dx} - \frac{dE_v}{dx} \right) + kT\mu_n \frac{dn}{dx} = n\mu_n \frac{dE_g}{dx} + kT\mu_n \frac{dn}{dx}$$

(12)

The net field driving electrons is the derivative of the bandgap of the material. For Al composition variation of $x = 0.1$ across a base region of thickness 1000 Å, the effective field is 12.5 kV/cm. Such a field is in excess of what causes velocity saturation in large samples. In this instance, electron velocity is not reduced due to scattering to the satellite valleys lying 0.3 V above the conduction band minimum, because the entire energy drop across the base is only 0.125 eV, and it is improbable that more than this energy will be imparted to any electron. The electron velocity reaches about 2×10^7 cm/s, as measured by picosecond optical pulse techniques [15].

Let us consider the amount of aluminum that must be incorporated in the emitter of an HBT. The collector current density J_c may be represented as

$$J_c = qn(b_0)v(n)$$

(13)

where $n(b_0)$ is the number of electrons at the emitter edge of the base, and $v(n)$ is their effective velocity toward the collector (which may be due to drift as described above, or more conventional diffusion). For the combined processes, Eq. (12) shows that $v(n)$ is given a very good approximation by

$$v(n) = \mu_n \mathscr{E}_{\text{eff}} + \frac{D}{w_{\text{b}}} \tag{14}$$

where $\mathscr{E}_{\text{eff}}$ is the effective quasi-field from bandgap grading and w_{b} is the width of the base region. Similarly, J_{b}, the hole current density at the edge of the emitter, is

$$J = q \cdot n \cdot \text{velocity}$$

$$J_{\text{b}} = qp(e_0)v(p) \tag{15}$$

where $p(e_0)$ is the number of holes at the base edge of the emitter and $v(p)$ their effective velocity. Typically, $v(p)$ is governed by diffusion in the presence of recombination, and is given by

$$v(p) = \frac{D_p}{L_p} \tag{16}$$

with L_p the hole diffusion length in the emitter. The current gain of the device (ignoring all other contributions to base current) is

$$\beta = \frac{J_{\text{c}}}{J_{\text{b}}} = \frac{n(b_0)}{p(e_0)} \frac{v(n)}{v(p)} \tag{17}$$

If we assume that quasi-Fermi levels of the electrons and holes are continuous across the emitter–base junction, then the carrier concentrations in Eq. (17) are[†]

$$n(b_0) = N_{\text{DE}} \exp\left(\frac{-V_n}{kT}\right)$$

$$p(e_0) = N_{\text{AB}} \exp\left(\frac{-V_p}{kT}\right) \tag{18}$$

where N_{DE} and N_{AB} are shallow doping concentrations in emitter and base, respectively, by inspection of the band diagram illustrated in Fig. 4.9. The diagram also shows

$$V_n = E_{\text{gb}} - V_{\text{appl}} - \epsilon_{\text{fb}} - \epsilon_{\text{fe}}$$

$$V_p = E_{\text{ge}} - V_{\text{appl}} - \epsilon_{\text{fb}} - \epsilon_{\text{fe}} \tag{19}$$

[†] (Editor Note) The symbol V in eqs (18) and (19) is defined as "potential energy" by authors in order to conform with Fig. 4.9.

where E_{gb} and E_{ge} are energy bandgaps in the base and emitter, respectively, ϵ_{fb} and ϵ_{fe} are the separations of the majority carrier quasi-Fermi levels from the band edge in these regions, and V_{appl}/q is the applied voltage. Combining the above equations leads to

$$\beta = \frac{v(n)}{v(p)} \frac{N_{DE}}{N_{AB}} \exp \frac{-(E_{gb} - E_{ge})}{kT} \tag{20}$$

In a typical case, $v(n) = 5v(p)$. If it is desired to obtain β as high as 100, and to have a base doping level 500 times that of the emitter, the bandgap difference between base and emitter must be $9.2kT$ or 240 meV at 27°C. This can be obtained by the inclusion of 20 mol % of Al in the emitter.

To obtain Eqs. (18–20), we assumed that quasi-Fermi levels were constant across the space charge region. In HBTs, this is a more drastic assumption than in the case of homojunction devices, for which it almost invariably holds. In HBTs, as noted above, a potential barrier may be formed in the conduction band if the Al concentration is not sufficiently graded. If the potential barrier is high enough, the flow of electrons across it will be restricted and may become the limiting factor in establishing overall collector current. Figure 4.10 illustrates the band diagram for this case, showing that transport across the base and transport across the heterojunction barrier may each contribute to limiting the electron current.

Analytical expressions for the general case may be written (based on integrals that must be determined for any particular doping distribution and applied bias) [16]. In most practical situations, however, numerical analysis of the above device equations may be used for device design. Solution of the drift–diffusion equations in one dimension is straightforward and can be accomplished by minor modification of programs used for Si device simulation [17, 18]. For a representative HBT, the calculated distributions of

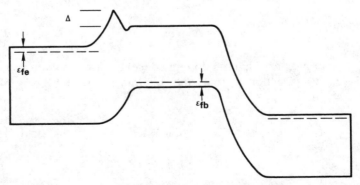

Figure 4.10   Band diagram of HBT under forward bias, in the case the emitter–base potential barrier contributes to limiting electron current.

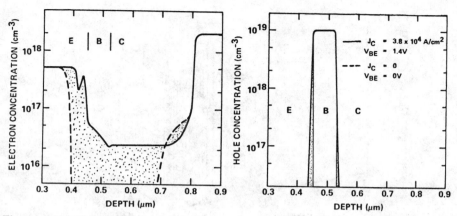

Figure 4.11   Calculated distribution of electrons and holes within a representative HBT under forward bias.

electrons and holes are shown in Fig. 4.11 for the equilibrium case and for a forward-biased emitter–base junction. The electron distribution in forward bias shows, for example, a constant density of carriers traversing the base–collector space charge layer (with constant saturated velocity); a decreasing electron concentration across the base, corresponding to diffusive flow; and considerable added carriers in the emitter–base space charge layer, with a complex profile governed by the Al concentration profile assumed for the device. The additional electrons in the transistor under forward bias correspond to stored charge that must be supplied by the input terminal (base). From Eq. (2), the added charge may be related to the

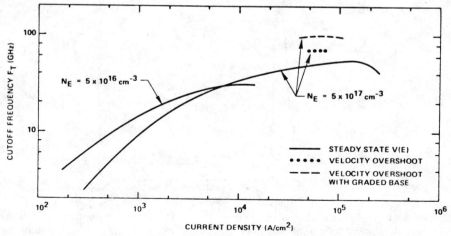

Figure 4.12   Calculated f_t versus current density for a representative HBT.

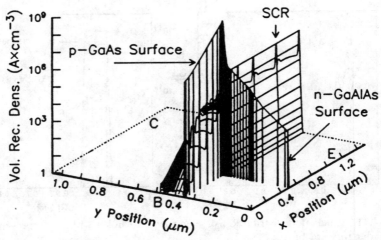

Figure 4.13 Calculated electron–hole recombination density in the vicinity of the emitter and base contacts of an HBT with exposed GaAs base, obtained from two-dimensional modeling. (After Tiwari [19].)

cutoff frequency f_t of the device. By numerically integrating the carrier concentrations, estimates of f_t (as well as junction capacitances) may be made. Figure 4.12 shows f_t calculated for a representative HBT, as a function of collector current density, using the drift–diffusion model.

To understand more fully the edge effects in transistors, simulations have

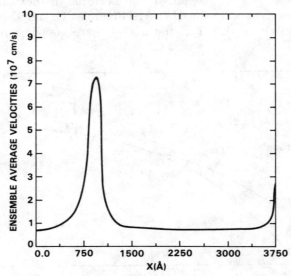

Figure 4.14 Calculated ensemble average electron velocity versus position within the base–collector region of an HBT, obtained from Monte Carlo simulations. (After Maziar et al. [21].)

been carried out using the drift–diffusion equations in two dimensions [19, 20]. Figure 4.13 shows, for example, the distribution of electrons in the vicinity of the base contact for a transistor of the type shown in Fig. 4.2, for which the surface of the base is exposed. Due to Fermi-level pinning at free surfaces, electrons in the vicinity of the exposed base are drawn to the surface, where they recombine. As detailed in Section 4.4, suppression of this edge recombination effect is necessary to have high current gain in small devices.

A number of studies have gone beyond the drift–diffusion equations to understand the effect of electron velocity overshoot and transport over short distances and times, using the Monte Carlo technique [13, 14, 21, 22]. In this technique, the paths of individual electrons through the device are simulated by taking into account the scattering probability for all important interactions. Calculated mean electron velocity at different positions within an HBT is shown in Fig. 4.14.

4.3 HBT FABRICATION

4.3.1 Epitaxial Growth

High-performance HBTs with realistic prospects for application within ICs have become possible only after the development of advanced epitaxial technology capable of the fabrication of sophisticated layer structures, such as that of Fig. 4.3, with adequate uniformity and control over composition, doping, thickness, and defect density in the different layers. The availability of the epitaxial layer structures is one of the most important factors that will govern the extent of HBT application. It is therefore appropriate to consider here (albeit briefly and incompletely) the principal HBT layer growth techniques, MBE and MOCVD.

In the MBE growth technique, gallium, arsenic molecules (As_2 or As_4), aluminum, and various dopant atoms are evaporated from heated crucibles within an ultrahigh vacuum chamber [22]. The resultant atomic or molecular beams come together at the site of a heated substrate of GaAs, where they react to form layers of GaAs or GaAlAs. Mechanical shutters are used to interrupt the beams as needed to produce variations in doping and composition. Growth rates are typically around 1 μm/h (approximately one monolayer per second). Since shutters can turn beams on or off in a fraction of a second, excellent abruptness of interfaces can be achieved. Using Si as n-type dopant and Be as p-type dopant, a large range of doping concentrations is possible (typically, 10^{14}–10^{19} cm^{-3}). The substrate is heated to a temperature of 550–700°C during growth (to provide sufficient energy to adsorbed species to allow them to migrate to proper, low-energy bonding sites on the growing surface). A representative MBE system is shown in Fig. 4.15. Various pieces of analytical equipment are attached to the chambers,

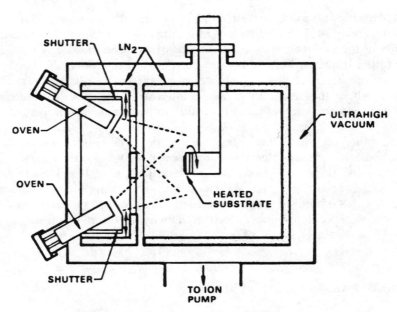

Figure 4.15   Representative MBE system.

such as reflection high-energy electron diffraction (RHEED), mass spec-
trometers, and Auger systems to provide in-situ characterization of the
growing layers. The RHEED technique has proved to be particularly
valuable, and can provide a precise measure of the growth rate (accurate to
1%). By analyzing growth rates for different incident beams (e.g., Ga, Al,
or In), the alloy composition of GaAlAs or GaInAs can be calibrated.
RHEED can also be used to monitor the relation between the Group III
and the Group V flux. Dopant fluxes must be calibrated by measurements
outside the chamber, but calibrated dopant oven settings drift only very
slowly, making frequent recalibration unnecessary. Moreover, the dopant
calibrations are not strongly dependent on other parameters such as substrate temperature, alloy composition, III/V ratio, etc.

As a result of the flexibility of MBE and its convenience for growth of
complicated structures, it has been widely applied in the growth of HBT
layers. The principal limitations of MBE at present in this application are
low throughput (up to 2 h growth time for a single wafer) and high capital
cost. Also, with MBE growth, there are frequent morphological defects on
the epitaxial layer (known as oval defects) which can in some cases limit
device yield.

Metallorganic chemical vapor deposition (MOCVD), also known as
organometallic vapor phase epitaxy (OMVPE), is a technique in which
GaAs or GaAlAs is produced by the reaction of arsine (AsH_3) with
trimethyl gallium ($Ga(CH_3)_3$) or trimethyl aluminum ($Al(CH_3)_3$) molecules

present in a hydrogen gas stream [23]. The reaction takes place on a GaAs substrate heated to 650–800°C by rf induction heating, within a cold-wall chamber. The gas pressure within the growth chamber may be atmospheric, or lower (typically 0.1 atm). To dope the layers appropriately, Zn (in the form of diethyl zinc $Zn(C_2H_5)_2$) is often used for p material, and silane (SiH_4) or hydrogen selenide (H_2Se) for n material. To vary the composition or doping, gas flows from the various sources are modulated. To achieve the most abrupt variations, it is necessary to have high gas velocities, small gas manifolds, appropriately designed chambers, and to switch gas streams between growth and vent lines. By incorporating these features, excellent abruptness may be obtained, down to changes in Al composition over 5–10 Å, as evidenced in quantum well structures. Figure 4.16 illustrates a representative MOCVD apparatus.

In MOCVD, layer growth rates are typically higher than in MBE; frequently, 5–25 μm/h are used. With proper attention to reactor design, good uniformity over large areas (multiple wafers) may also be achieved. These features improve the throughput over that of MBE. The principal disadvantages of MOCVD are associated with the difficulty of obtaining high-purity source gases; the need for extensive safety precautions in the handling of the AsH_3 and other gases; the relative difficulty of calibration, due to the considerable number of factors that affect doping and composition; and the fact that p-doping has been limited so far to below $10^{19} \, cm^{-3}$ (while higher values are desirable for use in the base of HBT structures). Various laboratories have demonstrated high-performance MOCVD-grown HBTs [24, 25].

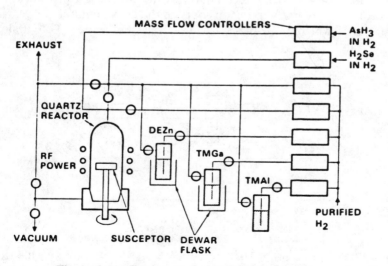

Figure 4.16 Representative MOCVD apparatus.

4.3.2 Device Processing

With an epitaxial structure such as that of Fig. 4.3, the principal goals of device processing are to make electrical contact to the different layers (emitter, base, and collector), and to isolate devices electrically by eliminating or rendering inactive the epitaxial layers outside of the transistors. Various techniques can be used to achieve each objective.

Contact to the emitter is made from the top surface of the wafer in the transistor of Fig. 4.2. Such a contact can be directly made with, for example, alloyed layers of Au, Ge, and Ni, as is usually done in GaAs MESFET technology. In recent work, epitaxial structures have ended with a thin layer of InAs or InGaAs on the wafer surface [26]. Since the Fermi level is pinned within the conduction band at the surface of these materials, making them degenerately n-type, excellent ohmic contacts may be obtained with arbitrary contact metals. This allows making contacts that will withstand higher temperature cycling than AuGeNi alloys and that have lower specific contact resistance (down to 2×10^{-7} Ω-cm^2).

Contact to the base can be made by etching away the top layers of the transistor structure and depositing metal films directly on the base layer. To accomplish this, however, it is necessary to control the etch depth precisely, because the base layer itself is very thin. Etchants whose etch rate is sensitive to Al fraction may be used for this purpose [27, 28]. Excellent uniformity of etching over 3 in. diameter wafers has been demonstrated using this approach.

An alternative method of making contact to the base is to convert the n-type top layers of the structure to p-type with the use of ion implantation [29] or diffusion [30] of acceptors. A representative structure of this type is pictured in Fig. 4.17. This technique has the advantage of increasing the lateral conductivity of the p regions outside the active emitter (or "extrinsic" base region), which contributes to lower base resistance. Potentially, the

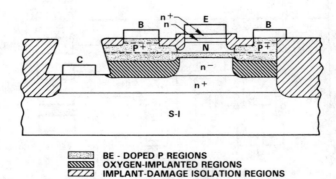

BE - DOPED P REGIONS
OXYGEN-IMPLANTED REGIONS
IMPLANT-DAMAGE ISOLATION REGIONS

Figure 4.17   Structure of an HBT fabricated by p+ implantation in the extrinsic base region.

implanted or diffused p-doping concentration can be made very high, allowing lower contact resistance to be obtained in the base contacts. The potential disadvantage of this technique is the requirement of thermally cycling the material to remove implant damage or to diffuse in the acceptors. During this heating cycle, the grown-in dopants, particularly the base dopant, may diffuse excessively. Rapid thermal annealing techniques are generally used for these heating cycles to minimize this problem.

If p-type implants or diffusions are made from the wafer surface, care must be taken to eliminate the p-n junction formed between the topmost GaAs or GaInAs n emitter contact regions and the adjacent p-converted material. The built-in potential for these junctions is lower than that of the intrinsic transistor base–emitter junction, and significant current might flow in the lateral parasitic junctions under forward bias. This current contributes directly to base current and reduces the transistor current gain. To remove the side junctions, the lower-bandgap top layers may be etched off in the region of the extrinsic base. The sidewall p-n junctions remaining in the AlGaAs are not a problem for base current, because the built-in potential in this wide-gap material is greater than that of the intrinsic transistor; the junctions do not turn on during transistor operation.

Contact to the collector region of the device is usually made by etching through the uppermost epitaxial layers, down to the n^+ subcollector region. AuGeNi films can then be deposited and alloyed to produce low contact resistance. The principal difficulty of this procedure is associated with the considerable etch depth required (up to $1\ \mu m$). To minimize problems with step coverage of subsequent metal layers connecting different device regions, it is possible to use orientation sensitive etching of the GaAs to ensure that the sidewalls of the etched steps have gradual slopes (corresponding to GaAs (111)B planes). Alternatively, etching vias with vertical sidewalls by reactive ion etching (RIE), and filling the resulting openings with metal has been demonstrated [31].

Isolation between devices can be achieved by using ion-implantation-induced lattice damage to make the epitaxial semi-insulating outside the active transistor area. Extensive work has been reported with boron, oxygen, and hydrogen implants to increase the resistivity of both n-type and p-type layers to the order of $10^7\ \Omega$-cm with appropriate doses. The increase in resistivity can be stable to at least 500°C. The detailed mechanism of the resistivity increase is not known; part of the effect is no doubt associated with deep levels due to lattice defects, while a portion of it may be related to the formation of complexes with dopants and to the gettering of impurities.

An alternative technique on isolating devices is by etching away the epitaxial layers in suitable regions. Individual transistors can be left as mesas on the semi-insulating GaAs substrate. Alternatively, narrow etched gaps at the device edges can be fabricated and backfilled with dielectric material to achieve a planar surface (trench isolation) [32].

With transistors fabricated in this fashion, ICs may be completed in a

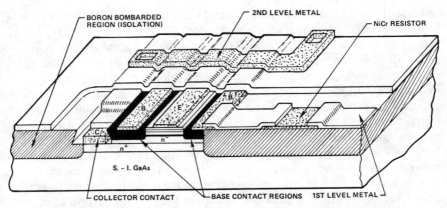

Figure 4.18   Representative cross section of an HBT-based IC.

straightforward manner using techniques employed widely in GaAs MES-FET circuit fabrication. These techniques differ from those used with Si in the extensive use of Au-based metallization rather than Al-based metallization to achieve compatibility with the AuGeNi contacts. Figure 4.18 shows a representative HBT IC cross section indicating metal-film resistors, and two levels of interconnect metal, separated by interlevel dielectric of SiO_2, Si_3N_4, or polyimide.

The techniques described are employed with the layer structure of Fig. 4.3 to fabricate HBTs that approximately correspond to the device of Fig. 4.2. However, a variety of HBT approaches that depart significantly from this structure have been reported. In the heterojunction integrated injection logic (HI^2L) structure [33], the collector region is on the wafer surface and the emitter region is buried beneath it, as shown in Fig. 4.19. All emitters are electrically connected to the n^+-GaAs substrate, simplifying the interconnection problem and contacting problem (but limiting the logic approach

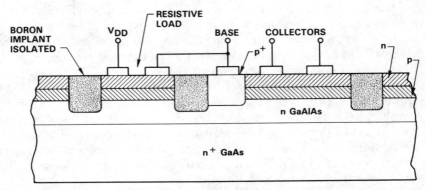

Figure 4.19   Structure developed for the implementation of heterojunction integrated injection logic (HI^2L). (After Yuan et al. [67].)

to I^2L). The base doping is introduced by implantation from the wafer surface.

In another approach, collector, base, and emitter dopings are all introduced by ion implantation [34]. This technique leads to considerable flexibility in combining devices of different kinds (e.g., MESFETs and HBTs) to achieve specific circuit goals. It also simplifies isolation and leads to easily formed collector contacts. The limited flexibility and control over the base doping in this and the last technique, however, limit the transistor speed that may be achieved.

An important goal during transistor fabrication is to minimize parasitic elements that degrade device speed. One of the most significant of these parasitic elements is extrinsic base–collector capacitance, the space charge layer capacitance between the base and the collector in the regions outside the active (emitter) area of the device. Although the extrinsic base region serves only to support the base contact, its area may be large compared with the active device area, and its associated capacitance may seriously degrade the transistor speed. In GaAs-based transistors, it has been possible to reduce the capacitance of this region by employing implants of oxygen or protons into the collector region under the base contact [35, 36]. The implants compensate the n-collector material and widen the effective spacing between base and collector layers, decreasing the capacitance per unit area, as shown in Fig. 4.20. The implants can be done from the wafer

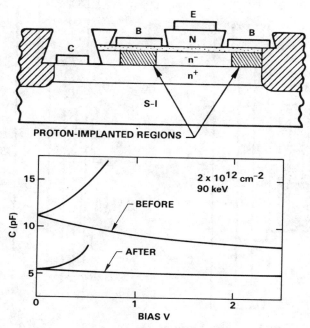

Figure 4.20 Capacitance versus bias voltage for base–collector junction diodes with and without implantation of protons into the n-collector region.

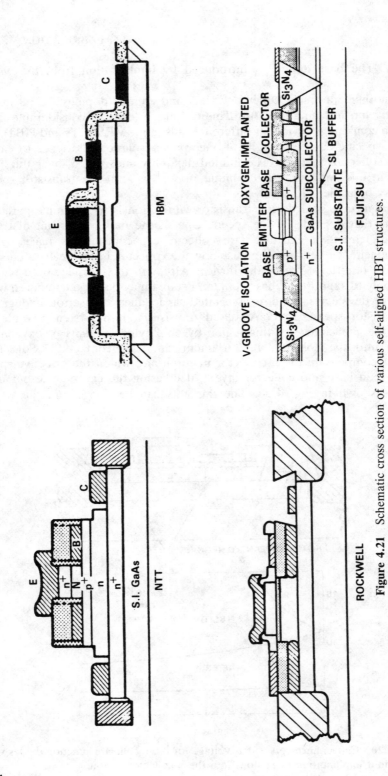

Figure 4.21 Schematic cross section of various self-aligned HBT structures.

surface, through the base layer, since it is possible to choose an implant dose that will compensate the lightly n-type collector drift region, yet not affect the conductivity of the heavily doped base.

4.3.3 Self-alignment Techniques

To decrease the influence of the extrinsic base capacitance further, and to minimize the base resistance contributed by this region, it is desirable to use self-aligned processing techniques. These techniques allow the definition of various device features with the same photoresist pattern, thereby avoiding the relatively wide spaces that must otherwise be included to allow for photolithography alignment tolerances. For example, with presently available projection lithography production systems, "3 sigma" conservative design spacings between regions defined on different mask layers are on the order of 0.5–0.8 μm. If, by advanced processing techniques, the different regions can be defined with the same resist, their spacing may be reduced to below 0.25 μm. In HBTs, it is desirable to self-align the edge of the base contact to the edge of the active emitter, the edge of the emitter contact to the active emitter edge, and in some cases, the full width of the extrinsic base to the active emitter. Development of appropriate self-aligned processing techniques is a principal focus of HBT process research, particularly since the high-speed HBT performance is limited in most cases by parasitic elements outside of the intrinsic device. Moreover, the parasitic elements become more severe as the emitter size is reduced unless the extrinsic areas can be scaled also.

Numerous self-alignment approaches have been reported for HBT processing [32, 37–40]. Figure 4.21 illustrates a number of the structures described. Among the process innovations used to accomplish self-alignment are (1) dielectric sidewall spacers defined by reactive ion etching; (2) self-aligned base implants, annealed after (refractory) emitter contacts have been defined; and (3) self-aligned dielectrics defined by lift-off. Further developments in this area are to be expected.

4.4 HBT CHARACTERISTICS

4.4.1 Dc Characteristics

Common-emitter curves of collector current I_c versus collector–emitter voltage V_{ce} for various values of base current I_b are shown in Fig. 4.22[†] for a representative HBT. Figure 4.23 shows the corresponding behavior of I_c and I_b versus base–emitter voltage V_{be} for a particular base–collector bias ($V_{bc} = 0$) on a logarithmic scale (known as the Gummel plot). Both sets of

---

[†](Editor's Note) For uniformity, authors of this chapter had adopted, in most cases, lower-case subscripts for symbols in the text. They are inter-changeable with upper-case subscripts as otherwise shown in the figures.

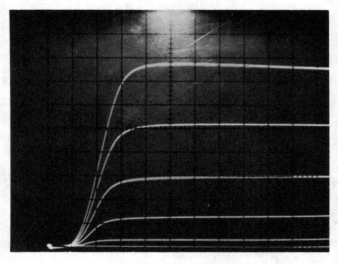

I_C: 0.5 mA/DIV

V_CE: 0.2 V/DIV

I_B: 10 μA/DIV

Figure 4.22 Collector current versus collector–emitter voltage for a representative HBT, with stepped base current. Emitter dimensions are $2 \times 3.5\ \mu m$.

curves are qualitatively similar to those obtained with homojunction bipolar transistors.

For devices of the type represented in Fig. 4.23, with a graded composition base–emitter junction, the forward current I_c varies exponentially with V_{be} with an ideality factor close to 1.0, up to a limit established by series resistance. The magnitude of the current is in close agreement with the result given by Kroemer for heterojunctions without conduction band energy barriers in the junction regions [41]:

$$J_c = \frac{qD_n n_{ib}^2}{p_b w_b} \exp\left(\frac{qV_{be}}{kT}\right) \tag{21}$$

Here D_n, p_b, w_b, and n_{ib} are the electron diffusion coefficient, hole concentration, thickness, and intrinsic carrier density in the base. A comparison between experimental I–V curves and other parameters in Eq. (21) leads to a determination of the quantity n_{ib} appropriate to the heavily doped p-type base regions. The effective value of n_i is slightly higher than in undoped material as a result of bandgap narrowing effects.

In a significant number of cases, J_c is lower than indicated by Eq. (21) by up to several orders of magnitude, and the ideality factor is larger than unity (up to 1.5). This corresponds to the case described in Section 4.2.4, in which

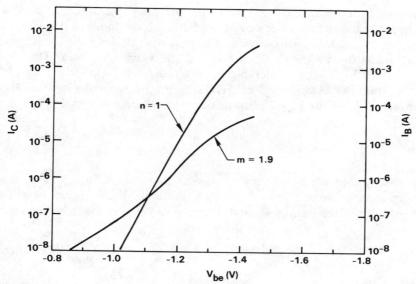

Figure 4.23   Collector current and base current versus base–emitter voltage at zero base–collector voltage for a representative HBT (Gummel plot).

a potential barrier in the conduction band near the emitter–base junction limits current flow. Analyses of the magnitude of J_c for such a case are presented in Ref. 16, for example.

The maximum J_c attainable in HBTs is typically very high, corresponding to a current density of over 10^5 A/cm^{-2}. In Si transistors, the current density is frequently limited by several effects: the onset of high-injection effects in the base, and the Kirk effect. In the first case, the injected electron concentration in the base becomes comparable to the base doping, which leads to a decrease in current gain. This condition is not attainable in HBTs, with their considerably higher base doping. In the Kirk effect, the electron concentration in the collector depletion layer becomes comparable to the donor doping level, causing a change in the field distribution at the base–collector junction, and the injection of a significant hole concentration into the collector (base "push-out") [4]. In HBTs, this effect is reduced by high electron velocity, and by relatively high doping of the collector drift region.

It can be expected from Eq. (21) that the value of V_{be} required to reach a given current J_c is highly reproducible (so that the transistor "turns on" at a precisely established input voltage). The required voltage to obtain collector current density J_c is given from Eq. (21) by

$$V_{be} = \frac{E_{gb}}{q} - \frac{kT}{q} \ln \frac{qD_n N_c N_v}{J_c p_b w_b} \tag{22}$$

expressed in terms of the bandgap of the base E_{gb} and the conduction and valence band densities of states N_c and N_v. V_{be} is dominated by the first term; the typical value of the second term is near 0.25 V, and it changes by only 2.6 mV for a 10% change of any of the variables within. The reproducibility of V_{be} is of considerable importance for IC applications.

The transconductance g_m of HBTs is given implicitly by the above behavior of I_c versus V_{be}. The intrinsic transconductance g_{m0} has the value

$$g_{m0} = \frac{qI_c}{kT} \tag{23}$$

as is well known for homojunction bipolar transistors. The extrinsic transconductance, measured from the device terminals, is influenced by emitter and base series resistances R_e and R_b, respectively, and is given by

$$\frac{1}{g_m} = \frac{1}{g_{m0}} + \frac{R_e}{\alpha} + \frac{R_b}{\beta} \tag{24}$$

Here, β is the incremental current gain and $\alpha = \beta/(\beta + 1)$ is the current transport factor. Experimentally, g_m can reach very high values on an absolute and a per-unit area basis, particularly in light of the high output current density attainable. Extrinsic transconductances as high as 10 mS/μm^2 have been demonstrated, limited essentially by emitter series resistance $(0.5–5 \times 10^{-6} \ \Omega\text{-cm}^{-2})$. Transconductance per unit device length is typically used as a figure of merit of FET performance, with values up to 400 mS/mm in exceptional cases; the corresponding figure of transconductance per unit emitter length in HBTs is 10,000 mS/mm.

As can be seen in Fig. 4.22, the output conductance (i.e., the change in I_c for changing V_{cb} at fixed I_b) is remarkably low for HBTs. The principal mechanism for V_{cb} to affect the output current is modulation of the base width w_b due to the penetration of the base–collector space charge region into the base. For a given increment of voltage dV_{cb}, the charge depleted from the base is

$$dQ = C_{bc}dV_{cb} = \frac{\epsilon}{w_c} \ dV_{cb} \tag{25}$$

where w_c is the depletion region thickness and C_{bc} its capacitance per unit area. From this relation and Eq. (21) above, it may be seen that

$$\frac{dI_c}{I_c} = -\frac{dw_b}{w_b} = \frac{C_{bc}}{qp_bw_b} \ dV_{cb} \tag{26}$$

The output conductance in bipolar transistors is typically expressed in terms of the Early voltage V_A with

$$V_A = I_c \left(\frac{dI_c}{dV_{cb}} \right)^{-1} = \frac{q p_b w_b}{C_{bc}} \tag{27}$$

For typical HBTs, this is in the range 100–400 V, up to 20 times higher than that obtained in Si bipolar devices, as expected due to the higher base doping employed in HBTs.

Another feature of the I–V curves evident in Fig. 4.22 is the fact that positive values of I_c are attained only after a nonzero value of V_{ce} is applied (0.2 V in the figure shown). This voltage, commonly termed "offset" voltage, corresponds in bipolar nomenclature to V_{cesat}, the output voltage measured under saturation conditions with $I_c = 0$. With a forward-biased base–emitter junction, the condition $I_c = 0$ can only be established if the forward current predicted by Eq. (21) is canceled by a current flowing out of the collector due to a forward-biased base–collector junction. V_{cesat} is, to lowest order, the difference between the values of V_{be} and V_{bc} needed to maintain such a current. In HBTs, non-negligible values of V_{cesat} arise because the effective turn-on voltage of the base–emitter heterojunction is typically greater than that of the base–collector homojunction. For the latter junction, in addition to electron flow from collector to base, there is a significant flow of holes injected from the base into the collector. For representative doping profiles, the hole injection component is on the order of 250 times greater than the electron current; a hole-diffusion-current-dominated offset voltage of $\ln(250)kT/q = 145$ mV is thus predicted. Additional components of V_{cesat} may result from the fact that base–collector junction area is often much larger than the base–emitter junction area; from excess recombination currents at the base–collector space charge layer; and from resistive voltage drops in the base layer between the contacts and the intrinsic device.

With increasing temperature, the value of I_c increases for a fixed V_{be}, in accord with Eq. (21). Equivalently, the value of V_{be} needed to maintain a given I_c decreases. The rate of decrease is approximately the same as in Si bipolar transistors, 1.0–1.5 mV/°C.

The maximum collector–emitter voltage that can be applied to HBTs is limited by breakdown resulting from avalanche carrier multiplication by impact ionization at the base–collector junction. Calculations show that in experimental HBTs, the breakdown voltages attained are close to those predicted on the basis of measured impact ionizations coefficients for electrons and holes in GaAs in a simple, one-dimensional approximation. Breakdown fields are higher than those obtained in Si bipolar transistors by 20–30%, due to the higher bandgap of GaAs. As in the case of Si devices, the measured breakdown voltage BV_{ceo}, measured between collector and emitter with the base open-circuited, is smaller than the corresponding

BV_{ces}, measured with the base short-circuited to the emitter; the difference arises because with an open-circuited base, impact-ionization-generated carriers flowing into the base add to the base current and are amplified through the gain mechanism of the HBT [42].

4.4.2 Current Gain

To understand the magnitude of the HBT current gain, it is necessary to consider the base current flow in detail. As depicted in Fig. 4.24, there are a number of contributions to the base current, corresponding to electron-hole recombination in the various device regions.

As discussed in Section 4.2.4, the contribution I_{b1} of holes injected into the quasi-neutral emitter may be reduced to any desired level (even in the presence of heavy doping in the base region) by the inclusion of a sufficient amount of Al in the emitter, increasing the barrier for hole flow across the base–emitter junction.

The component I_{b2} is due to recombination within the quasi-neutral base region. For this case, I_{b2}/I_c may be estimated to be the ratio τ_b/τ_{rec}, where τ_b is the average residence time of injected electrons within the base and τ_{rec} is their average recombination time in the base. The recombination lifetime for minority carriers in GaAs is smaller than that typically obtained in Si, and it is further reduced by the heavy base doping. For example, for a base doping of 10^{19} cm^{-3}, the radiative recombination lifetime is expected to be only 1 ns, and nonradiative processes further decrease τ_{rec}. Nonetheless, this component of base current is frequently not large, because of the very short transit time τ_b, on the order of 1 ps.

Both I_{b1} and I_{b2} vary exponentially with V_{be}, with an ideality factor of unity (matching the ideality factor for I_c). In devices whose base current is

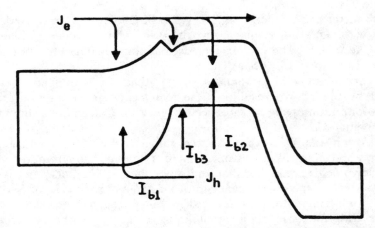

Figure 4.24 Schematic illustration of various contributions to base current in HBTs.

dominated by these contributions, current gain is independent of collector current density.

The current I_{b3} corresponds to recombination in the emitter–base space charge region. The net recombination rate U is given approximately by

$$U = N\sigma v \, \frac{pn}{p + n} \tag{28}$$

where N is the density of recombination centers, σ the cross section for capture of carriers (taken to be equal for electrons and holes), and v an average carrier thermal velocity. In the case of homojunction transistors, pn is constant across the emitter–base junction, given by

$$pn = n_i^2 \exp\left(\frac{qV_{be}}{kT}\right) \tag{29}$$

As a result, U has a strong maximum at the plane where $p = n$. The net recombination current is

$$I_{b3} = qAN\sigma v w_r n_i \exp\left(\frac{qV_{be}}{2kT}\right) \tag{30}$$

where w_r is an effective width of the recombination region (on the order of $kT/q\mathscr{E}_m$, with $\mathscr{E}_m$ the junction electric field), and A is the area of the junction [43].

In the case of heterojunction devices, pn is no longer constant across the junction. The intrinsic carrier density n_i (which depends strongly on the semiconductor bandgap) varies between base and emitter. Some variation may also occur in the electron quasi-Fermi level if a potential barrier is present near the junction. Under these circumstances, U again will typically reach a sharp maximum, not where $p = n$ but where $n/p = \mathscr{E}_n/\mathscr{E}_p$. Here, $\mathscr{E}_n$ and $\mathscr{E}_p$ are the effective drift fields (including both electrostatic fields and conduction and valence band gradients) for electrons and holes. A relation for recombination current similar to Eq. (30) arises, with an ideality factor only approximately equal to two, and with an intrinsic carrier concentration n_{ieff} that corresponds to a bandgap greater than that of the base but smaller than that of the emitter. The detailed value of the recombination current depends on the details of the composition grading between base and emitter, and the relative placement of the heterojunction and the p-n (doping) junction. It is possible to decrease I_{b3} by causing the p-n junction to occur in relatively wide-bandgap material, thereby decreasing significantly the value of n_{ieff} [44].

A final contribution to the base current, I_{b4}, is given by recombination at the edges of the emitter. This component is stronger in GaAs-based devices than in Si bipolar transistors, because the surface recombination velocity for most GaAs surfaces (or interfaces) is very high, on the order of 10^6 cm/s. Since the thermal equilibrium effective velocity of incidence of electrons on free surfaces is 1.5×10^7 cm/s, this corresponds to effective capture by the surface of 10% of the incident electrons. The centers responsible for the rapid recombination are not known, but are presumed to be related to those causing Fermi-level pinning near midgap. In HBTs, recombination can occur at the surface of regions of the base adjoining the emitter; in this case, electrons diffuse from the p-n junction to the surface of the base. Recombination may also occur at the surface of the base–emitter space charge layer.

An approximate picture of the magnitude of periphery-related base current effects may be obtained by assuming that half the electrons injected into the base within an electron diffusion length of the emitter edge are lost to surface recombination. For a diffusion length of 0.1 μm, then, the maximum current gain of a device with 1 μm emitter width is 9–10. Quantitative estimation of the electron distribution and of the magnitude of recombination may be done with two-dimensional modeling, as has already been shown in Fig. 4.13.

Emitter edge components of base current may be recognized experimentally by measuring I_b and I_c at a fixed V_{be} for a series of devices of different dimensions. As shown in Fig. 4.25, I_c varies linearly with emitter width w_e, as expected (for the emitter length L_e held constant). The corresponding I_b values also fit a straight line, but the data extraplate to a nonzero intercept

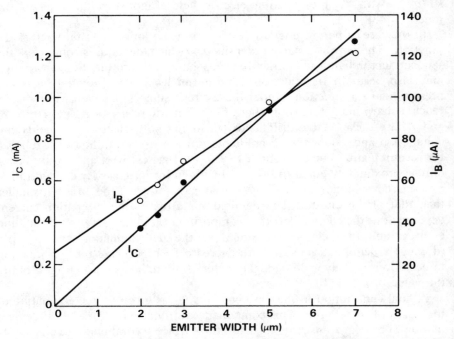

Figure 4.25 Experimental variation of base current and collector current with emitter width, for a case where emitter edge recombination is significant.

at $w_e = 0$. The intercept denotes the base current associated with a periphery of $2L_e$. On this basis, emitter edge recombination currents characterized by ideality factors of both one and two have been obtained.

A number of techniques have been developed to decrease the edge recombination component. One strategy is to minimize the diffusion of electrons to the surface of the base by incorporating high quasi-electric fields in the base by composition grading [45]. Another approach is to reduce the surface recombination velocity S by the application of a suitable passivation layer. A particularly convenient choice of passivation material is AlGaAs, since it has been shown that the resultant value of S is on the order of 10^3 cm/s. To prevent the AlGaAs from acting as a mere extension of the emitter, it may be converted to p-type, by implantation or diffusion. Alternatively, it can be thinned to the point where it is pinched off due to surface depletion [46]. A novel passivation approach is to treat the surface with spun-on layers of $Na_2S_{0.9}H_2O$ [47].

As temperature is increased, it is observed experimentally that I_b for a given V_{be} increases. In most cases, however, the rate of increase of I_b is lower than the corresponding rate of increase of I_c, so that the current gain decreases with increasing temperature. This phenomenon gives rise to differential negative output conductance, as seen in many measured HBT common-emitter I–V characteristics with stepped I_b (although the effect does not persist at high frequency). This phenomenon is opposite to what is commonly observed in Si devices. In that case, due to bandgap narrowing from heavy doping, the emitter bandgap is lower than that of the base. In the HBT, by varying the semiconductor composition, the bandgap is made narrower in the base than in the emitter.

4.4.3 High-frequency Characteristics

To study the high-frequency behavior of individual devices, a common procedure is to measure S-parameters of transistors with an automatic network analyzer. To eliminate complications to the data analysis from transistor fixturing, suitable coplanar transmission line high-frequency probes are used (such as those produced by Cascade Microtech). Devices are provided with large output pads for the common terminal to decrease inductance and small bond pads for the signal terminals to decrease capacitance. Figure 4.26 illustrates the layout of a high-frequency HBT.

From the S-parameters, the transistor gain versus frequency may be calculated, using expressions well known for two-port linear networks [48]. Figure 4.27 shows the behavior of current gain h_{21}; maximum stable power gain, MSG; maximum available power gain, MAG; and unilateral power gain, U, versus frequency calculated for a representative HBT. MAG is well defined only in the region in which the device is unconditionally stable (frequencies above 21 GHz in the example shown), and so it is plotted only

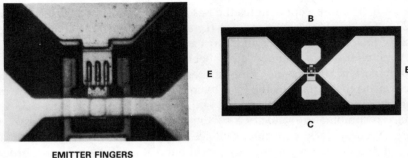

EMITTER FINGERS
1.2 μm x 9 μm

Figure 4.26 Photomicrograph of a fabricated high-frequency discrete HBT, connected for common-emitter operation.

for a portion of the abscissa; MSG is shown in the remaining portions of the curve.

The cutoff frequency f_t corresponds to the frequency for which the current gain drops to unity. Since available data commonly extend only to 26 GHz, for high f_t devices it is necessary to determine f_t by extrapolation.

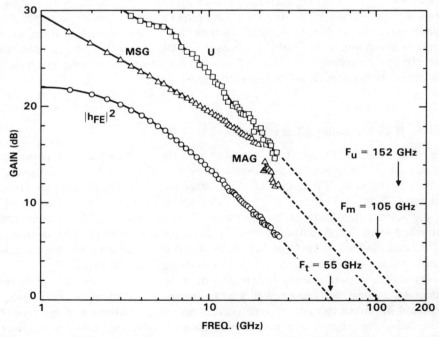

Figure 4.27 Frequency dependence of incremental short-circuit current gain, h_{21}; maximum stable power gain, MSG; maximum available power gain, MAG; and unilateral power gain U, obtained from S-parameter measurements.

This extrapolation is commonly done by assuming a 6 dB/octave roll-off of h_{21} with frequency, as would be obtained if the transistor current transfer function were dominated by a single pole. The device of Fig. 4.27 is thus inferred to have an f_t of 55 GHz.

Values of f_t up to 105 GHz have been obtained with HBTs [44]. HBT f_t's are considerably higher than those obtainable with Si bipolar transistors (for which 5–10 GHz is commonplace and 25 GHz is the maximum reported); they are also higher than those obtained with GaAs MESFETs (which achieve 15–25 GHz with 1 μm gate lengths, and have reached 60–80 GHz with 0.2 μm gate lengths). The reported HBT results were achieved with straightforward optical lithography, employing emitter widths greater than 1 μm.

The transistor f_t, as described in Sections 4.1.2 and 4.2.4, may be related to the emitter–collector transit time of electrons τ_{ec}. Within the charge control model of the device, τ_{ec} is equal to the ratio of stored hole charge Q_b to the output current I_c. To understand in detail the magnitude of τ_{ec}, contributions from different device regions must be considered. The different components are typically expressed in the relation [4]:

$$\tau_{ec} = \tau_e + \tau_b + \tau_{sclc} + \tau_c \tag{31}$$

Here, τ_e corresponds to hole charge stored at the emitter edge of the base, within the emitter–base space charge layer, or within the emitter itself. This lasts contribution is negligible for HBTs due to the wide emitter bandgap (although it is significant for homojunction devices). For the most part, the storage of holes corresponds to an expansion or contraction of the space charge region width, and the magnitude of the associated charge is the value inferred from the base–emitter space charge layer capacitance C_{be} times the change in V_{be} required to change the output current (g_m). Thus, the value of τ_e is

$$\tau_e = C_{be} \frac{kT}{qI_c} \tag{32}$$

According to this relation, to minimize τ_e, the transistor should be operated at the maximum possible current I_c. Additionally, C_{be} can be decreased by decreasing the emitter doping. Such a strategy can be effective only to the point where the emitter carrier concentration can no longer support the required current density due to electron velocity saturation. Furthermore, detailed consideration of τ_e in numerical calculations shows that the charge associated with electrons within the space charge layer is a contribution that cannot be neglected; this contribution depends on the details of composition grading near the junction.

The contribution τ_b corresponds to the charge of electrons traversing the base. According to drift–diffusion modeling, τ_b is given by

$$\tau_b = \frac{w_b^2}{2D_n} + \frac{w_b}{v_{edge}} \tag{33}$$

Here v_{edge} is the net velocity of carriers at the collector edge of the base. An upper limit to v_{edge} is provided by the effusion velocity v_e, the effective velocity of carriers incident upon a plane in thermal equilibrium (with $v_e = [kT/(2\pi m^*)]^{1/2} = 2 \times 10^7$ cm/s). Improved estimates of τ_b are obtainable from Monte Carlo calculations [21].

The contribution τ_{sclc} corresponds to the base charge induced by the electrons that are crossing the base–collector space charge layer. Under the simplifying assumption that the electrons travel at a constant, saturated drift velocity v_{sat}, the density of electrons in the space charge region is $n_c = J_c/qv_{sat}$. The corresponding hole charge in the base is $Q_b = \frac{1}{2}n_c w_c$, where a factor of $\frac{1}{2}$ is included to account for the fact that the fields associated with the traveling electrons are terminated only partly in the base; the remaining half of the fields are terminated in the quasi-neutral collector (for a one-dimensional geometry).

A more realistic calculation of τ_{sclc} must take into account the fact that the electron velocity reaches very high values at the base edge of the space charge layer, due to electron velocity overshoot; it decreases to the steady-state saturated velocity value only after traversing a non-negligible distance d_0. The value of d_0 is a strong function of the field in the base–collector region; to lowest order, d_0 is the distance taken by the electrons to acquire an energy of 0.3 eV, needed to be able to reach the high effective mass satellite valleys of the conduction band. To maximize the distance d_0, it has been shown that it is advantageous to tailor the doping in the space charge layer to decrease the electric field near the base edge while maintaining a high field at the collector edge [49, 50].

A final contribution to τ_{ec} results from the fact that a change in output current produces a change in base–collector voltage, principally through resistive drops. The change in junction voltage in turn induces a charge in the base, through the depletion layer capacitance. The corresponding τ_c is

$$\tau_c = \left(\frac{kT}{qI_c} + R_e + R_c\right)C_{bc} \tag{34}$$

Here, R_e and R_c are the parasitic emitter and collector resistances, respectively, and C_{bc} is the base–collector capacitance (including both intrinsic and extrinsic components). Whereas the previous contributions to τ_{ec} depended solely on the vertical structure of the device and could be analyzed in a one-dimensional model, this last term depends on the lateral structure of the device. The reduction of this contribution during the course of HBT process development has been the major factor in the increases reported in f_t.

The power gain of the devices at high frequency is specified through the figure of merit f_{max}, as described in Section 4.1.2. An experimental measure of f_{max} may be obtained by extrapolating the behavior of power gain versus frequency inferred from S-parameter measurements, to the point where the power gain drops to unity. It is theoretically demonstrable that MAG and U both must reach unity at the same frequency. Their behavior at lower frequency may be significantly different, however, which complicates the task of extrapolation. As shown in Fig. 4.27, for example, the value obtained by extrapolating U at 6 dB/octave is larger than that inferred from MAG with the same extrapolation rule. The results imply that MAG and U do not decrease with frequency at the same rate. Simulation based on small-signal equivalent circuits indicates that U decreases in best accord with 6 dB/octave slope, whereas MAG decreases somewhat more slowly. This indicates that f_{max} for the device of Fig. 4.27 is in the vicinity of 150 GHz.

Values of f_{max} attained with HBTs are significantly better than those demonstrated with Si bipolar transistors (limited to date to values on the order of 35 GHz). They are comparable to results obtainable in MODFET devices with gate lengths of 0.2 μm or lower [51], or those of permeable base transistors (PBTs) [52].

As presented in Section 4.1.2, f_{max} in HBTs is given approximately by

$$f_{max} \cong \sqrt{\frac{f_t}{8\pi R_b C_{bc}}} \tag{5}$$

To obtain the highest f_{max}, it is desirable to minimize the parasitic base resistance. To accomplish this in the device of Fig. 4.27, the base doping was raised to the value 10^{20} cm^{-3}. An additional, well-established technique is to use an interdigitated structure in the transistor layout, with narrow emitter fingers interspersed with base contacts. In such a way, the base contact periphery per unit emitter area is maximized. In Si bipolar technology, 0.35 μm emitters with a 2 μm pitch have been attained. In HBT technology to date, 1.2 μm emitters with a 2.8 μm pitch are the narrowest structures employed.

To obtain the highest possible value of f_{max}, it is additionally important to reduce C_{bc} as far as possible. The contribution to C_{bc} from the intrinsic device (under the active emitter) can be decreased only by increasing the width of the base–collector depletion region. This, in turn, causes a decrease in f_t due to the resultant increase in the space charge region transit time. For optimal rf performance, design trade-offs must thus be made (by taking into account all parasitic elements affecting transistor performance). The contribution to C_{bc} from the extrinsic base area (outside of the active emitter) is typically significant also. It was shown in Section 4.3.2 that with GaAs HBT technology, it is possible to reduce C_{bc} in this area by making use of ion implantation to introduce damage centers into the collector layer under the base, reducing its effective doping. Another approach, suggested

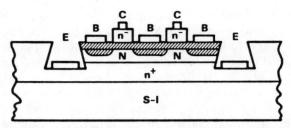

Figure 4.28 Schematic diagram of collector-up (inverted) HBT structure which minimizes base–collector capacitance.

by Kroemer and unique to HBTs [3], is based on an inverted transistor structure (Fig. 4.28). The device makes use of a wide-bandgap emitter buried beneath the base and collector, and a collector layer located on the wafer surface. In the extrinsic base regions, the semiconductor is etched away, decreasing the base–collector capacitance very extensively. For the transistor to have adequate current gain, however, it is necessary to suppress the injection of electrons from the emitter in the regions covered by the base contact (where they would recombine rather than be collected). In the transistor of Fig. 4.28, injection in these regions is suppressed by the fact that the base–emitter junction turn-on voltage is greater in the extrinsic base regions because of the wider bandgap of the material.

The small-signal rf characteristics of HBTs can be described in terms of an equivalent circuit model, such as that shown in Fig. 4.29a. This model is similar to the one used most frequently for Si bipolar transistors, although for both cases it should be noted that the models are not unique. The current generator used in the model, for example, can be voltage-controlled or current-controlled, and can connect base and collector or emitter and collector in various embodiments. With appropriate definitions, all the models are equivalent. The relationship between the circuit model of Fig. 4.29a and the transistor structure is illustrated in Fig. 4.29b. Identification of the various resistances and capacitances provides a way to estimate the different element values. A corresponding set of experimental circuit element values can be obtained by fitting measured S-parameter data at many frequencies. The element values change with bias condition, however, and for the most part scale with transistor size, so that care must be exercised in applying the equivalent circuit.

4.5 HBT APPLICATIONS

This section examines the potential performance obtainable with HBTs in a variety of applications and describes results obtained to date in these areas. It must be noted, however, that HBT technology is evolving rapidly, and performance limits are continually being extended.

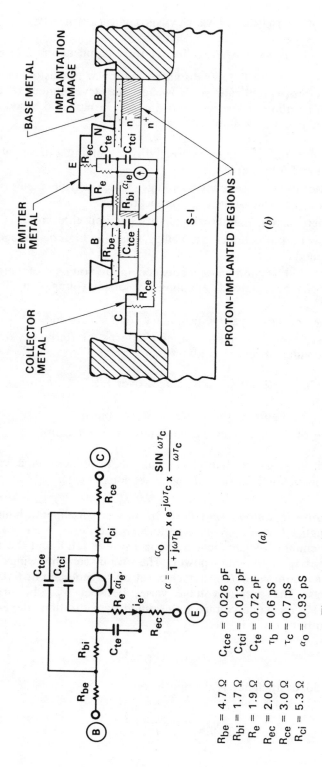

Figure 4.29 (a) Lumped element circuit model of an HBT, (b) schematic diagram showing relationship between circuit elements and the physical HBT structure.

$$\alpha = \frac{\alpha_o}{1 + j\omega\tau_b} \times e^{-j\omega\tau_c} \times \frac{\text{SIN } \omega\tau_c}{\omega\tau_c}$$

(a)

$R_{be} = 4.7\ \Omega$	$C_{tce} = 0.026$ pF
$R_{bi} = 1.7\ \Omega$	$C_{tci} = 0.013$ pF
$R_e = 1.9\ \Omega$	$C_{te} = 0.72$ pF
$R_{ec} = 2.0\ \Omega$	$\tau_b = 0.6$ pS
$R_{ce} = 3.0\ \Omega$	$\tau_c = 0.7$ pS
$R_{ci} = 5.3\ \Omega$	$\alpha_o = 0.93$ pS

4.5.1   Analog, Microwave, and Millimeter-wave Applications

In recent years, GaAs FETs have displaced Si bipolar transistors from many high-frequency linear applications. However, the development of HBTs is expected to bring about a resurgence of interest in bipolars, based on their advantage over FETs in a number of areas:

1. Their high f_t and f_{max}, and the possibility of obtaining these values with conventional lithography.
2. Their controllable breakdown voltage, which can be easily increased for high-power, high-efficiency applications.
3. Their high current handling capability per unit chip area.
4. Their low output conductance values, enabling high dc voltage gain to be obtained.
5. The fact that a good match between input voltages of neighboring devices can be obtained, enabling low offset voltages in differential pairs.
6. Their low noise in the low-frequency (or $1/f$) regime.
7. The absence of hysteresis, trapping effects, or frequency-dependent output conductance which affect many FETs.

By contrast, GaAs FETs have advantages over HBTs based on

1. Lower noise figure in the white noise regime.
2. Absence of dc input current (infinite dc current gain).

This comparison suggests that HBTs will be prominent in a variety of applications, as detailed in the following sections.

4.5.1.1   Microwave Power Amplifiers. In power amplifiers, high output power and high efficiency are key performance criteria. Figure 4.30 illustrates the schematic $I–V$ characteristics of a power HBT and the load line employed to extract maximum power. The output current swings between zero and a maximum value I_{max}, typically set by current gain or f_t roll-off, or heating effects. The output voltage varies between V_B, the breakdown voltage, and V_k, the knee voltage determined by saturation (dependent on I_{max}). The output power is given by

$$P_{out} = \tfrac{1}{8} I_{max}(V_B - V_k) \tag{35}$$

The power-added efficiency is defined to be

$$\eta_{add} = \frac{P_{out} - P_{in}}{P_{dc}} \tag{36}$$

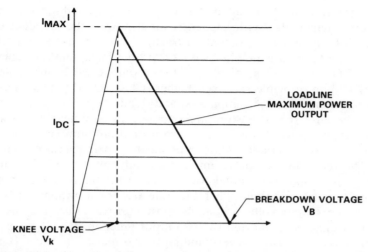

Figure 4.30 Simplified HBT *I–V* characteristics and load line used for power amplification.

where P_{in} is the input rf power. Transistor biasing, which establishes the input power P_{dc}, can be accomplished in several different ways. To maximize linearity and gain of the amplifier, a bias current and voltage of $I_{max}/2$ and $(V_B + V_k)/2$ can be selected (Class A amplifier). Thus, if the rf power gain is G, for the Class A amplifier, the maximum power added efficiency is

$$\eta_{addmax} = \frac{1}{2} \times \frac{G-1}{G} \times \frac{1 - V_k/V_B}{1 + V_k/V_B} \qquad (37)$$

Clearly, to maximize output power as well as efficiency, it is desirable to have a large value of breakdown voltage V_B. Unfortunately, this value cannot be increased indefinitely in microwave devices. To avoid avalanche breakdown of the base–collector junction, electric fields in that area must be kept below tolerable limits. To allow high voltages to be maintained, the width of the space charge region must be increased, with a corresponding increase in electron transit time and reduced f_t. An optimal space charge layer width should be selected, according to the maximum operating frequency desired. HBTs are advantageous in optimizing this voltage. The largely one-dimensional nature of the field distribution in the critical base–collector space charge region of HBTs, and the control of the doping and thickness of this region possible during epitaxial growth, allow the straightforward tailoring of the breakdown voltage. High values of breakdown voltage, e.g. BV_{cbo} above 25 V, can be easily achieved.

In Class A amplifiers, the efficiency is limited to values below 50%, as seen in Eq. (37). Class B and Class C amplifiers use alternative biasing approaches which permit greater efficiency. The transistor is biased at or close to cutoff, with the result that under quiescent conditions the power dissipation is negligible. With applied rf input power, the amplifier output current will be appreciable for only a fraction of the rf cycle. For Class B amplifiers, the fraction is nominally 50%, and the maximum efficiency rises to 78.5%. For Class C amplifiers, a lower duty cycle is used, and the efficiency may in principle rise even higher. Class B and C amplifiers have lower gain than Class A amplifiers and tend to have greater output distortion because the transistors are operated in nonlinear regimes. In a variety of applications, this is not a severe restriction, however, especially if the transistors are operated as a complementary pair (push–pull arrangement). The realization of these amplifiers is difficult in a FET technology because of the demands on breakdown voltage: the input voltage has large excursions to negative voltage when the transistor is cut off, and under these conditions the output voltage is at its most positive value, increasing the voltage that must be supported by the output junction. It may be expected that with HBTs superior power-handling performance in Class B and C amplifiers may be realized, just as has been found with Si bipolar transistors.

Microwave amplifiers may be designed based on common-emitter, common-base, or common-collector operation. The first two are in most frequent use. Only in the common-emitter mode is there current gain greater than unity. However, in common-base mode, appreciable power gain may be obtained by virtue of the impedance transformation offered by the amplifier: low input impedance and high output impedance. The gain in both configurations is limited by the same value of f_{max} (which is configuration invariant). At frequencies below f_{max}, however, the frequency dependence of gain differs slightly between the two modes, with somewhat higher gain resulting from common-base operation. The ease of impedance matching is better in common-emitter mode. A considerable advantage for power amplifiers of the common-base mode is the fact that the breakdown voltage BV_{cbo} is substantially higher than BV_{ceo}.

In experimental results obtained so far, output powers of 0.4 W have been obtained in CW common-emitter operation at 10 GHz, limited by heating effects, as shown in Fig. 4.31 [42]. This corresponds to an output power of 4 W/mm of emitter length, a considerably higher power density than demonstrated with FETs. The power-added efficiency reached 48%. In pulsed operation at 10 GHz, a corresponding value of 8 W/mm of emitter length has been achieved [53].

4.5.1.2 Millimeter-wave Amplifiers. A central requirement of millimeter-wave devices is high gain at millimeter-wave frequencies (above 30 GHz) or, equivalently, high values of f_{max}. HBTs have thus far attained a value of f_{max} of 170 GHz. Various alternative technologies have achieved values of this order or, in a few instances, even higher (above 200 GHz for pseudomorphic MODFET technology and permeable base transistor technology) [51, 52]. However, among candidate devices, the HBT is the only one that can be

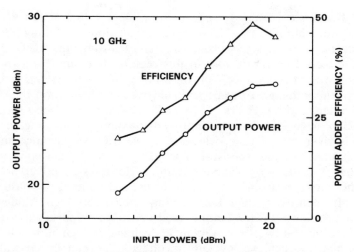

Figure 4.31 Experimental measurements of output power and power-added efficiency at 10 GHz for an HBT with 100 μm emitter length.

fabricated on the basis of straightforward optical lithography with lateral features on the order of 1 μm or greater. This is possible, since the critical small dimensions are established by epitaxial growth. Additionally, for millimeter-wave power applications, it is beneficial to achieve the required output capability in relatively compact devices; if the transistor is spread out over dimensions greater than 0.1 wavelength, then different sections of the device may be operating with different phases of the input signal, and their outputs will interfere. At 100 GHz, 0.1 wavelength (in GaAs loaded transmission lines) is on the order of 100 μm. The HBT and PBT have the best opportunity to achieve significant output power under this constraint.

4.5.1.3 Microwave Oscillators and Mixers. In these circuits, low noise is an important requirement. For linear circuits, the relevant noise is that associated with the operating frequency only. In the inherently nonlinear oscillators and mixers, the prime concern is often noise at low frequencies near dc (the $1/f$ regime), since the nonlinearity of the circuit causes intermixing of the noise spectral components. Noise near dc gives rise to phase noise of the rf output. For FETs, high values of $1/f$ noise are common [54]. These are believed to result from temporal variations in the channel conductance due to trapping and detrapping of electrons at sites along the channel surface and the channel–substrate interface. HBTs are relatively free from such effects because the intrinsic device is shielded from such surfaces. Accordingly, preliminary reports of $1/f$ noise in HBTs indicate a noise corner frequency (where $1/f$ and white noise contributions intersect) below 1 MHz [55]. A comparison of oscillators fabricated with HBTs and FETs has shown lower phase noise with the HBTs [56].

4.5.1.4 Wideband Analog Amplifiers. Amplifiers with frequency response from dc to several GHz have a variety of applications, including fiber-optics communications, radar signal processing, and instrumentation. GaAS FET-based amplifiers face a difficulty in this regime stemming from frequency-dependent output resistance (in the range of 1–100 MHz), arising on account of the influence of traps and substrate conduction on the channel–substrate interface [57]. HBTs are free of such trapping effects. At the same time, HBTs have significantly improved maximum voltage gain u. This quantity is given by $u = g_m r_o$, where g_m is the transconductance and r_o the output resistance of the transistor. For HBTs at dc, g_m can be significantly higher than for FETs of comparable dimensions, as described in Section 4.4.1. The output resistance, by virtue of the shielding of the base–emitter junction afforded by the high base doping, is also very high. Values of u in excess of 200 have been demonstrated. Also of importance for analog amplifiers is the ability to match input V_{be} voltages, which allows differential transistor pairs to achieve low offset voltage. The match achievable with bipolars tends to be better than that of FETs. Figure 4.32 shows, for example, histograms showing uniformity across a wafer of input offset voltage and gain for simple differential pair test patterns. This combination of advantages leads to the expectation of HBT application to operational, fixed gain feedback, AGC, and logarithmic amplifiers. Excellent performance has recently been demonstrated for a monolithic four-stage "true" logarithmic IF amplifier implemented with HBTs at TRW, which achieved a dc-3 GHz bandwidth [58].

4.5.2 Digital Applications

HBTs have a variety of advantages for use in digital ICs. Included among them is the fact that f_t of the transistors can be made very high, the highest among possible candidates for commercial digital circuits (which to date do not employ electron beam lithography). The high transconductance of HBTs is of importance in allowing circuit operation with small input voltage swings and for providing low output impedance driver stages for charging load capacitances rapidly. Additionally, the input voltage matching of the devices is compatible with low voltage swings as required for high-speed, low-power operation. In contrast, with FET-based digital circuits, threshold-voltage control is a significant problem in device fabrication. The threshold voltage depends directly on the doping and thickness of the channel, and variations of V_{th} among devices can prevent circuit operation in FET technologies.

Key performance issues for digital circuit technologies are propagation delay time, power dissipation per gate, and integration level. No analytical treatment exists to calculate directly the influence of device characteristics on these circuit parameters, inasmuch as (1) transistor operation in digital circuits is inherently nonlinear and thus difficult to treat; and (2) a great number of variables must be taken into account, including logic family and

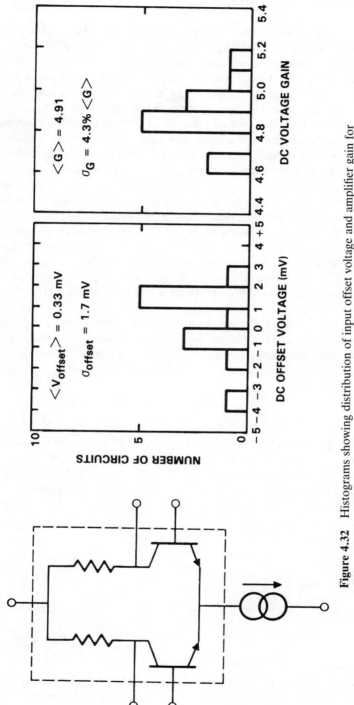

Figure 4.32 Histograms showing distribution of input offset voltage and amplifier gain for differential amplifier test structures on a fabricated HBT wafer.

215

circuit approach, noise tolerance required, interconnect characteristics of the technology (number of wiring levels of metal and the influence of device density on interconnect capacitance); required yield and thus cost, etc. In approximate terms, every technology may be characterized by a family of curves describing the propagation delay time of a gate versus the power dissipation of the gate. Representative curves for two technologies are shown in Fig. 4.33. At low power, the product of the propagation delay τ_d times the power dissipation P is approximately constant, with a value called the "power-delay product" $P\tau$, or switching energy of the technology. At higher levels of power dissipation per gate, the propagation delay typically reaches a minimum value, τ_{min}, which establishes a floor to the switching speed no matter what the available power to be expended. It is possible for two technologies to have different relative performance in the two different regions of the curves, as shown in Fig. 4.33 (e.g., the technology with the lowest $P\tau$ may have a greater value of τ_{min}). The regime of operation in specific ICs depends on the integration level and on the available power. In general, SSI or MSI (small- and medium-scale ICs) often operate in the regime of minimum delay, while LSI and VLSI circuits are in the $P\tau$-limited regime.

Semiquantitative understanding of the importance of various device characteristics on digital performance may be obtained by referring to the simplified prototype circuit of Fig. 4.34. The propagation delay τ_d (for an optimized choice of load resistance) is of the form

$$\tau_d = k_1 \frac{C_{device} + C_{interconnect}}{G_m} \tag{38}$$

where k_1 is a constant of order unity, G_m is the effective device transconductance, C_{device} is the total input capacitance of the devices loading the gate output (fan-out-dependent), and $C_{interconnect}$ is the associated wiring capacitance. The small-signal values of both device transconductance and input capacitance depend strongly on operating point, which varies dramati-

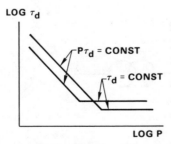

Figure 4.33 Relationship between propagation delay and power dissipation per gate for representative technologies.

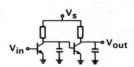

Figure 4.34 Elementary digital circuit used for the calculation of digital technology figures of merit.

cally over the voltage range used in switching. To apply Eq. (38) properly, appropriately averaged large-signal quantities should be used. For example

$$G_m = \frac{I_{out} \text{ (high)} - I_{out} \text{ (low)}}{V_{in} \text{ (high)} - V_{in} \text{ (low)}} \tag{39}$$

$$C_{device} = \frac{Q_{in} \text{ (high)} - Q_{in} \text{ (low)}}{V_{in} \text{ (high)} - V_{in} \text{ (low)}} \tag{40}$$

where appropriate values of the variables for logic high and logic low conditions are used. Equation (38) illustrates that high transconductance is important to achieve small switching delay time, even more so than high f_t, because of the impact of the interconnect capacitance.

The same prototype circuit can be used to evaluate the power-delay product. For one switching operation, charge is first transported from the positive power supply V^+ to the output node to raise its potential; in a subsequent phase, the charge is transported from the output node to the negative supply V^- (or ground, as in Fig. 4.34) to lower the output voltage. The amount of charge transported is given by the product of the node capacitance and its voltage swing. The overall energy per switching cycle, the charge transported times the voltage drop, is thus given by

$$P_\tau = k_2(C_{device} + C_{interconnect})(V^+ - V^-)V_{logic} \tag{41}$$

where k_2 is a constant of order unity. This equation accounts directly for the dynamic switching energy; an additional contribution related to dc power dissipation typically must be included. The equation illustrates the considerable importance of minimizing device and interconnect capacitance, as well as operation with low voltage swings (which requires high uniformity of switching threshold voltage and high transconductance).

Various logic circuit families have been explored for use with HBTs, adapted from work on Si-based bipolar transistors. The most widespread families may be generically referred to as ECL and I^2L, each with a number of variants.

4.5.2.1 ECL Circuits.

In emitter coupled logic circuits (ECLs), current is directed from a current source through one or more differential pairs of transistors, which steer the current to one or another load resistor in accordance with the logic state of the inputs. Emitter-follower buffer stages may be used to drive the subsequent stages, in "classical ECL," or they may be omitted in current mode logic (CML). Figure 4.35 illustrates the corresponding circuit diagrams. Typically, complementary outputs are produced from each ECL stage. A further increase in logic power is possible by "series-gating," that is, the use of multiple levels of differential switch transistor pairs using the same current source, as shown in Fig. 4.35c. This allows complex functions such as exclusive OR to be carried out with a

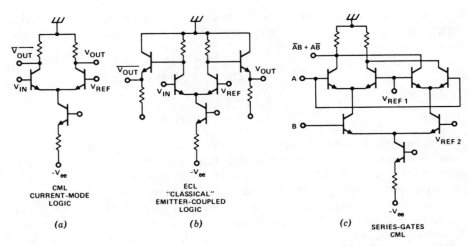

Figure 4.35 Circuit diagrams of various types of logic circuits within the ECL family.

single gate. ECL has a number of advantages for high-speed operation, including:

1. Transistors are kept out of saturation because the current source limits the voltage swing of the outputs.
2. Low output impedance is available to drive following logic stages, particularly if the emitter-follower buffers are used.
3. The differential structure tends to minimize the effect on logic output of power supply, temperature, and wafer-to-wafer device variations.
4. Low voltage swings are used, typically from 0.4 to 0.8 V, which tends to minimize power-delay product.
5. The current drawn from the supply lines is independent of logic state and relatively constant during switching, minimizing noise due to supply line resistance and inductance.
6. The logic capability available from complementary outputs, series gating, and emitter dotting (establishment of wired-OR functions by directly connecting emitter-follower outputs) allows efficient circuit implementation in many cases.

There are several disadvantages of using ECL as well. Among these are

1. High device count, particularly for the simplest logic operations (3–5 transistors and 3–5 resistors for a two-input OR).
2. High power dissipation. Whereas power and speed can be extensively traded off, the dissipation of ECL circuits tends to be greater than that

of saturating circuits. A fundamental reason for this is that base–collector voltages must be higher to avoid saturation under worst-case conditions.

Numerous demonstrations of ECL circuits implemented in HBTs have been reported. One of the most widespread circuits for technology evaluation is the frequency divider. In ECL technology, this is implemented as a master–slave flip–flop with its output inverted and fed back to its input. Figure 4.36 shows the corresponding logic and circuit diagram for a series-gated CML. The maximum frequency of operation corresponds to $\frac{1}{2}\tau_d$, where τ_d is the propagation delay of the compound (bi-level) gate. Frequency dividers have been made with HBTs [59, 60], with operating frequency up to 26.6 GHz. Figure 4.37 shows a fabricated divide-by-four circuit. The propagation delay for compound CML gates within these circuits was 20 ps, with a fan-out that varied between two and three. The power dissipation was 50 mW per compound gate, or on the order of 17 mW per equivalent NOR gate (assuming the logic power of one bi-level ECL gate is equivalent to that of three two-input NOR gates). The circuits are significantly faster than corresponding circuits implemented in Si [61] (the fastest of which has operated at 13 GHz), and somewhat faster than those implemented with GaAs MESFETs and MODFETs using electron beam

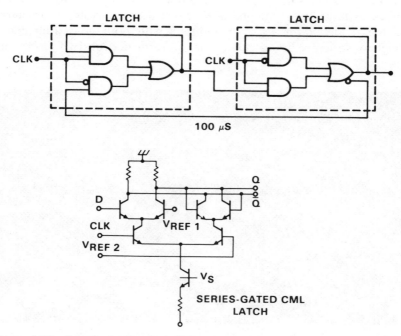

Figure 4.36 Schematic diagram of frequency divider circuit used with HBTs.

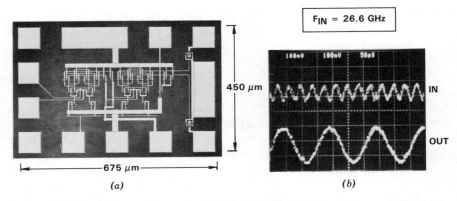

Figure 4.37   (a) Fabricated HBT divide-by-four circuit, (b) oscillograph illustrating divider operation at 26.6 GHz.

lithography to define $0.2\,\mu m$ gate lengths [62, 63]. By contrast, the HBTs employed to reach ultrahigh speed had emitter widths of $2\,\mu m$.

Several more complex ECL circuits have been made with HBTs. Multiplexers and demultiplexers for use in fiber-optics communication systems have been reported [64]. Figure 4.38 shows a fabricated 8-bit universal shift register that was capable of 3 GHz operation [65].

Theoretical studies have been carried out of the prospective performance of HBTs in large logic circuits. Tiwari has used transistor characteristics calculated in detail for $1 \times 2\,\mu m$ emitters with both nonself-aligned and self-aligned structure to calculate the performance of ECL logic under

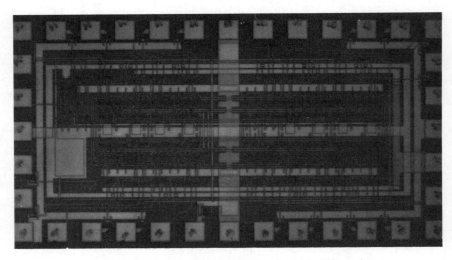

Figure 4.38   Fabricated 8-bit HBT universal shift register.

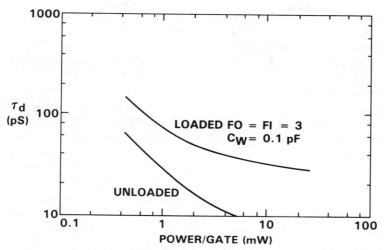

Figure 4.39 Calculated relation between propagation delay and power per gate for ECL circuits based on an optimized HBT technology with $1 \times 2 \, \mu m$ emitters, for unloaded and loaded conditions. (After Tiwari [66]).

various conditions of loading and power dissipation [66]. Figure 4.39 illustrates the calculated delay versus power both for fan-in = fan-out = 1 condition, and for fan-in = fan-out = 3 together with 0.1 pF wiring capacitance loading.

4.5.2.2 I^2L Circuits. In I^2L (integrated injection logic) circuits, logic gates contain a single transistor with multiple collectors, or a single collector with multiple Schottky diodes attached to it, as shown in Fig. 4.40. A current source is implemented to provide input base current to the switching transistor such that when the inputs to the gate are all high, the transistor is in saturation, providing a logic low output. If any input to the gate is low, the base input current is shunted away from the switching transistor, and the transistor is cut off, providing a logic high output. The logic function thus implemented is a multiple-input NAND.

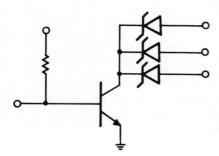

Figure 4.40 Basic multiple–output NAND gate used to implement heterojunction integrated injection logic (HI^2L).

Circuits of the I^2L family may be implemented with HBTs in a variety of ways. A particularly clever implementation, which makes use of numerous advantages of the GaAlAs/GaAs system, has been extensively developed by Texas Instruments (TI) [67]. Figure 4.19 illustrates a cross section of the circuit elements. The current source is provided by a resistor. The switching element is an HBT with its collector on the wafer surface and its AlGaAs emitter buried below. An n^+ substrate is used, connected to all the emitters of the transistors. This provides a low inductance reference voltage and eliminates the need to route one of the power supply lines (although it also reduces somewhat the flexibility of circuit that can be implemented). Implanted p-regions in the extrinsic base area are made; the implant is made deep enough that the p-n junction extends into the underlying AlGaAs emitter layer. The turn-on voltage of this p-n junction is greater than that of the intrinsic device, so that current injection in these parasitic areas of the device is avoided. This overcomes one of the significant problems in Si-based I^2L, since charge injected into the extrinsic base regions contributes both to a decrease in current gain and an increase in input capacitance.

In I^2L, the switching transistor is allowed to saturate. The logic voltage swings between the turn-on voltage of the emitter–base junction (logic high), and the saturation voltage V_{cesat} plus the forward drop of a GaAs Schottky diode (logic low). The low swing (about 0.3 V) contributes to a low power-delay product. During saturation, however, charge injection into the collector occurs as the base–collector junction becomes forward-biased. This tends to increase the propagation delay time. The charge disappears relatively rapidly, since the recombination lifetime is low in GaAs, especially when surfaces are near as in the TI design. To avoid the extra delay, however, it is also possible to design a logic family in which Schottky diodes are used to clamp the base–collector junctions of the switching transistor, preventing hard saturation.

The I^2L approach permits high packing density, above 750 gates per mm^2 with 2 μm design rules. Measured gate delay for fan-out of four is on the order of 200 ps with these design rules. At present, very large circuits approaching the VLSI regime are being processed at TI using this technology. Figure 4.41 shows a photograph of a 32 bit CPU, recently demonstrated by TI, containing 12,900 gates [68]. Fabrication of such complex circuits represents a milestone in HBT technology and in GaAs device technology in general.

4.5.3 Analog-to-Digital Conversion

As digital signal processing and data processing speeds advance, increasing pressures are placed on analog-to-digital (A/D) converters to interface these systems with the real, analog, world. HBTs are being actively explored for

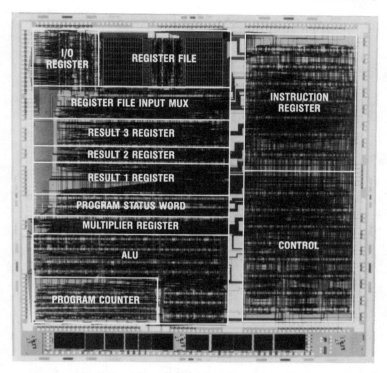

Figure 4.41 Microphotograph of a 32 bit CPU chip containing 12900 HI^{2}L gates developed at Texas Instruments [68].

this application [69]. HBTs combine high f_t and high transconductance with low offset voltages (since the turn-on voltages are well matched). In addition, trapping effects that have been identified in GaAs MESFETs as the cause of frequency-dependent output resistance or "current-lag" phenomena are absent in HBTs.

Key circuits for A/D converters are voltage comparators and sample-and-hold (S/H) circuits. Voltage comparators must recognize input signals that are within millivolts of a reference voltage and deliver outputs at proper logic levels within short times (down to fractions of a nanosecond). The large amplification necessary is conveniently done by a regenerative (or latching) circuit, as shown in Fig. 4.42 which uses a clock input to determine the instant of comparison and initiate the latching. Comparators of this type have operated at clock frequencies above 5 GHz. All-parallel ("flash") A/D converters consisting of a string of these comparators have recently been reported, with up to 4 bit complexity (16 comparators). The circuits have operated at up to 2 GSamples/s [70, 71]. Figure 4.43 illustrates an A/D converter of this type. In future work, it is expected that higher resolution

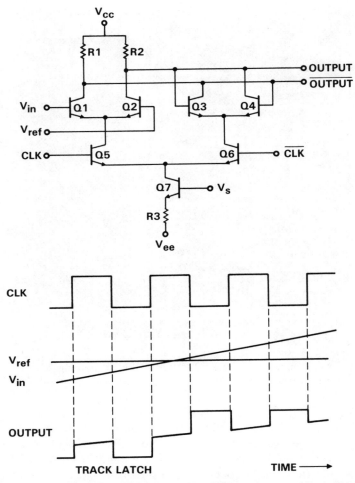

Figure 4.42 Circuit diagram of latching voltage comparator used in A/D converters based on HBTs, together with a timing diagram of its operation.

circuits will be fabricated by paralleling more comparators up to limits established by power dissipation and yield (e.g., an 8-bit converter of this type requires 255 comparators or 10,000–15,000 transistors, a significant increase in level of integration).

S/H circuits are required to increase the dynamic performance of flash A/D converters and to allow implementation of more complex A/D architectures (to hold the analog input signal constant over the relatively long time of conversion). These circuits are at present being designed in HBT technology; initial performance has recently been reported [72].

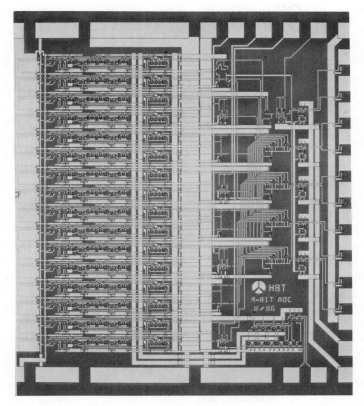

Figure 4.43   Fabricated 4 bit A/D converter containing 480 HBTs.

4.6 FUTURE DIRECTIONS

On a relative scale, HBT technology is in its early infancy. Future work on the application of HBTs to the areas discussed is expected to increase performance considerably. Furthermore, beyond the direct improvement expected from device scaling, layer structure optimization, and superior circuit design, a variety of more profound innovations may be expected. The inclusion of resonant tunneling structures within bipolar transistors has already been demonstrated [73] and is reported elsewhere in this book. Work is progressing on p-n-p devices [74, 75] and it may be possible soon to make use of monolithically integrated complementary transistors. The integration of HBTs and FETs (MESFETs or JFETs) on the same chip offers additional possibilities for improved circuit performance. Another avenue for progress is in the use of new materials systems that may offer additional advantages over AlGaAs/GaAs. HBTs made with AlGaAs/GaAs epilayers grown on Si substrates have been reported [76]. This technology offers the potential for larger wafers that are mechanically more rugged and have better thermal conductivity than ones based on GaAs alone; on a long-term basis, integration of GaAs and Si devices on the same chip may possibly become practical. The use of different III–V materials is also of considerable interest. HBTs in the InGaAs/InP or InAlAs system, lattice-matched to InP substrates, have been reported from a number of laboratories [77, 78]. This system has advantages for high speed (stemming from higher electron mobility and the potential for increased electron velocity overshoot effects); for monolithic integration with long-wavelength sources and detectors for fiber-optic communication systems; and, importantly, for lower power-delay products in digital circuits. This last advantage results from the fact that the bandgap of $In_{0.53}Ga_{0.47}As$ (which is lattice-matched to InP) is 0.75 eV at room temperature, significantly lower than the bandgap of GaAs (1.42 eV). From Eq. (22), it is apparent that the bandgap of the base layer is the prime determinant of the turn-on voltage V_{be} of the device. This turn-on voltage determines the power supply voltage that must be used. In turn, the power-delay product depends strongly on the power supply voltage, as expressed in Eq. (41). By decreasing the bandgap of the base, it may be estimated that $P\tau$ may be decreased by a factor of 1.6 times over the case of AlGaAs/GaAs.

The argument in favor of InGaAs base layers has considerable generality and may be used to compare potential VLSI technologies. With larger circuits, interconnect capacitance typically increases; with smaller devices, input capacitance decreases. In the VLSI limit with sufficiently advanced device technologies, the interconnect capacitance may dominate the overall load capacitance (to the point that device f_t no longer has an effect on $P\tau$). Within any bipolar technology, the logic swing required for device switching is the same, limited to values on the order of $5–10kT$ in principle; in practice, noise tolerance and logic signal and power supply distribution issues are additionally important. Then, the only factor that is open to optimization in the dynamic switching energy from Eq. (41) is the power supply voltage. In an FET technology, the threshold voltage is a convenient design variable, established by doping and thickness of the channel. The power supply voltage is reduced to values on the order of the logic swing. In bipolar technology, as explained above, the turn-on voltage is dependent on the material system (which leads to good reproducibility but little flexibility of design). Optimization of $P\tau$ requires a proper choice of material system. The optimal bandgap of the base layer depends on operating temperature; it may be estimated that at room temperature a value on the order of 0.5 eV should be used. With homojunction transistors, such a choice of bandgap would lead to undesirable leakage currents, particularly from the reverse-biased base–collector junctions. With heterojunction transistors, such leakage currents can be avoided by the use of a wide-bandgap collector region, even though a low-bandgap base layer is employed.

An additional factor in digital switching power-delay product is the static power dissipation, which is not included in a transparent way in Eq. (41). It is of great importance to reduce the power dissipation of circuits while they are not actively switching, as in CMOS technology. The search for practical methods of accomplishing this will be a continuing technology of HBT research for digital applications.

REFERENCES

1. W. Shockley, U. S. Patent 2,569,347 (1951).
2. H. Kroemer, *Proc. IRE* **45**, 1535 (1957).
3. H. Kroemer, *Proc. IEEE* **70**, 13 (1982).
4. S. M. Sze, "Physics of Semiconductor Devices," 2nd ed. Wiley, New York, 1981; a comprehensive reference for HBT high-frequency performance is M. B. Das, *IEEE Trans. Electron Devices* **ED-35**, 604 (1988).
5. R. L. Anderson, *Solid State Electron.* **5**, 341 (1962).
6. H. C. Casey, Jr., and M. B. Panish, "Heterostructure Lasers." Academic Press, New York, 1978.
7. M. O. Watanabe, J. Yoshida, M. Mashita, T. Nakanisi, and A. Hojo, *Tech. Dig., Int. Conf. Solid State Devices Mater.* p. 181 (1984).
8. H. Kroemer, W. Y. Chien, J. S. Harris, Jr., and D. D. Edwall, *Appl. Phys. Lett.* **36**, 295 (1980).
9. J. E. Sutherland and J. R. Hauser, *IEEE Trans. Electron Devices* **ED-24**, 363 (1977); also A. H. Marshak and C. M. Van Vliet, *Proc. IEEE* **72**, 148 (1984).
10. G. M. Martin, A. Mitonneau, and A. Mircea, *Electron. Lett.* **13**, 192 (1977).
11. J. G. Ruch and G. S. Kino, *Phys. Rev.* **174**, 921 (1968).
12. T. J. Maloney and J. Frey, *J. Appl. Phys.* **48**, 781 (1977).
13. K. Tomizawa, Y. Awano, and N. Hashizume, *IEEE Electron Device Lett.* **EDL-5**, 362 (1984).
14. C. M. Maziar, M. E. Klausmeier-Brown, S. Bandyopadhyay, M. S. Lundstrom, and S. Datta, *IEEE Trans. Electron Devices* **ED-33**, 881 (1986).
15. B. F. Levine, C. G. Bethea, W. T. Tsang, F. Capasso, K. K. Thornber, R. C. Fulton, and D. A. Klienman, *Appl. Phys. Lett.* **42**, 769 (1983).
16. A. Marty, G. E. Rey, and J. P. Bailbe, *Solid-State Electron.* **22** 549 (1979).
17. P. M. Asbeck, D. L. Miller, R. Asatourian, and C. G. Kirkpatrick, *IEEE Electron Device Lett.* **EDL-3**, 403 (1982).
18. J. Yoshida, M. Kurata, K. Morizuka, and A. Hojo, *IEEE Trans. Electron Devices* **ED-32**, 1714 (1985).
19. S. Tiwari, *Tech. Dig., IEEE Bipolar Circuits Technol. Meet.* p. 21 (1986).
20. K. Yokoyama, M. Tomizawa, and A. Yoshii, *IEEE Trans. Electron Devices* **ED-31**, 1222 (1984).
21. C. M. Maziar, M. E. Klausmeier-Brown, and M. S. Lundstrom, *IEEE Electron Device Lett.* **EDL-7**, 483 (1986).

22. A. Y. Cho and J. R. Arthur, *Prog. Solid State Chem.* **10**, 157 (1975).

23. *Proc. Intl. Conf. Metalorganic Vapor Phase Epitaxy*, G.B. Stringfellow, (ed.) *J. Cryst. Growth* **77**, 1–652 (1986).

24. C. Dubon-Chevallier, R. Azoulay, P. Desrousseux, J. Dangla, A. M. Duchenois, M. Hountondji, and D. Ankri, *Tech Dig.—Int. Electron Devices Meet.* p. 689 (1983).

25. P. J. Topham, R. C. Hayes, I. H. Goodridge, C. Tombling, and D. Benn, *Tech. Dig. IEEE Gallium Arsenide IC Symp.*, p. 167 (1986).

26. M. A. Rao, E. J. Caine, S. I. Long, and H. Kroemer, *IEEE Electron Device Lett.* **EDL-8**, 30 (1987).

27. R. P. Tijburg and T. van Dongen, *J. Electrochem. Soc.* **123**, 687 (1976).

28. K. Hikosaka, T. Mimura, and K. Joshin, *Jpn. J. Appl. Phys.* (1982).

29. P. M. Asbeck, D. L. Miller, E. J. Babcock, and C. G. Kirkpatrick, *IEEE Electron Device Lett.* **EDL-4** (4), 81–84 (1983).

30. D. Ankri, A. Scavennec, C. Courbet, F. Heliot, and J. Riou, *Appl. Phys. Lett.* **40**, 816 (1982).

31. K. Morizuka, M. Obara, K. Tsuda, M. Asaka, H. Tamura, M. Mashita, J. Yoshida, and A. Hojo, *Tech. Dig.*, *Int. Conf. Solid-State Mater. Devices*, p. LD-57 (1984).

32. T. Ohshima, K. Ishii, N. Yokoyama, T. Futatsugi, T. Fujii, and H. Nishi, *Tech. Dig.*, *IEEE Gallium Arsenide ICC Symp.*, pp. 53–56 (1985).

33. H. T. Yuan, W. V. Mclevige, H. D. Shih, and A. S. Hearn, *Dig. Tech. ISSCC* pp. 42–48 (1984).

34. J. W. Tully, W. Hant, and B. B. O'Brien, *IEEE Electron Device Lett.* **EDL-7**, 615 (1986).

35. P. M. Asbeck, D. L. Miller, R. J. Anderson, and F. H. Eisen, *IEEE Electron Device Lett.* **EDL-5**, 310 (1984).

36. O. Nakajima, K. Nagata, Y. Yamauchi, H. Ito, and T. Ishibashi, *Tech. Dig.—Int. Electron Devices Meet.* p. 266 (1986).

37. M. F. Chang, P. M. Asbeck, D. L. Miller, and K. C. Wang, *IEEE Electron Device Lett.* **EDL-7** (1), 8–10 (1986).

38. T. Ishibashi, Y. Yamauchi, O. Nakajima, K. Nagata, and H. Ito, *Tech. Dig.—Int. Electron Devices Meet.* pp. 86–89 (1986).

39. M. F. Chang, P. M. Asbeck, K. C. Wang, G. J. Sullivan, N. H. Sheng, J. A. Higgins, and D. L. Miller, *IEEE Electron Device Lett.* **EDL-8** (7) (1987).

40. S. Tiwari, *Tech. Dig.—Int. Electron Devices Meet.* pp. 262–265 (1986).

41. H. Kroemer, *Solid-State Electron.* **28**, 1101 (1985).

42. N. H. Sheng, M. F. Chang, P. M. Asbeck, K. C. Wang, G. J. Sullivan, D. L. Miller, J. A. Higgins, E. Sovero, and H. Basit, *Tech. Dig.—Int. Electron Devices Meet.* p. 619 (1987).

43. C. H. Henry, R. A. Logan, and F. R. Merritt, *J. Appl. Phys.* **49**, 3530 (1978).

44. S. C. Lee and G. L. Pearson, *J. Appl. Phys.* **52**, 275 (1981).

45. O. Nakajima, K. Nagata, H. Ito, T. Ishibashi, and T. Sugeta, *Jpn. J. Appl. Phys.* **24**, 1368 (1985).

46. H. H. Lin and S. C. Lee, *Appl. Phys. Lett.* **47**, 839 (1985).

47. C. J. Sandroff, R. N. Nottenburg, J. C. Bischoff, and R. Bhat, *Appl. Phys. Lett.* **51**, 33 (1987).

48. G. Gonzalez, "Microwave Transistor Amplifiers, Analysis and Design." Prentice-Hall, Englewood Cliffs, New Jersey, 1984.

49. T. Ishibashi and Y. Yamauchi, *IEEE Trans. Electron Devices* **ED-35**, 401 (1988).

50. C. M. Maziar, M. E. Klausmeier-Brown, and M. S. Lundstrom, *IEEE Electron Device Lett.* **EDL-7**, 483 (1986).

51. T. Henderson, M. I. Aksun, C. K. Peng, H. Morkoç, P. C. Chao, P. M. Smith, K. H. G. Duh, and L. F. Lester, *Tech. Dig.—Int. Electron Devices Meet.* p. 464 (1986).

52. M. A. Hollis, K. B. Nichols, R. A. Murphy, R. P. Gale, S. Rabe, W. J. Piacentini, C. O. Bozler, and P. M. Smith, *Tech. Dig.—Int. Electron Devices Meet.* p. 102 (1985).

53. B. Bayraktaroglu, N. Camilieri, H. D. Shih, and H. Q. Tserng, *Tech. Dig.—IEEE Int. Microwave Symp.*, p. 969 (1987).

54. B. Hughes, N. G. Fernandez, and J. M. Gladstone, *IEEE Trans. Electron Devices* **ED-34**, 733 (1987).

55. X. N. Zhang, A. van der Ziel, K. H. Duh, and H. Morkoç, *IEEE Electron Device Lett.* **EDL-5**, 277 (1984); see also P. M. Asbeck, A. K. Gupta, F. J. Ryan, D. L. Miller, R. J. Anderson, C. A. Liechti, and F. H. Eisen, *Tech. Dig.—Int. Electron Devices Meet.* p. 864 (1984).

56. K. K. Agarwal, *Tech. Dig.—IEEE Int. Microwave Symp.* p. 95 (1986).

57. M. Rocchi, *Physica B* (*Amsterdam*) **129B**, 119 (1985).

58. A. K. Oki, M. E. Kim, G. M. Gorman, and J. B. Camou, *Tech. Dig.—IEEE Microwave Millimeter-Wave Monolithic Circuits Symp.*, (1988).

59. T. Isibashi, Y. Yamauchi, O. Nakajima, K. Nagata, and H. Ito, *IEEE Electron Devices Lett.* **EDL-8**, 194 (1987).

60. K. C. Wang, P. M. Asbeck, M. F. Chang, G. J. Sullivan, and D. L. Miller, *IEEE Electron Devices Lett.* **EDL-8**, 383 (1987).

61. S. Konaka, Y. Yamamoto, and T. Sakai, *IEEE Trans. Electron Devices* **ED-33**, 526 (1986).

62. J. F. Jensen, L. G. Salmon, D. S. Deakin, and M. J. Delaney, *Tech. Dig.—Int. Electron Devices Meet.* p. 496 (1986).

63. J. F. Jensen, U. K. Mishra, A. S. Brown, R. S. Beaubien, M. A. Thompson, and L. M. Jelloian, *Tech. Dig.*, ISSCC p. 268 (1988).

64. K. C. Wang, P. M. Asbeck , M. F. Chang, G. J. Sullivan, and D. L. Miller, *Tech. Dig.*, *IEEE Bipolar Circuits Technol. Meet.* (1987).

65. K. C. Wang, P. M. Asbeck, M. F. Chang, D. L. Miller, and G. J. Sullivan, *Tech. Dig.—Int. Electron Devices Meet.* p. 159 (1986).

66. S. Tiwari, *Tech. Dig.*, *IEEE Gallium Arsenide IC Symp.*, p. 95 (1985).

67. H. T. Yuan, J. Delaney, H. D. Shih, and L. Tran, *Tech. Dig.*, ISSCC, p. 74 (1986).

68. D. A. Whitmire, V. Garcia, and S. Evans, *Tech. Dig.*, ISSCC p. 34 (1988).

69. K. de Graaf and K. Fawcett, *Tech. Dig.*, *IEEE Gallium Arsenide IC Symp.*, p. 205 (1986).

70. K. C. Wang, P. M. Asbeck, M. F. Chang, G. J. Sullivan, and D. L. Miller, *Tech. Dig.*, *IEEE Gallium Arsenide IC Symp.*, p. 83 (1987).

71. A. K. Oki, M. E. Kim, J. B. Camou, C. L. Robertson, G. M. Gorman, K. B. Weber, L. M. Hobrock, S. W. Southwell, and B. K. Oyama, *Tech. Dig.*, *IEEE Gallium Arsenide IC Symp.*, p. 137 (1987).

72. G. M. Gorman, J. B. Camou, A. K. Oki, B. K. Oyama, and M. E. Kim, *Tech. Dig.—Int. Electron Devices Meet.* p. 623 (1987).

73. F. Capasso and R. Kiehl, *J. Appl. Phys.* **58**, 1366 (1985).

74. N. Chand, T. Henderson, R. Fischer, W. Kopp, H. Morkoç and L. J. Giacoletto, *Appl. Phys. Lett.* **46**, 302 (11985).

75. D. A. Sunderland and P. D. Dapkus, *IEEE Electron Devices Lett.* **EDL-6**, 648 (1985); see also J. Hutchby, *ibid.* **EDL-7**, 108 (1986).

76. R. Fisher, J. Klem, J. S. Gedymin, T. Henderson, W. Kopp, and H. Morkoç, *Tech. Dig.—Int. Electron Devices Meet.* p. 332 (1985).

77. R. J. Malik, J. R. Hayes, F. Capasso, K. Alavi, and A. Y. Cho, *IEEE Electron Devices Lett.* **EDL-4**, 383 (1983).

78. H. Kanbe, J. C. Vlcek, and C. G. Fonstad, *IEEE Electron Devices Lett.* **EDL-5**, 172 (1984); see also R. N. Nottenburg, H. Temkin, M. B. Panish, R. Bhat, and J. C. Bischoff, *ibid.* **EDL-7**, 643 (1986).

5 Resonant Tunneling Diodes and Transistors: Physics and Circuit Applications

SUSANTA SEN[†], FEDERICO CAPASSO, and
FABIO BELTRAM

AT&T Bell Laboratories, Murray Hill, New Jersey

5.1 INTRODUCTION

The tunnel diode pioneered by Leo Esaki [1] in 1958 is the first example of a semiconductor device utilizing the quantum phenomenon of tunneling for its operation. Despite its great promise for high-speed circuit applications, after a decade or so this device became impractical. The main limitations of tunnel diodes are twofold. First the difficulty in reproducing the current–voltage (I–V) characteristics, since the latter is very sensitive to the doping levels. Because of the ultrahigh carrier concentrations required in tunnel diodes, the latter is difficult to control accurately. Second, its two-terminal nature makes it difficult to use in most logic applications. In addition, it has proven difficult to conceive transistors based on Zener tunneling.

In recent years, a new device, the resonant tunneling (RT) diode or double barrier (DB), first demonstrated in 1974 by Chang et al. [2] has emerged as a much more promising device than the tunnel diode for a variety of reasons. In RT diodes, the voltage position of the I–V peak is primarily controlled by the thickness of the quantum well between the barriers. This dimension can be precisely controlled (by approximately a few percent) using molecular beam epitaxy (MBE). More importantly, several transistors based on RT through DBs are possible, some of which are very promising for digital applications with reduced circuit complexity.

Since the first demonstration [2] of RT by Chang et al., the material quality has improved to the point that negative differential resistance (NDR) can be observed at room temperature [3]. Recently, peak-to-valley ratios of nearly 4:1 in the current through a single double-barrier quantum

[†]Present address: Institute of Radio Physics and Electronics, University of Calcutta, Calcutta-700 009, India.

well (DBQW) were reported at room temperature [4, 5]. At 77 K, the observed value of peak-to-valley is as high as 15:1 [5]. Further improvement in the peak-to-valley ratio (14:1 at room temperature) was obtained by the use of pseudomorphic structures [6]. Mendez et al. have also observed RT of holes [7].

Recently, Reed et al. [8] showed that the replacement of AlGaAs in the barriers with an AlAs/GaAs superlattice with the same average composition considerably improves the current–voltage characteristics of RT diodes by making it symmetric. Nakagawa et al. have also reported RT in triple barrier diodes [9]. Sen et al. have recently observed RT through parabolic quantum wells [10]. As many as 14 resonances were observed in the *I–V* of one such sample. Unlike rectangular quantum wells, the multiple resonances of the parabolic well were nearly equally spaced.

The first practical application of RT diodes was made by Sollner et al. at the MIT Lincoln Laboratories in 1983 [11]. They used the NDR of RT diodes in detectors and mixers at frequencies up to the terahertz range. With improvement of technology, the parasitic series resistance of RT diodes reduced and they could be used for microwave generation [12] as well. Recently, Brown et al. have also reported oscillation frequencies up to 200 GHz [13] using RT diodes. Apart from these early applications, RT structures have assumed great importance in recent years, as functional devices for circuit applications [14]. A variety of three-terminal devices (both unipolar and bipolar) have been proposed and implemented [15–28]. The RT approach to circuits is one of the several ones proposed to circumvent the limits imposed by scaling laws to the ever-increasing functional density [29]. In fact, Sen et al. have shown that through the use of RT devices, many circuits can be implemented with less devices per function [30]. In particular, RT devices with multiple negative resistance regions are of considerable interest for a variety of potential applications which could be realized with greatly reduced circuit complexity.

This chapter deals with the operation of some of these interesting RT devices and their circuit applications.

5.2 RESONANT TUNNELING DIODES

In this section, we first discuss the physics of RT through DBQW followed by some state-of-the-art results on RT diodes. Finally, the last part of the section is devoted to the discussion of some novel applications involving integration of RT diodes.

5.2.1 Physics of Resonant Tunneling

5.2.1.1 The Origin of Negative Differential Resistance. RT through a DB occurs when the energy of an incident electron in the emitter matches that

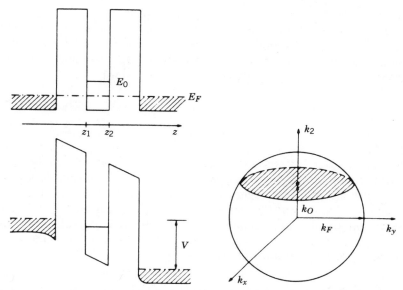

Figure 5.1   Illustration of the operation of a double-barrier RT diode. The top left part shows the electron energy diagram in equilibrium. The bottom left displays the band diagram for an applied bias V, when the energy of certain electrons in the emitter matches unoccupied levels of the lowest subband E_0 in the quantum well. The right illustrates the Fermi surface for a degenerately doped emitter. Assuming conservation of the lateral momentum during tunneling, only those emitter electrons whose momenta lie on a disk $k_z = k_0$ (shaded disk) are resonant. The energy separation between E_0 and the bottom of the conduction band in the emitter is given by $\hbar^2 k_0^2 / 2m^*$. In an ideal double-barrier diode at zero temperature, RT occurs in a voltage range during which the shaded disk moves down from the pole to the equatorial plane of the emitter Fermi sphere. At higher V (when $k_0^2 < 0$), resonant electrons no longer exist.

of an unoccupied state in the quantum well (QW) corresponding to the same lateral momentum. NDR arises simply from momentum and energy conservation considerations and does not require the coherence of the electron wave function (as is the case for the electronic analogous of the Fabry–Perot effect). In fact, NDR is a manifestation of the reduction in the dimensionality of the electronic states (in the present case, from 3D to 2D, but 2D to 1D was also considered [19]) in tunneling from the emitter to the quantum well. This has been clarified by Luryi [31] and is illustrated in Fig. 5.1.

Consider the Fermi sea of electrons in the degenerately doped emitter. Their energy can be expressed as

$$E_{3D} = E_c + \frac{\hbar^2 k_z^2}{2m^*} + \frac{\hbar^2 k_\perp^2}{2m^*} \tag{1}$$

where E_c is the bottom of the conduction band and $k_\perp^2 = k_x^2 + k_y^2$. In doing so, we are ignoring the quantization at the interface of the injecting electrode. This would simply add further structure to the peaks in the I–V without altering the main finding, that is, the NDR. The energy in the 2D target states in the QW, on the other hand, is given by

$$E_{2D} = E_n + \frac{\hbar^2 k_\perp^2}{2m^*} \tag{2}$$

where E_n is the bottom of the relevant subband in the QW. The RT of electrons into the 2D states requires the conservation of energy and of lateral momentum $k_\perp$. It is, of course, assumed that the barriers are free from impurities and inhomogeneities. The momentum conservation condition requires that the last term in Eqs. (1) and (2) be equal. So, from energy conservation we find that tunneling is possible only for electrons whose momenta lie in a disk corresponding to $k_z = k_0$ (shaded disk in the figure in which the case $n = 0$ is illustrated) where

$$k_0^2 = \frac{2m^*}{\hbar^2}(E_0 - E_c) \tag{3}$$

Only those electrons have isoenergetic states in the quantum well with the same $k_\perp$. This is a general feature of tunneling into a two-dimensional system of states. As the emitter–base potential rises, so does the number of electrons that can tunnel; the shaded disk moves downward to the equatorial plane of the Fermi sphere. For $E_n = E_c$, which corresponds to $k_0 = 0$, the number of tunneling electrons per unit area equals $m^* E_F / \pi \hbar^2$. When E_c rises above E_n, at $T = 0$ temperature, there are no electrons in the emitter which can tunnel into the quantum well while conserving their lateral momentum. Therefore, one can expect an abrupt drop in the tunneling current. This has been experimentally observed by Morkoç et al. [32]. Of course, similar arguments of conservation of lateral momentum and energy leading to NDR apply also to systems of lower dimensionality; for example, to tunneling of two-dimensional electrons through a quantum wire and to RT in one dimension.

5.2.1.2 Coherent (Fabry–Perot-type) Resonant Tunneling. Let us now consider the Fabry–Perot effect. In the presence of negligible scattering of the electrons in the well, the above NDR effect is accompanied by a coherent enhancement of the transmission, analogous to that occuring in an optical Fabry–Perot. In the case of a symmetric structure (such as the one discussed in Section 5.3), after a transient in which the electron wave function builds up in the quantum well, an equilibrium is reached where the portion of the incident wave reflected by the first barrier is exactly cancelled by the fraction of the electron wave function leaking from the QW to the left. The net effect is a total transfer of electrons from the left to the right throught the

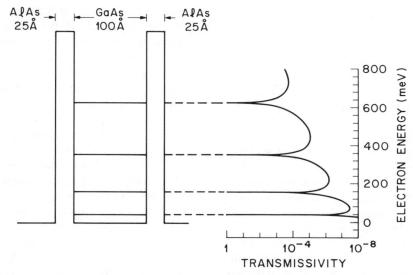

Figure 5.2 Schematic conduction band energy diagram of an AlAs/GaAs/AlAs RT double barrier with 25 Å barriers and 100 Å well (left) and the calculated transmissivity as a function of incident electron energy (right).

DB. This is shown in Fig. 5.2 by the unity transmissivity peaks. In this case of coherent RT, the peak transmission at resonance is given approximately by $T_{\min}/T_{\max}$, where $T_{\min}$ is the smallest among the transmission coefficients of the two barriers and $T_{\max}$ is the largest [33]. It is, therefore, possible to achieve unity transmission at the resonance peaks as discussed above, by making the transmission of the left and right barriers equal. This crucial role of symmetry has been discussed in detail by Ricco and Azbel [33]. Application of an electric field to a symmetric DB introduces a difference between the transmissions of the two barriers, thus significantly decreasing, below unity, the overall transmission at the resonance peaks. Unity transmission can be restored if the two barriers have different and appropriately chosen thicknesses; obviously, with this procedure, one can only optimize the transmission of one of the resonance peaks. This is because the exit barrier becomes lower under application of an electric field and therefore has a higher transmission than the input barrier. Unity transmission can be restored by making the exit barrier thicker. However, as discussed by Weil and Vinter [34] and Jonson and Grincwajg [35] and as recently elaborated by Luryi [36], these arguments do not strictly apply to the DB under the large biases required for RT operation. In fact, in order to calculate the current, one must average over the energy distribution of incoming electrons, which typically is much broader than the resonance width of the quantum state. It was shown [34–36] that in this case the peak current is proportional to $T_{\min}$ only and the above approach does not apply.

In a recent experiment, Choi et al. have obtained the first strong evidence of coherent tunneling [37], by measuring the photocurrent in two different bias polarities in a specially designed superlattice structure. Each period of the superlattice consisted of a thick GaAs QW (72 Å), a thin $Al_{0.33}Ga_{0.67}As$ barrier (39 Å), followed by a thin GaAs QW (18 Å) and a thick $Al_{0.33}Ga_{0.67}As$ barrier (154 Å). All the layers in the superlattice were undoped except the 72 Å thick QWs, which were n-type doped to $1 \times 10^{18}/cm^3$. For electrons photoexcited from the ground to the first excited state of the doped QW, Choi et al. [37] have calculated the transmission coefficients of the two barriers for opposite bias polarities. It was shown that under suitable bias condition in one polarity, the transmission coefficient of the two barriers are nearly equal, resulting in a large coherent enhancement of the photocurrent. In the other polarity of the applied bias, however, the transmission coefficient of the thicker barrier is always much smaller than the thinner one. Theoretical calculations [37] also indicate that the maximum ratio of the forward- and reverse-bias transmissivities would be 10 and 100, respectively, under sequential and coherent tunneling conditions, which was observed experimentally, confirming the presence of coherent RT.

5.2.1.3 The Role of Scattering: Incoherent (Sequential) Resonant Tunneling.

RT through a DB has been investigated experimentally by many researchers [2–13, 38–44]. All of these investigations assumed that a Fabry–Perot-type enhancement of the transmission was operational in such structures. However, as previously discussed, the observation of NDR does not imply a Fabry–Perot mechanism. Other types of tests are necessary to show the presence of the coherence of the wave function, such as the dependence of the peak current on the thickness of the exit barrier.

The presence of scattering gives rise to another physical picture for RT. Once the electrons are injected into the QW as discussed in Sec. 5.2.1.1, scattering events can randomize the phase of the electronic wave function; this considerably weakens the coherent enhancement of the transmission. RT is then a two-step process in which the electrons first tunnel into the well and then out of it through the second barrier. *The first step is the one that gives rise to NDR.* A lucid discussion of this point has recently been given by Stone and Lee [45] in the context of RT through an impurity center. Unfortunately, their work has gone unnoticed among researchers in the area of quantum well structures. Their conclusions can also be applied to RT through quantum wells and we shall discuss them in this context.

To achieve the resonant enhancement of the transmission (Fabry–Perot effect), the electron probability density must be peaked in the well. The time constant for this phenomenon, τ_0, is of the order of $\hbar/\Gamma_r$, where Γ_r is the full width at half maximum of the transmission peak. Collisions in the DB tend to destroy the coherence of the wave function, and therefore the electronic density in the well will never be able to build up to its full resonant value. If the scattering time τ is much shorter than τ_0, the peak transmission at resonance is expected to be decreased by the ratio τ_0/τ. The principal effects of collisions are to decrease the peak transmission by the ratio $\tau_0/(\tau_0 + \tau)$ and to broaden the resonance. In addition, the ratio of the number of electrons that resonantly tunnel without undergoing collisions to the number that tunnel after undergoing collisions is equal to τ/τ_0 [45]. To summarize, coherent RT is observable when the intrinsic resonance width ($\approx \hbar/\tau_0$) exceeds or equals the collision broadening ($\approx \hbar/\tau$). In the other limit, electrons will always tunnel through one of the intermediate states of the well, but they will do it incoherently without resonant enhancement of the transmission. We shall apply now the above criterion to RT through AlGaAs/GaAs DB investigated in many experiments.

Consider a 50 Å thick GaAs well sandwiched between two $Al_{0.30}Ga_{0.70}As$ barriers. Table 5.1 shows the ground-state resonance widths Γ_r (full width at half the maximum of the transmission curve) calculated for different values of the barrier thicknesses L_B (assumed equal). Note the strong dependence of Γ_r on L_B. This is due to the fact that Γ_r is proportional to the transmission coefficient of the individual barriers which decreases exponentially with increasing L_B. The case $L_B = 50$ Å corresponds to the microwave oscillator reported by Sollner et al. [12].

Because of dimensional confinement in the wells and because the wells are undoped, one can obtain a good estimate of the scattering time of electrons in the wells from the mobility of two-dimensional electron gas (2DEG) (in the plane of the layers), measured in selectively doped AlGaAs/GaAs heterojunctions [46]. For state-of-the-art selectively doped AlGaAs/GaAs heterojunctions, the electron mobility at 300 K is $\sim 7000 \, cm^2/V \cdot s$. From this value, we can infer an average scattering time $\cong 3 \times 10^{-13}$ s, which corresponds to a broadening of $\cong 2$ meV. In Table 5.1, the ratio of the resonance width Γ_r to the collision broadening Γ_c is also shown. For the 50 and 70 Å barrier case, the resonance width is much smaller than the collision broadening so that, by the previously discussed criterion, there is very little resonant enhancement of the transmission via the Fabry–Perot mechanism at 300 K. However, the latter effect should become visible in structures with thinner barriers (<30 Å), as seen from

TABLE 5.1 Resonance and Collision Widths of $Al_{0.30}Ga_{0.70}As$/GaAs RT Diode (at Zero Bias) for Different Barrier Thicknesses

L_W	L_B	Γ_r	Γ_r/Γ_c^a		
(Å)	(Å)	(meV)	300 K	200 K	70 K
50	70	1.28×10^{-2}	6×10^{-3}	1.93×10^{-2}	2.6×10^{-1}
50	50	1.5×10^{-1}	7.5×10^{-2}	2.26×10^{-1}	3.08
50	30	1.76	8.8×10^{-1}	1.32	3.62
20	50	6.03	3.02	4.56	124.02

[a]Resonance width/collision broadening.

Table 5.1. Consider now a temperature of 200 K; from the mobility ($\approx 2 \times 10^4$ cm^2/V·s) [46], one deduces $\tau \simeq 1$ ps, which corresponds to a broadening of $\simeq 0.67$ meV. This value is comparable to the resonance width for a barrier width of 50 Å. This implies that in Sollner's microwave oscillators [12] (which operated at 200 K), coherent RT effects were probably present. This is definitely not the case for the mixing and detection experiments performed up to terahertz frequencies [11] in DB RT structures with $L_W = L_B = 50$ Å, and Al mole fraction $x = 0.25$–0.30 at a temperature of 25 K. In this case, the well was intentionally doped to $\simeq 10^{17}$/cm^3, which would correspond to a mobility of ≈ 3000 cm^2/V·s, giving a collision broadening of 4 meV, significantly larger than the resonant width. Thus, in this case, electrons are tunneling incoherently (i.e., sequentially) through the DB.

Finally, an estimate of Γ_r/Γ_c for a temperature of 70 K is given in Table 5.1. State-of-the-art mobilities in selectively doped interfaces exceed 10^5 cm^2/V·s, so that scattering times are typically longer than 1 ps and the broadenings are less than 0.5 meV. Thus, coherent RT will significantly contribute to the current for barrier widths $\lesssim 70$ Å and dominate for $L_B \lesssim 30$ Å. The values of Γ_r/Γ_c at 70 K in Table 5.1 were obtained using a mobility of 3×10^5 cm^2/V·s [46].

The situation appears to be different in the case of AlAs/GaAs DB with well widths of 50 Å. The confining barriers in this case are much higher (≈ 1.35 eV) [22], and for barrier thicknesses in the 3–70 Å range, the resonance widths are $\lesssim 10^{-2}$ meV. Thus, coherent RT is negligible at room temperature but is expected to become dominant at 70 K for $L_B \lesssim 70$ Å in high-quality DBs.

Capasso et al. [47] have demonstrated sequential RT through a superlattice under strong electric fields. In a strong electric field, the miniband picture in a superlattice breaks down when the potential drop across the superlattice period exceeds the miniband width. When this condition is satisfied, the quantum states become localized in the individual wells. In this limit, an enhanced electron current will flow at sharply defined values of the external field, when the ground state in the nth well is degenerate with the first or the second excited state in the $(n + 1)$th well, as illustrated in Fig. 5.3a. Under such conditions, the current is due to electron tunneling between the adjacent wells with a subsequent dexcitation in the $(n + 1)$th well, by emission of photons. In other words, electron propagation through the entire superlattice involves sequential RT.

Experimental difficulties in studying this phenomenon are usually associated with the nonuniformity of the electric field across the superlattice and the instabilities generated by negative differential conductivity. To ensure a strictly and spatially uniform electric field, Capasso et al. [47] placed the superlattice in the n^- ($\lesssim 10^{14}$ cm^3) region of a reverse-biased p^+-n^--n^+ junction. This structure allowed observation of the sequential RT for the first time. Two NDR peaks observed in the photocurrent characteristics (Fig. 5.4) correspond to the resonances shown schematically in Fig. 5.3.

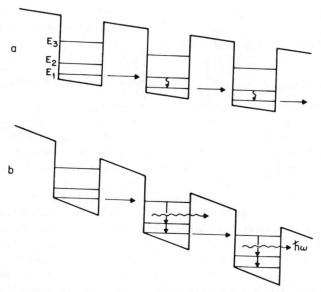

Figure 5.3 Band diagram of (a) sequential resonant tunneling and (b) far infrared laser using sequential resonant tunneling.

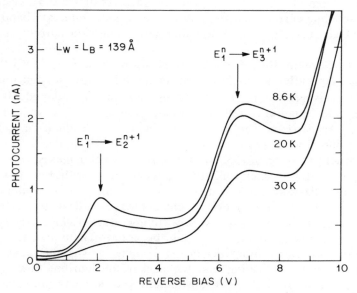

Figure 5.4 Photocurrent–voltage characteristics at $\lambda = 0.6328 \, \mu m$ (pure electron injection) for a superlattice of $Al_{0.48}In_{0.52}As / Ga_{0.47}In_{0.53}As$ with 138 Å thick wells and barriers and 35 periods. The arrows indicate that the peaks correspond to resonant tunneling between the ground state in the nth well and the first two excited states of the $(n + 1)$ well.

For the sequential RT regime, there is the possibility of a laser action at the intersubband transition frequency—an effect not yet observed experimentally in a superlattice (Fig. 5.3b).

5.2.2 $Ga_{0.47}In_{0.53}As/Al_{0.48}In_{0.52}As$ Resonant Tunneling Diodes

Many efforts have been directed toward achieving better performance , such as room-temperature operation and improved peak-to-valley ratio. So far, RT structures composed of the GaAs/AlGaAs material system have received major attention. The highest peak-to-valley current ratio at room temperature obtained so far in this material system is 3.9:1 [4]. Recently, Muto et al. have reported a systematic study of $Ga_{0.47}In_{0.53}As/Al_{0.48}In_{0.52}As$ RT diodes at low temperature (77 K) [43]. The same material system has been used by Yokoyama et al. in their resonant tunneling hot electron transistor (RHET) [48] to obtain improved performance at 77 K. Sen et al. have presented data on the room-temperature operation [5] of a $Ga_{0.47}In_{0.53}As/Al_{0.48}In_{0.52}As$ RT diode grown lattice-matched to InP substrate by MBE. The $I–V$ characteristics of these diodes exhibit peak-to-valley ratios of 4:1 at room temperature and 15:1 at 80 K. The improved peak-to-valley ratio compared with AlGaAs/GaAs RT diodes is due to the lower electron effective mass in the barriers ($0.075\,m_0$ for $Al_xIn_{1-x}As$ at $x = 0.48$ compared with $0.092\,m_0$ for $Al_xGa_{1-x}As$ at $x = 0.3$), resulting in higher tunneling current density, and to the larger conduction band discontinuity which strongly reduces the thermionic emission current across the barrier.

The structure in Ref. 5 consisted of $1\,\mu m$ thick n^+-($\sim 3 \times 10^{17}\,cm^{-3}$) $Ga_{0.47}In_{0.53}As$ buffer layer grown on an n^+ InP substrate. On top of the buffer layer is grown the RTDB, consisting of an undoped $50\,\text{Å}$ wide $Ga_{0.47}In_{0.53}As$ quantum well sandwiched between $50\,\text{Å}$ wide undoped $Al_{0.48}In_{0.52}As$ barriers. The growth ends with a $1\,\mu m$ thick $Ga_{0.47}In_{0.53}As$ cap layer doped to $n^+ \sim 3 \times 10^{17}\,cm^{-3}$.

The structures were processed into $50\,\mu m$ diameter mesas, using photolithography and wet chemical etching techniques, with top and bottom contact metallizations and tested in a Helitran dewar equipped with microprobes. The room-temperature characteristics indicate a peak-to-valley ratio of as high as 4:1 [5]. At low temperature (80 K), the peak-to-valley ratio increases to 15:1. It should be noted that though the peak-to-valley ratio increases dramatically on cooling down, the peak current remains the same, indicating that the background current flowing due to inelastic processes only changes substantially with temperature. The peak in the $I–V$ occurs at $\sim 600\,mV$ and does not change with temperature. An electron tunneling transmission calculation shows that the first resonance is at $E_1 = 126\,meV$ from the bottom of the quantum well. Note that the peak in the $I–V$ appears at a voltage greater than $2E_1/e = 252\,mV$. This can be explained by considering the voltage drop in the depletion and accumulation regions in the collector and emitter layers adjacent to the DB. Thus, a larger voltage must be applied across the entire structure to line up the first subband in the well with the bottom of the conduction band in the emitter in order to quench RT. A simple calculation, taking the above effects into account, indicates that the peak should occur at $\simeq 580$ mV applied bias, which is in reasonable agreement with the measured value.

The large peak-to-valley ratio observed at room temperature makes this device suitable for many circuit applications. A circuit with a $30\,\Omega$ load resistance in series with the device and a 3.0 V supply has two stable operating points, which are measured to be 0.47 V and 0.85 V, respectively, at room temperature. The corresponding load line drawn on the room temperature I–V characteristics indicates the stable operating points at 0.46 V and 0.84 V, respectively, which are in close agreement with the measured values. The circuit can thus be used as a static RAM cell involving only one device. Such a RAM cell is also suitable for integration in a large memory array, as discussed in Section 5.2.4.1, in connection with multistate memories.

5.2.3 Resonant Tunneling Through Parabolic Quantum Wells

Parabolic quantum wells have interesting possibilities for device applications [15], because the levels in such a well are equally spaced, unlike rectangular wells. The I–V characteristics of RT structures with parabolic wells therefore are expected to produce equally spaced peaks in voltage. The first experimental observation of such resonances was reported by Sen et al. [10] in 1987.

The samples in Ref. 10 were grown by MBE on silicon-doped (100) GaAs substrates at a substrate temperature of 680°C. The growth was computer-controlled and calibrated by ion-gauge flux measurement at the position of the substrate. Parabolically graded well compositions were produced by growth of short-period (~ 15Å), variable duty cycle GaAs/$Al_xGa_{1-x}As$ superlattices in which the Al content within each period of the superlattice corresponded to the Al content at the same point in a smooth parabolic well [49]. A cross-sectional transmission electron micrograph (TEM) of one such structure is shown in Fig. 5.5. The structure consists of a 439 Å parabolic quantum well of $Al_xGa_{1-x}As$, with x varying from 0.3 at the edges to 0 at the center, sandwiched between two 35 Å AlAs barriers. The parabolic part of the structure is composed of variable-gap superlattice with a period of nearly 10 Å, as discussed above. The brighter lines in the well part of the TEM picture represent $Al_{0.3}Ga_{0.7}As$ layers, while the darker lines represent GaAs layers. Notice how the relative widths of the bright and the dark lines change from the edges of the well to its center. The electrons, of course, "sense" the local average composition, since their de Broglie wavelength is much greater than the superlattice period.

Two types of structures were grown. In one, sample A, the 300 Å

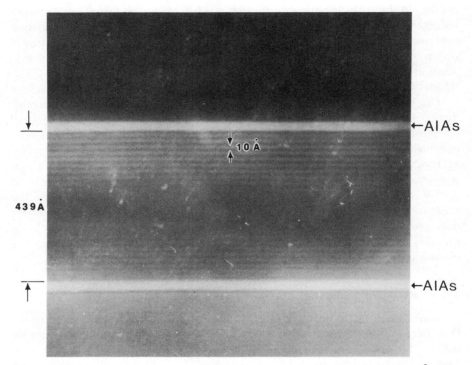

Figure 5.5 A cross-sectional transmission electron micrograph of a 439 Å wide parabolic quantum well composed of $Al_xGa_{1-x}As$, with x varying from 0.3 at the edges to 0 at the center, sandwiched between two 35 Å AlAs barriers.

undoped well is sandwiched between two 20 Å AlAs undoped barriers. The parabolic well composition is effectively graded from $x = 1$ at the edges to $x = 0$ at the center. The portion of the well from $x = 0.49$ to $x = 0$ is achieved by means of an $Al_{0.50}Ga_{0.50}As/GaAs$ superlattice, whereas the rest (from $x = 0.49$ to $x = 1$) is achieved using an AlAs/GaAs superlattice. Undoped 20 Å thick GaAs spacer layers were used between the barriers and the Si-doped ($n = 10^{18}$ cm^{-3}) 5000 Å thick GaAs contact layers. Systematic studies have shown that offsetting the doping in the regions adjacent to the barriers significantly improves the I–V of RT diodes [40].

In sample B the 439 Å undoped well is bounded by 35 Å AlAs undoped barriers, and the composition of the well is graded from $x = 0.30$ at the edges to $x = 0$ at the center using an $Al_{0.325}Ga_{0.675}As/GaAs$ superlattice. $Al_{0.02}Ga_{0.98}As$ 1000 Å thick layers Si-doped to 5×10^{17} cm^{-3} (with a doping offset of 50 Å from the barriers) were used as contact regions to the RTDB. The composition of these layers was chosen in such a way that the bottom of the conduction band in the emitter is nearly lined up with (but always below) the first energy level of the well, a technique successfully used in improving the peak-to-valley ratio of RT transistors and diodes [22, 41, 42]. These layers are followed by 1000 Å regions compositionally graded from $x = 0.02$ to $x = 0$ Si-doped to $n = 5 \times 10^{17}$ cm^{-3} and by 4000 Å thick Si-doped ($n = 1 \times 10^{18}$ cm^{-3}) GaAs. The thickness of the layers were determined by cross-sectional TEM.

The energy band diagrams at the Γ point [50] for samples, A and B are shown in Fig. 5.6. The structures were processed into 50 μm diameter mesa diodes with top and bottom ohmic contacts (Section 5.2.2). The diodes were tested at temperatures in the range from 7 to 300 K in a Helitran dewar equipped with microprobes. The I–V and the differential conductance were measured with an HP4145 parameter analyzer.

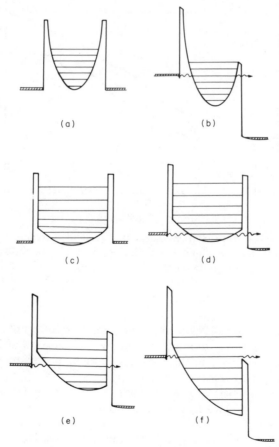

Figure 5.6 (a–b) Band diagram of sample A in equilibrium and under RT conditions. (c–f) Band diagram of sample B in equilibrium and under different bias conditions. The wells are drawn to scale; however, for the sake of clarity, only half the number of levels in an energy interval are shown.

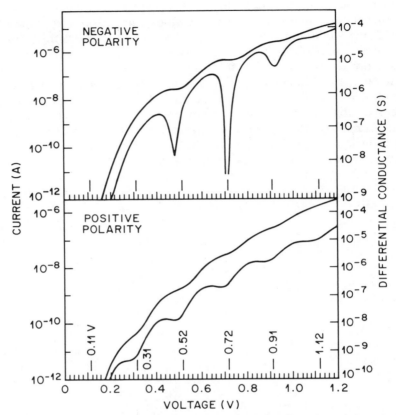

Figure 5.7 Current–voltage characteristics at 7.4 K and corresponding conductance for a representative diode of sample A under opposite bias polarity conditions. The vertical segments near the horizontal axis indicate the calculated positions of the resonances.

Figure 5.7 shows the I–V and the corresponding differential conductance di/dV for a representative diode from sample A measured at 7.5 K. Positive and negative polarity refer to the top contact being positively and negatively biased with respect to the bottom contact.

Consider first the positive polarity data. These display five equally spaced inflections in the I–V and five corresponding minima in the conductance. For negative polarity, four of the conductance minima occur at practically identical voltages as the corresponding minima for positive polarity, with the exception of the resonance at +0.3 V, which is not observed experimentally for negative polarity. The resonances for negative polarity are more pronounced, and the one at $\simeq$0.72 V actually exhibits negative differential resistance.

Energy levels in the wells under bias were determined by electron tunneling transmission calculations for the grown layer sequence; thus, the effect of the superlattice grading is directly included. The effects of depletion and accumulation in the emitter and collector layers are also taken into account. A conduction band offset of 0.60 times the direct energy gap differences was assumed [49]. Electron effective mass dispersion with Al content was included. From these calculations, the voltage position of the transmission peaks can be directly obtained. These voltages are indicated by the vertical segments at the bottom of Fig. 5.7 and correspond, respectively, to the first six energy levels of the well. The first resonance is not seen experimentally because the corresponding current is below the detection limit of the apparatus (~1 pA). The other calculated positions of the resonances (E_2-E_6) are in good agreement with the observed ones. The calculations show that not only the energy levels for a given bias are nearly equally spaced, but also that the spacing ΔE between the quasi-bound states depends little on the electric field as the voltage is varied from 0 to 1 V. For example, at zero bias, $\Delta E \simeq 90$ meV, while at 1 V, $\Delta E \simeq 80$ meV. The latter effect is easily understood if one considers that the application of a uniform electric field to a parabolic well (Fig. 5.6a) preserves the curvature of the parabola and therefore the spacing ΔE, while shifting its origin to the right (Fig. 5.6b).

Figure 5.8 shows the I–V and corresponding conductance for sample B for opposite bias polarities. The band diagrams at different voltages are shown in Fig 5.6c–f.

It is interesting to note that the group of resonances from the 5th to the 11th are the most pronounced ones and actually display negative differential resistance. A total of 14 resonances is observed in the sample of Fig. 5.8, for positive polarity. In a few diodes, two additional resonances were also observed. The resonances were observed up to temperatures $\simeq 100$ K, but are considerably less pronounced. The vertical segments near the horizontal axis indicate the calculated positions of the transmission peaks. Overall good agreement with the observed minima in the conductance is found.

The overall features of the I–V can be interpreted physically by means of the band diagrams of Fig. 5.6c–f and of the calculations. At zero bias, the first six energy levels of the well are confined by a parabolic well 225 meV deep, corresponding to the grading from $x = 0$ to $x = 0.30$; their spacing is $\simeq 35$ meV. When the bias is increased from 0 to 0.3 V, the first four energy levels probed by RT (Fig. 5.6d) remain confined by the parabolic portions of the well, and their spacing is practically independent of bias, for reasons identical to those discussed in the context of sample A. This gives rise to the calculated and observed equal spacing of the first four resonances in the I–V characteristic (Fig. 5.8). Consider now the higher energy levels confined by the rectangular part of the well (>230 meV) at zero bias. When the voltage is raised above 0.3 V, these levels become increasingly confined on the emitter side by the parabolic portion of the well and on the collector side by

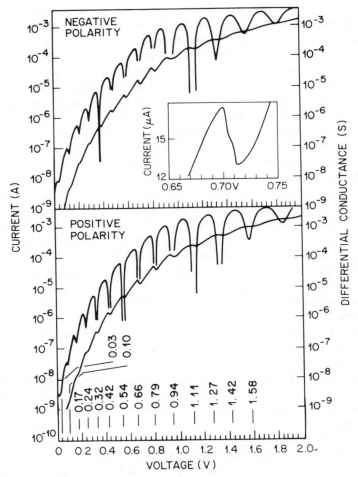

Figure 5.8 Current–voltage characteristics at 7.1 K and conductance for a representative diode of sample B under opposite bias polarity conditions. The inset shows the eight resonance on a linear scale. The vertical segments near the horizontal axis indicate the calculated positions of the resonances.

a rectangular barrier, thus becoming progressively more separated, although retaining the nearly equal spacing (Fig. 5.6e). This leads to the observed gradual increase in the voltage separation of the resonances as the bias is increased from 0.3 to 1.0 V. Above 1 V, the electrons injected from the emitter probe the virtual levels in the quasi-continuum above the collector barrier (Fig. 5.6f). These resonances result from electron interference effects [51] associated with multiple quantum mechanical reflections at the well–barrier interface for energies above the barrier height. It should be noted that these reflections give rise to the existence of quasi-2D states in the well region. The observed NDR is due to tunneling into these states. These interference effects give rise to the four resonances observed above 1 V and must be clearly distinguised from the ones occurring at lower volatages, which are due to RT through the DB.

A simple physical explanation of why the resonances above the 4th (5th–12th) are the most pronounced ones, leading to negative differential resistance, can be easily given in terms of the calculated voltage dependence of the transmission.

Up to 0.3 V tunneling out of the well (Fig. 5.6d) occurs through the thick parabolic part of the collector barrier, and the resulting widths of the transmission resonances are very small ($<$1 μeV). As the bias is increased above 0.3 V, not only is the barrier height further reduced but electrons tunnel out of the well through the thin (20 Å) rectangular part of the barriers. This greatly enhances the barrier transmission and the resonance widths (Fig. 5.6e). This behavior is clearly observed in the calculation of the total transmission versus bias. For example, the calculated energy width of the level corresponding to the 10th resonance (0.9 V) is 1 meV, which is not negligible compared with the width of the incident energy distribution in the emitter. As the bias voltage is further increased, the competing effect of the decrease of the peak-to-valley ratio takes over, as shown by the calculations. This explains why the highest resonances (above 1 V) become less pronounced.

5.2.4   Integration of Resonant Tunneling Diodes and Their Circuit Applications

A variety of potential applications could be realized with devices exhibiting multiple peaks in the I–V. These include ultrahigh-speed analog-to-digital converters, parity bit generators, and multiple valued logic, [14, 15]. RT devices offer interesting possibilities toward realizing these circuits. One way of obtaining multiple peaks in the I–V naturally, is by using the multiple resonances of a quantum well. However, this approach suffers from the difficulty that the peaks corresponding to the excited states generally carry significantly higher current than that associated with the ground state for a variety of structural and material reasons. On the other hand, the above-mentioned circuit applications require nearly equal peak currents. A novel approach for obtaining multiple peaks in the I–V of RT devices, introduced by Capasso et al. [52], is the integration of a number of RT diodes. In this method, a single resonance of the quantum well is used to generate the multiple peaks. Hence, they occur at almost the same current level and exhibit similar peak to valley ratios. There are two possible methods of integrating RT diodes to achieve this characteristic. One is to integrate them horizontally so that the diodes come in parallel in the equivalent circuit [30, 52]; the other is to integrate them vertically where they are in series [53–55].

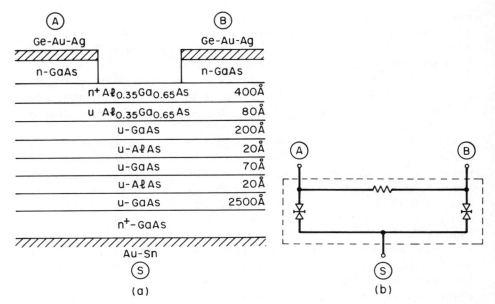

Figure 5.9 (a) Schematics of the integrated RT diode structure. (b) Equivalent circuit: two resonant tunneling (RT) diodes in parallel connected by the resistance R of the 200 Å GaAs channel between A and B. The resistance of the $Al_{0.35}Ga_{0.65}As$ layer between A and B is much higher than R due to carrier depletion by electron transfer into the GaAs channel. The choice of the circuit symbol for the RT diode (two back-to-back tunnel diodes) is motivated by the symmetry of the current–voltage characteristics of the RT diode).

5.2.4.1 Horizontal Integration of RT Diodes.

The device structure, grown on a $\langle 100 \rangle n^+$ Si-doped GaAs substrate, is shown in Fig. 5.9a. An undoped GaAs layer and is followed by the RTDB. The latter consists of a 70 Å GaAs quantum well sandwiched between two 20 Å AlAs barriers. A modulation doped $Al_{0.35}Ga_{0.65}As$/GaAs heterojunction is then grown on top of the DB; the 200 Å thick GaAs is undoped, while the 480 Å thick $Al_{0.35}Ga_{0.65}As$ layer is doped with Si to 2×10^{18} cm^{-3} except for an 80 Å spacer region adjacent to the GaAs channel. The channel contains a high density ($\simeq 10^{18}$ cm^{-3}) high-mobility electron gas, spatially separated from the parent donors in the AlGaAs layer. As a result, the AlGaAs layer is completely depleted. The growth ends with an n^+-GaAs 1400 Å contact layer doped to $n \simeq 2 \times 10^{17}$ cm^{-3}.

The prototype device fabricated has two rectangular (240×80 μm) contact pads separated by a distance of 6.5 μm along the long side, defined by successively evaporating Ge(120 Å), Au (270 Å), Ag (1000 Å), and Au (1500 Å) and using lift-off techniques. The metallizations were then alloyed at 380°C for 10 s and used as a mask for wet chemical etching. A selective stop etch (H_2O_2 and NH_4OH, pH = 7.2) was used to reveal the $Al_{0.35}Ga_{0.65}As$ barrier. Note that the thickness of the cap layer and the composition of the two top contacts and the alloying temperature and time were the same used for the fabrication of the charge injection transistor, which is structurally similar to this device [56]. This ensures that the contacts to the electron gas in the GaAs layer beneath the AlGaAs barrier do not penetrate through the RTDB.

This is thus an integration of two RT diodes in parallel (Fig. 5.9b) that exhibits two peaks of nearly equal current at 100 K. The device has been used to demonstrate for the first time a number of practical circuits using devices with multiple peaks in the I–V. The scheme may, however, be extended to the integration of more than two RT diodes, as discussed in the following section.

Device operation. The operation of the device can be understood from the equivalent circuit shown in Fig. 5.9b. This device consists essentially of two monolithically integrated RT diodes in parallel. The resistance shown in Fig. 5.9b is that of the GaAs 200 Å channel connecting the two diodes and its measured value is $\approx 12\,\Omega$. In the two-diode case discussed in this section, this resistance is not essential for the operation of the device. But the scheme can be obviously extended to more than two RT diodes; in the latter case, the resistance of the channel linking the devices provides the useful function of a monolithically integrated voltage divider. For proper biasing of this voltage divider, the structure should be suitably designed so that the current in the divider network is sufficiently large compared with that through the RT diodes.

The use of the modulation doped heterojunction allows the formation of a low-resistance ohmic contact to the RT diodes while keeping the dopants away from the DB. In addition, the AlGaAs passivates the GaAs channel between the two metallizations.

The substrate current (i.e., the one through terminal S) is measured as a function of positive bias applied between terminals S and A (which is grounded) for different values of the potential difference V_{BA} applied between B and A. The substrate current consists primarily of the sum of the two RT currents flowing through the two RT diodes. For zero potential difference V_{BA}, the structure behaves like a conventional RT diode and the I–V displays one peak (Fig. 5.10a and b). The negative conductance region is, of course, due to the quenching of RT through the two DBs under terminals A and B, respectively. When terminal B is biased negatively with respect to terminal A, the I–V characteristics (see Fig. 5.10a) develops an additional peak at lower voltages; the position of one peak remains unchanged while that of the other moves to lower bias as the potential difference V_{BA} between B and A is made more negative. Note that by appropriate choice of the bias between B and A the two peak currents can be made nearly equal.

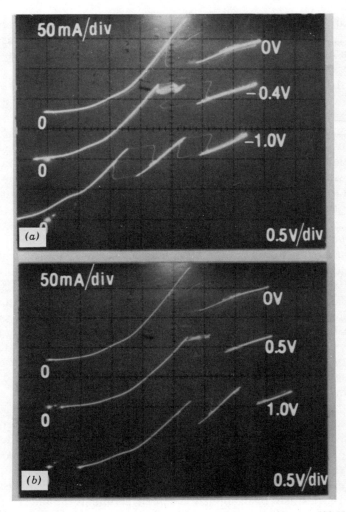

Figure 5.10 Substrate current versus positive substrate bias at 100 K with (a) negative and (b) positive potential difference V_{BA} between terminals B and A as the parameter. Terminal A is grounded.

This effect is explained as follows. As a result of the bias applied between A and B, the potential differences across the two DBs are different and for B negatively biased with respect to A, RT through the DB under B is quenched at a lower substrate bias than in the DB under terminal A, leading to two peaks in the I–V.

The peak that does not shift with varying V_{BA} is, naturally, associated with quenching of RT through diode A. Note also that, as expected, the separation between the peaks is nearly equal to the bias applied between A and B. Finally, if terminal B is positively biased with respect to A, a higher voltage is required to quench RT through the DB B, leading to a second peak which shifts to higher voltages as V_{BA} is increased (Fig. 5.10b). Similar results are obtained with negative bias applied to S.

The characteristics of Fig. 5.10 were obtained at an operating temperature of 100 K. With improved processing and material quality, it should be possible to operate the device at room temperature. This has already been achieved in RT diodes and transistors.

Circuit Applications. The I–V of Fig. 5.10 has been used in a variety of circuit applications ranging from frequency multipliers to multiple valued logic elements [30, 52]. Such applications for devices with multiple peaks in the I–V had been predicted before [14, 15] but could not be implemented for the lack of a practical device exhibiting such characteristics. The circuits constructed using the new device are discussed and the experimental results are presented in this section.

Frequency Multiplier. The I–V characteristic with two peaks is used to design a frequency multiplier. Its operation is understood from Fig. 5.11a and b, which show the I–V for a typical bias voltage V_{BA} between the terminals A and B of the device. Figure 5.11a and b shows the operation with a sawtooth and a sinewave input, respectively. Let us consider the operation with a sawtooth input first. The substrate bias V_{SS} is adjusted to select the quiescent operating point at A_2 of the I–V. As the sawtooth input voltage increases from A_1 to B_1, the operating point shifts from A_2 to B_2 along the I–V, with the substrate current I_s increasing almost linearly. The output voltage across a resistance in series with the structure is proportional to I_s and hence also increases from A_3 to B_3 linearly. As the input increases beyond B_1, the current I_s suddenly drops to the valley point B_2' resulting in a sudden drop in the output voltage from B_3 to B_3'. Between B_3 and C_3, the output continues to rise again, followed by a second drop at C_2, and rise thereafter as the input continues to rise up to D_1. At D_1, the input returns to zero to start a new cycle and the operating point also shifts back to A_2, with a drop in the output as well. Thus, the frequency of the sawtooth input signal has been multiplied by a factor of three. It should be noted here that the multiplier circuit described is independent of the input signal frequency.

The operation of the circuit with a sinewave input is shown in Fig. 5.11b. The output waveform in this case can also be explained following similar arguments and is found to be rich in the fifth harmonic of the input. Figure 5.12a and b show the experimental results for a sawtooth and a sinewave input, respectively, with the device biased at $V_{BA} = 1$ V and $V_{SS} = 2.3$ V. The efficiency of this device in generating the fifth harmonic is therefore found to be much better than in conventional devices, such as step-recovery diode, used in frequency multiplier circuits. It should also be noted here that if V_{BA} is adjusted to produce a single peak in the I–V instead of two, the sawtooth

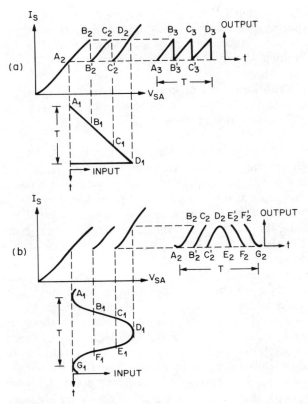

Figure 5.11 The schematic operation of the frequency multiplier for (a) sawtooth and (b) sinewave inputs.

will be multiplied by a factor of two instead of three and the output for sinewave will be rich in the third harmonic. Sinewave multiplication by a factor of two using a single peak in the $I-V$ of a RT hot-electron transistor has been demonstrated before [16].

MULTIPLE VALUED LOGIC. The circuit shown in the inset of Fig. 5.13 can be used as a memory element in a three-state logic system. The bias voltage V_{BA} between the terminals A and B is again adjusted to produce the $I-V$, as shown in Fig. 5.13, with two nearly equal peaks at the same current level. For a suitable supply voltage V_{SS} and load resistance R_L, the load line intersects the $I-V$ at five different points, three of which (Q_1, Q_2 and Q_3) are in the positive slope parts of the curve and are hence stable operating points. The output voltage of the circuit corresponding to the three operating points Q_1, Q_2, and Q_3 is represented, respectively, by V_1, V_2 and V_3, as shown in Fig. 5.13. The circuit can stay indefinitely on any one of the three points, thus retaining the last voltage information impressed on it. It can

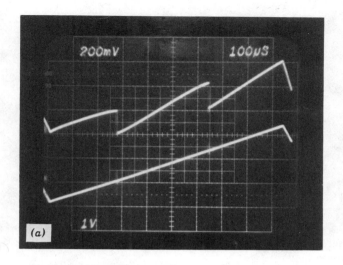

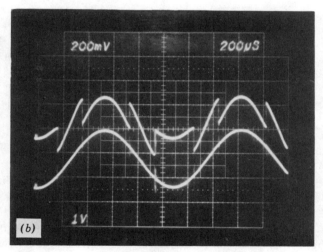

Figure 5.12 Experimental results of frequency multiplication: (a) sawtooth input, (b) sinewave input.

therefore be used as a memory element in a three-state logic circuit, with V_1, V_2 and V_3 being the voltages corresponding to the three logic states. This is a significant component reduction over the existing three-state logic circuits which require four conventional transistors and six resistors to construct a memory cell [57]. The circuit can be switched from one stable state to another by applying a short voltage pulse. In the experimental studies, [30, 52], the operating point was shifted from one state to another by momentarily changing the supply voltage V_{SS}, which has the same effect as

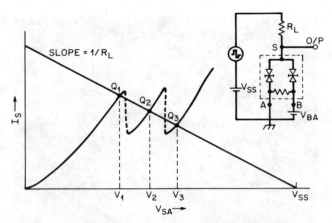

Figure 5.13 The current–voltage characteristic and schematic of a three-state memory cell using the integrated RT diodes. The associated load line shows three stable operating points, Q_1, Q_2 and Q_3.

applying a short voltage pulse. With a supply voltage $V_{SS} = 16$ V, load resistance $R_L = 215\,\Omega$, and the device biased to $V_{BA} = 0.7$ V, the three stable states were measured to be at 3.0 V, 3.6 V, and 4.3 V. The corresponding load line drawn on the measured I–V characteristic of the device at $V_{BA} = 0.7$ V intersects at 2.8 V, 3.4 V, and 4.1 V, respectively, which are in close agreement with the measured values of the three stable operating points.

The three-state memory cell discussed above is also suitable for integration in memory ICs with read/write and decoding network laid out as shown in Fig. 5.14. The memory cells are placed in a matrix array and a particular element in the array is addressed by activating the corresponding row and column select lines. A row select connects each device in that row to the corresponding column lines. The column select finally connects the selected column to the data bus. Consider the element (i, j) of the memory matrix shown in Fig. 5.14. When the row select line is activated, it turns the driving switch Q_1 on. It also turns on the switches for every element in the ith row. The column select logic now connects the jth column only to the data bus. The ternary identity cell [58] T acts as the buffer between the memory element and the external circuit for reading data. For reading data from the memory, the identity cell is activated with the read enable line and data from the element number (i, j) in the matrix goes, via the data bus, to the in/out pin of the IC. When the write enable line is activated, data from external circuit is connected to the data bus and is subsequently forced on the (i, j)th element in the array and is written there.

PARITY GENERATOR. A 4 bit parity generator circuit using the new device is shown in Fig. 5.15a. The operation of the circuit can be understood from

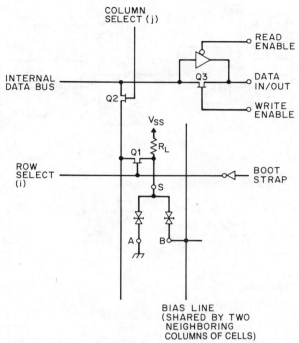

Figure 5.14 Typical layout of an integrated circuit (IC) using the three-state memory cells of Fig. 5.13.

Fig. 5.15b which shows the $I-V$ of the device with the bias voltage V_{BA} properly adjusted and the resultant voltage waveforms at various points in the circuit for different input conditions. The four digital inputs are added in the inverting summing amplifier A_1 to produce five distinct voltage steps at its output corresponding to the number of digital bits being in the high state. Normally, the output of A_1 would be negative for positive input voltages. The addition of a suitable negative offset voltage V_{off} at the input shifts the whole waveform up, to produce the A_1 output as shown at the bottom of Fig. 5.15b. The substrate bias voltage V_{SS} is adjusted to select the operating points of the device at the five dots shown in the $I-V$ curve, corresponding respectively to the five different voltage levels at A_1 output. The substrate current of the device generates a voltage across the $7.5\,\Omega$ resistor, which is picked up by the buffer amplifier A_2. Note that the output is high when the number of input bits set high is odd and vice versa. The circuit can thus be used as a 4 bit parity generator. The two different amplifiers in the circuit can be constructed using three transistors each. It is thus found that there is considerable reduction in the number of components compared with a conventional circuit, which needs three exclusive-OR gates, each requiring eight transistors.

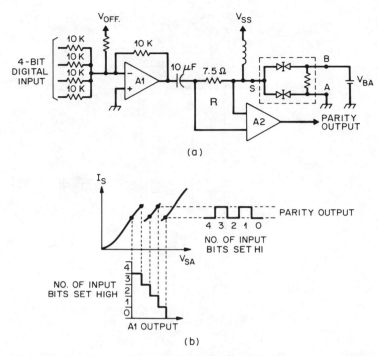

Figure 5.15 (a) The 4 bit parity generator circuit, (b) the current–voltage characteristic of the device and the waveform at various points in the circuit for different input conditions.

Typical experimental output from the circuit of Fig. 5.15a is shown in Fig. 16. The four digital inputs are driven by the outputs of a 4 bit binary counter. The top trace of Fig. 5.16 shows the output of amplifier A_1, and the bottom trace, that of A_2. Considering the dotted line as the reference level of a logic circuit, we find that the output is high for the second and the fourth voltage levels of A_1, whereas it is low at the first, third, and the fifth levels. It should be noted, however, that there are considerable decoding spikes in the output whenever the operating point of the device is shifted across a negative resistance region. This is believed to be due to the inherent oscillations of a circuit involving a negative resistance device, and may be taken care of by proper choice of the resistance R (7.5 Ω in the present experiments). The oscillations can be suppressed if this resistance is made larger than the magnitude of the negative resistance in the $I–V$. However, the choice of this resistance is very important for the operation of the parity generator circuit. Too large a resistance will lead to a much flatter load line, producing hysteresis [59] in the circuit operation whenever the operating point is moved across a peak. This is the case in the memory circuit described before, where multiple stable points are obtained at a given bias.

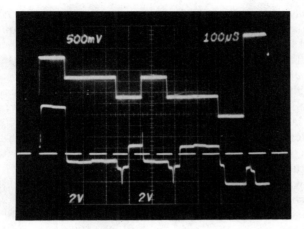

Figure 5.16 Experimental results of the parity generator circuit. The top trace shows the output of the amplifier A_1, and the bottom trace, the parity output. The dashed line is the threshold-voltage level of the output logic.

For the successful operation of the parity checker, on the other hand, there must be only one stable operating point at a given bias voltage. This is obtained with a nearly vertical load line, which can be freely shifted back and forth between any of the peaks and the following valley, without encountering any hysteresis.

Summary. The structure also lends itself to the realization of multiple peaks, using the same operating principle and a series of metallizations. The resistance of the channel in this case performs the useful function of a monolithically integrated voltage divider with the bias V_{BA} being applied between the pads at the two extreme ends. Multiple peaks in the $I–V$ characteristics could, of course, also be realized by connecting in parallel a series of tunnel diodes with resistors in between, which amounts basically to the circuit of Fig. 5.9b in the case of two diodes. The approach presented here, however has, two clear potential advantages: (1) it is a monolithic integration of RT devices and the voltage divider with resulting reduced parasitic resistances and capacitances (note that the monolithic integration of tunnel diodes would be a much more demanding task); and (2) reproducibility of the $I–V$'s, as discussed in Section 5.1.

Following the above demonstration, Söderström and Andersson [60] have demonstrated a new way of combining RT diodes in parallel, having different series resistances with each of them to separate the peaks in voltage, instead of the external bias. This is advantageous in that the external bias source is not required. However, the integration of different series resistances is a more demanding task, making the fabrication of the device difficult.

5.2.4.2   Vertical Integration of RT Diodes. Another way of obtaining multiple peaks in the I–V using only the ground-state resonance of quantum wells is the vertical integration of RT diodes [53–55]. The structure consists of a number n of RTDBs, each composed of 50 Å undoped $Ga_{0.47}In_{0.53}As$ QWs sandwiched between two undoped 50 Å $Al_{0.48}In_{0.52}As$ barriers, connected in series through 1000 Å thick heavily doped ($\sim 5 \times 10^{17}$ cm^{-3}) n-type $Ga_{0.47}In_{0.53}As$ cladding layers. The individual RTDBs were separated from the cladding layers on the two sides by 50 Å thick undoped $Ga_{0.47}In_{0.53}As$ layers to offset the effect of dopant diffusion during the high-temperature growth process. The entire structure was grown lattice-matched to an n^+-InP substrate, on which first a 5000 Å n^+-(1×10^{18} cm^{-3}) $Ga_{0.47}In_{0.53}As$ buffer layer was deposited. The growth ended with another 5000 Å thick n^+-(1×10^{18} cm^{-3}) $Ga_{0.47}In_{0.53}As$ contact layer. The devices were processed into 50 μm diameter mesas. A Au/Ge/Ag alloy was used for the top contact to avoid penetration of the metal beyond the 5000 Å contact layer. The bottom metallization is noncritical, and a Au/Ge/Ni alloy was used. The devices were tested in a Helitran dewar equipped with microprobes.

It has been previously shown that the GaInAs/AlInAs material system is suitable for the room-temperature operation of RT diodes with high peak-to-valley ratios [5, 43]. The present device is equivalent to a series combination of n such RT diodes. The DBs are designed so that the ground state in the QW is substantially above the Fermi level in the adjacent cladding layers. When a voltage is applied, the electric field is higher at the anode end of the device (Fig. 5.17a), because of the screening of the applied field by the charge accumulated in the QWs under bias. Quenching of RT is thus initiated across the DB adjacent to the anode first and then sequentially propagates to the other end, as shown in Fig. 5.17a and b. Once RT has been quenched across a DB, the voltage drop across it quickly increases with bias because of the increased resistance. The nonresonant tunneling component through this DB is large enough to provide continuity for the RT current through the other DBs closer to the cathode. An NDR region is obtained in the I–V, corresponding to the quenching of RT through each diode. Therefore, with n diodes, we observe n peaks in the I–V. This expansion of the high-field region from one DB to the next is somewhat similar to the phenomenon observed by Choi et al. [61] in sequential tunneling through a superlattice. Generating multiple peaks by combining tunnel diodes in series is well known [62]. However, the mechanism in that arrangement is different. The tunnel diodes used in such a combination must have different characteristics with successively increasing peak currents, so that each of them can go into the NDR region only when the corresponding current level is reached [62]. Besides, the present structure has the usual advantages over tunnel diodes discussed before.

Devices consisting of three and five RT structures in series were tested by Sen et al. [55]. The resulting I–V characteristics taken in both polarities of the applied bias are shown in Figs. 5.18 and 5.19, respectively. Positive

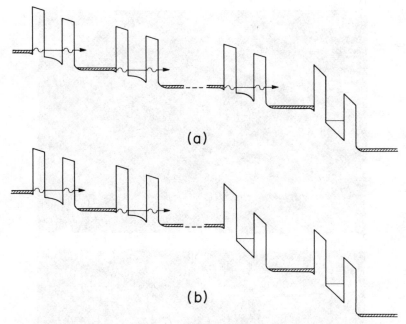

(a)

(b)

Figure 5.17 Band diagram under applied bias (a) with resonant tunneling quenched through the DB adjacent to the anode and (b) after expansion of the high-field region to the adjacent DB with increasing bias. The arrows indicate the resonant tunneling component of the current.

polarity here refers to the top of the mesa being biased positive with respect to the bottom and vice versa. A systematic study of the *I–V*'s at various temperatures from 300 K down to 77 K was made. The results presented in Figs. 5.18 and 5.19 are for room-temperature (top) and 77 K (bottom). At 77 K, the devices with three RTDBs indicate three peaks (Fig. 5.18, bottom) and the one with five indicates five peaks (Fig. 5.19, bottom) in both polarities, as expected. Note that the respective peaks in both polarities occur at nearly the same voltages. At 300 K, Fig. 5.18 shows three distinct peaks in the positive polarity, while in the negative polarity the third peak is not observed because of the rapidly increasing background current. The difference in the observed characteristics in the two polarities may be due to structural asymmetry, unintentionally introduced during growth and processing. The device with five DBs shows five clear peaks in both polarities also at 300 K (Fig. 5.19, top). The best peak-to-valley ratio observed at room temperature is 5:1 in both structures. At 77 K, the highest peak-to-valley ratios are, respectively, 9:1 and 18:1 in the structures with three and five peaks.

Note that there is considerable hysteresis associated with nearly all the peaks in the *I–V*'s. This occurs whenever there is a significantly large

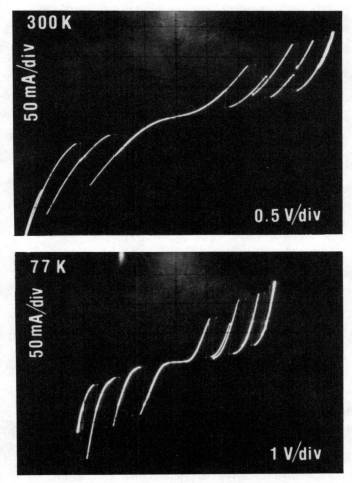

Figure 5.18 Current–voltage characteristics of the device with three vertically integrated RT double barriers taken for both bias polarities at 300 K (top) and 77 K (bottom).

parasitic resistance in series with the devices [3, 39]. In the present devices, when one of the RT diodes is active, all the other diodes contribute to this parasitic resistance. This hysteresis [3, 39] is different in nature from the load line effect [59] discussed previously. In the present case, hysteresis effect cannot be avoided even with an absolutely vertical load line ($R_L = 0\,\Omega$) because of the multiple-valued-current nature of the device characteristic around the peak. The effect, however, may be reduced with proper optimization of the device design. The series resistance also pushes the peaks in the I–V's toward higher voltages [39]. Systematic studies at different temperatures indicate that the positions of the peaks gradually shift

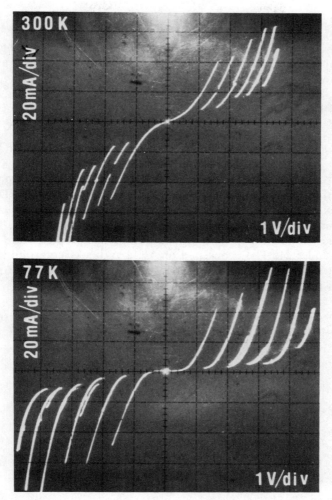

Figure 5.19 Current–voltage characteristics of the device with five vertically integrated RT double barriers taken for both bias polarities at 300 K (top) and 77 K (bottom).

toward lower voltages with increasing temperature. This is due to reduction in the parasitic resistance with increasing temperature.

In conclusion, *vertical integration* is another technique of obtaining multiple peaks in the *I–V* of RT devices, which combines *growth and processing simplicity and does not require an auxiliary bias source* to generate the peaks. The prototype devices demonstrated so far exhibit three and five peaks with high peak-to-valley current ratios at room temperature. These devices can be used in most of the applications involving devices with multiple peaks, demonstrated in the previous section. However, the hysteresis in the characteristics makes it difficult to use if for parity bit generation, since in that application a nearly vertical load line has to be moved back and forth between any of the peaks and the following valley by changing the bias voltage. The hysteresis will prevent the operating point from returning to the peak without going through the previous valley. The multiple-valued-current nature of the characteristic on the other hand, can be utilized in other applications, such as a Schmitt trigger circuit.

5.3 RESONANT TUNNELING BIPOLAR TRANSISTOR (RTBT)

The negative differential resistance of RTDBs showed enough potential to be included in three-terminal devices to take advantage of the NDR as well as the transistor action. The first of these kinds of structures was the RT bipolar transistor (RTBT), proposed by Capasso and Kiehl from Bell Laboratories in 1985 [15]. The structures initially proposed are shown in Figs. 5.20 and 5.21. They consisted of AlGaAs/GaAs heterojunction bipolar transistors with a quantum well in the p-type base layer. In order to always satisfy the condition of tunneling through symmetric DBs, discussed in Section 5.2.1, these structures employed high-energy or ballistic injection of minority carriers into the base to achieve RT through the DB, rather than applying a field across the latter. This method does not alter the transmis-

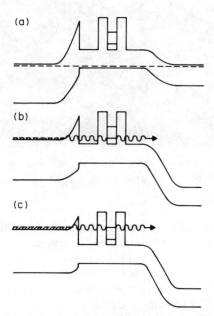

Figure 5.20 Band diagram of RTBT with tunneling emitter under different bias conditions: (a) in equilibrium, (b) RT through the first level in the well, (c) RT through the second level. (Not to scale.)

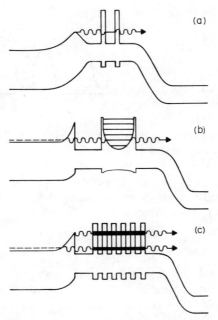

Figure 5.21 (a) Band diagram of RTBT with graded emitter (at resonance). Electrons are ballistically launched into the first quasi-eigenstate of the well. (b) RTBT with a parabolic quantum well in the base and tunneling emitter. A ballistic emitter can also be used. (c) RTBT with superlattice base. (Not to scale.)

sion of the two barriers and consequently should lead to near-unity transmission at all resonance peaks and to larger negative conductance and peak-to-valley ratios than conventional RT structures.

Figure 5.20 shows the band diagram of one of these devices. The structure is a heterojunction bipolar transistor with a degenerately doped tunneling emitter and a symmetric DB in the base. The collector current as a function of base–emitter voltage V_{BE} should exhibit a series of peaks corresponding to RT through the various quasi-stationary states of the well. Multiple negative conductance in the collector circuit can therefore be achieved.

An alternative injection method is the abrupt or nearly abrupt emitter which can be used to launch electrons ballistically into the quasi-eigenstates with high momentum coherence. As V_{BE} is increased, the top of the launching ramp eventually reaches the energy of the quasi-eigenstates so that electrons can be ballistically launched into the resonant states (Fig. 5.21a).

To achieve equally spaced resonances in the collector current, the rectangular quantum well in the base should be replaced by a parabolic one (Fig. 5.21b). RT through parabolic quantum wells was discussed in Section

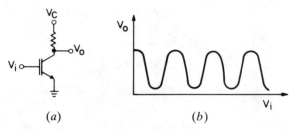

(a)   (b)

Figure 5.22 Common-emitter amplifier circuit using the RTBT and the corresponding multiple-valued voltage transfer characteristics.

5.2.3. Finally, in Fig. 5.21c, we illustrate another method, that of high-energy injection and transport in the minibands of a superlattice, using ballistic launching or tunnel injection.

5.3.1 Circuit Applications of RTBT

These new functional devices, because of their multiple resonant characteristic, can have many potential applications, leading to tremendous reduction in circuit complexity and size. These are discussed in this section.

5.3.1.1 Multiple Valued Logic. Consider the common-emitter circuit shown in Fig. 5.22a. For an input voltage V_i in the base for which the electrons undergo RT, the transistor strongly conducts and the output voltage V_o is low. Off resonance, instead, the device basically does not conduct and the output voltage is high. This results in the multiple-valued voltage transfer characteristic of Fig. 5.22b, having as many peaks as the number of resonances in the well. The output voltage V_o takes on one of the two values in accordance with the level of the input voltage V_i. Thus, the device provides a binary digital output for an analog input, or a multiple valued digital input.

5.3.1.2 Parity Generator. The multiple valued characteristics of Fig. 5.22b can be used to design the parity generator circuit shown in Fig. 5.23. In this circuit, the binary bits of a digital word are added in the resistive network at the input of the RTBT. With proper weighting of the resistors R_o and R_B,

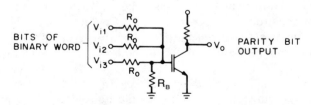

Figure 5.23 The parity generator circuit using RTBT

the operating point would be placed either on a peak or a valley of the I–V depending on whether the total number of 1's at the input is even or odd, respectively. The advantage of this approach over conventional circuits is that the RTBT implementation, apart from being smaller in size and simpler, should also be extremely fast, since it uses a single high-speed switching device. Conventional transistor implementation requires complex circuitry involving many logic gates with a consequent reduction in speed.

Compare this implementation of the parity generator with that discussed in Section 5.2.4.1. The advantage of using a transistor structure, as shown in this section, is further simplification of the circuit. The function of adding the voltages corresponding to the digital input bits, as performed by the amplifier A_1 in Fig. 5.16, is performed by the transistor in this case. Also, since the input and the output in the transistor are isolated, the amplifier A_2 is not necessary, as the output automatically comes ground-referenced.

5.3.1.3   *Analog-to-Digital Converter.* The circuit shown in Fig. 5.24 can be used as an analog-to-digital converter. In this application, the analog input is simultaneously applied to an array of RTBT circuits having different voltage scaling networks. To understand the operation of the circuit, consider the simplest system comprising of only the two transistors Q_1 and Q_2. The voltages at different points of this circuit are shown in Fig. 5.25a for various input voltages V_i. Consider that the resistances R_0, R_1 and R_2 are so chosen that the base voltages of the transistors Q_1 and Q_2 vary according to the

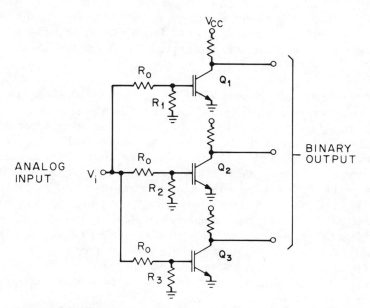

Figure 5.24   The analog-to-digital converter circuit using RTBT.

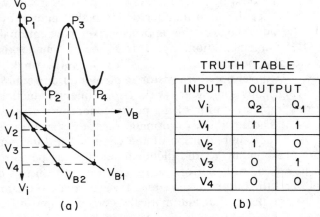

Figure 5.25 The schematic operation of the analog-to-digital converter circuit of Fig. 5.24, involving only 2 bits: (a) the voltages at different points of the circuit at various input voltages, (b) the truth table.

curves V_{B1} and V_{B2} respectively with V_i. With the input voltage at V_1, the output of both the transistors will be at the operating point P_1 (Hi-state). With the input changing to V_2, the output of Q_1 will become low (P_2), while that of Q_2 will still remain high (closer to P_1). Applying this logic to the input voltages V_3 and V_4, it can be easily shown that circuit indeed follows the Truth Table of Fig. 5.25b. The outputs of the RTBT array thus constitute a binary code representing the quantized analog input level. The system can be extended to more bits with larger number of peaks in the I–V. Again, the circuitry involved in this approach is simple and should be very fast.

5.3.1.4 Multiple-state Memory.

This application takes advantage of the ability to achieve a multiple-valued negative differential resistance characteristic. This type of characteristic is achieved at the emitter–collector terminals by holding the base–collector junction at fixed bias V_{BC}, as shown in the inset of Fig. 5.26. With V_{BC} fixed, variations in V_{CE} produce variations in V_{EB} which cause the collector current to peak as V_{EB} crosses a tunneling resonance (Fig. 5.26). When connected to a resistive load R_L and voltage supply V_{CC}, as shown in the circuit of Fig. 5.26, the resulting load line intersects the I–V at N stable points where N is the number of resonant peaks. The circuit thus acts as an N-state memory element, providing the possibility of extremely high-density data storage. Such an element can be latched onto any one of the stable states by momentarily applying a voltage close to the desired state. It can therefore be integrated in a large memory array, as discussed in Section 5.2.4.1.

An N-state latch such as this can also be used to build a counter in the

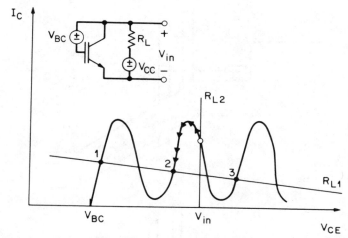

Figure 5.26   The current–voltage characteristics with multiple-valued negative differential resistance when the base–emitter voltage of the RTBT is held fixed, as shown in the circuit in the inset of the figure. The load line corresponding to R_{L1} demonstrates its operation as a multiple valued memory element. The solid circles denote stable states. The load line R_{L2} and the path indicated by the arrows show how such a memory can be pulsed from one stable state to another.

N-state logic system, as the circuit can be switched from one stable state to the immediately adjacent one by a voltage pulse of height V_{LS} to the circuit, forcing the operating point to that of the open circle on the unstable part of the characteristics in Fig. 5.26. It can be easily shown that as the input pulse is removed, the operating point will move along the indicated trajectory, finally latching onto the State 2. Circuits such as these and others, such as multipliers and dividers in the N-state logic system, had been of considerable interest for some time [57]. However, since no physical device providing multiple valued negative resistance previously existed, such circuits were possible only with combinations of binary devices. In order to achieve N states, N two-state devices were connected, resulting in a complex configuration with reduced density and speed.

5.3.2   Resonant Tunneling Electron Spectroscopy

The RT bipolar transistor structures initially proposed and discussed so far relied heavily on the ability to transfer the electrons ballistically through the p-type base region up to the DB. Recent experiments with resonant tunneling spectroscopy [63, 64] of hot electrons injected at high energy into a p-type GaAs well indicate that it may be extremely difficult to achieve that goal.

Figure 5.27 illustrates the energy band diagram of the structure used for resonant tunneling electron spectroscopy. It consists basically of a reverse

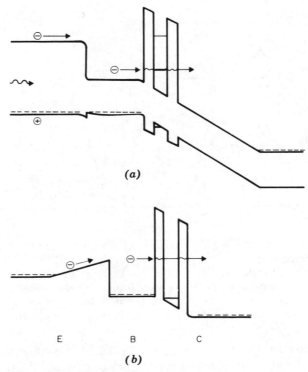

(a)

E   B   C

(b)

Figure 5.27 (a) Band diagram of heterojunction diode used for resonant tunneling spectroscopy of hot minority-carrier electrons. By measuring the photocurrent as a function of the reverse bias, the hot-electron energy distribution can be directly probed. (b) Unipolar transistor structure for resonant tunneling spectroscopy of hot majority-carrier electrons in the n$^+$ base layer.

biased p-i-n heterojunction and can be used to investigate hot minority-carrier transport. Low-intensity incident light is strongly absorbed in the wide-gap p$^+$ layer. Photogenerated minority-carrier electrons diffuse to an adjacent low-gap layer. Upon entering this region, electrons are ballistically accelerated by the abrupt potential step and gain a kinetic energy $\cong \Delta E_c$ and a forward momentum $p_\perp \simeq \sqrt{2m_e^* \Delta E_c}$. Collisions in the low-gap layer tend to randomize the injected, nearly mono-energetic distribution. Hot electrons subsequently impinge on the DB in the collector. From simple considerations of energy and lateral momentum $p_\parallel$ conservation in the tunneling process, it can be shown that only those electrons with a perpendicular energy $e_\perp$ ($p_\perp^2/2m_e^*$ for a parabolic band) equal (within the resonance width) to the energy of the bottom of one of the subbands of the quantum well resonantly tunnel through the quantum well and give rise to a current [31]. Thus, by varying the applied bias (i.e., changing the energy difference between the resonance of the quantum well and the bottom of the conduction band in the low-gap p^+ layer) and measuring the current, one directly probes the electron energy distribution $n(E_\perp)$. One has therefore

$$E_\perp = E_n - \frac{e(V + V_{bi})(L_B + L_W/2 + L_{sp})}{L_c} \qquad (4)$$

where V is the reverse-bias voltage, V_{bi} the built-in potential of the p-i-n diode, L_c the total collector layer thickness, L_B and L_W the barrier and well layer thicknesses, respectively, and L_{sp} the thickness of the undoped spacer layer (20 Å) between the p-type region and the DB (see later in text). E_n is the energy of the bottom of the nth subband measured with respect to the bottom of the center of the well. (Note that E_n is assumed to be independent of the electric field F, which is a good approximation as long as E_n is significantly greater than eFL_W). Identical arguments apply to the case of the unipolar transistor structure of Fig. 5.27b, which can be used to analyze the electron distribution in the base layer by measuring the collector current as a function of the collector–base voltage. In the above arguments, it is assumed that thermionic currents over the DB can be minimized. This can be done by operating the structure at sufficiently low temperature and suitably designing the DB [3]. To obtain the actual energy distribution $n(E_\perp)$ from the current, the latter must be properly normalized by taking into account the field dependence of the resonant tunneling probability (integrated over the resonance width). This procedure does not alter, of course, the position of the peaks in the current–voltage characteristic (see Fig. 5.28), since the above probability varies monotonically with the electric field, irrespective of whether electrons resonantly tunnel sequentially or coherently through the DB [31]. The main features of the electronic transport can therefore still be obtained directly from the current, without normalization.

The structures were grown by molecular beam epitaxy on a $\langle 100 \rangle$ p^+-GaAs substrate and consist of p-i-n heterojunction diodes. Their band diagram is shown in Fig. 5.27a at a given reverse bias. The growth starts with a 2000 Å thick $n^+ = 2 \times 10^{17}$ cm^{-3} buffer layer followed by an undoped ($|N_D - N_A| \simeq 10^{14}$ cm^{-3}) 5000 Å GaAs layer and an AlAs/GaAs/AlAs DB, with barrier and well thicknesses of 20 Å and 80 Å, respectively. A 20 Å undoped GaAs spacer layer separates the DB from the p^+-($=3 \times 10^{18}$ cm^{-3}) GaAs layer, in which electrons are launched. Different thicknesses were used for this region (250 Å, 500 Å, 1800 Å), keeping everything else the same. The last layer consists of 2 μm thick Al$_{0.3}$Ga$_{0.7}$As doped to p = 3×10^{18} cm^{-3}. This provides a launching energy $\cong 225$ meV, that is, the conduction band discontinuity between GaAs and Al$_{0.3}$Ga$_{0.70}$As (obtained from $\Delta E_c = 0.6\Delta E_g$). The depletion width on the p^+ side of the junction is negligible with respect to the p^+-GaAs well thickness, up to the highest applied bias (-10 V), due to the high doping. The parameters of the DB were chosen in such a way that over the applied voltage range (0–10 V) the

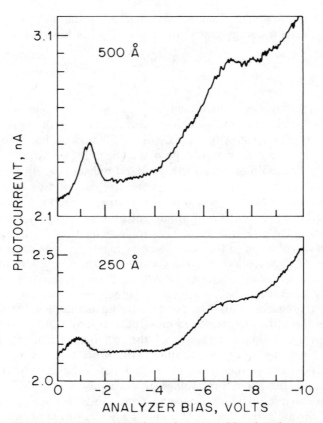

Figure 5.28 Photocurrent as a function of reverse bias for the structure of Fig. 5.27a, with GaAs p^+ layer thickness of 500 Å (top) and 250 Å (bottom).

electron energy distribution is probed essentially by one resonance at a time. For this DB, the first resonance is at $E_1 = 60$ meV from the well bottom and the second at $E_2 = 260$ meV, with full widths at half maximum of ≈ 0.1 meV and ≈ 1 meV, respectively [3]. It is easily shown from Eq. (4) that over the range of applied bias (Fig. 5.28) the first resonance samples the $E_\perp$ energy range from 37 meV to 0 meV, whereas the second resonance samples the energy range from 225 meV to 80 meV. The thickness of the collector layer L_c was made much greater than that of the DB to enhance the energy resolution of the spectrometer.

The samples were processed into mesa devices, using standard photolighographic, wet etching, and metallization techniques. The photosensitive area of these detectors is 10^{-4} cm^2. Light from a He–Ne laser ($\lambda = 6328$ Å) heavily absorbed in the $Al_{0.3}Ga_{0.7}As$ region (absorption length $\simeq 5000$ Å) was used to achieve pure electron (minority-carrier) injection and the dc photocurrent was measured with an HP4145 parameter analyzer as a function of reverse bias at low temperature in a Helitran dewar. Figure 5.28 illustrates the measured photocurrent at 9.2 K for the structure with a 500 Å thick GaAs p^+ layer. At these current levels and higher (up to 10 μA), space charge effects are negligible, as shown by varying the light intensity and monitoring the photocurrent–voltage characteristic. The dark current was completely negligible ($\leq 10^{-13}$ A) in the same voltage range. Two distinct features are present at 1.3 V and 7 V, respectively. Using Eq. (4), one can easily see that the first peak corresponds to electrons with perpendicular energy of a few tens of meV ($\simeq 17$ meV) that have resonantly tunneled through the first resonance of the quantum well. The second peak is much broader and corresponds to incident electrons with energy $E_\perp \simeq 130$ meV which have resonantly tunneled through the second resonance of the well. It is therefore clear that the energy distribution of the electrons in the p^+-GaAs layer, following high-energy injection, consists of two parts. One has relaxed close to the bottom of the conduction band; the other has considerably higher perpendicular kinetic energy $E_\perp$. The distribution is thus strongly non-Maxwellian, similar to what has been found in the case of majority-carrier electrons in the base of hot-electron planar doped barrier transistors [65]. Note that the peaks in the photocurrent were observed at temperatures as high as 70 K and did not appreciably shift with temperature.

Similar results are found by decreasing the GaAs p^+ layer thickness from 500 Å to 250 Å (Fig. 5.28). The peaks are located at somewhat lower voltages (corresponding to 10–20% higher energies), implying that the relaxation of carriers is somewhat less, due to the thinner layer, as expected. Overall, however, the shape of the energy distribution has not changed significantly, which implies that already over a length of a few hundred angstroms, the near-ballistic injected distribution has been strongly randomized by scattering and has reached a quasi-steady state. Additional manifestation of strong scattering comes from the fact that no evidence is found in the data of the quantized subbands of the 250 Å p^+ well into which electrons are injected. Since this subband structure should be reflected on the electron distribution, one would expect to observe peaks in the photocurrent at such voltages that the resonances of the DB would coincide in energy with the resonances of the 250 Å thick layers. The fact that this is not observed implies that the collisional broadening $\hbar/\tau$ must be comparable to or greater than the typical energy separation between the resonances of the p^+ layer. The latter varies roughly from 30 meV (between the first two resonances of the p^+ well) to 80 meV (between the highest two quasi-bound resonances of the p^+ well). This implies that the scattering time (avaraged over the hot-electron distribution) is $\leq 10^{-14}$ s. This estimate is consistent with recent studies of electron dynamics in p-type GaAs. Hopfel et al. [66] have investigated minority-electron transport in GaAs quantum wells following picosecond photoexcitation of small carrier densities ($n \ll p$). Their results show that the high-density hole plasma in the wells ($p = 1.5 \times 10^{11}$ cm^{-2}) induces strong electron-hole scattering. At very low electric fields (a few tens of V/cm) and lattice temperatures ≈ 15 K, their measured minority-electron mobilities give a total momentum relaxation time of 4×10^{-14} s. Such time is expected to be $\leq 10^{-14}$ s at higher hole densities $p \approx 5 \times 10^{12}$ cm^{-2}), comparable to that of the p$^+$ well in these samples. Previous work by Hopfel, Shah, and Gossard [67] had also shown that the energy relaxation rate of minority electrons in quantum wells is considerably higher than that of majority electrons. Although their results were obtained at a lattice temperature of 300 K, this is expected to be valid at low temperatures as well.

These results clearly show that quasi-ballistic or ballistic transport of a significant fraction of the electrons injected in p$^+$ GaAs does not occur even over distances ≤ 500 Å. Indeed, recent electron spectroscopy measurements in heterojunction bipolar transistors with base thicknesses as short as 400 Å have also demonstrated that electrons undergo strong relaxation in the p$^+$ ($= 2 \times 10^{18}$ cm^{-3}) base [68].

On the other hand, previous work on hot-electron transistors [69, 70] has shown that a significant fraction of the electrons injected at similar energies (≈ 0.2 eV) in n$^+$-GaAs of comparable doping density ($\approx 10^{18}$ cm^{-3}) traverses the base quasi-ballistically or ballistically for base thicknesses in the 300–500 Å range. In fact, recent experimental determinations of the scattering rates of nonequilibrium injected electrons in n$^+ = 1 \times 10^{18}$ cm^{-3} GaAs give a scattering times in the 3×10^{-14}–5×10^{-14} s range [65]. These values are considerably longer than the ones estimated in the present experiment, for p$^+$-GaAs, as comparable injection energies and doping levels.

Measurements have also been performed in similar structures with a thicker p$^+$ region (1800 Å). From the photocurrent–voltage characteristics, it is found that electrons undergo RT through the DB starting from a few meV energy from the bottom of the conduction band in the p$^+$-GaAs layer, and that there is not a hot-electron distibution at higher energy. This is to be expected, since by making the p$^+$ layer much thicker, electrons have had time to thermalize to the bottom of the band before undergoing RT.

In summary, the results show that nonequilibrium energetic electrons injected in p$^+$-GaAs undergo very strong scattering. This implies that the implementation of ballistic heterojunction bipolar transistor [71] and ballistic RT transistors [15] will likely not be possible.

5.3.3 Resonant Tunneling Bipolar Transistors Operating at Room Temperature

The first operating resonant tunneling bipolar transistor (RTBT), demonstrated by Capasso et al. [22] in 1986, was designed to have the minority electrons *thermally* injected into and transported through the base, rather than utilizing *hot-electron* or *quasi-ballistic* transport. This made the operation of the device much less critical, and the structure implemented in the AlGaAs/GaAs material system showed resonance peaks even at *room temperature*.

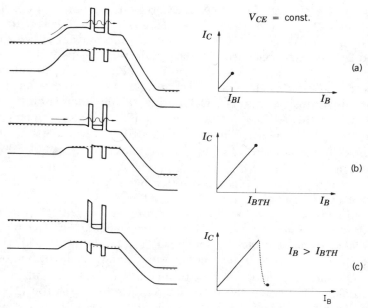

Figure 5.29 Energy band diagrams of the RTBT with thermal injection and corresponding schematics of collector current I_C for different base currents I_B at a fixed collector–emitter voltage V_{CE} (not to scale). As I_B is increased, the device first behaves as a conventional bipolar transistor with current gain (a), until near flat-band conditions in the emitter are achieved (b). For $I_B > I_{BTH}$, a potential difference develops across the AlAs barrier between the contacted and uncontacted regions of the base. This raises the conduction band edge in the emitter above the first resonance of the well, thus quenching resonant tunneling and the collector current (c).

The band diagram of this transistor is shown in Fig. 5.29. Thermal injection is achieved by adjusting the alloy composition of the portion of the base adjacent to the emitter in such a way that the conduction band in this region lines up with or is slightly below the bottom of the ground-state subband of the quantum well (Fig. 5.29a). For a 74 Å well and 21.5 Å AlAs barriers, the first quantized energy level is $E_1 = 65$ meV [3]. Thus, the Al mole fraction was chosen to be $x = 0.07$ (corresponding to $E_g = 1.521$ eV) so that $\Delta E_c \simeq E_1$. This equality need not be rigorously satisfied for the device to operate in the desired mode, as long as E_1 does not exceed ΔE_c by more than a few kT. The quantum well is undoped; nevertheless, it is easy to show that there is a high-concentration ($\cong 7 \times 10^{11}$ cm^{-2}) two-dimensional hole gas in the well. These holes have transferred from the nearby $Al_{0.07}Ga_{0.93}As$ region, by tunneling through the AlAs barrier, in order to achieve Fermi level line-up in the base.

The structures were grown by MBE on an n$^+$Si-doped GaAs substrate. A 2100 Å n $= 3 \times 10^{17}$ cm^{-3} GaAs buffer layer is followed by a 1.6 μm thick GaAs n-type ($= 1 \times 10^{16}$ cm^{-3}) collector. The base layer starts with a Be doped p^+ ($=1 \times 10^{18}$ cm^{-3}) 1900 Å GaAs region adjacent to the collector, followed by a 210 Å undoped GaAs set-back layer. The DB is then grown. It consists of a 74 Å undoped GaAs quantum well sandwiched between two undoped 21.5 Å AlAs barriers. The last portion of the base is 530 Å thick $Al_{0.07}Ga_{0.93}As$, of which 105 Å adjacent to the DB is undoped and the rest is p-type doped to $\simeq 1 \times 10^{18}$ cm^{-3}. The purpose of the two set-back layers is to offset Be diffusion into the DB during the high-temperature growth ($T = 680°C$) of the AlGaAs graded emitter [72]. The latter consists of a 530 Å $n^+ \approx 3 \times 10^{17}$ cm^{-3} region linearly graded between $x = 0.07$ and $x = 0.24$, adjacent to the base, and of 3200 Å thick $Al_{0.25}Ga_{0.75}As$ doped to $\cong 3 \times 10^{17}$ cm^{-3}. The growth ends with a 1000 Å $n^+ = 3 \times 10^{18}$ cm^{-3} GaAs contact layer separated from the emitter by a 530 Å $n^+ = 3 \times 10^{18}$ cm^{-3} region linearly graded from $x = 0.24$ to $x = 0$.

Test structures with 7.5×10^{-5} cm^2 emitter area were fabricated using photolithography and wet and anodic etching techniques. The base layer was revealed by anodic etching in H_3PO_4/H_2O. The portion of the base ($Al_{0.07}Ga_{0.93}As$) adjacent to the emitter was also anodically etched off at 12 Å/V, while the mesa height was continuously monitored with a Dektak depth profiler. The rest of the base was contacted using AuBe (1% Be by weight, 400 Å); Au(1100 Å) alloyed at 400°C in a H_2 flow for 2 s. Au(500 Å)/Sn(250 Å)/Au(2000 Å) alloyed at 450°C for 1 s was used as the n-type contact to the emitter and collector.

The present structure consists, therefore, of a HBT with a DB *in the base region*. In an alternative RBT design, the DB can be placed *between a wide-gap graded emitter and a GaAs base*, giving rise to RT transistor action similar to that discussed in the present paper.

To understand the operation of this device, consider a common-emitter bias configuration. Initially, the collector-emitter voltage V_{CE} and the base current I_B are chosen in such a way that the base–emitter and the base–collector junctions are respectively forward- and reversed-biased. If V_{CE} is kept constant and the base current I_B is increased, the base–emitter potential also increases until flat-band condition in the emitter region is reached (Fig. 5.29b, left). In going from the band configuration of Fig. 5.29a to that of Fig. 5.29b, the device behaves like a conventional transistor, with the collector current linearly increasing with the base current (Fig. 5.29a–b, right). The slope of this curve is, of course, the current gain β of the device. In this region of operation, electrons in the emitter overcome, by thermionic injection, the barrier of the base–emitter junction and undergo RT through the DB. If now the base current is further increased above the value I_{BTH} corresponding to the flat-band condition, the additional potential difference drops primarily across the first semi-insulating AlAs barrier (Fig. 5.29c), between the contacted and uncontacted portions of the base, since the highly doped emitter is now fully conducting. This pushes the conduction band edge in the $Al_{0.07}Ga_{0.93}As$ above the first energy level of the well, thus quenching the RT. The net effect is that the base transport factor and the current gain are greatly reduced. This causes an abrupt drop of the collector current as the base current exceeds a certain threshold value I_{BTH} (Fig. 5.29c, right). This is the most important manifestation of the inherent negative transconductance of this device. It should be noted that although the base metallization penetrates into the AlAs barrier, the latter can still sustain the potential drop in the region under the emitter, as discussed above, since the base contact is placed away from the emitter mesa and the barrier is undoped (semi-insulating). Nevertheless, it should be noted that RT transistor action can also be achieved if the p-type ohmic contact is made *only* to the portion of the base adjacent to the collector and not to the quantum well. Obviously, the band diagram configuration in the base region under operating conditions will be somewhat different. In this case, although there is no direct ohmic contact to the quantum well, electrical connection between the 2D hole gas in the well and the part of the base adjacent to the collector is established via tunneling of holes through the AlAs barrier.

The devices were biased in a common-emitter configuration at 300 K and the I–V characteristics were displayed on a curve tracer. For base currents ≤ 2.5 mA, the transistor exhibits normal characteristics, whereas for $I_B \geq$ 2.5 mA the behavior previously discussed was observed. Figure 5.30 shows the collector current versus base current at $V_{CE} = 12$ V, as obtained from the common-emitter characteristics. The collector current increases with the base current and there is clear evidence of current gain ($\beta = 7$ for $I_C >$ 4 mA). As the base current exceeds 2.5 mA, there is a drop in I_C because the current gain mechanism is quenched by the suppression of RT. The transistor characteristics were also measured in a pulsed mode using 300 μs pulses. No changes were detected and a behavior identical to that of Fig. 5.30 was observed, ruling out heating effects.

The devices exhibited similar behavior in all the investigated temperature range from 100 to 300 K, although the negative conductance effects were more pronounced at lower temperatures, because the quenching of RT is more abrupt than at higher temperatures. Figure 5.31 illustrates the common-emitter characteristics at 100 K. At relatively low V_{CE}, both the emitter–base and the collector–base junctions are forward-biased (saturation region) and the collector current is negative, corresponding to a net flow of electrons from the collector contact into the collector layer. This gives rise to a collector–emitter offset voltage, which is typical of an assymmetrical heterojunction bipolar transistor [73]. In the present case, this offset is large due to the relatively small α of this structure at low temperatures and to the relatively large emitter resistance and ideality factor ($n \approx 2$) of the base–emitter p-n junction [73]. This offset is greatly reduced at room temperature. As V_{CE} is further increased, the collector–base junction becomes sufficiently reverse-biased, and for base currents $I_B \leq$ 4 mA the characteristics are similar to that of a conventional bipolar (Fig. 5.31). For $I_B > 4$ mA, at sufficiently high V_{CE}, the collector current instead

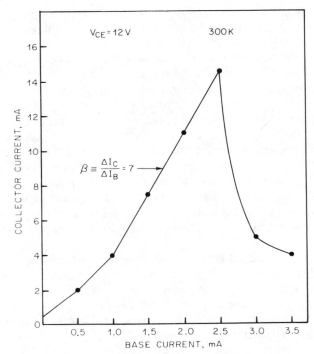

Figure 5.30   Collector current versus base current in the common-emitter configuration of the RTBT of Fig. 5.29, at room temperature with the collector–emitter voltage held constant. The line connecting the data points is drawn only to guide the eye.

decreases with increasing base current. It is apparent from Fig. 5.31 that in addition to this behavior previously discussed, there is also a large negative conductance in the I_C versus V_{CE} curve for base currents in excess of the threshold value ($=4$ mA). This is easy to understand by noting that in order to reach the band configuration of Fig. 5.29c and quench RT (at a fixed $I_B > I_{BTH}$) the collector–emitter voltage V_{CE} should be large enough for the collector–base junction to be reverse-biased and draw a significant collector current. In the present structure, the higher I_{BTH} at lower temperatures (4 mA at 100 K compared with 2.5 mA at 300 K) is a consequence of the larger collector–emitter offset voltage.

These bipolars in the common-emitter configuration act as oscillators when biased in the negative conductance region of the characteristic. The current oscillation in the collector circuit was picked up by a loop and displayed. Fig. 5.32 shows the oscillation in the time domain at room temperature and in the frequency domain at both 300 K and 100 K, at an operating point $V_{CE} = 12$ V, $I_B = 6$ mA. Note the high spectral purity (near single frequency response), particularly at low temperature. The oscillation

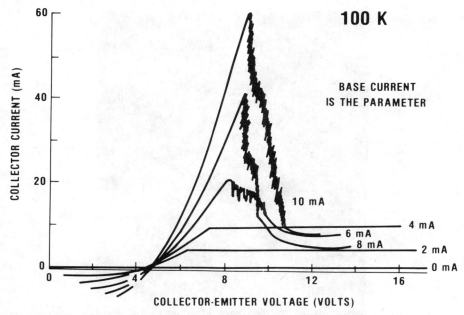

Figure 5.31 Common-emitter characteristics of the RTBT of Fig. 5.29 at 100 K.

frequency ($\approx$20 MHz) is limited at present by the probe stage used, and the collector bias circuit. Much higher oscillation frequencies (>10 GHz) may be ultimately expected.

By changing the bias conditions (I_B, V_{CE}), the oscillation frequency can be tuned over a few MHz range. At room temperature, the device ceases to oscillate for base currents $\lesssim$2.5 mA, since at such base currents the device is out of the negative conductance region (RT is not quenched; see Fig. 5.30). A simple, physical picture of the oscillations can be given. As the dc base bias current is increased above the threshold value, RT is suppressed. Thus, the collector current is reduced, implying a reduction of the emitter current, since the bias current I_B is kept fixed. This in turn implies a decrease of the voltage applied across the emitter, followed by a restoration of the flat-band conditions (since $I_B > I_{BTH}$), and the cycle is repeated.

5.3.4 Alternative Designs of RTBT

An alternative RTBT design, with the DB in the base–emitter junction rather than in the base region, was reported by Futatsugi et al. [23] of Fujitsu Laboratories. The operation of this device is very similar to the one discussed in the previous section. With the emitter–base junction forward-biased, resonance peak could be observed at 77 K in the emitter and hence in the collector current also as a function of base–emitter voltage V_{BE} at a

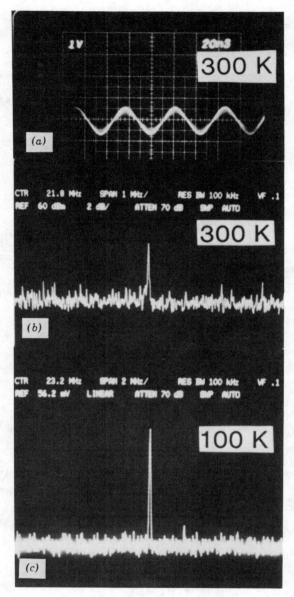

Figure 5.32 (a) Oscillations and (b) spectral characteristics for a resonant tunneling transistor operating as an oscillator at room temperature (center frequency 21.8 MHz; span 1 MHz). (c) Frequency response at 100 K of the same device (center frequency 23.2 MHz; span 2 MHz). The oscillation frequency is limited by the external circuit.

fixed collector–emitter voltage V_{CE}. However, in this design, the tunneling probability becomes a strong function of the base–emitter voltage V_{BE}. This makes it difficult to observe NDR at room temperature. Recently, the Fujitsu group had implemented an RTBT in the AlInAs/GaInAs material system, grown lattice-matched to an InP substrate, with the DB in the emitter layer [74]. This structure also suffers from the drawback that the tunneling probability is a strong function of the base–emitter voltage. However, the parameter advantages of this material system, discussed earlier, made it possible to observe NDR at room temperature [74].

5.4 RESONANT TUNNELING UNIPOLAR TRANSISTORS

Following the announcement of the RTBT from Bell Laboratories in 1984, several unipolar three-terminal devices were proposed and implemented that utilized the RT structure as electron injectors to generate voltage–tunable NDR and negative transconductance characteristics. The first of these unipolar structures was the resonant tunneling hot-electron transistor (RHET) [16], developed at the Fujitsu Laboratories in 1985. The same year, Luryi and Capasso proposed the quantum wire transistor [19], a novel device in which the quantum well is linear rather than planar and the tunneling is of two-dimensional electrons into a one-dimensional density of states. In 1987, the resonant tunneling gate field effect transistor (RT-FET) [24, 26, 27] was developed at Bell Laboratories. Recently, a device exhibiting the Stark effect [17] and the quantum capacitance effect [75] has been demonstrated at Bell Laboratories [76]. Among other proposals of unipolar three-terminal devices, the one involving integration of RT diodes and FETs, developed by California Institute of Technology and Xerox PARC [18, 25, 77] and their circuit applications [78], deserve mention. Some of these devices are discussed in this section.

5.4.1 Resonant Tunneling Hot-electron Transistor (RHET)

The schematic band diagram of the resonant tunneling hot-electron transistor (RHET), demonstrated by Yokoyama et al. of Fujitsu Laboratories [16], is shown in Fig. 5.33. The structure consisted of a RTDB placed between GaAs base and emitter layers. The RT structure was made of a 56 Å thick GaAs quantum well sandwiched between two 50 Å thick $Al_{0.53}Ga_{0.67}As$ barriers. The RTDB between the base and the emitter simply served the purpose of injecting, through RT, high-energy electrons into the base region. The high-energy electrons are transported ballistically through the 1000 Å thick n^+ base region before being collected at the 3000 Å thick $Al_{0.20}Ga_{0.80}As$ collector barrier. The barriers and the quantum well were undoped; the emitter, base and the collector layers were n-type doped to 1×10^{18} cm^{-3}.

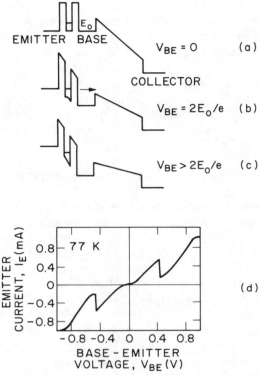

Figure 5.33 The band diagrams of the RHET at (a) $V_{BE} = 0$, (b) $V_{BE} = 2E_0/e$ (RT established), and (c) $V_{BE} > 2E_0/e$ (RT quenched), illustrating the operating principle of the device, and (d) the base–emitter current–voltage characteristics measured at 77 K.

The operation of the device in the common-emitter configuration with a fixed collector–emitter voltage V_{CE} is schematically shown in the band diagrams of Fig. 5.33. When the base–emitter voltage V_{BE} is zero (Fig. 5.33a), there is no electron injection, and hence the emitter as well as the collector currents are zero even with a positive V_{CE}. A peak in the emitter and the collector current occurs when V_{BE} is nearly equal to $2E_0/e$ (where E_0 is the energy of the first resonant state in the quantum well; Fig. 5.33b). With further increase in V_{BE}, RT is quenched (Fig. 5.33c) and hence there is a sudden drop in the emitter current (Fig. 5.33d) and a corresponding drop in the collector current as well [16].

This device could in principle be used for the same applications as discussed in Section 5.3.1. The Fujitsu group has demonstrated its application as an exclusive-NOR gate [16], which is essentially the parity generator circuit of Fig. 5.23 with two inputs. The base–emitter characteristics was also used to design a flip–flop [79], as shown in the inset of Fig. 5.34. The

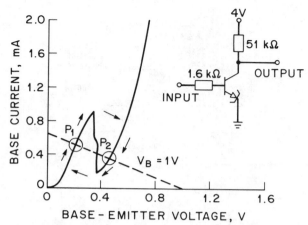

Figure 5.34 Schematic operation of the flip–flop circuit (shown in the inset) using the RHET.

circuit consists of a RHET with a 51 kΩ collector resistance and a 1.6 kΩ resistance in series with the base. The collector and the base supply voltages are 4.0 V and 1.0 V, respectively. It should be noted that the base current versus base–emitter voltage also exhibits a resonance peak. The load line corresponding to the 1.6 kΩ base resistance and 1.0 V base supply voltage, drawn on the I_B versus V_{BE} characteristic, has two stable points P_1 and P_2 (Fig. 5.34). The circuit can stay in either of these two points indefinitely and may also be shifted from one to the other by applying suitable pulses. At P_1, the transistor is heavily conducting and therefore the output at the collector node will be low, whereas at P_2, since RT has been quenched, the transistor is essentially in the poorly conducting state, resulting in a high state at the output.

The RHET is, however, operable at a low temperature (77 K) only and possesses very little current gain, since it is an unipolar device. A large fraction of the hot electrons injected into the base cannot surmount the collector barrier, because of large phonon scattering in the n^+ base region. Recently, the RHET has been implemented in the $Al_{0.48}In_{0.52}As/$ $Ga_{0.47}In_{0.53}As$ material system [48] with improved characteristics, as was expected.

5.4.2 Resonant Tunneling Gate Field Effect Transistor (RT-FET)

5.4.2.1 Structure and Processing. Figure 5.35a shows the schematics of the device grown by MBE on an LEC semi-insulating substrate. The structure consists of an n-type GaAs channel, 1 μm thick, and doped to 4×10^{17} cm^{-3}. The DB, consisting of a 70 Å undoped GaAs well layer sandwiched between two 25 Å undoped AlAs barriers, was grown on top of the channel. The

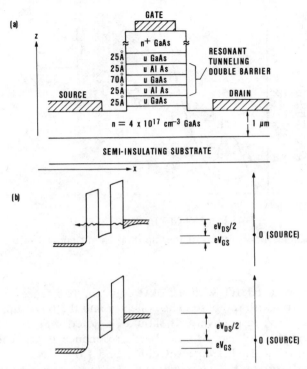

Figure 5.35 (a) Schematic cross section of the resonant tunneling gate FET. (b) Band diagram showing RT of electrons injected from the channel through double barrier in the gate, and quenching of RT. Note the combined action of the gate and drain voltages in biasing the double barrier.

gate contact layer is $0.4\,\mu m$ thick GaAs and doped to $5 \times 10^{17}\,cm^{-3}$. The undoped GaAs spacer layers ($\sim 25\,\text{Å}$) are left on the two sides of the undoped DB, to offset the effect of Si diffusion from the adjacent n-type layers, during the high-temperature growth process.

The devices have a $4.65\,\mu m$ gate length, $308\,\mu m$ gate width, $2.15\,\mu m$ source–gate spacing, and $9.8\,\mu m$ source–drain spacing. The asymmetry between the source–gate and drain–gate spacing was unintentionally introduced during processing.

The devices were processed by evaporating the gate electrode followed by mesa etching. The gate mesa was formed by wet etching using the stop etches $H_2O_2 + NH_4OH$ (7.2 pH) for the GaAs and 1:1 $HCl + H_2O$ for the AlAs layers alternately. The height of the gate mesa was $5500\,\text{Å}$. The source and drain electrodes were then deposited on the exposed channel. The gate contact was composed of Ge(60 Å)/Au(135 Å)/Ag(500 Å)/Au(750 Å) alloy. Ni(50 Å)/Au(385 Å)/Ge(215 Å)/Au(750 Å) alloy was used as the source and drain contacts. All the contacts were alloyed together at 400°C for 10 s in a hydrogen flow.

5.4.2.2   Principle of Operation. The operation of the device is based on the quenching of RT through the DB between the gate electrode and the channel. The biasing of the DB, when a gate-to-source (V_{GS}) as well as a drain-to-source (V_{DS}) voltage is applied, is shown in Fig. 5.35b. Note the combined action of the gate voltage and half the drain voltage in controlling RT. The voltage appearing across the DB is, to a first approximation, given by

$$V_{DB} = V_{GS} - r \cdot V_{DS} \tag{5}$$

where r is a factor determined by the ratio of the gate-to-source and the drain-to-gate spacings. Ideally, r should be 0.5. In the devices tested, however, this factor was 0.42, due to structural asymmetry unintentionally introduced during processing. Thus, RT through the DB may be controlled by either the gate or the drain voltage to produce peaks in the gate current characteristics. This in turn generates structures in the drain current as well, since the gate current adds to the drain-to-source current in the channel. Thus, in addition to negative differential resistance, negative transconductance was obtained as well, by controlling structures in the drain current with the variaiton of the gate voltage.

I_D *and* I_G *versus* V_{GS}. To understand the current–voltage characteristics quantitatively, it is necessary to discuss the potential distribution inside the device. Let us first consider the gate (I_G) and the drain (I_D) current variation with V_{GS} when $V_{DS} = 0$ V. Under this condition, the electrons tunneling from the gate to the channel and vice versa form two equal (in the ideal case, when $r = 0.5$) and opposite current flows in the channel, between its center and the source and drain electrodes. Half the gate current thus forms the drain current. The resulting potential distribution is shown in Fig. 5.36a and b for positive and negative gate voltages, respectively. The conduction band energy diagram across the DB is shown on the left of Fig. 5.36; the electron potential energy distribution due to ohmic drop along the channel in the region immediately adjacent to the DB is shown on the right. The interesting points to note from these diagrams are listed below:

1. For positive gate bias, an electron accumulation layer is formed at the heterointerface between the channel and the bottom AlAs barrier while the gate side of the DB is depleted of carriers. The density of electrons in the thin (100–200 Å) accumulation layer substantially exceeds the average carrier concentration in the channel, thus strongly reducing, by screening, the source-to-drain electric field in the immediate vicinity of the DB channel interface, compared with the rest of the channel, as shown at the right of Fig. 5.36a. The near equipotential at the DB channel interface allows RT to be quenched everywhere at the same gate bias, resulting in an abrupt drop of the current with voltage, as observed in the experimental data (Fig. 5.37a). The important impact of the accumulation and depletion layers adjacent to the DB on the I–V characteristics of RT diodes has been

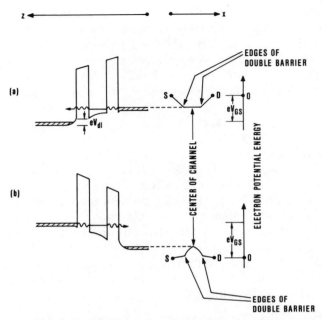

Figure 5.36 Conduction band energy diagram across the double barrier (left) and electron potential energy distribution along the channel in the region immediately adjacent to the double barrier (right) at $V_{DS} = 0$ V for (a) $V_{GS} > 0$ V and (b) $V_{GS} < 0$ V.

discussed previously by Goldman et al. [80]. These data provide the first direct evidence of such effect.

2. For negative gate bias, on the other hand, the accumulation layer is formed on the gate side while the region of the channel adjacent to the DB is depleted of carriers. As a result, the flow of electrons causes a near triangular potential variation in the channel, with the center being at the highest energy, and the edges of the DB being at the lowest (Fig. 5.36b). Therefore, quenching of RT is initiated at the edges of the DB and gradually proceeds to the center with increasing magnitude of negative gate voltage. The I–V thus exhibits a gradual drop, as opposed to the abrupt drop with positive gate bias (Fig. 5.37a). The same effect is also observed for finite voltages applied externally between the drain and the source (Fig. 5.37b and c), since the accumulation layer in the channel also screens the applied drain-to-source field in the immediate vicinity of the DB, rendering it near equipotential. For negative gate bias in this latter case, the electron potential energy diagram in the channel (which is left to the reader as an exercise) implies that RT is quenched initially at either the drain or the source end (depending on the polarity of V_{DS}) of the DB and gradually proceeds to the other end with increasing magnitude of gate voltage.

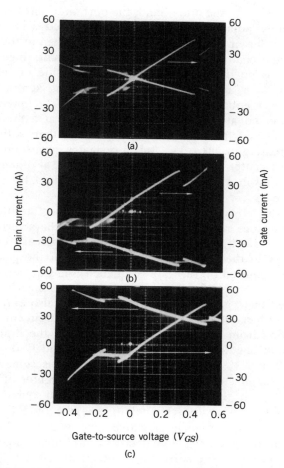

Figure 5.37 Drain and gate currents versus gate-to-source voltage V_{GS} measured at 100 K for (a) $V_{DS} = 0$ V, (b) $V_{DS} = -0.2$ V, and (c) $V_{DS} = 0.2$ V.

3. For positive gate voltage (Fig. 5.36a), there is a significant potential difference between the equipotential region under the DB and the source and drain electrodes (which are at the same potential, since $V_{DS} = 0$). This is due to the flow of electrons in the channel that come from the gate by RT. Because of this ohmic drop, only a part of the applied V_{GS} appears across the DB. Also, there is a potential drop V_{dl} (~0.1 V) in the depletion layer adjacent to the DB on the gate side (Fig. 5.36a, left). Consequently, the voltage appearing across the DB is substantially less than that applied between the gate and source electrodes externally. Therefore, a larger V_{GS} (0.4 V) is required to quench RT than what is expected from a simple tunneling calculation ($2E_1 \simeq 0.2$ V).

For negative V_{GS}, on the other hand, the portion of the channel under the gate immediately adjacent to the DB is depleted. Hence, the ohmic potential drop is larger in this part of the channel than in the rest, bringing the center of the channel at the highest energy point (Fig. 5.36b, right). Also, as there is no depletion layer in the gate side of the barrier, the potential drop V_{dl} is absent. The voltage appearing across the DB at its edges and almost equal to the applied voltage V_{GS}. Quenching of RT, therefore, is iniated at the edges of the DB and at a much lower $|V_{GS}|$ (≈ 0.2 V, as expected from tunneling calculations) than with positive gate bias. With increasing negative V_{GS}, this propagates inward until RT is finally quenched at the center. The absence of the drop V_{dl} also helps it to quench RT totally at a lower V_{GS} (0.3 V), as opposed to the 0.4 V required in the case of positive V_{GS}.

When the drain is negatively biased with respect to the source, the negative conductance region and the overall I–V characteristics shift to a lower V_{GS} (Fig. 5.37b). This is a result of the increase of the electron potential energy in the region of the channel under the gate, so that less positive bias is required on the gate to quench RT. For negative gate bias, however, a larger voltage is necessary to obtain the same effect. Note that as RT is quenched there is a change not only in I_G but also in I_D as in the case with $V_{DS} = 0$. In fact with $V_{DS} < 0$ and $V_{GS} > 0$, the drain current consists of electrons flowing from the drain to the source and the drain to the gate. When RT is quenched, the latter flow is reduced and I_D drops.

For $V_{DS} < 0$ and $V_{GS} < 0$, however, the picture is somewhat different. With $V_{GS} \simeq 0$, the electron potential energy in the center of the channel is higher than that in the gate due to the drain-to-source current flow. Electrons thus flow from the channel to the gate by RT, resulting in a positive gate current for small negative gate voltages. With increasing negative bias on the gate, this flow decreases and finally changes direction, giving rise to a negative gate current as a sufficiently large V_{GS}. Before RT is quenched, electrons flowing from the gate to the channel by RT add up to those coming from the drain by ohmic conduction, and flow out in the source electrode, causing an ohmic potential drop in the source–gate part of the channel. The electron potential energy in the region under the gate is consequently higher than what would have been expected in the absence of the gate current. When RT is quenched, the potential drop in the source–gate part, therefore, is decreased. But, since the drain-to-source voltage is fixed, the potential drop in the drain–gate part of the channel has to increase, resulting in an increase in the drain current flow. This is observed as a sudden increase in the magnitude of the negative drain current.

For $V_{DS} > 0$ (Fig. 5.37c), a negative gate current is produced even at $V_{GS} = 0$ by the electrons resonantly tunneling from the gate to the channel, since the positive V_{DS} has brought the electron potential energy in the middle of the channel below that in the gate. These electrons also add to the drain current. With V_{GS} going positive, the RT gate current is first reduced to zero and then changes direction with a corresponding reduction in the drain current, as some of the electrons coming from the source are directed to the gate. When RT is finally quenched with a drop in the gate current, there is a corresponding abrupt rise in the drain current because all the electrons coming from the source are now directed to the drain. For negative V_{GS}, the resonantly tunneling electrons from the gate always add to the drain current, and hence the latter increases with increasing $|V_{GS}|$. A drop in the gate as well as the drain current is observed when RT is quenched.

I_D and I_G versus V_{DS} for Different V_{GS}. From the foregoing discussions, we find that a negative drain bias has the same effect as a positive gate bias and vice versa. It is therefore easy to understand that the drain I_D and the gate I_G current variation with V_{DS} at different V_{GS} will show similar NDR characteristics, with the suppression of RT being abrupt at negative V_{DS} and gradual with positive V_{DS}. Also, for a change ΔV_{GS} in the gate bias, the NDR region will shift by $\Delta V_{DS} \approx 2\Delta V_{GS}$.

Figure 5.38 shows I_G and I_D as a function of the drain-to-source bias V_{DS} for $V_{GS} = 0$ (top) and $V_{GS} = +0.2$ V (bottom). Let us consider the curves for $V_{GS} = 0$ first. Note that for positive V_{DS}, the region of the channel adjacent to the DB is depleted and electrons resonantly tunnel from the gate contact layer into the channel. For negative V_{DS}, on the other hand, electrons resonantly tunnel from the accumulation layer at the DB channel interface into the gate contact layer. It is therefore easy to explain the differences in the steepness of the *I–V* (Fig. 5.38) in the NDR regions for different bias polarities, in a manner similar to those in Fig. 5.37. Following similar arguments, it is also understood why the NDR region occurs at a significantly lower $|V_{DS}|$ when the drain is positively biased than when it is negatively biased. Application of a positive gate bias ($V_{GS} = +0.2$ V; Fig. 5.38, bottom) shifts the NDR region to higher V_{DS} by nearly 0.4 V, as expected. Also, at $V_{DS} = 0$, electrons now tunnel from the channel into the gate, giving rise to positive gate current. A positive V_{DS} is then required to suppress this tunneling first ($I_G = 0$ at $V_{DS} \approx 0.4$ V for $V_{GS} = 0.2$ V) and then reverses the direction of RT electrons.

An even more interesting situation occurs when the applied gate bias is large enough to quench RT even at $V_{DS} = 0$. The conduction band energy diagram across the DB under this condition is shown in Fig. 5.39 (left). Even if RT is suppressed, there will be some flow of electrons from the channel to the gate by thermionic emission over the barrier and inelastic tunneling, resulting in a positive gate current (Fig. 5.40). These electrons are supplied by the drain and the source contacts. The resulting ohmic drop will place the drain and the source points at a higher energy than the region under the DB, as shown in the right half of Fig. 5.39a. As a positive drain voltage is applied, the electron flow from the drain end is reduced, resulting in a reduction in the negative drain and positive gate currents until a

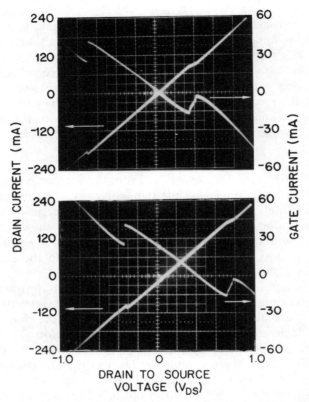

Figure 5.38   Drain and gate currents versus drain-to-source voltage V_{DS} measured at 100 K for $V_{GS} = 0$ V (top) and $V_{GS} = 0.2$ V (bottom).

condition is reached when the drain is at the same potential as the middle of the channel and the drain current is reduced to zero. As the drain voltage is further increased, a positive drain current starts flowing. The positive drain voltage also pulls down the middle of the channel in the energy diagram, and the gate current continues to decrease until RT is restored at some V_{DS} (Fig. 5.39b). At this point, a large fraction of the electrons are again transferred to the gate by RT, resulting in a sudden increase in the gate current and corresponding decrease in the drain current, as observed experimentally (Fig. 5.40). Note that the onset of RT is quite abrupt in this case because of the accumulation layer in the channel (Fig. 5.39b). Beyond this point, the gate current continues to decrease with a decreasing RT flow of electrons and the drain current increases.

At larger V_{DS}, the drain current is large compared with the gate current, and hence the potential in the channel can be solely determined by V_{DS}. So, at $V_{DS} \simeq 2V_{GS}$, the gate is at the same potential as the center of the channel (Fig. 5.39c) and no gate current flows. The experimental data taken at $V_{GS} = 0.5$ V (Fig. 5.40) in fact indicate that $I_G = 0$ at $V_{DS} \simeq 1.0$ V.

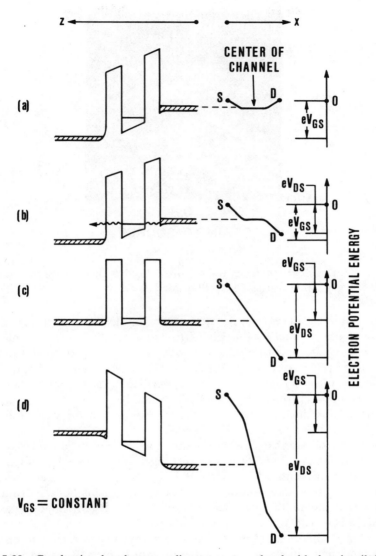

Figure 5.39 Conduction band energy diagram across the double barrier (left) and electron potential energy distribution along the channel in the region immediately adjacent to the double barrier (right) at a large positive V_{GS} (constant): (a) $V_{DS} = 0$ V, RT quenched; (b) small positive V_{DS}, RT established; (c) larger V_{DS}, gate current is zero; and (d) large positive V_{DS}, RT quenched again.

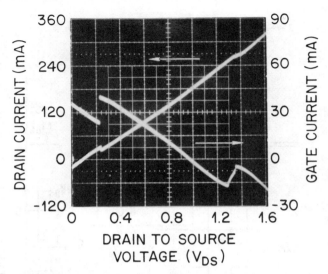

Figure 5.40 Drain and gate currents versus positive drain-to-source voltage at $V_{GS} = 0.5$ V, exhibiting two peaks, as illustrated in Fig. 5.39.

With a further increase in the drain voltage, electrons start flowing from the gate to the channel by RT, giving rise to a negative gate current. These electrons also add up in the channel with those coming from the source to the positive drain current further increase (Fig. 5.40). When RT is quenched again at a sufficiently large V_{DS} (Fig. 5.39d), there is a drop in the gate as well as the drain currents (Fig. 5.40). It should be noted that the quenching of RT is gradual in this case, as expected, because of the depletion of the channel adjacent to the gate (Fig. 5.39d).

A similar situation can also be obtained with large negative bias applied to the gate and the characteristics taken against negative V_{DS}. Figure 5.41 shows the experimental data taken at $V_{GS} = -0.3$ V. These curves can be explained quantitatively, following similar arguments as before. It may also be noted that, unlike the previous one, the first NDR region ($V_{DS} = -0.2$ V) in this case is broad, whereas the second one ($V_{DS} = -1.3$ V) is abrupt. With $V_{GS} = -0.3$ V, the channel is at a lower energy than the gate when $V_{DS} = -0.2$ V, and hence its depletion region adjacent to the DB leads to a gradual suppression of RT. For $V_{DS} = -1.3$ V, on the other hand, an accumulation layer is formed in the channel, and thus the quenching is abrupt.

This part of the characteristics may be important in applications involving multiple peaks [15, 30, 52], since the two peaks are obtained by a suitable manipulation of only the ground-state resonance of the quantum well.

In another attempt to integrate RTDB with field effect transistors (FETs), developed jointly by California Institute of Technology and Xerox PARC [18, 25, 77], the DB was placed in the channel, adjacent to the source electrode. The operation of this class of structures, called DB/FETs by their

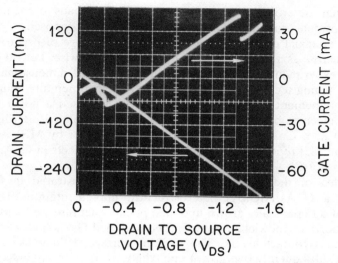

Figure 5.41 Drain and gate currents versus negative drain-to-source voltage at $V_{GS} = -0.3$ V, exhibiting two peaks.

developers, is quite different from the RT-FET discussed above. In principle, the DB/FET is electrically equivalent to a RT diode with a resistance (that of the channel) in series with it. The role of the resistance is to push the NDR region in the *I–V* characteristics to higher voltages, since a significant fraction of the applied voltage drops across the same. In the DB/FET, since the gate voltage changes the channel resistance, it effectively controls the position of the peak. Therefore, a voltage-controllable negative resistance characteristic is obtained [18]. The DB/FETs have been implemented both in the vertical FET [25] and the MESFET [77] geometry. The *I–V* characteristics in both geometries illustrate the same basic nature, namely, gate-controllable negative resistance, although there are differences in details arising out of the differences in the field distributions in the two geometries. The DB/FETs have been used to construct frequency multipliers and flip–flops [78].

5.5   QUANTUM WIRE TRANSISTOR

This device, proposed by Luryi and Capasso [19], uses a linear rather than a planar quantum well as the active region. In this device, electrons resonantly tunnel from 2D to 1D (quantum wire) states. The properties of 0- and 1-dimensional systems are receiving increasing attention as new techniques are developed to realize them. Sakaki [81] discussed the possibility of obtaining an enhanced mobility along quantum wires, because of the suppression of the ionized-impurity scattering. He proposed a V-groove etch of a planar heterojunction quantum well as a means of achieving the one-dimensional confinement. Chang et al. [82] proposed a technique involving epitaxial overgrowth of a vertical $\langle 110 \rangle$ edge of pregrown $\langle 100 \rangle$ heterostructrue as a means of obtaining the quantum wire. Petroff et al. [83] reported experimental attempts of implementing one-dimensional confinement by etching techniques. Cibert et al. [84] experimentally demonstrated carrier confinement in 1 and 0 dimensions. They used a novel technique involving electron beam lithography and laterally confined interdiffusion of aluminum in a GaAs/AlGaAs heterostructure grown by MBE. Recently, Reed et al. [85] reported dimensional quantization effects in GaAs/AlGaAs "quantum dots."

The idea of the quantum wire transistor is illustrated in Fig. 5.42, assuming a GaAs/AlGaAs heterostructure implementation. The device consists of an epitaxially grown undoped planar quantum well and a double AlGaAs barrier sandwiched between two undoped GaAs layers and heavily doped GaAs contact layers. The working surface defined by a V-groove etching is subsequently overgrown epitaxially with a thin AlGaAs layer and

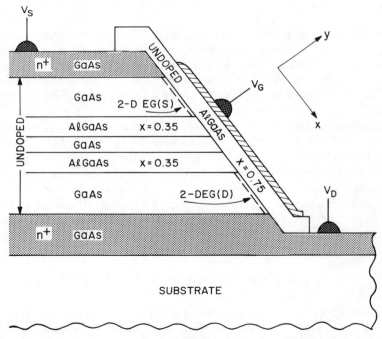

Figure 5.42 Schematic cross section of the proposed surface resonant tunneling device, the quantum wire transistor structure. A "V-groove" implementation of the quantum wire is assumed. Thicknesses of the two undoped GaAs layers outside the double barrier region should be sufficiently large ($\gtrsim 1000$ Å) to prevent the creation of a parallel conduction path by the conventional (bulk) RT.

gated. The thickness of the gate barrier layer $(d \gtrsim 100 \,\text{Å})$ and the Al content in this layer $(x \gtrsim 0.5)$ should be chosen so as to minimize gate leakage. The thickness of the quantum well barrier layers are chosen so that their projection on the slanted surface should be $\gtrsim 50 \,\text{Å}$ each. The Al content in these layers should be typically $x \lesssim 0.45$. Application of a positive gate voltage V_G induces 2D electron gases at the two interfaces with the edges of undoped GaAs layers outside the quantum well. These gases will act as the source (S) and drain (D) electrodes. At the same time, there is a range of V_G in which electrons are not yet induced in the quantum wire region (which is the edge of the quantum well layer) because of the additional dimensional quantization.

To understand the operation of the device, consider first the band diagram in the absence of a source-to-drain voltage, $V_{DS} = 0$ (Fig. 5.43a). The diagram is drawn along the x direction (from S to D parallel to the surface channel.). The y direction is defined as the one normal to the gate, and z that along the quantum wire. Dimensional quantization induced by the gate results in a zero-point energy of electronic motion in the y direction, represented by the bottom E_0 of a 2D subband corresponding to the free motion in the x and z directions. The thicknesses of the undoped S and D layers are assumed to be large enough $(\gtrsim 1000 \,\text{Å})$ so that the electronic motion in x direction in these layers can be considered free. On the other hand, in the quantum well region of the surface channel, there is an additional dimensional quantization—along the x direction—which defines the quantum wire [81]. If t is the x projection of the quantum well layer thickness, then the additional zero-point energy is approximately given by

$$E_0' - E_0 = \pi^2 \hbar^2 / 2m^* t^2 \tag{6}$$

This approximation, of course, is good only when the barrier heights substantially exceeds E_0'.

Application of a gate voltage V_{GS} moves the 2D subband E_0 with respect to the (classical) bottom of the conduction band E_C and the Fermi level E_F. The operating regime of this device with respect to V_{GS} at $V_{GS} = 0$ corresponds to the situation when E_F lies in the gap $E_0' - E_0$. RT condition is set in by the application of a positive V_{DS}, as illustrated in Fig. 5.43b. In this situation, the energy of certain electrons in S matches unoccupied levels in the quantum wire (Fig. 5.43c). Compare this with the tunneling of 3D electrons into 2D density of states discussed in Section 5.2.1.1. In the present case, the dimensionality of both the emitter and the base is reduced by one. Hence, the emitter Fermi sea of Fig. 5.1 has become a disk in this case and the Fermi disk of the previous case is replaced by a resonant segment, as shown in Fig. 5.43c. Since both k_x and k_y are quantized in the quantum wire, RT requires conservation of energy and the lateral momentum k_z. This is true only for those electrons whose momenta lie in the

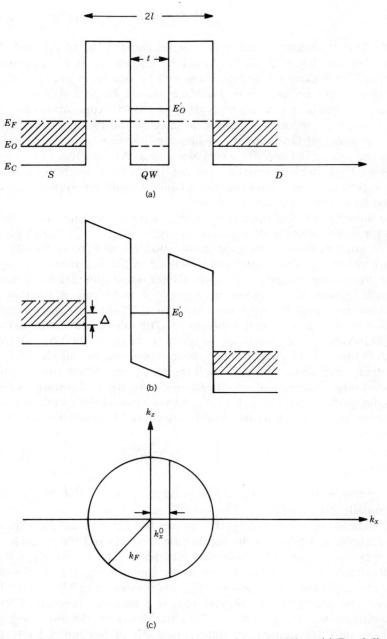

Figure 5.43 Illustration of the quantum wire transistor operation. (a) Band diagram along the channel in "equilibrium," that is, in the absence of a drain bias. (b) Band diagram for an applied bias V_{DS} when the energy of certain electrons in the source (S) matches unoccupied levels of the lowest 1D subband E'_0 in the quantum wire. (c) Fermi disk corresponding to the 2D degenerate electron gas in the source electrode. Vertical chord at $k_x = k_x^0$ indicates the momenta of electrons that can tunnel into the quantum wire while conserving their momentum k_z along the wire.

segment $k_x = k_x^0$ (Fig. 5.43), where

$$\frac{\hbar^2 (k_x^0)^2}{2m^*} = \Delta \tag{7}$$

It should be noted that the energies of all electrons in this segment $(k_x = k_x^0)$ lie in the band $E_0 + \Delta \leq E \leq E_F$. However, only those electrons in this energy band that satisfy the momentum conservation condition are resonant. As V_{DS} is increased, the resonant segment moves to the left (see Fig. 5.43c), toward the vertical diameter $k_x = 0$ of the Fermi disk, and the number of tunneling electrons grows, reaching a maximum $[2m^* (E_F - E_0)]^{1/2}/\pi\hbar$ per unit length in the z direction when $\Delta = 0$. At higher V_{DS}, when $\Delta < 0$, there are no electrons in the source that can tunnel into the quantum wire while conserving their lateral momentum. This gives rise to the NDR in the drain circuit.

In the present device, apart from obtaining high electron mobility, as discussed by Sakaki [81], additional flexibility is achieved throught the gate electrode. The gate voltage in this structure not only determines the number of electrons available for conduction but also controls the position of the E_0' level in the quantum wire with respect to E_0 in the source. This latter control is effected by the fringing electric fields and gives rise to the interesting possibility of negative transconductance. Luryi and Capasso [19] have solved the corresponding electrostatic problem by suitable conformal mappings and have shown that in the operating regime of the device, an increasing $V_{GS} > 0$ *lowers* the electrostatic potential energy in the base (quantum wire) with respect to the emitter (source) nearly as effectively as does an increasing V_{DS}.

5.6   CONCLUSION

Resonant tunneling has been found to be very promising in the design of new functional devices [86]. With the development of band gap engineering and modern crystal growth techniques, such as MBE, many new devices have been demonstrated in recent years. Practical circuits involving less complexity were also constructed using some of these devices. An up-to-date review of the research done in the field is given in this chapter. The future also looks very promising. Without trying to predict the future, it may be said that the scope of the area is wide open. This chapter is intended to stimulate the readers and make them aware of the challenges that lie ahead. It is only left to the designer's imagination to conceive new devices with increased functionality. At the time of going to press, resonant tunneling bipolar transistors with multiple peaks in the characteristics and their application in various circuits have been demonstrated [87–91].

REFERENCES

1. L. Esaki, New phenomenon in narrow germanium p-n junctions. *Phys. Rev.* **109** 603, (1958).
2. L. L. Chang, L. Esaki, and R. Tsu, Resonant tunneling in semiconductor double barriers. *Appl. Phys. Lett.* **24** 593, (1974).
3. M. Tsuchiya, H. Sakaki, and J. Yoshino, Room temperature observation of differential negative resistance in an AlAs/GaAs/AlAs resonant tunneling diode. *Jpn. J. Appl. Phys.* L**24**, 466 (1985).
4. C. I. Huang, M. J. Paulus, C. A. Bozada, S. C. Dudley, K. R. Evans, C. E. Stutz, R. L. Jones, and M. E. Cheney, AlGaAs/GaAs double barrier diodes with high peak-to-valley current ratio, *Appl. Phys. Lett.* **51**, 121 (1987).
5. S. Sen, F. Capasso, A. L. Hutchinson, and A. Y. Cho, Room temperature operation of $Ga_{0.47}In_{0.53}As/Al_{0.48}In_{0.52}As$ resonant tunneling diodes. *Electron. Lett.* **23**, 1229 (1987).
6. T. Inata, S. Muto, Y. Nakata, S. Sasa, T. Fujii and S. Hiyamizu, A pseudo-morphic $In_{0.53}Ga_{0.47}As/AlAs$ resonant tunneling barrier with a peak-to-valley current ratio of 14 at room temperature. *Jpn. J. Appl. Phys.* **26**, L1332 (1987).
7. E. E. Mendez, W. I. Wang, B. Ricco, and L. Esaki, Resonant tunneling of holes in AlAa-GaAs-AlAs heterostructures. *Appl. Phys. Lett.* **47**, 415 (1985).
8. M. A. Reed, J. W. Lee, and H.-L. Tsai, Resonant tunneling through a double GaAs/AlAs superlattice barrier, single quantum well heterostructure. *Appl. Phys. Lett.* **49**, 158 (1986).
9. T. Nakagawa, H. Imamoto, T. Kojima, and K. Ohta, Observation of resonant tunneling in AlGaAs/GaAs triple barrier diodes. *Appl. Phys. Lett.* **49**, 73 (1986).
10. S. Sen, F. Capasso, A. C. Gossard, R. A. Spah, A. L. Hutchinson, and S. N. G. Chu, Observation of resonant tunnelling through a compositionally graded parabolic quantum well. *Appl. Phys. Lett.* **51**, 1428 (1987).
11. T. C. L. G. Sollner, W. D. Goodhue, P. E. Tannenwald, C. D. Parker, and D. D. Peck, Resonant tunneling through quantum wells at frequencies up to 2.5 Thz. *Appl. Phys. Lett.* **43**, 588 (1983).
12. T. C. L. G. Sollner, P. E. Tannenwald, D. D. Peck, and W. D. Goodhue, Quantum well oscillators. *Appl. Phys. Lett.* **45**, 1319 (1984).
13. E. R. Brown, T. C. L. G. Sollner, W. D. Goodhue, and C. L. Chen, High-speed resonant-tunneling diodes. *Proc. SPIE—Int. Soc. Opt. Eng.* **943**, 2 (1988).
14. F. Capasso, New high speed quantum well and variable gap superlattice devices. in *"Picosecond Electronics and Optoelectronics"* (G. A. Mourou, D. M. Bloom, and C. H. Lee, eds.), p. 112. Springer-Verlag, Berlin and New York, 1985.
15. F. Capasso and R. A. Kiehl, Resonant tunneling transistor with quantum well base and high-energy injection: A new negative differential resistance device. *J. Appl. Phys.* **58**, 1366 (1985).
16. N. Yokoyama, K. Imamura, S. Muto, S. Hiyamizu, and H. Nishi, A new funct5onal resonant tunneling hot electron transistor (RHET). *Jpn. J. Appl. Phys.* **24**, L853 (1985).

17. A. R. Bonnefoi, D. H. Chow, and T. C. McGill, Inverted base-collector tunnel transistors. *Appl. Phys. Lett.* **47**, 888 (1985).

18. A. R. Bonnefoi, T. C. McGill, and R. D. Burnham, Resonant tunneling transistors with controllable negative differential resistance. *IEEE Electron Device Lett.* **EDL-6**, 636 (1985).

19. S. Luryi and F. Capasso, Resonant tunneling of two dimensional electrons through a quantum wire: A negative transconductance device. *Appl. Phys. Lett.* **47**, 1347 (1985); also erratum: *ibid.* **48**, 1693 (1986).

20. Y. Nakata, M. Asada, and Y. Suematsu, Novel triode device using metal insulator superlattice proposed for high speed response. *Electron. Lett.* **22**, 58 (1986).

21. F. Capasso, K. Mohammed, and A. Y. Cho, Resonant tunneling through double barriers, perpendicular quantum transport phenomenon in superlattices, and their device applications. *IEEE J. Quantum Electron.* **QE-22**, 1853 (1986).

22. F. Capasso, S. Sen, A. C. Gossard, A. L. Hutchinson, and J. H. English, Quantum well resonant tunneling bipolar transistor operating at room temperature. *IEEE Electron Device Lett.* **EDL-7**, 573 (1986).

23. T. Futatsugi, Y. Yamaguchi, K. Ishii, K. Imamura, S. Muto, N. Yokoyama, and A. Shibatomi, A resonant-tunneling bipolar transistor (RBT); A proposal and demonstration for new functional devices with high current gains. *Tech. Dig.— Inter. Electron Devices Meet.* p. 286 (1986).

24. F. Capasso, S. Sen, F. Beltram, and A. Y. Cho, Resonant tunneling gate field-effect transistor. *Electron. Lett.* **23**, 225 (1987).

25. T. K. Woodward, T. C. McGill, and R. D. Burnham, Experimental realization of a resonant tunneling transistor. *Appl. Phys. Lett.* **50**, 451 (1987).

26. S. Sen, F. Capasso, F. Beltram, and A. Y. Cho, The resonant tunneling field-effect transistor: A new negative transconductance device. *IEEE Trans. Electron Devices* **ED-34**, 1768 (1987).

27. F. Capasso, S. Sen, and A. Y. Cho, Negative transconductance resonant tunneling field effect transistor. *Appl. Phys. Lett.* **51**, 526 (1987).

28. F. Capasso, S. Sen, and A. Y. Cho, Resonant Tunneling: Physics, new transistors and superlattice devices. *Proc. SPIE—Int. Soc. Opt. Eng.* **792**, 10 (1987).

29. G. H. Heilmeir, Microelectronics: End of the Beginning or beginning of the end? *Tech. Dig—Int. Electron Device meet.* p. 2 (1984).

30. S. Sen, F. Capasso, A. Y. Cho, and D. Sivco, Resonant tunneling device with multiple negative differential resistance: Digital and signal processing applications with reduced circuit complexity. *IEEE Trans. Electron Devices* **ED-34**, 2185 (1987).

31. S. Luryi, Frequency limit of double-barrier resonant-tunneling oscillators. *Appl. Phys. Lett.* **47**, 490 (1985).

32. H. Morkoç, J. Chen, U. K. Reddy, T. Henderson, and S. Luryi, Observation of a negative differential resistance due to tunneling through a single barrier into a quantum well. *Appl. Phys. Lett.* **49**, 70 (1986).

33. B. Ricco and M. Ya. Azbel, Physics of resonant tunneling. The one dimensional double-barrier case. *Phys. Rev. B: Condens. Matter* [3] **29**, 1970 (1984).

34. T. Weil and B. Vinter, Equivalence between resonant tunneling and sequential tunneling in double-barrier diodes, *Appl. Phys. Lett.* **50**, 1281 (1987).

35. M. Jonson and A. Grincwajg, Effect of inelastic scattering on resonant and sequential tunneling in double barrier heterostructures. *Appl. Phys. Lett.* **51**, 1729 (1987).

36. S. Luryi, Coherent versus incoherent resonant tunneling and implications for fast devices. *Superlattices Microstruct.* (to be published).

37. K. K. Choi, B. F. Levine, C. G. Bethea, J. Walker, and R. J. Malik, Photoexcited coherent tunneling in a double-barrier superlattice. *Phys. Rev. Lett.* **59**, 2459 (1987).

38. M. Tsuchiya and H. Sakaki, Precise control of resonant tunneling current in AlAs/GaAs/AlAs double barrier diodes with atomically-controlled barrier widths. *Jpn. J. Appl. Phys.* **25**, L185 (1986).

39. M. Tsuchiya and H. Sakaki, Dependence of resonant tunneling current on well widths in AlAs/GaAs/AlAs double barrier diode structures. *Appl. Phys. Lett.* **49**, 88 (1986).

40. S. Muto, T. Inata, H. Ohnishi, N. Yokoyama, and S. Hiyamizu, Effect of silicon doping profile on I-V characteristics of an AlGaAs/GaAs resonant tunneling barrier structure grown by MBE. *Jpn. J. Appl. Phys.* **25**, L577 (1986).

41. H. Toyoshima, Y. Ando, A. Okamoto, and T. Itoh, New resonant tunneling diode with a deep quantum well. *Jpn. J. Appl. Phys.* **25**, L786 (1986).

42. M. A. Reed and J. W. Lee, Resonant tunneling in a GaAs/AlGaAs barrier/InGaAs quantum well heterostructure. *Appl. Phys. Lett.* **50**, 845 (1987).

43. S. Muto, T. Inata, Y. Sugiyama, Y. Nakata, T. Fujii, H. Ohnishi, and S. Hiyamizu, Quantum well width dependence of negative differential resistance of $In_{0.52}Al_{0.48}As/In_{0.53}Ga_{0.47}As$ resonant tunneling barriers grown by MBE. *Jpn. J. Appl. Phys.* **26**, L220 (1987).

44. M. Tsuchiya and H. Sakaki, Dependence of resonant tunneling current on Al mole fractions in $Al_xGa_{1-x}As$-GaAs-$Al_xGa_{1-x}As$ double barrier structures. *Appl. Phys. Lett.* **50**, 1503 (1987).

45. A. D. Stone and P. A. Lee, Effect of inelastic processes on resonant tunneling in one dimension. *Phys. Rev. Lett.* **54**, 1196 (1985).

46. C. W. Tu, R. Hendel, and R. Dingle, Molecular beam epitaxy and the technology of selectively doped heterostructure transistors. *in* "Gallium Arsenide Technology" (D. K. Ferry, ed.), p. 107. Howard & Sams, Indianapolis, Indiana, 1985.

47. F. Capasso, K. Mohammed, and A. Y. Cho, Sequential resonant tunneling through a multiquantum well superlattice. *Appl. Phys. Lett.* **48**, 478 (1986).

48. N. Yokoyama, K. Imamura, H. Ohnishi, T. Mori, S. Muto, and A. Shibatomi, Resonant tunneling hot electron transistor (RHET). *Proc. Int. Conf. Hot Carriers Semicond. 5th, 1987.*

49. A. C. Gossard, R. C. Miller, and W. Wiegmann, MBE growth and energy levels of quantum wells with special shapes *Surf. Sci.* **174**, 131 (1986).

50. Recent systematic studies in Ref. 44 have shown that electron RT through GaAs/$Al_xGa_{1-x}As$ diodes with thin barriers (30 Å) is dominated by the barrier height at the Γ point also in the indirect gap region ($0.45 \le x \le 1$).

51. M. Heiblum, M. V. Fischetti, W. P. Dumke, D. J. Frank, I. M. Anderson, C. M. Knoedler, and L. Osterling, Electron interference effects in quantum wells: Observation of bound and resonant states. *Phys. Rev. Lett.* **58**, 816 (1987).

52. F. Capasso, S. Sen, A. Y. Cho, and D. Sivco, Resonant tunneling devices with multiple negative differential resistance and demonstration of a three-state memory call for multiple-valued logic applications. *IEEE Electron Device Lett.* **EDL-8**, 297 (1987).

53. R. C. Potter, A. A. Lakhani, D. Beyea, H. Hier, E. Hempfling, and A. Fathimulla, Three-dimensional integration of resonant tunneling structures for signal processing and three-state logic. *Appl. Phys. Lett.* **52**, 2163 (1988).

54. A. A. Lakhani, R. C. Potter, and H. S. Hier, Eleven-bit parity generator with a single, vertically integrated resonant tunneling device. *Electron. Lett.* **24**, 681 (1988).

55. S. Sen, F. Capasso, D. Sivco, and A. Y. Cho, New resonant tunneling devices with multiple negative resistance regions and high room temperature peak to valley ratio. *IEEE Electron Device Lett.* **9**, 402 (1988).

56. S. Luryi, A. Kastalsky, A. C. Gossard, and R. H. Hendel, Charge injection transistor based on real-space hot-electron transfer. *IEEE Trans. Electron Devices* **ED-31**, 832 (1984).

57. C. Rine, ed., *"Computer Science and Multiple Valued Logic"*. North-Holland Publ., Amsterdam, 1977.

58. A. Heung and H. T. Mouftah, An all-CMOS ternary identity cell for VLSI implementation. *Electron. Lett.* **20**, 221 (1984).

59. "General Electric Tunnel Diode Manual, *1*st ed., p. 66 (1961) (A GE internal publication).

60. J. Söderström and T. G. Andersson, A multiple-state memory cell based on the resonant tunneling diode. *IEEE Electron Device Lett.* **9**, 200 (1988).

61. K. K. Choi, B. F. Levine, R. J. Malik, J. Walker, and C. G. Bethea, Periodic negative conductance by sequential resonant tunneling through an expanding high-field superlattice domain. *Phys. Rev. B: Condens. Matter* **35**, 4172 (1987).

62. S. P. Gentile, "Basic Theory and Application of Tunnel Diodes," p. 156, Van Nostrand, Princeton, New Jersey, 1962.

63. F. Capasso, S. Sen, A. Y. Cho, and A. L. Hutchinson, Resonant tunneling electron spectroscopy. *Electron. Lett.* **23**, 28 (1987).

64. F. Capasso, S. Sen, A. Y. Cho, and A. L. Hutchinson, Resonant tunneling spectroscopy of hot minority electrons injected in gallium arsenide quantum wells. *Appl. Phys. Lett.* **50**, 930 (1987).

65. J. R. Hayes and A. F. J. Levi, Dynamics of extreme nonequilibrium electron transport in GaAs. *IEEE J. Quantum Electron.* **QE-22**, 1744 (1986).

66. R. A. Hopfel, J. Shah, P. A. Wolff, and A. C. Gossard, Negative absolute mobility of minority electrons in GaAs quantum wells. *Phys. Rev. Lett.* **56**, 2736 (1986).

67. R. A. Hopfel, J. Shah, and A. C. Gossard, Nonequilibrium electron-hole plasma in GaAs quantum wells. *Phys. Rev. Lett.* **56**, 765 (1986).

68. J. R. Hayes, A. F. J. Levi, A. C. Gossard, and J. H. English, Base transport

dynamics in a heterojunction bipolar transistor. *Appl. Phys. Lett.* **49**, 1481 (1986).

69. A. F. J. Levi, J. R. Hayes, P. M. Platzman, and W. Wiegmann, Injected-hot-electron transport in GaAs. *Phys. Rev. Lett.* **55**, 2071 (1985).

70. M. Heiblum, M. Nathan, D. C. Thomas, and C. N. Knoedler, Direct observation of ballistic electron transport in GaAs. *Phys. Rev. Lett.* **55**, 2200 (1985).

71. D. Ankri and L. F. Eastman, GaAlAs-GaAs ballistic heterojunction bipolar transistor. *Electron. Lett.* **18**, 750 (1982).

72. R. J. Malik, F. Capasso, R. A. Stall, R. A. Kiehl, R. W. Ryan, R. Wunder, and C. G. Bethea, High gain, high frequency AlGaAs/GaAs graded band-gap base bipolar transistors with a Be diffusion setback layer in the base. *Appl. Phys. Lett.* **46**, 600 (1985).

73. D. Ankri, R. A. Zoulay, E. Caquot, J. Dangla, C. Dubon, and J. Palmier, Analysis of D. C. characteristics of GaAlAs/GaAs double heterojunction bipolar transistors. *Solid-State Electron.* **29**, 141 (1986).

74. T. Futatsugi, Y. Yamaguchi, S. Muto, N. Yokoyama, and A. Shibatomi, InAlAs/InGaAs resonant tunneling bipolar transistor (RBTs) operating at room temperature with high current gains. *Tech. Dig. IEEE Int. Electron Device Meet.* p. 877 (1987).

75. S. Luryi, Quantum capacitance devices. *Appl. Phys. Lett.* **52**, 501 (1988).

76. F. Beltram, F. Capasso, S. Luryi, S. N. G. Chu, A. Y. Cho, and D. L. Sivco, Negative transconductance via gating of the quantum well subbands in a resonant tunneling transistor. *Appl. Phys. Lett.* **53**, 219 (1988).

77. T. K. Woodward, T. C. McGill, H. F. Chung, and R. D. Burnham, Integration of a resonant-tunneling structure with a metal-semiconductor field-effect transistor. *Appl. Phys. Lett.* **51**, 1542 (1987).

78. T. K. Woodward, T. C. McGill, H. F. Chung, and R. D. Burnham, Applications of resonant-tunneling field-effect transistors. *IEEE Electron Device. Lett.* **EDL-9**, 122 (1988).

79. N. Yokoyama and K. Imamura, Flip-flop circuit using a resonant-tunneling hot electron transistor (RHET). *Electron. Lett.* **22**, 1228 (1986).

80. V. J. Goldman, D. C. Tsui, and J. E. Cunningham, Resonant tunneling in a magnetic field: Evidence for space-charge build-up. *Phys. Rev. B: Condens. Matter* **35** 9387, (1987).

81. H. Sakaki, Scattering suppression and high-mobility effect of size-quantized electrons in ultrafine semiconductor wire structures. *Jpn. J. Appl. Phys.* **19**, L735 (1980).

82. Y. C. Chang, L. L. Chang, and L. Esaki, A new one-dimensional quantum well structure. *Appl. Phys. Lett.* **47**, 1324 (1985).

83. P. M. Petroff, A. C. Gossard, R. A. Logan, and W. Wiegmann, Toward quantum well wires: Fabrication and optical properties. *Appl. Phys. Lett.* **41**, 636 (1982).

84. J. Cibert, P. M. Petroff, G. J. Dolan, S. J. Pearton, A. C. Gossard, and J. H. English, Optically detected carrier confinement to one and zero dimensions in GaAs quantum well wires and boxes. *Appl. Phys. Lett.* **49**, 1275 (1986).

85. M. A. Reed, R. T. Bate, K. Bradshaw, W. M. Duncan, W. R. Frensley, J. W.

Lee, and H. D. Shih, Spatial quantization in GaAs-AlGaAs multiple quantum dots. *J. Vac. Sci. Technol., B* **4**, 358 (1986).

86. F. Capasso, F. Beltram, and S. Sen, New functional devices based on quantum confinement and band-gap engineering. *Proc. Int. Conf. Electron. Mater. 1988.* (to be published).

87. S. Sen, F. Capasso, A. Y. Cho, and D. L. Sivco, Stacked double barriers and their application in novel multi-state resonant tunneling bipolar transistor. *Conf. Ser.—Inst. Phys.*, (to be published).

88. F. Capasso, S. Sen, A. Y. Cho, and D. L. Sivco, Multiple negative transconductance and differential conductance in a bipolar transistor by sequential quenching of resonant tunneling. *Appl. Phys. Lett.* **53**, 1056 (1988).

89. S. Sen, F. Capasso, A. Y. Cho, and D. L. Sivco, Multiple state resonant tunneling bipolar transistor operating at room temperature and its application as a frequency multiplier. *IEEE Electron Device Lett.* **9**, 533 (1988).

90. S. Sen, F. Capasso, A. Y. Cho, and D. L. Sivco, Parity generator circuit using a multi-state resonant tunneling bipolar transistor. Electron. Lett. **24**, 1506 (1988).

91. S. Sen, F. Capasso, A. Y. Cho, and D. L. Sivco, New resonant tunneling bipolar transistor (RTBT) with multiple negative differential resistance characteristics operating at room temperature with large current gain. *Proc. IEEE. Int. Electron Device Meet. 1988*, p. 834 (1988).

6 JFET Technology

WILLIAM A. GEIDEMAN

McDonnell Douglas Electronic Systems Company
Huntington Beach, California

6.1 INTRODUCTION

The junction field effect transistor (JFET) is fabricated in a semi-insulating GaAs substrate using a double implantation of an n-type channel and a p-type gate structure with an ohmic gate contact. A cross section of the JFET is shown in Fig. 6.1. The JFET differs from the more common section of the MESFET only in that the gate is formed inside the bulk material by the p-n junction and not at the surface by a Schottky barrier. The p-n junction allows for tighter control of the channel height, since this is controlled by the difference between two implants and not by a single implant. This control of the channel height promoted the use of enhancement mode devices, as the shallow channel is formed more easily in the JFET. These devices are fabricated by reducing the channel height so that a normally-off enhancement mode FET (E-JFET) results, with a positive pinch-off or threshold voltage. Another important advantage of these JFET circuits is that the logic swing of a circuit using MESFETs is limited by the forward turn-on voltage of the Schottky-barrier transistor gate to about 0.5 V, whereas for the E-JFET this voltage is increased to about 1 V. This gives a larger noise immunity for the E-JFET circuits and places stringent fabrication requirements on threshold-voltage control and uniformity for the E-MESFET to obtain good processing yield [1].

Although the E-JFET intrinsically has a higher noise margin than the E-MESFET because of the higher possible logic swing, the demonstrated complementary design of E-JFET circuits with enhancement mode n- and p-channel devices offers even further increase in noise margin and very low power dissipation per gate [2]. It should be noted that complementary circuit design is only practical in GaAs with the E-JFET, since the E-MESFET has a low Schottky-barrier height for the p-channel device that prohibits practical circuit designs and operation. This complementary design lends itself to the fabrication of a low-power static random access memory (SRAM) of high-yield capability owing to the high noise margin and excellent threshold-voltage control by a double ion-implanted p-n junction gate.

302

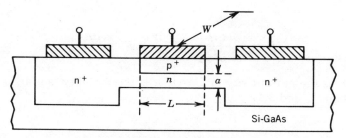

Figure 6.1 Cross section of an ion-implanted enhancement mode JFET: L is the gate length, W is the gate width, and a is the channel height.

The speed–power performance of the homojunction GaAs E-JFET technology will be presented and discussed. Low-power operation of E-JFET gates in direct coupled FET Logic (DCFL) is advantageous for the design of gate arrays and microprocessors. The low power per gate of the E-JFET DCFL allows larger circuit integration than can be expected in competing technologies. Indeed, VLSI circuits of greater than 10, 000-gate complexity levels are currently being constructed. A 32 bit microprocessor chip using the reduced instruction set architecture (RISC) was implemented in enhancement mode JFET technology [3]. With improvements in the GaAs JFET currently in development, the goal for this microprocessor is an execution cycle of 5 ns, which corresponds to a clock rate of 200 MHz.

In this chapter, the voltage–current and temperature characteristics of the JFET are presented. The equivalent circuit for modelling, as described, is utilized in circuit performance predictions. The all ion-implantation planar process for integrated circuit fabrication is described. This chapter also documents the LSI circuits that have been fabricated to date. The issue of radiation hardness is also included.

6.2 JFET THEORY AND MODEL

6.2.1 Voltage – Current Characteristics

The steady-state I–V characteristics for a uniformly doped channel have been derived by Shockley [4], assuming that electron drift velocity is proportional to the electric field. According to this theory, the current saturates for

$$V_D \gtrsim V_{SAT} = V_G - V_T \tag{1}$$

when the channel is pinched off at the drain. For ion-implanted channels, the MOSFET-like relation exists and we have

$$I_D = 2K\left[(V_G - V_T)V_D - \frac{V_D^2}{2}\right] \tag{2}$$

with

$$K = \frac{W\mu\epsilon_r\epsilon_0}{2a_{\mathrm{eff}}L} = K'\left(\frac{W}{L}\right); \; K' = \frac{\mu\epsilon_r\epsilon_0}{2a_{\mathrm{eff}}} \tag{3}$$

where $\epsilon_r\epsilon_0$ is the permittivity of GaAs, μ is the mobility, and $a_{\mathrm{eff}} = a$ is the channel height, and

$$a_{\mathrm{eff}} = R_p + \frac{2\sigma_p\sqrt{2}}{\pi} \tag{4}$$

appears more appropriate [5], since the width of the depletion layer varies little along most of the channel and also with applied gate voltage, its boundary being pinned more or less to the peak of the dopant concentration. K' in Eq. (3) is a gain parameter related to device structure and material used. It is one of the important circuit design parameters to be discussed in Section 6.2.4. As in Shockley's theory, the current saturates for $V_D \gtreqqless V_{\mathrm{SAT}} = V_G - V_T$, at

$$I_{\mathrm{SAT}} = K(V_G - V_T)^2 \tag{5}$$

The parameters W, L, and $a_{\mathrm{eff}} = a$ are defined in Fig. 6.1, which presents a cross section of a JFET.

Deviations of voltage–current characteristics from the square-law relations are encountered, however, when series resistance values of the source contact and the region between the source and gate edge are appreciable [6]. Equation (5) must then be modified, because the effective gate voltage is now

$$V_G' = V_G - R_S I_{\mathrm{SAT}} \tag{6}$$

where R_S is the total ohmic series resistance between source and gate. The reduced drain saturation current I_{DS}^* is approximately

$$I_{\mathrm{DS}}^* \simeq \frac{I_{\mathrm{SAT}}}{1 + 2R_S K(V_G - V_T)} \tag{7}$$

A uniform reduction of saturation current arising from equal resistances for all transistors would be tolerable. However, nonuniform alloying over a large wafer area, causing resistance variations, would be objectionable, since it would result in a spread of saturation current.

Optimization of device performance characteristics is accomplished through tailoring of ion-implantation profiles. Smallest channel height gives the highest drain saturation current I_{DS} for a given gate voltage V_G, according to Eqs. (3) and (5). This requires rather large channel doping, however, to maintain the desired threshold voltage. The design goal is

$a = 1000$ A, so that for a channel doping of 1.5×10^{17} cm^{-3}, $V_T \cong V_{bi} - V_P =$ 0.2 V (see Eqs. [8–10]). With $L = 2.5\,\mu$m and $\mu = 4000$ cm^2/V·s, one obtains $I_{DS}/W = 80\ (V_G - V_T)^2$ (mA/mm) and $g_m/W = (1/W)(dI_{DS}/dV_G) =$ 160 $(V_G - V_T)$ (mA·V/mm). Thus, for $V_G = +1$ V, we have $I_{DS} =$ 51 (mA/mm) and $g_m = 128$ (mS/mm). Since the gate capacitance per unit length of this device is $C_G/W = \epsilon_r \epsilon_0 L/a \simeq 2.5$ pF/mm, the estimated intrinsic RC time constant response is $\tau = 2.2 C_G/g_m \simeq 50$ ps.

For a JFET with $W = 10\,\mu$m and a 2.5 KΩ load resistance, the extrinsic RC time constant response is then $\tau = 2.2 C_G R_L = 150$ ps.

Experimental values of 150–200 ps have been obtained with complex integrated circuits.

The important parameters in controlling the current–voltage characteristics are primarily channel doping N and channel height a, with W and L being well defined by photolithographic delineations. The channel doping affects I–V characteristics through the threshold voltage

$$V_T \cong V_{bi} - V_P \tag{8}$$

where

$$V_P = \frac{q}{\epsilon_r \epsilon_0} \int N(x)x\,dx \tag{9}$$

is the pinchoff voltage and V_{bi} is the gate junction built-in potential. For a uniformly doped epitaxial layer

$$V_P = \frac{Nqa^2}{2\epsilon_r \epsilon_0} \tag{10}$$

Equation (3) with (5), (8) and (10) then shows that

$$\frac{\partial I_{DS}}{I_{DS}} = \frac{2V_P}{(V_G - V_T)}\frac{\partial N}{N} + \left(\frac{4V_P}{V_G - V_T} - 1\right)\frac{\partial a}{a} \tag{11}$$

For the typical values $V_P = 1.0$ V, $V_G = +1.0$ V, and $V_T \simeq 0$ V, this gives numerically

$$\frac{\partial I_{DS}}{I_{DS}} = 2\frac{\partial N}{N} + 3\frac{\partial a}{a} \tag{12}$$

Variations of built-in voltage and of mobility are neglected, being only second-order effects.

With a 10% change in N and 10% change in a, the fractional change of drain saturation current I_{DS} is 0.5, or 50%. Thus, tight control of a and N is required over large wafer areas to achieve good uniformity of electrical device characteristics and reasonable yields in integrated circuit fabrication.

Control of N and a to less than 10% over small wafer areas can be obtained with standard techniques of diffusion and epitaxy. It is, however, not the yield of devices and circuits per wafer, but rather the yield of good wafers, that renders the standard technology marginal in getting good overall yields in the integrated circuit fabrication. Implantation of silicon ions is now standard practice for channel regions, and measurements on GaAs devices indicated a drain saturation current variation of only 10% over large wafer area. This amounts to a control of N and a to within a 2% variation. These tolerances can be maintained for an optimized enhancement mode structure, and ion implantation offers the necessary tolerances to assure device uniformity and thus provides better yields in integrated circuit fabrication than diffusion and epitaxy. Uniformity of performance characteristics for ion-implanted devices depends primarily on the capping and annealing procedure and secondarily on the ohmic contact and series resistance control obtainable with the presently used fabrication technologies.

Effects in GaAs E-JFETs arising from neutron fluences and total doses of ionizing radiation have been assessed [7]. It was established that the degradation of the electrical characteristics of enhancement mode gallium arsenide junction field effect transistors exposed to fast neutrons ($E > $ 10keV) or ionizing radiation (Co^{60}) is caused substantially by changes in mobility and free-carrier concentration. Circuit operation of devices with channel impurity concentrations of about 10^{17} cm^{-3} will not be impaired by fast neutron fluences of 10^{15} neutrons/cm^2 and ionizing radiation doses of 10^8 rad (GaAs). In addition, single event upset (SEU) to heavy ions and protons and logic upset to pulsed ionizing radiation were measured with E-JFET SRAMs and are available from the literature [8].

6.2.2   Temperature Characteristics

A temperature-dependent semiempirical GaAs JFET device model was developed to interpret the experimental data. This model, as implemented in a circuit simulator, incorporates the temperature dependence of the threshold voltage, transconductance, subthreshold current, substrate leakage current, gate current, and implanted load resistor conductivity [9].

The dominant factor affecting operation at low and high temperatures is the threshold-voltage shift of the JFETs, which is approximately -1.5 mV/ °C, as shown in Fig. 6.2. The performance factor K and threshold voltage V_T are both temperature-dependent and will affect R_{ON}, the on-resistance in the linear region, as well as I_{DS} of the FET, which are expressed by the relations

$$I_{DS}(T) = \frac{\epsilon_r \epsilon_0 \mu(T)}{2a(T)} \left(\frac{W}{L}\right)[V_G - V_T(T)]^2 \tag{13}$$

$$R_{ON} = \frac{1}{2K(T)[V_G - V_T(T)]} \tag{14}$$

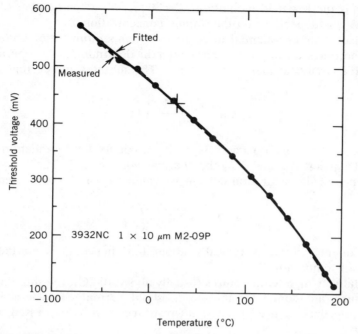

Figure 6.2 Threshold voltage V_T as a function of temperature for a GaAs JFET.

An accurate physical model [10] for temperature dependence of V_T requires determining the channel-to-substrate built-in voltage, which involves an iterative method, and such parameters as trap concentration and occupation as function of temperature, which are not easily measurable.

For the temperature under consideration here, however, the temperature dependence of the threshold voltage can be accurately approximated by a linear function with the addition of a second-order term as follows:

$$V_T(T) = V_{T0} + \Delta T + bT^2 \tag{15}$$

where V_{T0} is the nominal threshold voltage at $T = 300$ K, typically $\Delta \simeq -1.33$ mV/°C and $b \simeq 1.2$ μV/°C^2 for n-channel JFETs and $\Delta \simeq -2.33$ mV/°C for p-channel JFETs. Figure 6.2 compares the measured and calculated threshold voltages for an n-channel JFET.

The performance parameter K of Eq. (3) is temperature-dependent. In Eq. (3), mobility is the most important parameter determining the temperature dependence of K. The effective channel height is also a function of temperature, but its effect is less significant. For moderate doping conditions, the electron mobility in the channel can be described by [11]:

$$\frac{1}{\mu} = \frac{1}{\mu_0} \left(\frac{T}{300}\right)^\alpha + 4 \times 10^{-21} N_d \left(\frac{300}{T}\right)^{1.5} \tag{16}$$

where μ_0 is the low-field mobility at $T = 300$ K, N_d is the number of donor sites, and α is dependent on the doping concentration.

Because of its exponential increase with temperature, the subthreshold current becomes an important parameter in determining the operation of GaAs FET circuits at high temperatures. The subthreshold current is given by

$$I_{DS} = I_0 \exp\left(\frac{V_{GS} - V_T}{mV_{th}}\right)\left(1 - \frac{V_{DS}}{V_{se}}\right) \tag{17}$$

where m is the emission coefficient. V_{se} accounts for modulation of the potential barrier V_{DS} and V_{th}, the thermal voltage [12]. The temperature dependence of the saturation current parameter I_0 can be expressed as

$$I_0 = I_0(T = 300 \text{ K})\left(\frac{T}{300}\right)^{1.5 - \alpha} \tag{18}$$

The exponent $1.5 - \alpha$ is typically about 0.3, hence I_0 is an increasing function of temperature.

At relatively high temperatures (usually above 100°C), the shunting effect of the substrate dominates the subthreshold current characteristics. The temperature dependence of the shunt resistance can be expressed approximately as

$$R_{sh} = R_{sh0}\left(\frac{300}{T}\right)^{3/2}\left(\frac{T}{300}\right)^{2.3}\exp\left(\frac{E_g}{2kT}\right) \tag{19}$$

where R_{sh0} is a parameter determined by the geometry of the device.

The resistance of ion-implanted resistors may be modeled as a parallel combination of the implanted region resistance and the substrate resistance. For lightly doped resistors, the carrier mobility expression, Eq. (16), can be simplified to

$$\mu(T) = \mu_0\left(\frac{300}{T}\right)^{\alpha} \tag{20}$$

where α is dependent on the doping concentration ($\alpha = 2.3$ for intrinsic GaAs) and μ_0 is the mobility at $T = 300$ K. Then, the temperature dependence of the implanted region resistance R_i may be described approximately by

$$R_i = R_0\left(\frac{T}{300}\right)^{\alpha} \tag{21}$$

where R_0 is the nominal resistor value at 300 K. The substrate resistance R_s may be roughly expressed as

$$R_s = R_{s0} \exp\left(\frac{E_g}{2kT}\right) \tag{22}$$

Neglecting the temperature dependence of $R_{s0}(R_{s0} \propto 1/T)$, the total resistance may be rewritten as

$$R_T = \frac{R_0(T/300)^\alpha R_{s0} \exp(E_g/2kT)}{R_0(T/300)^\alpha + R_{s0} \exp(E_g/2kT)} \tag{23}$$

The calculated curve R_T versus T using Eq. (23) is shown in Fig. 6.3. As can be seen, R_T has a maximum value at $T = T_m$. Thus, the value of R_{s0} can be obtained from Eq. (23) by letting $\partial(R_T)/\partial T = 0$ at $T = T_m$, which yields

$$R_{s0} = \frac{E_g R_0}{kT_m(\alpha + 1)} \left(\frac{T_m}{300}\right)^\alpha \exp\left(\frac{-E_g}{2kT_m}\right) \tag{24}$$

Hence, the input parameters are α, R_0, and T_m, which can be easily extracted from the measured data.

The temperature performance of logic levels, that is, logic output low (V_{OL}) and high (V_{OH}) for the three JFET DCFL gate configurations (complementary, resistive load, and depletion JFET load), was simulated over the temperature range of $-55°C$ to $125°C$. The results are presented in Fig. 6.4. While the first two versions offer a tolerable reduction of logic swing up to $125°C$, the E/D inverter suffers a severe degradation above $100°C$ and therefore has a limited range for high-temperature operation.

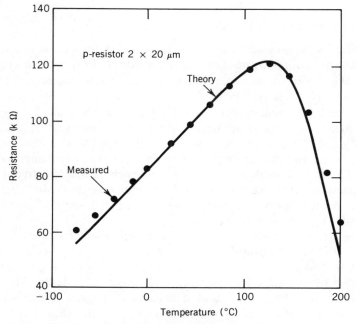

Figure 6.3 GaAs implanted resistor temperature characteristic.

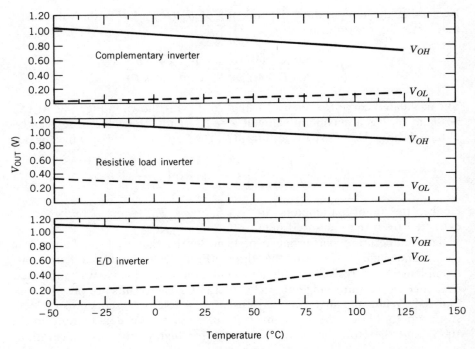

Figure 6.4 Logic levels of E-JFET inverters as function of temperature.

6.2.3 JFET Fabrication Process

The JFET process uses undoped GaAs wafers that are 3 in. in diameter and are grown by liquid-encapsulated Czochrolski (LEC) techniques. After cleaning the wafer surface to remove surface contaminants and polishing damage, a Si_3N_4 layer is deposited to protect the surface during processing and to form the annealing cap. Since dopants are implanted through this silicon nitride layer to form the n- and p-type regions, nitride thickness uniformity and repeatability are essential to the control of device properties.

Figure 6.5 shows the ion-implant sequence for the complementary JFET process. Photoresist is used as the mask for each of the selective implants. The implant species are Si^+ for n-type implants and Mg^+ for p-type implants. Annealing of the wafers to remove the implant damage and activate the dopants is done in a furnace.

For the n-channel E-JFETs, Si^+ ions are implanted at 150 keV and doses up to 1×10^{13} cm^{-2} to form the channel. Mg^+ ions are implanted at energies near 50 keV and doses of 5×10^{13} cm^{-2} for the p^+ gate. Figure 6.6 shows typical impurity profiles for the gate and channel implants. The built-in voltage generated by the p^+ and n regions depletes the carriers in the channel, causing the device to be normally-off. The threshold voltage of the device can be adjusted to the desired value by changing the Mg^+ gate

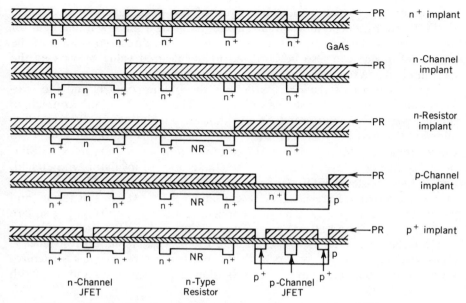

Figure 6.5 Ion-implant sequence for complementary E-JFET process.

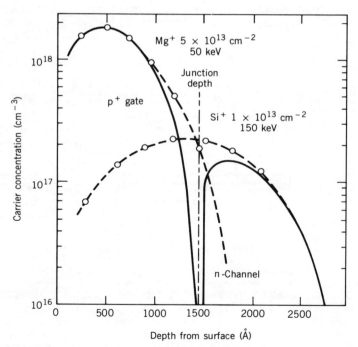

Figure 6.6 Carrier concentration profiles of gate and channel implants for n-channel E-JFET.

implant energy, which varies the depth of the p^+-n junction. p-Channel E-JFETs are produced in a similar manner by using a shallow n^+ implant for the gate and a deep p implant for the channel. The threshold voltage of the p-channel device can also be adjusted to the desired value by varying the gate implant energy.

The remainder of the process defines the ohmic contacts and the two-level interconnects and is shown in Fig. 6.7. The GeAuNi ohmic contacts to the n^+ regions are patterned by nitride-assisted photoresist lift and then alloyed. The first interconnect layer, which also forms the ohmic contact to the p^+ region, is sputtered TiPtAu patterned with ion milling. Next, a SiO_2 or Si_3N_4 dielectric layer is deposited for the crossover insulator, vias etched, and the second-level interconnect sputtered and ion-milled. Typical values of contact resistance are $1 \times 10^{-6} \ \Omega/cm^2$ and $2 \times 10^{-6} \ \Omega/cm^2$ for the n^+ and p^+ contacts, respectively. The interconnect resistivity is 150 mΩ per square for the first level and 60 mΩ per square for the second level.

No isolation implants are used during the process. Isolation resistance is measured on processed wafers with the n^+–n^+ and p^+–p^+ test structures shown in Fig. 6.8. The gap between the doped layers is 3 μm and the width

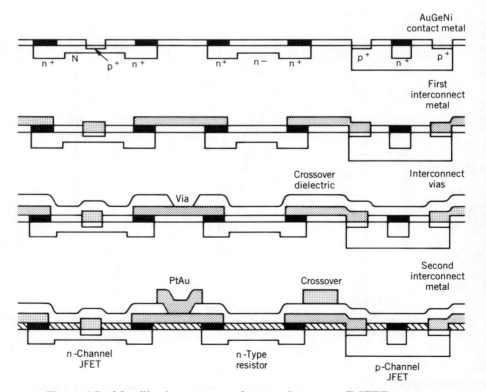

Figure 6.7   Metallization sequence for complementary E-JFET process.

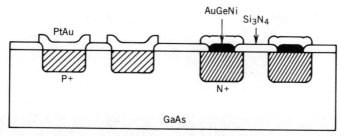

Figure 6.8 Substrate isolation test structures.

is 100 μm. Typical resistance values are $10^9 \, \Omega$ for the n^+–n^+ pattern and $10^7 \, \Omega$ for the p^+–p^+ patterns.

Typical *I–V* characteristics of $1 \times 10 \, \mu m$ n-channel and p-channel E-JFETs are shown in Fig. 6.9. The transconductance of n-channel JFETs, measured between $V_{GS} = 0.9 \, V$ and $V_{GS} = 1.0 \, V$, is greater than $100 \, mS/mm$ with $V_T = 0.2 \, V$. The highest value of g_m measured under these conditions is $152 \, mS/mm$. The transconductance of p-channel JFETs, measured between $V_{GS} = -0.9 \, V$ and $V_{GS} = -1.0 \, V$, is greater than $5 \, mS/mm$ with $V_T = -0.2 \, V$. The highest value of p-channel g_m measured under these conditions is $12 \, mS/mm$. Gate diode forward *I–V* curves for p^+-n diodes and n^+-p diodes, presented in Fig. 6.10, show the 1 V turn-on voltage of p-n junctions that allows DCFL circuits to operate with 1 V logic levels.

NOR gates, NAND gates, and cross-coupled inverter pair flip–flops are the basic building blocks of E-JFET logic. The NOR gate is the preferred gate circuit for two reasons: better switching speed and higher fan-in capability. A NAND gate of the same power dissipation and logic voltage levels requires larger inverter devices to maintain the logic low level; this results in lower switching speed. For this reason, NAND gates with more than three inputs are rarely used in E-JFET logic.

6.2.4 Device Circuit Modeling

An overview of semiconductor device modeling is shown in Fig. 6.11. This flowchart encompasses the entire modeling activity as applied to the JFET technology. Extensive modeling is used in integrated circuit practice to predict and optimize circuit performance and chip yield. The model can be based on theoretical physics or it can be an empirical model that is derived from curve fitting of experimental data and data extraction from device scaling experiments. As described in the previous section, the GaAs JFET voltage–current characteristics and their temperature dependence have been analyzed using a semiempirical model. With the aid of these voltage–current characteristics and the equivalent circuit of Fig. 6.12, a modified SPICE2 simulation was developed. The two-diode model was selected; the additional circuit elements with respects to FET terminology are self-explanatory.

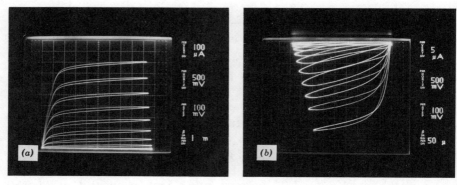

Figure 6.9 Curve tracer plots of room-temperature drain current–voltage relations: (a) n-channel ($V_{GS} = 0$–1 V), (b) p-channel ($V_{GS} = 0$ to -1 V).

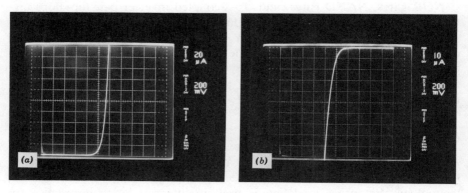

Figure 6.10 Room-temperature current–voltage plots of junction diodes in E-JFETs: (a) n-channel, (b) p-channel.

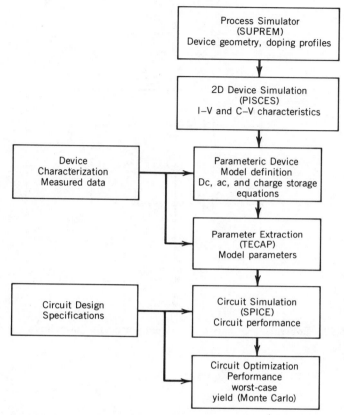

Figure 6.11   Overview of semiconductor device modelling.

Special analytical treatments were found mandatory for R_{SH}, the shunt resistance, in parallel with the channel conductance and the current generators I_B and I_{SUB}, which represent the breakdown current and leakage current through the semi-insulating substrate material.

The utility of the model is verified by a comparison of the measured and simulated performance of the circuit. Figure 6.13 shows the data for two values of load resistance in the 4 bit arithmetic logic unit (ALU) circuit. With a load resistance value of 2.5 KΩ, the circuit of Lot 1 measured an add time of 2.2 ns for a power dissipation of 55 mW. This compares with the simulated values for this load resistance of 2.1 ns for a 50 mW power performance.

The effect of changing the process parameters on the performance of the circuit is shown in these data. Lot 2 of the ALU circuit was processed at a different time from Lot 1 and the threshold-voltage and resistance values were significantly different. The model was able to predict the performance differences when the new parameters were used. The trade-off between

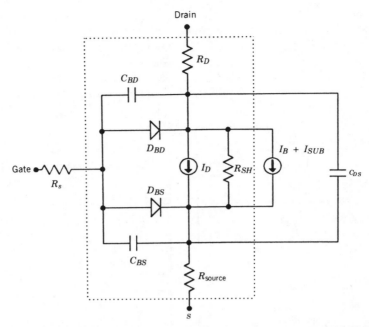

Figure 6.12 Equivalent circuits and elements used for modeling of E-JFETs.

	Lot 1	Lot 2	Simulation
Add Time – nsec	2.2	2.9	2.1
Power – mW	55	32	50
Load Value – kΩ	2.5	4.0	2.5

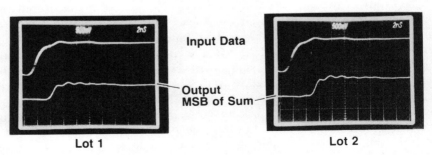

Figure 6.13 Comparison of simulated and measured speed–power performance of 4 bit ALU. Output is most significant bit (MSB) of sum.

speed and power is evident from a circuit from Lot 2. With increased load resistance, that is, 4 KΩ, the power dissipation was reduced to 32 mW with a corresponding increase in the add time to 2.9 ns. The good agreement of simulation with experimental results demonstrates the accuracy of our modified JFET SPICE 2 model and the voltage- and current-dependent circuit elements of Fig. 6.12 that were used for this simulation. The result establishes a confidence level for application of this model to predict the performance of circuits with higher complexity.

The output low voltage of a resistive load E-JFET inverter is given by

$$V_{OL} = R_{ON} \cdot I_{DS} \tag{25}$$

where $R_{ON} = \frac{1}{2} K (V_G - V_T)$.

Commonly, the load resistor (or active load element) is sized to produce a V_{OL} equal or close to the threshold voltage V_T of the FET, thus

$$V_{OL} = V_T \tag{26}$$

which, for worst-case conditions, should be

$$V_{OL1} \lesseqgtr V_{T2} \tag{27}$$

where V_{OL1} is the output voltage of inverter 1 and V_{T2} the V_T of the EJFET driven by it (i.e., second stage).

The value of V_{OH}, the output high voltage when the inverter FET is turned off, is given by

$$V_{OH} = V_{DD} - I_R \cdot R_R \cong V_{GS(ON)} \tag{28}$$

where $V_{GS(ON)}$ is the gate–source voltage of the FET driven by the inverter. V_{OH} can be determined graphically by plotting I_G versus V_{GS} and a load line given by the load resistor of the inverter. Figure 6.14 shows the intercept of the load line with the I_G versus V_{GS} curve for fan-outs of 1, 3, 5, 10, and 30. The load lines represent a nominal resistor value and a 25% lower- and 25% higher-than-nominal resistor values.

The change of V_{OH} as function of fan-out is an important parameter to be considered in circuit design; the V_{OH} "window" is widened when JFET parameters are given ranges of values rather than fixed, nominal values. The standard deviations of these JFET parameters derived from empirical data are $\sigma V_T = 50$ mV, $\sigma K' = 8 \, \mu A/V^2$, and $\sigma R = 5\%$ where K' is given in Eq. (3)). The mean values of these three parameters may vary ± 100 mV for V_T, from 75 to 110 $\mu A/V^2$ for K', and $\pm 10\%$ for the resistor (R).

An illustration of the effect of parametric variations on V_{OL} is given in Fig. 6.15, where V_{OL} is plotted as a function of FET threshold voltage. In this plot, the gate-voltage–drain current gain parameter K' is automatically

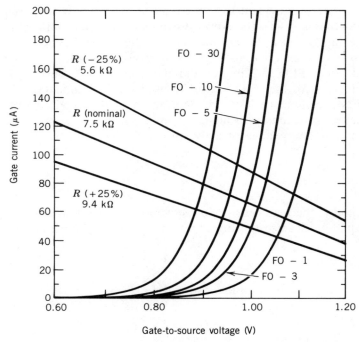

Figure 6.14 Characteristic curve of E-JFET inverter as a function of fan-out (FO) and load resistor (R) values.

adjusted as a function of V_T. K' increases as V_T increases and compensates for the increase in the value of the drain-source on-resistance R_{ON}. If a flat K' is used, V_{OL} changes are in the order of 10 mV for a 200 mV change in V_T. The range of load resistor value stipulated in the example shown in Fig. 6.16 is nominally 7.5 Ω, the lower limit is 5.625 Ω, and the high limit is 9.375 Ω.

The four paramaters that represent gate switching performance are rise time t_r, fall time t_f, delay from rising input edge to falling output edge at 50% of logic swing t_{pdRF}, and delay from falling input edge to rising output edge t_{pdFR}. A good circuit design keeps t_r and t_f equal, with ±10% an acceptable tolerance goal, and t_{pdRF} and t_{pdFR} equal within 20%. These goals can be readily accomplished for a small circuit by iterative circuit design which establishes a nominal case first and then examines several worst cases where the device parameters are at the edge of their respective design "windows." The windows are based on experimental results. Limits are commonly based on the maximum variation of the mean value plus or minus three times the standard deviation over the chip area. Windows typically used in E-JFET circuit design are 0.05–0.55 V for V_T, 50–125 μA/V^2 for K', and 0.75–1.25 times the nominal resistor value. The compatibility of differ-

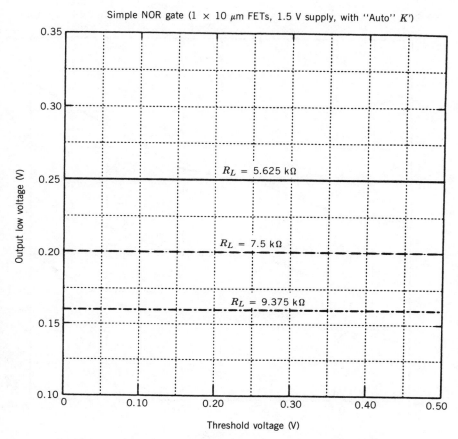

Figure 6.15 $V_{\rm OL}$ as a function of $V_{\rm T}$ for various load resistors.

ent circuits to interface with each other and guarantee functionality can be analyzed by establishing an input–output protocol that each circuit must obey.

The need to model the effects of statistical parameter variations of $V_{\rm T}$, K' and resistances led to the development of MONTE. This program uses Monte Carlo methods to vary parameters in a MDSPICE simulation with a desired distribution function for multiple runs. MDSPICE and MONTE simulations are run on the VAX 8600, and the results are reviewed on VT240 graphics terminals with a graphical postprocessor called VIEW-SPICE, which was developed at McDonnell Douglas. Another circuit design tool, OPTIMA II, allows the linear variations of device parameters as well as circuit input voltage to be assessed through multiple runs of MDSPICE. These circuit design tools make the accurate and efficient design of GaAs circuits possible.

Simple NOR gate (1 × 10 μm FETs, 1.5 V supply, without "Auto" K')   Output low voltage (V)

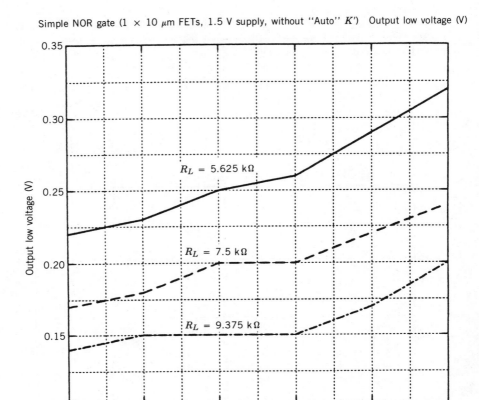

Figure 6.16   Effect of K' variation on V_{OL} versus V_T relation.

For complex circuits, there are many worst cases which are difficult to derive by hand. The best design tool here is the Monte Carlo analysis, where certain fail criteria are specified. The parameter values of the failing circuits are saved and subsequently examined, and corrections or redesigns of the circuit are made. This is an iterative procedure that can predict the circuit performance distribution and circuit yield. In contrast to MOS circuit design, where parameter windows are relatively small, GaAs designs require a more detailed and statistical analysis to obtain high circuit yield.

When designing a circuit for a temperature range, the design procedure discussed above is repeated for the two end-points of the range. If required, the design is changed or adjusted and then reconciled with the 25°C design. As discussed previously, device parameters change with temperature, the dominant one being V_T, which decreases approximately 1.5 mV/°C. Correct subthreshold drain–source current modeling is very important at elevated temperatures in gate array circuits with large fan-in or fan-out and in low-current circuits such as low-power memory cells, registers, and linear circuits that have high impedance nodes.

To complete the design cycle for GaAs JFET circuits, McDonnell Douglas makes full use of commercially available software. Using a library of standard parts that the company has developed for Mentor Graphics workstations, schematics are entered. Layout is done on a Calma GDSII IC design system. Layout design rule checks (DRCs) are performed with the Calma and MASKAP software. Layout verification is done with MASKAP or with DRACULA II. Simultaneous display of schematic and layout information is possible on the Mentor Graphics workstations. Using results from DRACULA II, circuit layout discrepancies are found quickly and interactively. As layouts are completed, initial estimates of layout capaci-

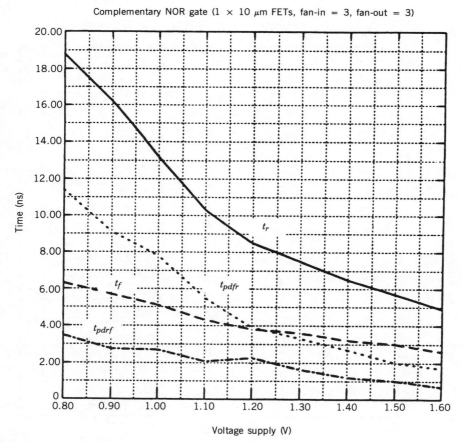

Figure 6.17 Response time of complementary NOR gate as a function of supply voltage.

tance are replaced with values extracted from the completed layouts with MASKAP, and circuits are resimulated with MDSPICE. After all design specifications have been met and the layout is fully verified, a mask set is procured for the fabrication of the circuits.

6.3 COMPLEMENTARY CIRCUITS

Circuit designs can also be implemented with p-channel E-JFETs which allow design analogs to silicon CMOS with the difference that E-JFETs consume gate current when forward-biased. However, the gate current, typically $1\,\mu A$ per micrometer of FET width at $V_{GS} = 1.0\,V$, is small and makes it possible to construct useful logic gates with power dissipation of $1.0-40\,\mu W$, depending on switching speed required.

Figure 6.17 shows simulated two-input complementary gate switching

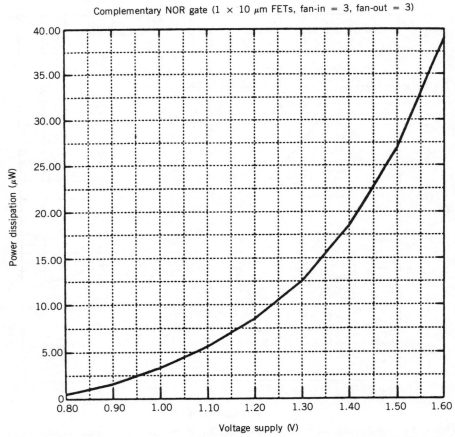

Figure 6.18 Power dissipation of complementary NOR gate as a function of supply voltage.

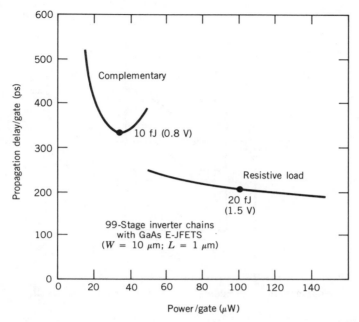

Figure 6.19 Switching performance of complementary and resistive load inverters.

characteristics as function of supply voltage, whereas Fig. 6.18 shows the power dissipation as function of power supply voltage. The switching performance of DCFL with GaAs E-JFETs for the complementary and resistive load inverter is shown in Fig. 6.19. These speed–power relations were obtained with 99-stage inverter chains and fan-in = fan-out = 1. With 0.8 V supply voltage, the complementary circuitry displays a switching energy of 10 fJ, corresponding to 330 ps and 30 μW, while the resistive load inverter with 1.5 V supply voltage performs with 20 fJ, that is, 200 ps and 100 μW.

The K' of p-channel E-JFETs is approximately one-tenth of the n-channel K', but device capacitances are the same for p- and n-channel FETs; thus, the switching speed of complementary inverters and gates is lower by a factor of 2–10, depending on power dissipation. The low switching speed of complementary E-JFET (C-EJFET) logic has restricted its use to circuits in which low-power dissipation is mandatory, such as static memory cells or registers. Figure 6.20 shows the circuit schematic of a C-EJFET memory cell.

Complementary E-JFET circuits with enhancement mode n- and p-channel devices offer a further increase in noise margin over that associated with the high logic swing of the E-JFET. The simulated noise margin for a DCFL complementary inverter is shown in Figure 6.21. Sensitivity of the complementary inverter to subthreshold leakage is greatly reduced because one transistor is in the off state.

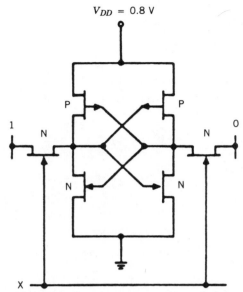

Figure 6.20 Complementary E-JFET static RAM cell.

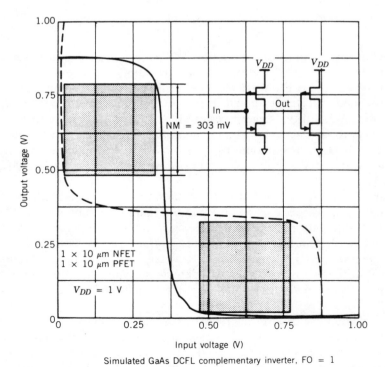

Simulated GaAs DCFL complementary inverter, FO = 1

Figure 6.21 Simulated noise margin (NM) for a complementary DCFL inverter.

6.4   JFET CIRCUIT APPLICATIONS

Since the inception of E-JFET logic in 1978, steady progress has been made in the design and development of digital integrated circuits. E-JFET IC development is carried out by the McDonnell Douglas Astronautics Co. in the United States and the Sony Corp. in Japan. Initial demonstration circuits included frequency dividers, inverter chains, 16 bit shift registers, and small SRAMs, 32×1 and 64×1 bit. In 1983, the first 256 bit SRAM was fabricated and tested. The next step was the development of a 1 kbit SRAM, which debuted in 1984 [13]. While the peripheral circuitry consisting of column and row decoding, I/O circuits, and read/write control were designed with n-channel E-JFETs and resistive loads, the six-transistor memory cell used a complementary flip–flop design. The complementary cell offers higher noise margin and circuit yield than the corresponding resistive load version.

The design of a high-speed circuit generally involves a trade-off between speed and sensitivity to device parameter uniformity across the area of the chip. The most critical parameter typically is the uniformity of V_T.

To tolerate large variations in V_T without sacrificing functional yield, logic gates are designed to operate at higher values of V_T, 0.4–0.5 V, with large load resistances for optimum noise margin, but also low speed. When high speed is one of the prime requirements, however, the mean value of V_T is lower, generally 0.2–0.3 V, with lower values of load resistance. The resulting noise margin is lower than the low-speed case, requiring a tighter tolerance on V_T and all other device parameters for the circuit to be functional.

For GaAs direct coupled FET logic (DCFL) [5], the interconnect capacitances are generally much larger than the device capacitance, and the transconductance does not saturate within the voltage swing utilized. Using the above assumptions, the logic gate delay can be approximated by [14]:

$$t_{pd} = \frac{2C_L}{K(V_{TURN-ON} - V_T)} \qquad (29)$$

where C_L is the line capacitance, $V_{TURN-ON}$ is the turn-on voltage of the p-n junction gate (i.e., 1.2 V), and $(V_{TURN-ON} - V_T)$ is the voltage swing (i.e., 1 V) with a threshold voltage $V_T = 0.2$ V.

The current through the depletion load is $\frac{1}{2}[K(V_{TURN-ON} - V_T)^2]$, which is half the maximum current through the enhancement mode device, so that rise and fall times are equal. The K value is defined by Eq. (3). Optimization of device performance is accomplished through tailoring of ion-implant profiles. Speed can be increased in two ways. The first is by reducing the channel height a, which requires a large channel doping to maintain the threshold voltage. Second, the increased channel doping and shallow ion implants give an increase in the transconductance of the JFET, thus further increasing the speed.

Currently, E-JFET logic gates are sensitive to circuit loading, which limits gate fan-outs, and other types of gate loading. The loading on a particular net is composed mainly of the interconnect resistance, the capacitance between the interconnects, interconnect crossovers, and the gate capacitance of the transistors connected to the net. This has a significant effect on the circuit designs developed and demands that the architecture, logic, circuit, and layout designers work as a closely knit team to produce desirable results. The loading effects will be reduced in the near future with self-aligned gate structures, which will reduce the sidewall capacitance of the gate. The effects of interconnect are being reduced by lowering the sheet resistance of the interconnect, and by decreasing the dielectric constant of the insulating material between the interconnect layers, lowering the interconnect capacitance.

The schematic for both the complementary and resistive load inverters is shown in Fig. 6.22. Both of these circuits can be manufactured on the same wafer, giving the circuit designer the ability to save power where speed is not critical. These are simple, low transistor count devices with no level translation required between any of the different circuit combinations, as is the case with many other GaAs fabrication processes, particularly those using depletion mode transistors. This allows the use of direct coupled FET logic (DCFL), with a voltage swing of about 0.8 V and a noise margin of 200 mV [2].

The low-power dissipation, simple transistor configurations, low transistor count per logic gate, and ease of processing make the E-JFET GaAs technology a good candidate for large-scale integrated circuits. This has been demonstrated with the successful design and fabrication of a 1 kbit static RAM (SRAM) with a wafer yield of 73%. The access and write time for the 1 kbit SRAM is typically in the range of 5–10 ns. Power dissipation at 100 MHz operation is 50 mW. A 4 kbit SRAM was developed at McDonnell Douglas during 1986 with fully functional parts showing access times in the 6–12 ns range. This chip contains over 27,000 transistors in a 4.1 ×

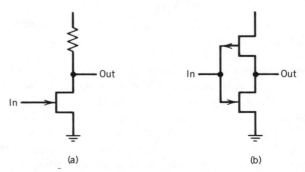

(a)   (b)

Figure 6.22   GaAs inverter circuits: (a) resistive load inverter, (b) complementary inverter.

Figure 6.23   Photomicrograph of complementary 1 kbit SRAM (2.8 × 2.4 mm).

5.0 mm area. A 16 kbit SRAM chip has been designed and is currently being fabricated on the pilot line. This SRAM contains 105,096 JFETs in a chip that is 8.2 × 5.5 mm. Figure 6.23 shows the 1 kbit SRAM, Fig. 6.24 shows the 4 kbit SRAM. These two circuits were processed with the same design rules. The 16 kbit SRAM shown in Fig. 6.25 was processed with tighter design rules.

The memory circuits have been fabricated with great success in recent months. The number of fully functional SRAMs per wafer (yield) has increased dramatically with minor process improvements. The best wafer yields measured are 71% for 1 kbit SRAM, 43% for 4 kbit SRAM, and 3% for 16 kbit SRAM. These data were obtained in April 1988 and will be described in future publications.

The development of E-JFET gate arrays was initiated in 1985 with an array of 1500 three-input NOR gates. The 1500-gate configurable cell array was personalized by the Mayo Foundation team to include a 5 × 5 multiplier in parallel with a 5 bit ALU. The yield of fully functional chips was 37% for the best wafer. The personalized 1500-gate configurable cell array is shown in Fig. 6.26.

A larger version with 6080 three-input NOR gates has been designed and the first personalization has been completed with full functionality. The chip contains 6080 equivalent three-input quasi-complementary NOR gates and was personalized with a large number of circuit elements, including an 8 × 8 bit multiplier, a 16 bit arithmetic logic unit, two 16 × 8 bit register stacks, an 8 bit binary counter, and two 243 stage inverter chains. The gate utilization of this personalization was about 77%. The yield of this circuit was 5% for

Figure 6.24 Photomicrograph of complementary 4 kbit SRAM.

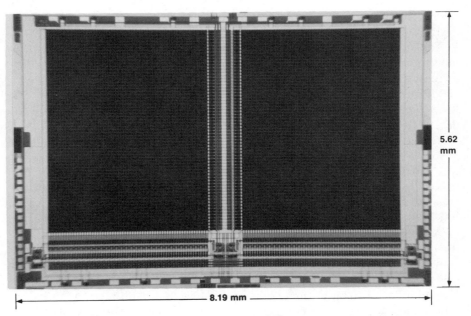

Figure 6.25 Photomicrograph of complementary 16 kbit SRAM.

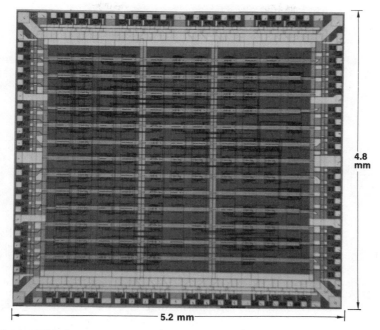

Figure 6.26   1500-Gate array personalized as 5 bit ALU in parallel with a 5×5 multiplier.

the best wafer. The floor plan of this personalization is shown in Fig. 6.27. Schematics for the major circuit elements are shown in Fig. 6.28. The completed chip is pictured in Fig. 6.29. Sony [6] demonstrated an 8×8 bit multiplier in 1986, which is shown in Fig. 6.30.

Probably the most ambitious development effort is the design and fabrication of a 32 bit RISC microprocessor under way at McDonnell Douglas. A preliminary chip demonstrated the feasibility of a 32 bit ALU in 1986.

Three circuits have been designed which demonstrate the GaAs technology for large-scale integrated circuits. The three circuits are called the MD2901, Demo I, and Demo II. The MD2901 emulates the silicon AM2901A chip, which is used in many processing applications. Demo I and Demo II are stepping stones to the realization of the GaAs single-chip 32 bit CPU. The MD2901 processor was chosen as the first demonstration chip. One of the chief reasons for this is the diversity of circuitry types involved. Multiplexers, a dual-port register file, registers, a shifter, and bidirectional I/O lines, as well as standard combinatorial logic, are included in its design. It is a GaAs functional equivalent of the Advanced Micro Devices AM2901 4 bit microprocessor slice. During 1985, a fully functional MD2901 was built by MDC and tested. It showed the feasibility of producing GaAs E-JFET

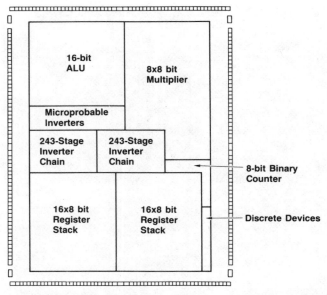

Figure 6.27   Floor plan of a personalization layout of the 6080-gate array.

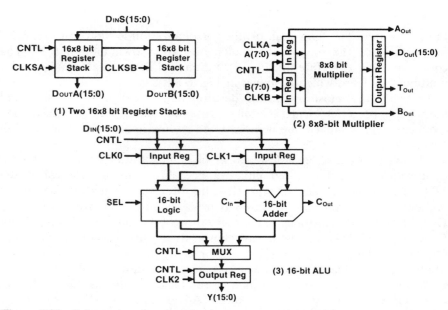

Figure 6.28   Schematics of major circuits on personalization layout of the 6080-gate array.

- Routed with 16 bit ALU, 8 x 8 multiplier, 16 x 16 registers and inverter chains (79% utilization)

- 6080 equivalent 3-input NOR gates

- 192 input/output pins

- 48 power/ground pins

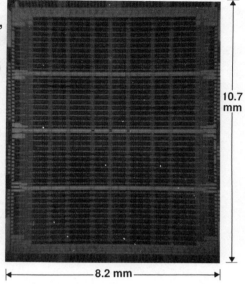

Figure 6.29   Photomicrograph of 6080-gate array.

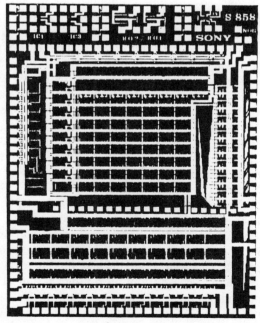

Figure 6.30   Sony 8×8 bit multiplier/accumulator (4.2×3.7 mm).

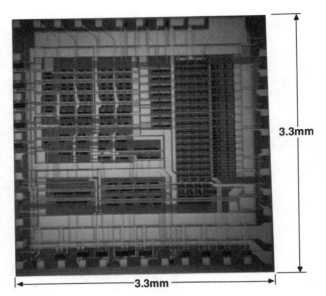

Figure 6.31 Photomicrograph of MD2901 4 bit GaAs microprocessor.

circuits with complexities of several thousand transistors. The 4 bit slice processor chip is shown in Fig. 6.31.

This circuit was realized with enhancement mode n- and p-channel JFETs with a gate length of 1 μm. With the exception of its 16 × 4 bit register stack, which has been implemented using complementary load devices, the chip was designed with resistive load devices. This chip was implemented with 1860 transistors on a 3.3 × 3.3 mm die and had a power dissipation of 135 mW. The measured circuit delays for the MD2901 were above the expected delay. Analysis revealed the cause to be larger-than-anticipated parasitic interconnect capacitance. Interconnect capacitances are now extracted directly from the layout and fed back into the simulator, giving accurate simulation results.

Demo I and Demo II were designed to provide intermediate results before implementing the entire 32 bit CPU. Demo I was used to demonstrate the full 32 bit ALU which will be used in the CPU. Demo II contains the ALU used in Demo I and a large portion of the CPU data path. This progression is illustrated in Fig. 6.32.

Demo I contains a 32 bit ALU, several registers, and multiplexers. The block diagram of this chip is shown in Fig. 6.33, and a photomicrograph is contained in Fig. 6.34. The ALU contains a 34 bit adder onto which a 32 bit logical unit was overlaid. The addition of the logical unit affected the speed of the adder minimally, only adding one gate delay. The logical unit required only five transistors per bit to implement and very little interconnect, since the adder and logical unit have some common gates and

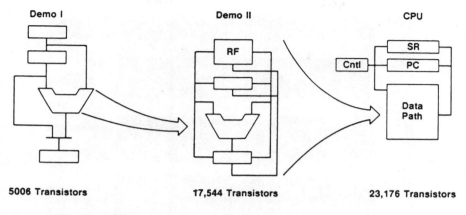

Figure 6.32   Progression of demonstration circuits leading to 32 bit CPU.

interconnect. The ALU uses only a small portion of the die. The rest of the circuitry was added to aid in determining the speed of the ALU. Demo I was fully functional on the first pass.

Demo I was implemented in 5006 transistors, on a die 5.5×4.5 mm, with a total power dissipation of 1.1 W. For the ALU only, the power was 0.89 W. The shortest ALU propagation delay measured was 9.7 ns.

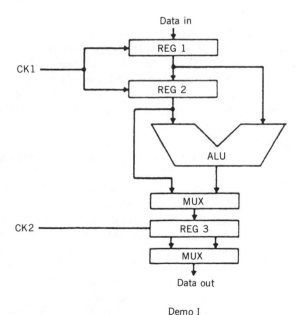

Demo I

Figure 6.33   Block diagram of 32 bit ALU demonstration chip.

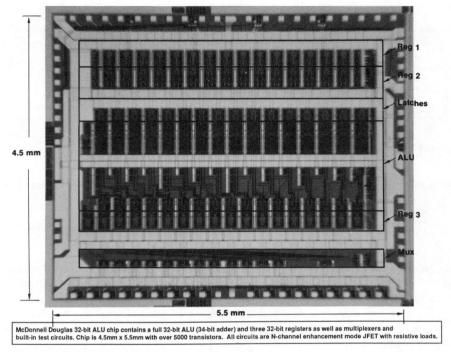

McDonnell Douglas 32-bit ALU chip contains a full 32-bit ALU (34-bit adder) and three 32-bit registers as well as multiplexers and built-in test circuits. Chip is 4.5mm x 5.5mm with over 5000 transistors. All circuits are N-channel enhancement mode JFET with resistive loads.

Figure 6.34 Photomicrograph of the 32 bit ALU demonstration circuit.

Demo II contains almost all of the components of the data path of the CPU and will be used to measure the cycle time expected for the CPU. This circuit contains the full ALU from Demo I, althought the ALU has been modified slightly to increase speed. The dual-port register file contains all 17 general-purpose registers and the null register used in the ALU. All of the divide and multiply support circuitry is included for evaluation. The operand memory interface also exists on Demo II. Demo II contains over 17,000 transistors, representing 76% of the total transistors in the CPU. The total power was calculated as 4.44 W. The functional block diagram is in Fig. 6.35.

While the Demo II chip contains most of the circuitry required for the final microprocessor chip, is a test vehicle for some of the new concepts and circuitry that will be needed. These circuits include dual-port register file memory that provides two 32 bit words simultaneously to the A and B buses; a precharged bus system uniquely constructed to provide shortest path and minimum load performance; an enhanced multiplexer that reduces area and power and increases speed; and spherical buffer circuits, including a tristate pad buffer, for driving high capacitance output loads.

The most important objective in designing this chip was to obtain the greatest speed possible. This translates into minimizing the propagation

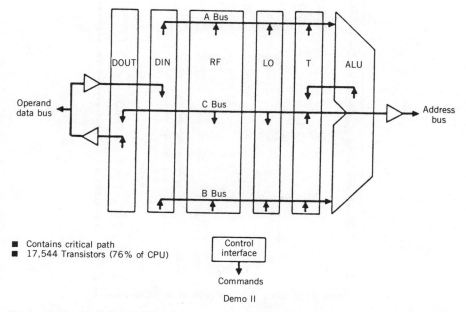

- ■ Contains critical path
- ■ 17,544 Transistors (76% of CPU)

Figure 6.35 Functional block diagram of the 32 bit data path demonstration chip.

delay in the critical path. A function of great importance in GaAs circuit design is the proper sizing of power buses: large enough to carry the supply and ground currents with minimum voltage drop, but not excessively large that they create excessive capacitance loads to signal lines that cross them.

Figure 6.36 shows the overall floor plan showing the positions of the major components. The layout architecture was carefully planned to take advantage of the 32 bit structures involved in the data path. Figure 6.37 shows the overall layout of Demo II. Each component (I/O, LO, ALU) is laid out with the exact same pitch, or distance between bit positions. This allows each component to match perfectly with the component above or below it, without having to fold the bus lines. The bus lines are distributed, in that each bit position in a component has direct access to the A or B bus without interference (the floor plan shows solid bus lines for clarity only). Taking this approach optimizes the circuit performance and eases partitioning of the layout work.

The control circuitry was placed along the right side of the chip to reduce the effect of control and clock skew. Since a finite delay exists in the interconnect between the least significant bit and most significant bit control or clock lines that run horizontally, it is preferable to generate them from the same side of the chip so that they are "in sync." Also, the least significant bit is clocked first, which is important for providing data to the ALU.

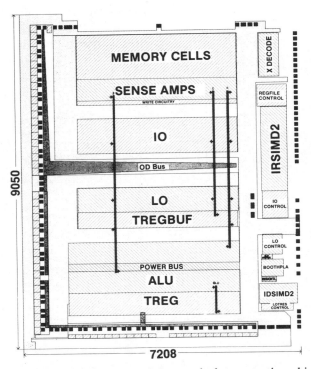

Figure 6.36 Floor plan of data path demonstration chip.

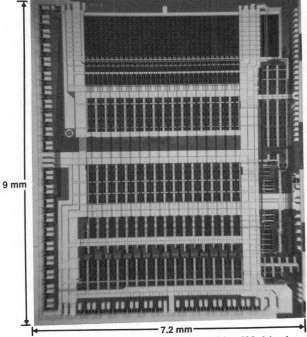

Figure 6.37 Layout of data path demonstration chip (32 bit data path; 17, 544 transistors).

336

Another feature of the Demo II layout includes the location of the power buses for the output pad drivers along the left side and bottom of the chip. This isolates the internal power buses from noise caused by power surges that occur when large external capacitive loads are driven high. This power arrangement also reduces the amount of power crossovers that the signal lines must take on their way to the output pads.

Demo II circuits were fabricated and tested. The test results proved the basic circuits and logic cells were functional, and provided the confidence that the CPU design could be fabricated to produce functional chips. Indeed, fully functional CPU chips have been fabricated. The best speed obtained by August 1988 was 60 MHz, or a complete calculation every 16 ns. The CPU data will be described in future publications.

6.5   RADIATION HARDNESS

The radiation encountered in certain military and space applications and in the nuclear industrial field creates one of the most demanding environments to which semiconductor devices may be exposed. Since GaAs FET devices and integrated circuits are now finding applications in systems that must function in a nuclear environment, it is appropriate to assess their radiation hardness. Electronic circuits are affected by neutrons, protons, gamma rays, cosmic rays, and electrons. Single event upsets (SEU) or soft errors are caused by cosmic rays and protons and are of particular importance for space electronics. These radiation types induce three fundamental effects: displacement of atoms from their lattice sites, ionization of atoms, and internal energy changes.

Radiation dosage is expressed in rad (GaAs) or as particle fluence. A rad (GaAs) is defined as the amount of radiation that deposits 100 ergs/g in GaAs material. The conversion factor from rad (Si) to rad (GaAs) is 1.06. Fluence, for example for neutrons, is defined as the time integral of neutron flux, and is expressed in units of n/cm^2 ($E > 10$ keV). The radiation energy range is specified to exclude thermal neutrons. For SEU in integrated circuits, the linear energy transfer (LET) of heavy ions is commonly used and is expressed in $MeV/mg \cdot cm^2$. The conversion factor from LET into linear charge deposition is 57 $MeV/mg \cdot cm^2 = 1$ pC/mm for GaAs, and, for comparison, is 98 $MeV/mg \cdot cm^2 = 1$ pC/mm for Si. This assumes an electron-hole pair generation of 4.8 eV for GaAs and 3.8 eV for Si, with corresponding densities of 5.32 g/cm^2 and 2.32 g/cm^2, respectively.

In this section, the four major nuclear and space radiation threat categories, namely, neutron effects, total dose effects, dose rate effects, and single particle phenomena are discussed. The effects due to electromagnetic pulse (EMP) and X-rays will not be addressed. Reference 8 provides a more detailed coverage of radiation hardness in GaAs circuits.

6.5.1 Total Dose Effects

Heavily doped GaAs FETs with p-n junction or Schottky-barrier gate structures are not susceptible to prolonged exposure to ionizing radiation. Up to a total dose of about 10^7 rad (GaAs), the changes are hardly noticeable. This is quite contrary to Si MOSFETs, which undergo a serious change in threshold voltage V_T at 10^6 rad (Si) due to charge buildup in the oxide. This effect is not present in GaAs MESFETs and JFETs, but would be encountered in GaAs metal–insulator–semiconductor FETs.

GaAs E-JFETs with a low channel doping of 10^{16} cm^{-3} exhibit discernible changes at 10^7 rad (GaAs) and pronounced changes at 10^8 rad (GaAs) [15]. Devices with a higher channel doping of 10^{17} cm^{-3} are expected to have improved tolerances to ionizing radiaiton. Figure 6.38 shows $I_{DS}^{1/2}$ versus gate voltage for an epitaxial GaAs E-JFET with channel doping of 10^{17} cm^{-3} before and after exposure to an ionizing radiation of 10^8 rad (GaAs) and after exposure to 1.7×10^{15} n/cm^2 $(E > 10 \text{ keV})$. From the square-law relationship

$$I_{DS} = \frac{\epsilon_r \epsilon_0 \mu W}{2aL} (V_G - V_T)^2 \tag{30}$$

mobility and threshold-voltage changes can be assessed.

The large total dose of 10^8 rad (GaAs) causes only a slight increase in V_T of 5 mV (see Fig. 6.38) and a negligible decrease of mobility in devices of 10^{17} cm^{-3} doping level. This can be seen from the shift of intersect and slope of the lines in Fig. 6.38. The voltage–current characteristics of a planar,

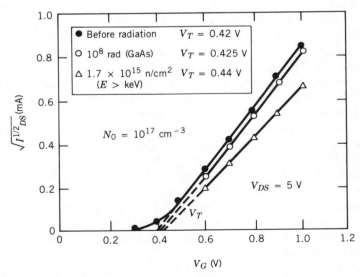

Figure 6.38 Effect of total dose on E-JFET characteristic curve.

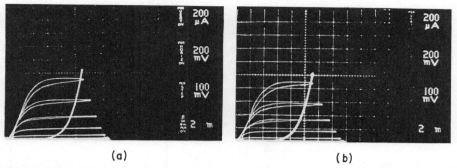

Figure 6.39 *I-V* characteristics of GaAs E-JFETs before and after exposure to 10^7 rad (GaAs) ionizing radiation.

ion-implanted device before and after exposure to a Co^{60} source are shown in Fig. 6.39. Only a very small reduction ($<50\ \mu A$) in the drain saturation current at $V_{DS} = 1$ V and $V_G = 1$ V is observed.

Because the discrete GaAs FETs are not susceptible to total doses of ionizing radiation up to 10^8 rad (GaAs), the same performance is expected for all GaAs planar integrated circuits; indeed, ICs of ring oscillators, flip–flops, buffers, frequency dividers, and static RAMs of other designs, from SSI to VLSI complexity, have operated without serious deterioration of electrical performance up to total doses of 10^8 rad (GaAs) [16, 17].

6.5.2 Transient Response to Pulsed Ionizing Radiation

To determine the logic upset of integrated circuits due to ionizing radiation, it is not sufficient to know only the shunt currents of the FETs; the shunt currents between all active elements must be included. As a first-order approximation for logic upset, the substrate photocurrent I_{DS} will be equated with the FET drain current, which yields a logic upset dose rate of

$$\dot{\gamma}_{UPSET} = \frac{K'(V_G - V_T)^2}{2.3 \times 10^{-11} V_A L_G \ln(8L_0/L)} \quad [\text{rad(GaAs)/s}] \quad (31)$$

With appropriate values for the E-JFET device geometry, $L_0/L = 2.5$, and operating conditions of $K' = 2 \times 10^{-4}$ A/V^2, $V_G - V_T = V_A = 1$ V, one obtains

$$\dot{\gamma}_{UPSET} \simeq \frac{0.5 \times 10^{11}}{L_G} \quad [\text{rad(GaAs)/s}] \quad (32)$$

when the channel length L_G is expressed in μm. The upset dose as a function of channel length is plotted in Fig. 6.40 according to Eq. (31) (solid line). The primary photocurrent upset prediction is shown by the dashed line

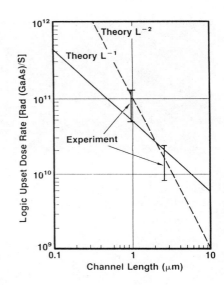

Figure 6.40 Logic upset tolerance as a function of channel length.

[18]. The ranges of experimental data from Ref. 18, including MSI and LSI circuits of 2.5 and 1.0 μm channel length, have been bracketed. The first-order theory (Eq. (31)) crosses through the ranges of logic upset and yields a satisfactory agreement with experiment. It is emphasized that, to predict logic upset accurately in an integrated circuit, a model has to be developed to take consideration of circuit layout and functions. This could move the displayed theoretical curve up and down, and could signal hope for reaching a logic upset dose rate of 10^{11} rad (GaAs)/s for optimized GaAs DCFL with 1 μm channel length FETs.

Circuit modeling could commence by merely including the shunting substrate currents to the GaAs FETs with prompt response, but it is necessary to consider the substrate shunting currents of other circuit components as well, such as resistors and widely spaced contacts with applied potentials. In contrast to the prompt response of closely spaced contact pads (≤ 50 μm), it was demonstrated by Zuleeg et al. [19] that widely spaced contact pads to semi-insulating GaAs show long time-constant transients. Ionizing radiation will cause shunt currents to flow in the semi-insulating substrate between any two conductors in integrated circuits provided sufficient potential difference exists and the conductors or interconnects are placed directly on the substrate material.

Placement of the first-level interconnect metallization on the semi-insulating substrate should be avoided when optimizing digital or analog GaAs ICs for gamma dot performance. The absence of long-term output voltage transients in GaAs E-JFET circuits is demonstrated in Fig. 6.41 for a divide-by-two circuit. The 20 ns wide ionizing radiation pulse at a dose rate of 1.2×10^{10} and 5.0×10^{10} rad (GaAs)/s does not upset the logic, whereas at 1×10^{11} rad (GaAs)/s, a logic upset takes place in the form of a missing

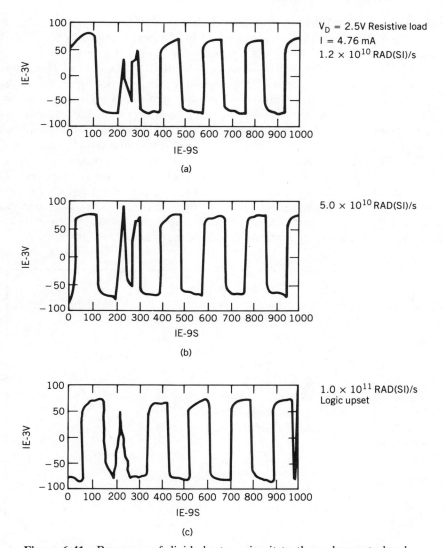

Figure 6.41 Response of divide-by-two circuit to three dose rate levels.

cycle. In all cases, the amplitude of the output signal remains constant and indicates the absence of long time-constant transient effects. It should be noted that the output buffer uses a positive voltage. If a source follower were designed into the circuit with a negative supply in addition to the positive supply, then an output voltage level shift with a long time-constant recovery would be observed. Latch-up, as observed in Si CMOS bulk circuits, is not present in GaAs ICs.

The first modeling of GaAs ICs, including the shunt substrate currents and the long-term ionizing radiation transients to predict logic upset, was made by Notthoff et al. [16] and applied to the performance of the 256 bit static RAM fabricated by GaAs enhancement mode JFETs in DCFL with resistive loads. The first cells—weak cells—failed at 8×10^9 rad (GaAs)/s, but the hard core cells (about 40% of the total cells) did not flip at dose rates of 3×10^{11} rad (GaAs)/s. Inferring from the first-order logic upset prediction of 5×10^{10} rad (GaAs)/s from Eq. (31) for 1 μm channel length devices, a general agreement with the memory logic upset behavior is obtained, but a more detailed modeling is required to predict integrated circuit logic upset.

The experimental results for a 4 kbit SRAM with complementary E-JFETs are shown in Fig. 6.42 for various cell power dissipations. For all power levels, the weakest cell shows a logic upset or bit-flip at 10^{10} rad (GaAs)/s. With increasing power dissipation of the memory cell, one can achieve a logic upset dose rate of 10^{11} rad (GaAs)/s at 100 μW/cell, which is demonstrated for an experimental E-JFET memory in Fig. 6.43.

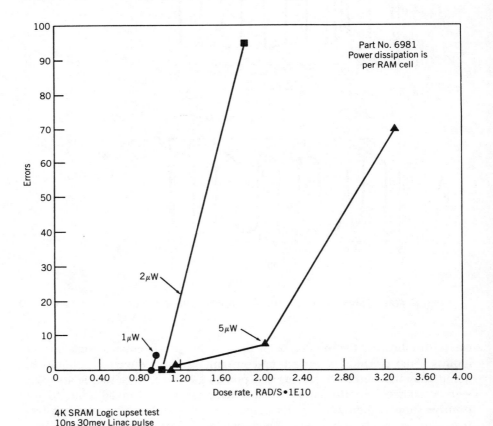

Figure 6.42 Logic upset of a 4 kbit SRAM as a function of dose rate and cell power dissipation.

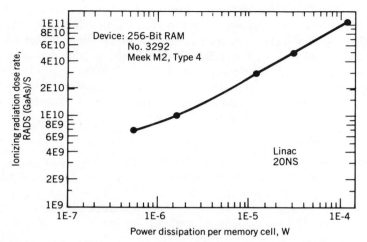

Memory cell power required for zero error operation versus ionizing radiation dose rate

Figure 6.43 Dose rate tolerance of 256 bit SRAM as a function of cell power dissipation.

6.5.3 Single Event Upset (SEU)

For space electronics, single event upsets (also referred to as soft errors) impose performance restrictions on integrated circuits. SEUs are caused by charge collection at sensitive nodes produced by incident radiation, such as cosmic rays, alpha particles, and protons. Cosmic rays include the nuclei of heavier elements, that is, heavy ions. The kinetic energies are in the range of 100 MeV to 100 BeV per nucleon. Charged particles are trapped by the earth's magnetic field in the Van Allen belt. In the region close to the earth, one encounters predominantly protons, while electrons extend out to about 20,000 km. The electrons and protons of the Van Allen belt are the primary origin of semiconductor device and IC failure in space applications. The maximum electron flux is near 10^{10} electrons/cm^2/S, which translates into about 100 rad/h. GaAs ICs with a total dose capability of 10^7 rad (GaAs) could operate for 11.4 yr in this environment without electronic failure. At the worst altitude, the proton fluxes are around 10^4 protons/cm^2/s, which amounts to a few rad/h. Solar flares can increase the proton flux by several orders of magnitude over periods of days.

While Si circuits have been studied extensively in these radiation environments and simulated tests of cosmic rays with ground-based equipment have been devised, this section discusses and analyzes the experimental results of prototype GaAs ICs with respect to SEU performance. Shapiro et al. [20] reported the upset cross section of resistive load 256 bit static RAMs to 40 MeV protons. A comparison of this result with Si technology was given by Simons [21], placing the hardness of this GaAs memory in the same range as Si NMOS circuits but above bipolar Si memory circuits. Si CMOS circuitry, however, outperforms the standard GaAs memory circuit by orders of magnitude. The concept of resistive decoupling of memory cells [22] was successfully applied to a Si CMOS SRAM to provide immunity to SEU [23]. Zuleeg et al. [2] demonstrated the realization of complementary GaAs integrated circuitry; this design is now being explored and holds the same promise as the Si complementary circuitry for immunity to SEU.

6.5.4 Neutron Degradation

The small degradation of g_m and almost unnoticeable change in drain saturation current I_{DS} of two GaAs JFETs with a channel doping of 1×10^{17} cm^{-3}, exposed to a neutron fluence of 1×10^{15} n/cm^2 ($E > 10$ keV), is demonstrated in Fig. 6.44. Drain saturation current in the hot-electron range is determined by a velocity-limited current component I_0 through the relation [24]

$$I_{DS} = I_n(1 - \mu_m) \tag{33}$$

where $I_n = qNv_m$ and μ_m is a normalized drain voltage that is related to z. Since μ_m and v_m are assumed constant, the degradation of I_{DS} is only related to the change in N, which is less than 5% for this fluence. For the Shockley case, I_{DS} is proportional to μN^2, and large changes would be expected, particularly due to degradation of μ. Figure 6.45 presents the predicted degradation of normalized transconductance versus neutron fluence for three different channel doping concentrations of GaAs E-JFETs operating in the hot-electron range. Experimental points are shown that present the average value of 18 GaAs JFETs fabricated on material with a channel doping of approximately 1×10^{17} cm^{-3} and exposed to successively higher fluence levels.

Zuleeg and Lehovec [25] analyzed the transconductance degradation of ion-implanted E-JFETs by neutron fluence and compared the results with those for uniformly doped channel devices. It was shown that the saturation

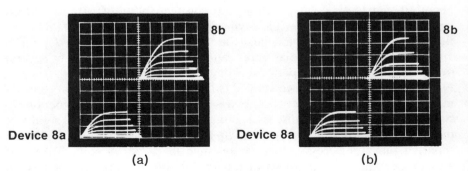

(a)   (b)

Figure 6.44 E-JFET device characteristics before and after exposure to neutrons.

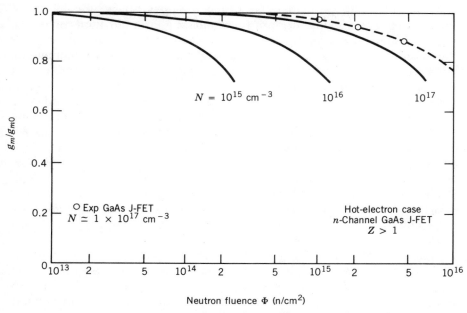

Figure 6.45 Degradation of normalized transconductance with neutron exposure.

transconductance degrades linearly with neutron fluence at about the same rate for devices with widely varying implant profiles and junction location but equal gate geometry and similar threshold voltage.

6.5.5 Summary

The experimental test data and theoretical analyses presented in these sections demonstrate the tolerance of GaAs discrete JFETS and planar integrated circuits to fast neutron and ionizing radiation, both under transient and cumulative conditions. Outstanding total dose ionizing radiation behavior is attributed to the p-n junction gate structure, which is free of charge buildup, the detrimental failure mechanism in the dielectric metal–oxide–semiconductor gate structure. 10^8 rad (GaAs) is an absolutely safe exposure with unnoticeable changes. 10^9 rad (GaAs) is a possible performance goal with functionality of most ICs, but undergoing minor changes. These changes are mostly due to surface charge and potential variations and are primarily extrinsic and peripheral to the intrinsic GaAs FET. Introduction of point defects are by no means ruled out at a total dose of 10^8 rad (GaAs) and require a more detailed study.

The logic upset dose rate to ionizing radiation pulses of 20 ns or less was demonstrated to fall into the range of 10^{10}–10^{11} rad (GaAs)/s, as predicted for 1 μm channel length devices and taking into account the substrate shunt currents. GaAs EJFETs in DCFL design with supply voltage of less than 2 V display a prompt response and are free from long-term radiation transients. The distinct behavior of this technology is ascribed to the forward-bias operation of the gate and the low and single supply voltage. Experimental investigations are warranted to establish the performance limitations of LSI and especially VLSI parts.

The SEU mechanism in GaAs logic designs needs additional studies to clarify the charge collection in the devices fabricated on semi-insulating substrates. The preliminary results of enhancement mode JFET and depletion mode MESFET resistive load static memory circuits reveal a high sensitivity to SEU and optimization procedures could improve the performance considerably. In analogy to Si CMOS SRAMs, the standard complementary GaAs JFET SRAM design has achieved SEU rates of 10^{-6} errors/bit-day for heavy ions. The complementary GaAs EJFET memory cell offers the prospects for SEU immunity with further research of device scaling and resistive decoupling in memory cells.

Permanent degradation due to fast neutron exposures has only thus far been investigated with discrete devices. The results can be applied to predict integrated circuit function failure by simulation. However, it should be emphasized that integrated circuits with passivation and two-level interconnections could impose additional circuit parameter changes not present in unpassivated devices utilized for some of the experimental investigations. Measurements on passivated and selectively ion-implanted MESFETs and JFETs have revealed increased degradation of parameters when compared with epitaxial channel JFETs, but still maintain only a 20% reduction in transconductance at a fluence of 1×10^{15} n/cm^2. With increased channel doping to 10^{18} cm^{-3} instead of the present 1.5×10^{17} cm^{-3} peak value, considerable improvement of neutron fluence tolerance is predicted.

6.6 CONCLUSION

The JFET technology in GaAs has been developed to the extent that promising VLSI circuits can be expected in the near future. The major advantages of the JFET are the higher noise margin as compared with MESFETs and the ability to fabricate true complementary circuits, which leads to low-power consumption and increased radiation resistance. The existing memory circuits and microprocessor chips will lead to the fabrication of an all-GaAs single board computer that will surpass current silicon designs in speed, lower power consumption, and increased radiation resistance for application in harsh environments such as space.

ACKNOWLEDGEMENTS

The author acknowledges the contribution of the GaAs processing and design groups at the McDonnell Douglas Astronautics Co., Huntington Beach, CA. The development of the E-JFET technology was performed under the direction of Dr. Rainer Zuleeg. Mr. J. K. Notthoff and Mr. C. Vogelsang were responsible for the design of the GaAs circuits and the models described in this chapter. The processing work was directed by Dr. Gary Troeger and the radiation hardness activities were led by Dr. Zuleeg. This work was sponsored by the Defense Advanced Research Projects Agency under the management of Mr. Sven Roosild, DARPA/DSO.

REFERENCES

1. R. Zuleeg, J. K. Notthoff, and G. L. Troeger, "A JFET GaAs Logic," Gallium Arsenide Technology, Howard W. Sams, Indianapolis, Indiana, (to be published).

2. R. Zuleeg, J. K. Notthoff, and G. L. Troeger, *IEEE Electron Device Lett.* **EDL-5**, 21–23 (1984).

3. T. L. Rasset, R. A. Niederland, J. H. Lane and W. Geideman, *Computer* **19**, 60–68 (1986).

4. W. Shockley, *Proc. IRE* **40**, 1374–1382 (1952).

5. R. Zuleeg, J. K. Notthoff, and K. Lehovec, *IEEE Trans. Electron Devices* **ED-25**, 628–639 (1978).

6. R. C. Eden, B. M. Welch, and R. Zucca, *IEEE J. Solid-State Circuits* **SC-131**, 419–425 (1978).

7. R. Zuleeg, J. K. Notthoff, and K. Lehovec, *IEEE Trans. Nucl. Sci.* **NS-24**, 2305–2308 (1977).

8. R. Zuleeg, *in* "VLSI Electronics (N. G. Einspruch and H. Huff, eds.), Chapter 11, Academic Press, Orlando, Florida, 1985.

9. C. H. Hyun, C. H. Vogelsang, and J. K. Notthoff, *Tech. Dig., IEEE Gallium Arsenide IC Symp.*, pp. 119–122 (1986).

10. C. H. Chen, M. Shur, and A. Peczalski, *IEEE Trans. Electron Devices* **ED-33** (6), 792–798 (1986).

11. J. S. Blackmore, *J. Appl. Phys.* **53** (10), R123–R181 (1982).

12. R. J. Brewer, *Solid-State Electron.* **18**, 1013–1017 (1975).

13. T. P. Nicalek and J. K. Notthoff, *Proc. Custom Integr. Circuits Conf.*, pp. 434–435 (1985).

14. B. J. van Zeghbroeck, W. Patrick, Heinz Meier, Peter Vettiger, *IEEE Electron Device Lett.* **EDL-8**, 118 (1987).

15. R. Zuleeg and K. Lehovec, *Solid-State Electron.* **13**, 1415 (1970).

16. J. K. Notthoff, R. Zuleeg, and G. L. Troeger, *IEEE Trans. Nucl. Sci.* **NS-30**, 4113 (1983).

17. S. I. Long, F. S. Lee, and P. Pellegrini, *IEEE Electron Device Lett.* **EDL-2**, 173 (1981).

18. R. Zuleeg, J. K. Notthoff, and G. L. Troeger, *IEEE Trans. Nuc. Sci.* **NS-29**(6), 1656 (1982).

19. R. Zuleeg, J. K. Notthoff, and G. L. Troeger, *IEEE Trans. Nuc. Sci.* **NS-30**(6), 4151 (1983).

20. P. Shapiro, A. B. Campbell, J. C. Ritter, R. Zuleeg, and J. K. Notthoff, *IEEE Trans. Nucl. Sci.* **NS-30**(6), 4610 (1983).

21. M. Simons, *Tech. Dig., IEEE Gallium Arsenide IC Symp., 1983*, p. 124 (1983).

22. J. L. Andrews, J. E. Schroeder, B. L. Gingerich, W. A. Kolasinski, R. Koga, and S. E. Diehl, *IEEE Trans. Nucl. Sci.* **NS-29**(6), 2040 (1982).

23. S. E. Diehl, A. Ochoa, Jr., P. V. Dressendorfer, R. Koga, and W. A. Kolasinski, *IEEE Trans. Nucl. Sci.* **NS-29**(6), 2032 (1982).

24. K. Lehovec and R. Zuleeg, *Solid-State Electron.* **13**, 1415 (1970).

25. R. Zuleeg and K. Lehovec, *IEEE Trans. Nucl. Sci.* **NS-25**, 1444 (1978).

7 Design of High-speed GaAs MESFET Digital LSI Circuits

CHRISTOPHER T. M. CHANG

Texas Instruments Incorporated, Dallas, Texas

7.1 INTRODUCTION

Future military and commercial computing system components must be capable of operating at gigabit data rates with power dissipation below a few hundred microwatts per gate. In the last few years, many gallium arsenide (GaAs) integrated circuits (ICs) were developed to meet this requirement. Among them, GaAs metal–semiconductor field effect transistor (MESFET) static random access memory (SRAM) with 1 to 16 kbit [1] capacity, and 8×8 bit to 16×16 bit parallel multipliers [2] have been demonstrated. Read access times of 1 ns have been achieved on 1 and 4 kbit designs [3, 4]. The power dissipation for these designs is 300 mW and 1.6 W, respectively. A multiplication time of 10.5 ns was reported for a 16×16 bit multiplier, with less than 1 W power dissipation [2]. Also, GaAs microprocessors with 10–15 kbit gate complexity are being developed using heterojunction bipolar technology [5]. The goal is to operate this processor at 200 MHz. Commercial 1 kbit SRAMs are available at GigaBit Logic, and more products are planned both in the United States, and overseas. While many fabrication facilities have been built in anticipation of the developing GaAs market, attention has gradually focussed on designing LSI products, which are required to fill the capacity of the manufacturing lines. This chapter introduces the digital GaAs MESFET circuit technologies, discusses the design issues for achieving high-speed operation and with large-scale integration (LSI) complexity, presents a model and methodology for GaAs LSI design, and describes a number of typical applications for GaAs MESFET LSI.

7.2 GENERAL CONSIDERATIONS FOR HIGH-SPEED CIRCUIT DESIGN

The design of high-speed digital circuits can be considered as the charging and discharging of a set of circuit nodes. To achieve high circuit speed, it is important to bring in and to take away a large amount of charge at a given node quickly. Since these charges are brought in or removed through digital switches (MESFETs in this case), one requirement for these switches is that they have high intrinsic transconductance g_m. Owing to the high electron mobility, GaAs MESFETs generally have transconductances in the range of hundreds of mS/mm, approximately 10 times larger than those of the corresponding silicon devices. However, since the extrinsic transconductance of a switch is degraded by the series resistor R_{gs}, it is a common practice for device designers to minimize R_{gs}. Several methods are used to accomplish this in GaAs MESFET design, including the use of recessed gate, n^+ implantation, or smaller source-to-drain spacings [6] and, ultimately, the use of self-aligned, n^+ implantation schemes. The last one combines the features of all the preceding techniques. The second way to achieve high circuit speed is to reduce the charges associated with these circuit nodes. The nodal charges can be reduced by reducing either nodal capacitances or the voltage swings associated with them. Capacitance can be decreased by using insulators with small dielectric constant (e.g., SiO_2) or by reducing physical dimensions of the device and circuit elements. The effects of both on speed have been demonstrated in silicon circuits; they are also applicable to the design of GaAs MESFET circuits.

7.3   MESFET IC FABRICATION

GaAs MESFET ICs with good device characteristics have been fabricated by several methods. A complete discussion of all fabrication methods is beyond the scope of this chapter; only the three most commonly used processes are discussed here. The recessed-gate process was evolved from the highly successful microwave MESFET fabrication process. Gate recess lowers the series resistances (i.e., R_{gs} and R_{gd}) and allows adjusting the threshold voltage while maintaining thin channels for high transconductance. However, both the transconductances and their variations across a wafer were limited by the alignment accuracy of the lithographic steps between n^+ implantation and by the gate definition. Variations in R_{gs} are also introduced by recess etching. The two self-aligned processes were introduced to lower the series resistances and their variations, so that uniform high transconductances across a wafer can be achieved. In one of these processes, refractory metal gates are used as mask for n^+ implantation. The refractory metals (e.g., tungsten silicide) were chosen to withstand the high-temperature ($\sim 800°C$) activation annealing after n^+ implantation. However, it is generally recognized that with the relatively large background doping and defect density in GaAs substrates, extensive device and material development is required to achieve the threshold control from slice-to-slice using this process. In another self-aligned scheme, multilayer sandwich structure is used to fabricate the mask for n^+ implantation. The gate pattern is preserved by SiN deposition afterward. Since gate metal is deposited after

1. ENHANCEMENT IMPLANT

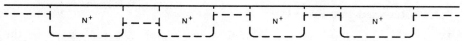

2. N$^+$ AND DEPLETION IMPLANTS

3. ISOLATION IMPLANTATION

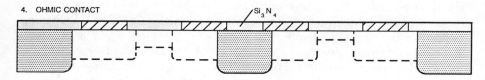

4. OHMIC CONTACT

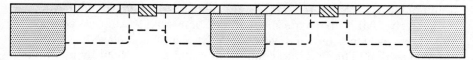

5. SCHOTTKY CONTACT

6. FIRST VIA ETCH AND FIRST-LEVEL METAL

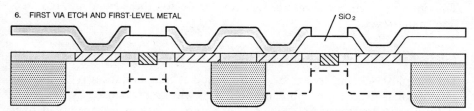

7. SECOND VIA ETCH AND SECOND-LEVEL METAL

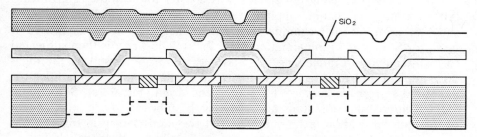

Figure 7.1 Cross-sectional views of recess-gate process.

the high-temperature activation annealing, commonly used TiPtAu metal can be used for gate metal. Some form of threshold adjustment through gate etching can also be implemented with this process.

7.3.1 Recessed-gate Process

Figure 7.1 shows the cross-sectional views of the different steps in the recessed-gate IC fabrication process now used for E/D-MESFET IC at Texas Instruments (TI) [7]. The process has 11 masking levels, with two noncritical ones for alignment-mark etching and bond-pad opening. It begins with sheet implantation of Si for E-MESFET channel. This is followed by a selective Si implantation for n^+ contacts. The alignment marks are then etched in the GaAs substrate. Another Si implantation increases the depth and dosage at selected areas to form the D-MESFET channel. Following the three implants, the slices are annealed using a proximity arsenic overpressure system. Planar isolation between devices is achieved through boron-implantation damage. The ohmic and Schottky metals are formed by evaporation with SiN-assisted lift-off. A single lithographic and recess step is used to define both the D-MESFET and E-MESFET gates simultaneously. The gate and ohmic metals are TiPtAu and AuGeNi, respectively. After the patterning of the ohmic and Schottky contacts, a SiO_2 layer is deposited and vias are etched using a plasma process for interconnection. The first interconnection metal is patterned by ion milling. This interconnection process is repeated again for the second interconnection metal level.

7.3.2 Refractory Metal Gate Self-aligned n^+ Implantation Process

The simplest scheme to achieve self-aligned n^+ implantation is to use the gate metal as mask for n^+ implantation. This was first proposed by the group at Fujitsu Laboratories [8]. Several refractory metals were examined over the last few years. Tungsten silicide was chosen for Schottky gate metal. After forming the channel implant, the gate metal is defined on the GaAs substrate using lift-off technique (Fig. 7.2a). The gate metal is then used as mask for Si (i.e., n^+) implantation (Fig. 7.2b). The wafer is then subjected to high-temperature annealing to remove the damage caused by the high-dose implant and to activate the dopant at the n^+ region. Ohmic metal is deposited to provide contacts for source and drain (Fig. 7.2c). Subsequent via and interconnection steps necessary for IC fabrication are independent of the gate fabrication technique. The last two steps shown in Fig. 7.1 can also be applied to both this and the process described in the next subsection.

7.3.3 Substitutional Gate Self-aligned n^+ Implantation Process

This process was first proposed by Nippon Telephone & Telegraph. It is known as self-aligned implantation for n^+ layer technology (SAINT) process

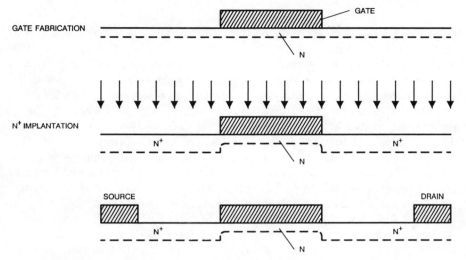

Figure 7.2 WSi gate self-aligned n$^+$ implantation process.

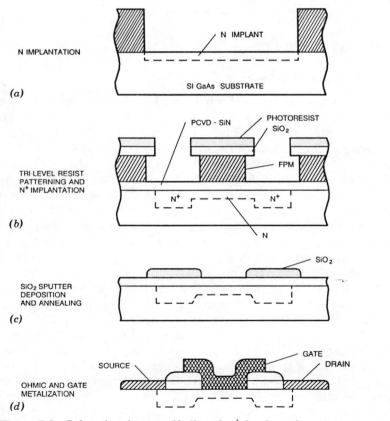

Figure 7.3 Substational gate self-aligned n$^+$ implantation process.

[9]. Cross-sectional views of different processing steps are shown in Fig. 7.3. In Fig. 7.3a, a selective Si implantation over a thin layer of sputtered SiN film forms the MESFET channel on the semi-insulating substrate. This is done to avoid the channeling of Si implant in GaAs crystal, which is deemed necessary for achieving good V_t control. This step can be repeated if more than one V_t is required. After a layer of SiN is deposited, an FPM/SiO$_2$/photoresist sandwich structure is put on the GaAs substrate and patterned. This dummy gate pattern is used as the mask for n$^+$ implantation (Fig. 7.3b). A layer of sputtered SiO$_2$ is deposited afterward. The dummy-gate structure is later removed (Fig. 7.3c) and the wafer is then subjected to high-temperature annealing. The gate metal is then deposited and patterned. This is followed by the deposition and formation of the ohmic metal, as shown in Fig. 7.3d. Owing to the slight undercut of the FPM resin during plasma etching, gate lengths for devices fabricated by this process are shorter than the design values. Therefore, it is possible to fabricate ~0.8 μm gate length devices using 1 μm design rules to achieve high transconductance. Note that the larger gate metal overlapping creates additional parasitic capacitances. In an improved SAINT process, higher-speed operation was achieved by removing the excessive gate metals.

7.4 MESFET CIRCUITS

7.4.1 Buffered FET Logic

The earliest attempt to establish GaAs MESFET for high-speed ICs was described by R. Van Tuyl and C. A. Liechti [10]. In a work published in 1974, they presented a circuit configuration that has since been named Buffered FET Logic (BFL). They reported a 60 ps gate delay with no output loading, and a 105 ps gate delay when output was loaded with three similar logic gates. A schematic diagram of this gate is shown in Fig. 7.4a. The basic gate consists of a switching branch and a level-shifting branch. Operation of the switching branch is shown in Fig. 7.5, where both the I_d versus V_{ds} characteristic of the switching FET and the nonlinear load are plotted. When a large negative voltage is applied to the gate of the switching FET to cut it off, the switching node is operated at point A. An increase in V_{gs} turns the switching FET on and provides a large current through the switching FET (see point B at $V_{gs} = 0$). The width of the load FET is kept to approximately 60% of the switching FET to ensure that the load line intersects the $V_{gs} = 0$ curve at the linear region and to yield a small turn-on voltage. However, because the switching FET is a depletion FET, a level-shifting branch is inserted to provide a negative logic level at its output voltage, so that a similar BFL gate connected to its output can be cut off completely. The level-shifting branch is also a buffer between the input and output of the gate. Thus, the speed of a BFL gate is comparatively insensitive to its output loading. Since the level-shifting branch is always

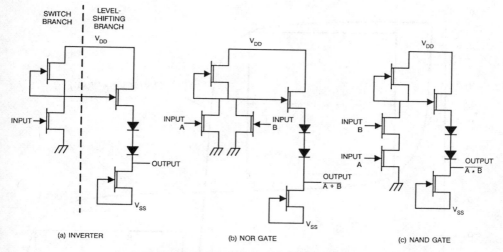

Figure 7.4 The Buffered FET Logic.

conducting, approximately 80% of the power is dissipated by this branch. On the other hand, BFL gates are relatively insensitive to the noise at the input terminal. Figure 7.6 shows a typical noise margin plot for a BFL gate. One of the curves shown in Fig. 7.6 is the dc-transfer curve. The other is the same transfer with the input and output axes interchanged. These two

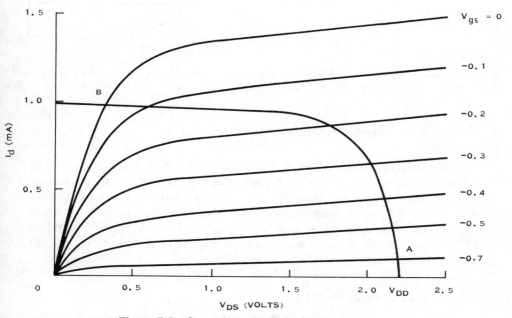

Figure 7.5 Operation of BFL switching branch.

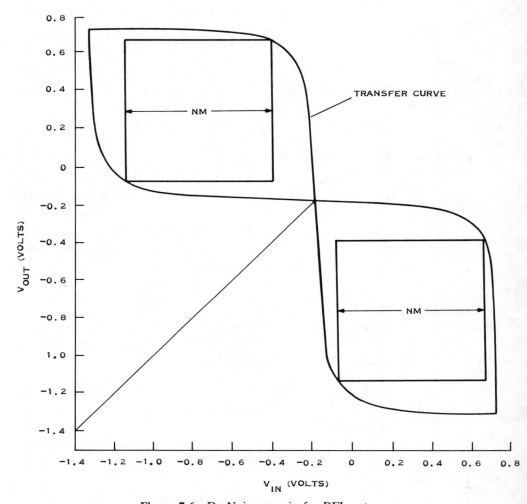

Figure 7.6 Dc Noise margin for BFL gate.

curves represent the operation of any two cascaded BFL gates (inverters) in a long inverter chain. The noise margin is defined as the side of the largest square circumscribed by these two transfer curves. In this case, the noise margin is 0.8 V, more than enough for LSI circuit design. More complex logic functions such as NOR and NAND gates can be achieved by adding FETs in parallel or in series with the switching FET, as shown in Fig. 7.4b and c, respectively.

7.4.2 Schottky-Diode FET Logic

To reduce the gate dissipation of the GaAs circuits, a new type of depletion FET logic, Schottky-Diode FET logic (SDFL), was proposed by Eden et al.

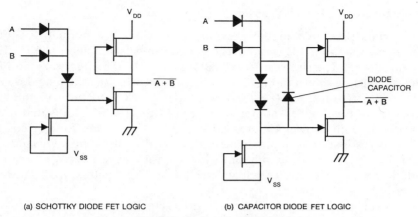

(a) SCHOTTKY DIODE FET LOGIC   (b) CAPACITOR DIODE FET LOGIC

Figure 7.7 Diode Coupled FET Logic.

at Rockwell in 1978 [11]. A schematic diagram of the SDFL gate is shown in Fig. 7.7a. In one sense, this circuit is a modification of the BFL gate. In SDFL, the output is taken from the drain of the switching FET. The level-shifting branch is now connected on the input side. The source-follower FET of the BFL circuit is eliminated. With this arrangement, the current for an SDFL gate is determined by the size of the pull-up FET; therefore, the power is much lower than that of BFL. The diode inputs allow implementation of the OR function, which becomes the NOR function after an inversion. To minimize the loading on the switching node and the gate dissipation, the current in the level-shifting branch is greatly reduced by using a small pull-down FET. However, since the input and output of the gate are no longer buffered, the speed of the SDFL gates is generally more sensitive to circuit loading. Therefore, the widths of the transistors in the switching branch are generally larger compared with those of BFL circuits.

7.4.3 Capacitor-diode FET Logic

A modification of SDFL gate, the capacitor-diode FET logic (CDFL), was recently introduced [12]. A reverse-bias Schottky diode is added across the level-shifting diodes of the SDFL gate (Fig. 7.7b). This diode serves as a speed-up capacitor, similar to the ac-coupling scheme described by several authors earlier [13, 14]. To provide driving power to the circuit, the value of the speed-up capacitor is set to at least 10 times the capacitive loading at the gate output. Therefore, the cell area for CDFL gates is generally large compared with those of other gates. To increase the stored charges futher, CDFL circuits were often designed with higher V_{dd} so that the diode capacitor will be connected over two or more serial diodes. However, the reverse Schottky-diode leakage from the capacitor diode can be substantial and can lead to circuit failures at high temperature. This circuit is used extensively by GigaBit Logic in design of its PicoLogic line of ICs.

7.4.4   Direct-coupled FET Logic

The circuits presented in the previous subsections use normally-on FET (i.e., depletion mode FET) only. The circuit can be simplified when E-FETs are introduced. The simplest E/D-MESFET circuit, the direct coupled FET logic (DCFL), is shown in Fig. 7.8a [15]. In this case, the switching device is a normally-off transistor (i.e., E-FET). It is possible to turn off the gate with a positive voltage. Another advantage of the DCFL gate is that it can be operated with a lower V_{dd} (typically, 1–2 V). Therefore, for a given gate width, its dissipation is approximately that of 10% BFL gates. Owing to its simplicity, DCFL gates usually have low parasitic capacitances and can be made to operate at extremely high speed. The NOR and NAND gates are shown in Fig. 7.8b and c, respectively.

However, when the gate is turned off in actual circuits, current from the load FET passes through the Schottky diode at the gate-to-source junction of the switching FET of the next logic stage. The high-output voltage for DCFL is clamped to ~0.7 V by TiPtAu Schottky-barrier diodes (with approximately 0.7 V barrier height). The logic swing and noise margin for these DCFL gates are limited to approximately 0.5 and 0.2 V, respectively. To limit the gate conduction, DCFL circuits are usually designed to operate at comparatively low supply voltage (i.e., V_{dd}). On the other hand, the direct coupled nodes are more sensitive to circuit loading. Since LSI chips using DCFL usually operate with large currents, special care is needed to minimize ohmic drops and the switching noises associated with the power buses.

At present, most GaAs LSI circuits fabricated in R&D laboratories were built using DCFL. However, it is not completely clear that good production yield can be achieved with current technology. Many efforts were made to improve the noise margin of the DCFL circuits. The most common techniques are to use higher barrier-height gate materials or to use new circuit

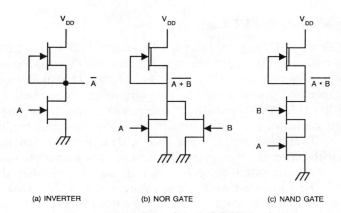

| (a) INVERTER | (b) NOR GATE | (c) NAND GATE |

Figure 7.8   Direct Coupled FET Logic.

techniques to circumvent the noise margin problem. Perhaps the simplest method to improve noise margin is to connect a serial diode at the DCFL gate [16]. The disadvantage of this circuit is that the serial diode resistance tends to lengthen the gate delay, resulting in a slower circuit.

7.4.5 Source-coupled FET Logic

One method to achieve high circuit speed and larger noise margin is to use SCFL. A schematic diagram of this circuit is shown in Fig. 7.9. This circuit is a version of current mode logic (CML) and is the GaAs MESFET equivalent to the well-known silicon emitter coupled logic (ECL). The input voltage V_{in} is applied at node 1. This voltage is compared to a reference voltage V_{ref} at node 2. When V_{in} is higher than V_{ref}, most current for the current source (i.e., F3) flows through F1, making node 3 low and node 4 high. Conversely, when the input is lower, most current is diverted to F2, making node 3 high and node 4 low. The voltages at nodes 3 and 4 are shifted through a pair of level-shifting branches to produce the appropriate voltage levels to drive another SCFL gate. Since the switching threshold is determined by V_{ref} and does not depend on the threshold voltage V_t of the FETs, this circuit is deemed to be more tolerant to process variations than DCFL if the characteristics of F1 and F2 can be made nearly identical. The circuit margin for this gate is similar to that of BFL. However, because the circuit consists of approximately two BFL gates, the dissipation is approximately twice that of BFL. The gate also provides a pair of complementary outputs. Since current is merely steered from one branch to the other and

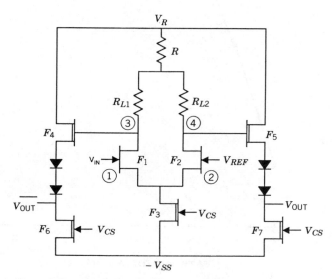

Figure 7.9 Source Coupled FET Logic.

since there is little change in the power supply current, the switching noise is, for this circuit, kept to a minimum. This circuit was first reported by Katsu et al. at Matsushita Electrons Corp. [17], and was subsequently used by Takahashi et al. (NEC) for a memory design. [18].

7.5 DESIGN ISSUES FOR LSI CIRCUITS

Early digital GaAs efforts were directed mainly toward high-speed applications and were built with several hundred gate circuits. To achieve high

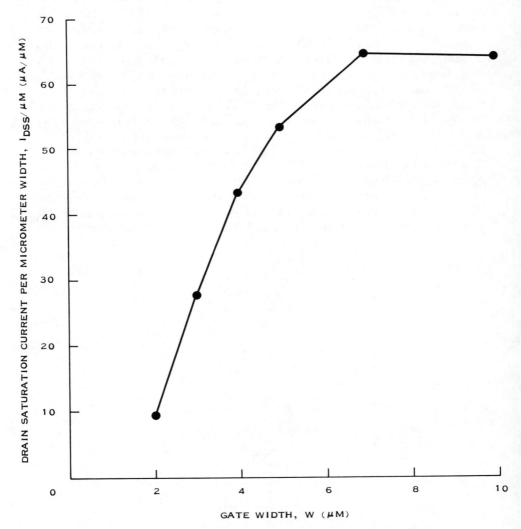

Figure 7.10 Width dependence for a GaAs enhancement MESFET.

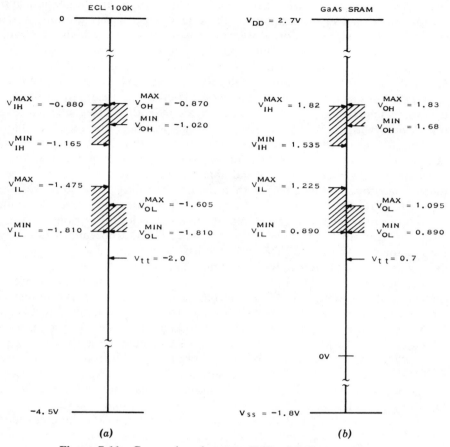

Figure 7.11   Conversions between ECL and GaAs voltages.

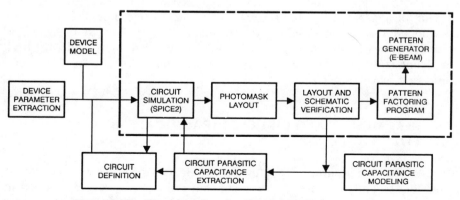

Figure 7.12   Flowchart for the GaAs LSI design process.

circuit speed, the device widths in these circuits were usually large, and chip dissipation of 1–2 W is common for these medium-scale integration (MSI) circuits. Since a large portion of the chip delay was associated with the pads and bonds of the input/output circuitries, it is always advantageous to integrate more functions on a chip to exploit the small gate delay of the GaAs devices. The inevitable result of going to LSI is that chip power dissipated rises as the gate count increases. Therefore, in LSI circuit design, it is important to keep power dissipation for individual gates small and to use devices with small gate widths whenever possible. The most common method for gate power reduction is to use D-FETs with smaller (less negative) V_t, since this reduces currents through logic gates and, hence, their dissipation. Recently, it was found that saturation current I_{dss} decreases much faster for small gate widths than the linear scaling assumed previously [19, 20]. This effect is plotted in Fig. 7.10. The drastic decrease of drain current per micrometer of gate width at small gate width is attributed to the lateral straggle of boron at isolation implant [19] and potential caused by gate-metal extension [20]. This effect must be incorporated into the LSI design process to yield a realistic design.

Another issue for the GaAs LSI circuit design is the problem of circuit interface. At present, no GaAs interface standard exists in the industry. This is partially the result of few GaAs products being available. Most early GaAs parts were designed for insertion into high-speed silicon systems with an ECL interface. One way to provide an ECL interface is shown in Fig. 7.11, where the −4.5 V ECL power supply is divided to provide the V_{dd} (2.7 V) and V_{ss} (−1.8 V) for a BFL circuit. The logic levels for the ECL input/output voltages (Fig. 7.11a) are shifted correspondingly to match the GaAs supply settings.

The third aspect of LSI design is its heavy reliance on computer-aided design (CAD) tools. These tools are required to ensure the first-pass success of a design and therefore eliminate the tedious tasks of debugging circuits on wafers. They also provide more efficient use of the designer's time. Figure 7.12 shows a flow of the GaAs LSI design process now used for MESFET circuit designs at TI. The boxes inside the dashed line are the steps developed for silicon design and are adapted to GaAs LSI design. The remaining boxes are steps unique to GaAs digital design. The MESFET modeling is discussed in Section 7.6. The remaining design steps are briefly discussed in Section 7.7.

7.6   GaAs MESFET AND CIRCUIT PARASITIC CAPACITANCE MODELS

The need for an accurate device model for GaAs MESFETs was recognized since the beginning of its development. Several models were developed in the past few years and were used for designing GaAs MESFET LSI circuits.

One of the most commonly used dc device models for the above-threshold region for GaAs MESFETs is the hyperbolic tangent model first proposed by Curtice [21]. A subthreshold-device conduction model was recently developed by Chang et al. [22] to account for the leakage of MESFETs when they were operated below cutoff. Subthreshold conduction is important for devices operated at high temperature (e.g., approximately 100°C or higher). Because of the high operating speed of the GaAs digital circuits, many factors deemed unimportant for the slower silicon circuits become more important for high-speed GaAs circuits. In this section, two important capacitance models and their effects on circuit speed will be discussed briefly. They are the device-capacitance model and the circuit parasitic capacitance model.

7.6.1 GaAs MESFET Model

7.6.1.1 Above-threshold Region (*The Hyperbolic Tangent Model*).
Owing to the similarity between the silicon and GaAs MESFETs, the early designs of GaAs MESFETs digital circuits used the SPICE2 JFET model that was previously applied to designing silicon MESFET circuits [23]. An equivalent circuit for the JFET model is shown in Fig. 7.13. However, it was soon discovered that, while the JFET model can describe the GaAs MESFET current in the saturated region with reasonable accuracy, the voltage where current saturation occurs was lower than in silicon MESFET circuits because of the much higher low-field mobility in GaAs MESFET circuits. This, in turn, caused errors in dc simulation. One way to correct this source of inaccuracy was to modify the SPICE2 JFET saturated current with a

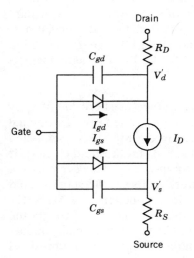

Figure 7.13   Device model for GaAs MESFET.

hyperbolic tangent multiplier; the additional parameter μ in the hyperbolic tangent multiplier allows the model to fit currents in the lower-field region. The equivalent circuit for the hyperbolic tangent model can also be represented by Fig. 7.13, with the current source described by

$$I_d = W\beta(V'_{gs} - V_t)^2(1 + \lambda V'_{ds}) \tanh(\mu V'_{ds}) \tag{1}$$

for $V'_{gs} - V_t > 0$, where

$$V'_{gs} = V_g - V'_s$$
$$V'_{ds} = V'_d - V'_s$$

and the parameters are

W = device width in μm,
β = transconductance factor in $A/V^2 - \mu$m,
V_t = threshold voltage in V,
λ = channel-length-modulation parameter in $1/V$, and
μ = hyperbolic tangent multipliers in $1/V$.

Other parameters of the equivalent circuit are

R_s = source ohmic resistance in Ω-mm (of device width),
R_d = drain ohmic resistance in Ω-mm,
C_{gs0} = zero-bias gate-to-source junction capacitance in pF/mm, and
C_{gd0} = zero-bias gate-to-drain junction capacitance in pF/mm,

for 1 μm gate length devices. Since the resistances do not scale linearly with gate length for approximate 1 μm gate length devices, different resistance measurements are required if the gate lengths of some devices on chips are substantially different from 1 μm. The measured data and calculations, based on Eq. (1), for both the D- and E-mode MESFETs, are compared in Fig. 7.14a and b for both D- and E-mode MESFETs. The agreement is excellent.

7.6.1.2 Below Cutoff Region (The Subthreshold Model). The value of I_d by Eq. (1) assumes an abrupt current cutoff at $V_{gs} = V_t$. However, in actual measurement, a nonzero residual current exists in the MESFET even when it is operated below its cutoff. This current is called subthreshold current. It is generally small and can be ignored in many instances. However, as the circuit density and/or the operating temperature increase, the subthreshold current becomes more important. Subthreshold conduction in a MESFET can be traced to two sources: (1) the residual conduction through the channel (i.e., between drain and source), and (2) the reverse-bias conduction through the gate-to-drain and gate-to-source diodes. The subthreshold

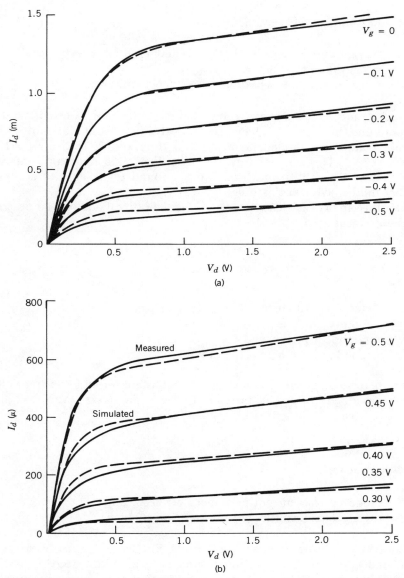

Figure 7.14 Comparison between the measurement and calculation based on eq. (1): (a) D-mode FET, (b) E-mode FET.

channel conduction can be described by

$$I_d = WJ_{s0}e^{[(q/N_s kT)(1-\alpha V_{ds})(V_{gs}-V_t+\gamma V_{ds})]} \tag{2}$$

where the subthreshold parameters are

> J_{s0} = drain current density at $V_{ds} = 0$ and V_t in ampere per μm^2,
> N_s = Schottky ideal factor,
> α, γ = coefficients for Schottky-barrier height lowering,

and q, k, and T are the electronic charge, Boltzmann constant, and the temperature (in degrees Kelvin), respectively. It is the major subthreshold current component immediate below the threshold voltage e.g., $-1\,V < V_{gs} < -0.7\,V$). This can be easily seen from Fig. 7.15a. The same data were replotted in Fig. 7.15b, and Eq. (2) was found to be able to accurately predict the behavior in this range.

The forward I_d–V_d characteristic of the Schottky diode can be described by the classical diode model

$$I_d = AJ_f[e^{(qV_d/N_f kT)} - 1] \tag{3}$$

for $V_d > 0$, where

> J_f = diode forward current in ampere per μm^2,
> A = Schottky diode area in μm^2,
> N_f = diode ideal factor, and
> V_d = applied voltage across the diode.

The reverse diode current does not stay constant for large reverse voltage as it is described in Eq. (3). Instead, it rises as the diode reverse voltage is lowered. It can be described by

$$I_d = AJ_r V_d e^{(-qV_d\delta/kT)} \tag{4}$$

for $V_d < 0$, where

> J_r = diode reverse current in ampere per μm^2, and
> δ = reverse-bias Schottky-barrier-lowering coefficient.

Both models were implemented on TI SPICE and were used for the simulated design of large MESFET circuits (e.g., 1 kbit SRAM, 8×8 bit multipliers, etc). They have provided simulated designs with first-pass functional bars on GaAs wafers.

7.6.1.3 *Device-capacitance Modeling.* Since both the gate-to-source and gate-to-drain junctions of an MESFET are Schottky-diode junctions, the

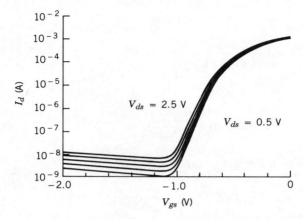

Measured I_d versus V_{gs} for V_{ds} = 0.5–2.5 V, in 0.5 V steps

(a)

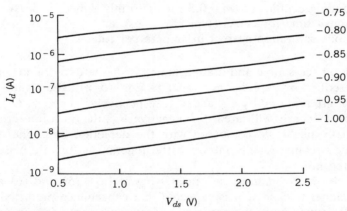

Data in (a) with -1 V $< V_{gs} < -0.75$ V replotted with I_d versus V_{ds}

(b)

Figure 7.15   Subthreshold characteristics of GaAs MESFETs.

most widely used capacitance model for MESFET transient analysis is the diode model [23]. They are represented by the C_{gs} and C_{gd} shown in Fig. 7.13. It is well known that the diode capacitance has two components, the depletion capacitance and the diffusion capacitance. They can be represented for gate-to-source junction as

$$C_{gs}(\text{depletion}) = \frac{C_{gs0}}{(1 - V_{gs}/\phi)^m} \qquad (5)$$

for $V_{gs} < FC\phi$,

$$C_{gs}(\text{depletion}) = \left[\frac{C_{gso}}{(1 - FC)^{1+m}}\right]\left[1 - (1 + m)FC + \frac{mV_{gs}}{\phi}\right] \tag{6}$$

for $V_{gs} > FC\phi$,

$$C_{gs}(\text{diffusion}) = \left[\frac{T_t J_s Q}{(N_f kT)}\right]e^{qV_{gs}/N_f kT} \tag{7}$$

and the total gate-to-source capacitance is

$$C_{gs} = C_{gs}(\text{depletion}) + C_{gs}(\text{diffusion}) \tag{8}$$

where the parameters are

ϕ = Schottky barrier (typically 0.7 V),
FC = forward-biased, nonideal junction capacitance coefficient,
m = grading coefficient (m = 0.5 for uniformly doped devices),
T_t = transit time for GaAs (typically 10 ps), and
J_s = diode saturation current in ampere per μm^2.

Because of the source and drain symmetry, the gate-to-drain capacitance can be described by expressions similar to Eqs. (5–8), where the variables C_{gs} and V_{gs} are replaced by C_{gd} and V_{gd}, respectively. The C_{gs0}, C_{gd0}, and J_s are all scaled linearly with area. For negative V_{gs}, the diffusion term in Eq. (8) is usually negligible compared with the depletion one. The diffusion capacitance becomes significant when the junction is forward-biased for a few hundred mV.

Since the approximation used in arriving at Eqs. (5) and (6) assumed infinite channel thickness, it is expected that capacitance predicted by this model will deviate from the experimental measurements. The solid line in Fig. 7.16 shows the measured capacitance of a 2×4 square-mil FET with the source and drain tied together. The broken line is the calculation based on Eq. (5) with the value of capacitance fitted at $V_{gs} = 0$. However, this curve begins to deviate from those measured near the threshold voltage for the FET. (i.e., V_{gs} approximates 0.63 V). A variable-capacitance model incorporating the effect of finite-channel thickness was presented by Takada et al. [24]. It provided a reasonable physical explanation for the shape of the MESFET C–V curve and is considered to be more accurate than the simple diode model. However, because of the complexity of the model, it is more difficult to incorporate into the SPICE2 simulator.

Another source of error for speed simulation comes from the partition of the measured capacitances between the gate-to-source and gate-to-drain junctions. Generally, the capacitance values for these junctions are expected

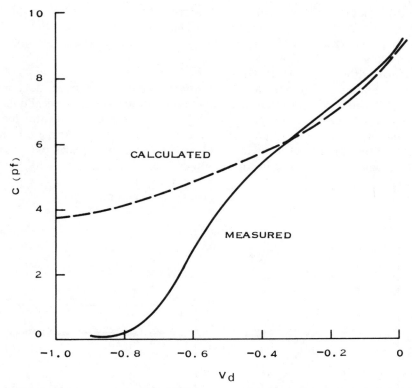

Figure 7.16 Measured and calculated diode capacitances for a 2×4 mil Schottky diode.

to depend on the voltage across the source–drain junction (i.e., V_{ds}). For FETs with symmetrical source–drain construction, the gate capacitance is expected to split equally between the two junctions if V_{ds} is zero. On the other hand, for large positive V_{ds}, a major portion of the junction capacitance is expected to be at the gate-to-source junction. The precise change of the capacitance partition from these two extremes is not understood. Most common practices at present assume both C_{gs0} and C_{gd0} to be independent of V_{ds} and are divided according to a fixed ratio (e.g., $C_{gs0}/C_{gd0} = 9$).

7.6.2 Circuit Parasitic Capacitance Modeling

It is well known that parasitic capacitances associated with device contacts and circuit interconnections can have considerable adverse effect on circuit speed. For silicon MOSFET circuits, these parasitic capacitances can be extracted using parallel plate capacitance approximation and accuracy is reasonable for large-conductor width-to-spacing ratio. However, for GaAs

MESFET fabricated on semi-insulation substrates, the virtual ground is placed far from the circuit (8 mils). The characteristics of the charge couplings become very different. Instead of having charge couplings between conductors and substrate ground (represented by parallel plate capacitors), the couplings are now between neighboring conductors.

A technique to model these parasitic capacitors has been developed [25, 26]. It is based on the Green's function method. The numerical computation was done on a TI program, MAXCAP20, which computes the charge distribution for a system of parallel conducting stripes placed at the interface of two different dielectric media with an upper ground plane over them. This geometry is shown in the inset of Fig. 7.17. The results of the computation, expressed in terms of a Maxwellian matrix, are subsequently converted into coupling capacitors between conductors.

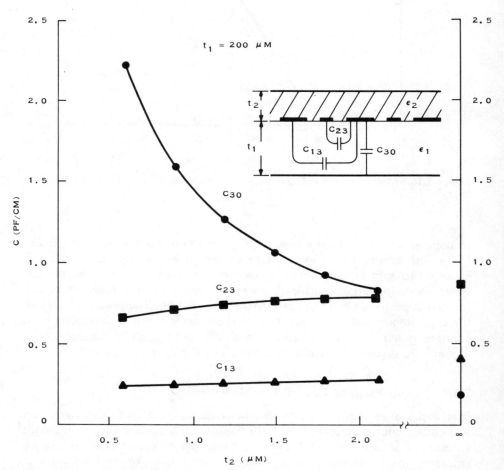

Figure 7.17 Calculated coupling capacitances for different values of dielectric thickness t_2.

Figure 7.17 shows the computed coupling capacitances C_{30} (between a 3 μm wide source contact and the virtual ground), C_{23} (between a 1 μm gate and drain), and C_{13} (between two 3 μm wide source-and-drain contacts). The separations between the metal stripes are 0.75 μm. The data points at $t_2 = \infty$ correspond to the case of no upper ground plane (i.e., microstrip). The graph shows that placing a metal layer approximately 0.6 μm from the contact metals on an SiO_2 insulator can increase C_{30} from 0.17 pF per cm to 2.2 pF per cm, an approximate 13-fold increase. The technique has been applied to find the parasitic on BFL inverters and larger circuits (e.g., memory cell, etc.). By incorporating the parasitic capacitances in the ring oscillator simulation, the simulated BFL gate delays agree to within 10% with those measured [25].

7.7 DESIGN METHODOLOGY FOR LSI CIRCUITS

This section describes the essential steps used for the digital GaAs circuit design shown in Fig. 7.12. Because of the diversity of available CAD tools, this discussion will be limited to the design tools now used for digital GaAs MESFET design at TI. Discussion of the specific algorithm used for the CAD tool is beyond the scope of this chapter. The readers are referred to the standard reference for detailed discussion, [27]. The specific models for GaAs MESFET and circuit parasitic capacitances were described in Section 7.6. The functions of the remaining steps are briefly described below.

7.7.1 Device Parameter Extractions

As the device model becomes more and more complex, it is more difficult to obtain the specific device parameters that will yield accurate representation of the devices produced from the processing facilities. This is especially true when the models are nonlinear and cannot be extracted by simple graphic methods. The standard method to extract the model parameters is to minimize the accumulated least-square error between the measured and calculated data points by optimizing different model parameters, [28, 29]. The accumulated square error can be expressed as

$$\text{Error} = \Sigma (I_{dm} - I_{dc})^2 \tag{9}$$

where I_{dm} and I_{dc} are the measured and calculated (based on the model) currents, respectively. Because the model is generally nonlinear, this minimization process is usually achieved using computer programs. One of the most commonly used programs is TECAP provided by Hewlett-Packard. When the proper model expression is entered, the computer will optimize the parameters to achieve the minimum least-square error. This set of parameters is the best parameter set for the measured data.

7.7.2 SPICE Modeling and Circuit Simulation

The most common design tool for circuit simulation is the SPICE general-purpose simulation program, first developed at the University of California at Berkeley [23]. The program can perform nonlinear dc, nonlinear transient, and linear ac analyses. Circuits elements (e.g., resistor, capacitors, FETs, diodes, current, and voltage sources, etc.) are connected to a set of nodes. The program begins with a set of initial nodal voltages. Then, the nodal equations are solved by matrix inversion for a new of nodal voltages. The values of the two successive trial voltages are then compared to see if they are within the specified error tolerances. The specific problem for GaAs design is to implement the physical models to the specific circuit simulation program so that the designer can use existing facilities. This has been implemented at TI by modifying the voltage-dependent current source for the SPICE2 JFET model according to Eqs. (1) and (2). From this simulation, the sizes of various devices are determined.

7.7.3 Circuit Layout

This task converts the device sizes obtained from the circuit simulation into the proper device geometries for circuit fabrication. Circuits are generally laid out on graphic terminals such as Calma or Apollo, where geometries from different layers can be created quickly and displayed by special graphic software (e.g., graphic programming language or GPL for Calma, and ICE for Apollo at TI). A plot of the memory cell layout for TI's 1 kbit SRAM is shown in Fig. 7.18. Each of these layers contains geometries for one of the masking steps shown in Fig. 7.1. A set of design rules are built to guide the layout. The rules will generally include tolerances for mask registration as well as dimensional variations for the lines and spacings. They are also designed to ensure good fabrication yield. The complex device geometries are later broken down into simple geometric figures (e.g., rectangles, triangles, and trapezoids, etc.), converted into a fixed format, and stored in a data file. Both Calma and Apollo terminals were used for GaAs layout work at TI. Communication lines were provided to link both systems with the IBM 370 and VAX computers, where the data conversions from GDSII (for Calma) or LAFF (for Apollo) to SBDF (e.g., symbolic bar data file, one of the data-file formats for design verification and photomask generation).

7.7.4 Design Rule Verification

Because of the complexity of the circuit layout, many layouts will inadvertently include violations of the design rules, which can cause failures on GaAs wafers or limit circuit yield. Special softwares have been developed to check if the layout conforms to the prescribed rules. A program called VER76 was developed at TI for design rule verification and device extrac-

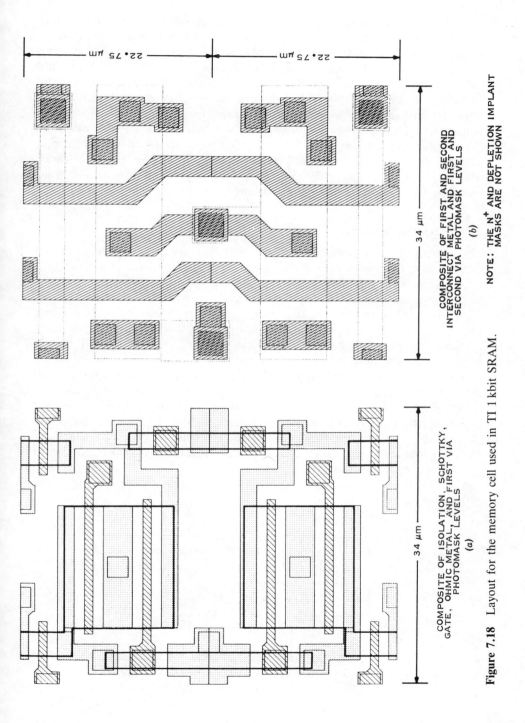

COMPOSITE OF ISOLATION, SCHOTTKY,
GATE, OHMIC METAL, AND FIRST VIA
PHOTOMASK LEVELS

(a)

22.75 μm 22.75 μm

34 μm

COMPOSITE OF FIRST AND SECOND
INTERCONNECT METAL AND FIRST AND
SECOND VIA PHOTOMASK LEVELS

(b)

NOTE: THE N⁺ AND DEPLETION IMPLANT
MASKS ARE NOT SHOWN

Figure 7.18 Layout for the memory cell used in TI 1 kbit SRAM.

tion. To perform the design rule check on a layout data base, a rule deck with input-level definition, extract definition, and the rules for both geometric dimensions and intra- and interlevel spacings is constructed. The extracts are composite levels used to separate figures from the same level into sublevels (or pseudolevels) so that they can be checked with different rules. For instance, in TI's 1 kbit SRAM design, the spacings between nonoverlapping first and second vias are 1 μm. However, for overlapping vias on large bond pads, the spacings are 4 μm. This difference is carried out with overlapping vias extract. The VER76 will first separate all figures from a design into different levels and pseudolevels, apply the rules to check whether there are any rule violations, and report all errors and their locations (i.e., coordinates) for error correction.

7.7.5 Schematic Verification

The schematic verification is used to check the errors in device sizes and connections. Figure 7.19 shows the steps for schematic verification at TI. It is based on the comparison of two schematics. One is constructed from the layout data base using VER76. The devices and its connections are extracted and placed in the EXTracted file in the schematic description data base (SDDB), as it is indicated in the upper path. The same logic diagram (schematics) are drawn on the Apollo terminal and converted into a hardware description language data base (HDLDB). This data base is further converted into the REFerence file in the SDDB, as is indicated by the lower path in Fig. 7.19. The two SDDB files are compared, and the result of the comparison is sent for corrections. It should be pointed out that many reported errors in schematic verification (SVER) runs are caused by the inability of the software to resolve the matching between the two data files; in some instances, they are false errors. Nevertheless, it is important to eliminate them from the error report to achieve an error-free comparison run, and not rely on human judgment which, in many instances, has been proven to be inaccurate and costly.

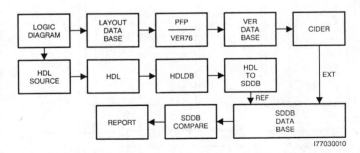

Figure 7.19 Steps for schematic verification.

7.8 APPLICATIONS

Although many applications were proposed for GaAs digital work, this discussion is limited to high-speed applications in which GaAs MESFET circuits will have a speed advantage. Since GaAs MESFET gates are sensitive to the output loading, they are most suited for applications that are highly repetitive and have minimum capacitive wire loadings. Three GaAs MESFET ICs are described in this section. They are 1 kbit SRAM, a 16×16 bit ripple-through multiplier, and the digital variable time-delay line. They represent the three areas of high-speed digital LSI applications for GaAs MESFETs: memory, structured logic, and memory and logic combinations.

7.8.1 Static Random Access Memory

One of the most important LSI digital GaAs MESFET applications is the SRAM. To date, there are over a dozen companies involved in developing the GaAs SRAMs. This is partly because the speed of the computing system is usually limited by the speed of its cache memory; an improvement in SRAM speed is perceived as one of the most direct means to improve the performance of a computing system. Figure 7.20 shows a photomicrograph of a 1 kbit SRAM design at TI [7]. The chip uses the recess-gate process described in Section 7.3.1. The unique feature of this design is derived from using DCFL gates for memory cells and BFL gates for peripheral circuits. The DCFL memory cells are a simple six-transistor cell that is small in cell area ($22.57 \times 34 \ \mu m$; Fig. 7.18) had have good memory-cell stability for a wide temperature range. The BFL peripheral circuits have large current drive capability and dc noise margin, which enable the SRAM design to operate at high speed and with high yield. The design uses three supply voltages (i.e., $V_{dd} = 2.7$ V, $V_{ss} = -1.8$ V, and $V'_{dd} = 0.7$ V) to achieve the ECL 100 K interface. The V'_{dd} is used to power the memory-cell array. A fourth supply, external to the memory chip, V_{tt}, was added for the 50 Ω ECL output resistor termination. A photomicrograph of the memory chip is shown in Fig. 7.20. Measuring 100×90 mils, it is among the smallest in the industry. Measurements indicated that read access times are as short as 1.3 ns (Fig. 7.21) and are typically between 1.5 and 3 ns. The total chip dissipation for this design is 1.4 W. A low-power version of the same design with 6 ns access time and 600 mW dissipation has achieved over 50% functional yield on a wafer and 22% yield on a complete lot of seven wafers. Since only active FET loads were used on this design, no mismatch in temperature coefficients between the acitve loads and the switching FETs was observed. This is verified by temperature measurements on eight packaged SRAMs. One of the SRAM packages operated over the complete military temperature range of -55 to $125°C$. The failures of the remaining packages were caused by a few random bit failures at high temperature.

Figure 7.20   Photomicrograph of a 1 kbit SRAM.

7.8.2   Ripple-through Multiplier

Parallel multipliers are one of the most important logic elements in computing and digital signal-processing systems. The speed of these systems is often determined by multiplier speed. Many GaAs multiplier designs were reported since the 8×8 bit parallel multiplier work of Lee et al. in 1982 [30]. Among them are the 16×16 bit DCFL multiplier by Nakayama et al. from Fujitsu [2] who reported a 10.5 ns multiplication time using a 2 μm gate length device with 952 mW power dissipation. Since the speed of MESFET gates is more sensitive to fan-out and wiring loadings, most MESFET multiplier designs used simple parallel organization with carrier ripple-through the final rows of the adders. A logic diagram of this multiplier is shown in Fig. 7.22. Note that this chip organization is highly structured. Wiring loading can be kept to a minimum and estimated with reasonable accuracy. Figure 7.23 is a photomicrograph of the 16×16 bit parallel multiplier.

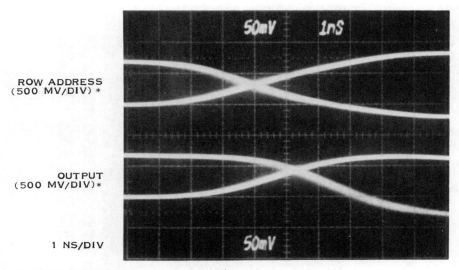

ROW ADDRESS
(500 MV/DIV) *

OUTPUT
(500 MV/DIV)*

1 NS/DIV

*MEASURED WITH A 0. 1 X PICOPROBE

Figure 7.21 1.3 ns Read access time measurement for a 1 kbit SRAM.

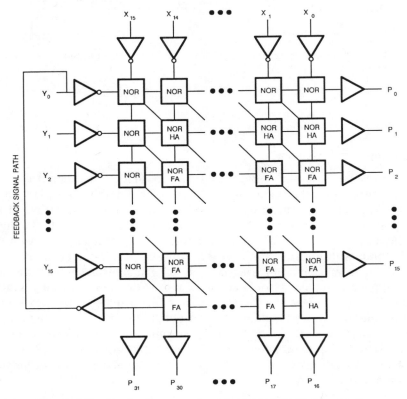

Figure 7.22 Logic diagram of a 16 × 16 bit parallel multiplier.

377

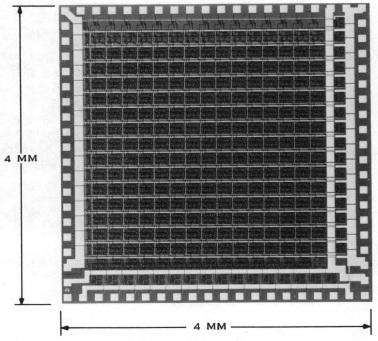

4 MM

4 MM

Figure 7.23 Photomicrograph of a 16×16 bit parallel multiplier (Courtesy of Yoshiro Nakayama of Fujitsu, Ltd.).

7.8.3 Digital Rf Memory

In many instances, it is advantageous to have both logic and memory circuits to be integrated on the same chip to shorten the delays associated with the input/output circuits and device packages. One of the applications of this type is a digital rf memory [31]. Figure 7.24 is a photomicrograph of a digital rf memory that is used as a variable time-delay line. Ordinarily, a delay line can be built with a serial shift register. However, the number of stages for the shift register can become excessively long if long delay time and fine time steps are required for such a system. Also, the clocking scheme required to synchronize this circuit can be difficult. One method to achieve the variable time-delay function is, instead of shifting the data at high speed, shifting the address of a SRAM serially, storing the information on memory cells, and accessing them later. An added advantage for this approach is that the delay time in this circuit can be altered by changing an offset between the write and read cycles.

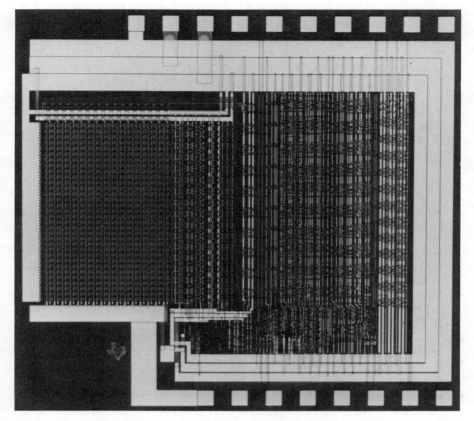

Figure 7.24   Photomicrograph of a digital rf memory chip.

ACKNOWLEDGMENTS

The author thanks Mr. Phil Congdon and Mr. R. Carroll for their encouragement during the preparation of this chapter. He is also grateful to many of his colleagues at Texas Instruments Incorporated, both in the Central Research Laboratories and in the Defense Systems & Electronics Group, for their technical assistance, as well as stimulating discussions in the past years that resulted in many of the insights in this chapter.

REFERENCES

1. M. Hirayama, M. Togashi, N. Kata, M. Suzuki, Y. Matsuoka, and Y. Kawasaki, A GaAs 16-kbit static RAM using dislocation-free crystal. *IEEE Trans. Electron Devices* **ED-33**(1), 104–110 (1986).

2. Y. Nakayama, K. Suyama, H. Shimizu, N. Yokoyama, H. Ohnishi, A. Shibatomi, and H. Ishkawa, A 16×16 bit parallel multiplier. *IEEE J. Solid-State Circuits* **SC-18**(5), 599–603 (1983).

3. N. Yokoyama, H. Onodera, T. Shinoki, H. Ohnishi, H. Nishi, and A. Shibatomi, A 3-ns GaAs $4k \times 1$ b SRAM. *Dig. Tech. Pap.—ISSCC 1984* pp. 44–45 (1984).

4. H. Tanaka, H. Yamashita, N. Masuda, N. Matsunaga, M. Miyazaki, H. Yanazawa, A. Masaki, and N. Hashimoto, A 4k GaAs SRAM with 1-ns access time, *Dig. Tech. Pap.—ISSCC 1987* pp. 138–139 (1987).

5. D. Whitmire, V. Garcia, and S. Evans, A 32 bit GaAs RISC microprocessor, *Dig. Tech. Pap.—ISSCC 1988*, pp. 34–35 (1988).

6. F. Katano, K. Takahashi, K. Uetake, K. Ueda, R. Yamamoto, and A. Higashisaha, Fully decoded GaAs 1-kb static RAM using closely spaced electrode FETs. *Tech. Dig.—Int. Electron Devices Meet.* pp. 336–339 (1983).

7. W. V. McLevige, C. T. M. Chang, and A. H. Taddiken, An ECL compatible GaAs MESFET 1-kbit static RAM. *IEEE J. Solid-State Circuits* **SC-22**(2). 262–267 (1987).

8. N. Yokoyama, T. Ohnishi, H. Onodera, T. Shinoki, A. Shibatomi, and H. Ishikawa, A GaAs 1 k static RAM using tungsten silicide gate self-aligned technology. *IEEE J. Solid-State Circuits* **SC-18**(5), 520–524 (1983).

9. K. Yamasaki, K. Asai, T. Mizutani, and K. Kurumada, Self-align implantation for N^+ layer technology (SAINT) for high-speed GaAs ICs. *Electron. Lett.* **18**(3), 119–121 (1982).

10. R. L. Van Tuyl, and C. A. Liechti, High-speed integrated logic with GaAs MESFETs *IEEE J. Solid-State Circuits* **SC-9**, 269–276 (1974).

11. R. C. Eden, B. M. Welsh, and R. Zucca, Planar GaAs IC technology: Application for digital LSI. *IEEE J. Solid-State Circuits* **SC-13**, 419–426 (1978).

12. R. Eden, Capacitor diode FET logic (CDFL) approach for GaAs D-MESFET ICs. *Tech. Dig., IEEE Gallium Arsenide IC Symp. 1984* pp. 11–14 (1984).

13. A. W. Livingston, and P. J. T. Mellor, Capacitor coupling of GaAs depletion-mode FETs. *Proc. Inst. Electr. Eng.* **127** (Part I, No. 5), 297–300 (1980).

14. M. R. Namordi, and W. A. White, A novel low-power static GaAs MESFET logic gate. *IEEE Electron Device Lett.* **EDL-3**(9), 264–267 (1982).

15. H. Ishikawa, H. Kusakawa, and M. Fukuta, Normally off type GaAs MESFET for low-power, high-speed logic circuits. *Dig. Tech. Pap.—ISSCC 1979*, pp.200–201 (1979).

16. L. Yang, A. Yuen, and S. I. Long, A simple method to improve the noise margin of III–V DCFL digital circuit, Coupling Diode FET Logic. *IEEE Electron Device Lett.* **EDL-7**(3), 145–148 (1986).

17. S. Katsu, S. Numbu, S. Shimano, and G. Kano, A GaAs monolithic frequency divider using Source-Coupled FET Logic. *IEEE Electron Device Lett.* **EDL-3**(8), 197–199 (1982).

18. K. Takahashi, T. Maeda, F. Katano, T. Furatsuka, and Higashisaka, A CML GaAs 4kb SRAM. *Dig. Tech. Pap.—ISSCC 1985* pp. 68–69 (1985).

19. A. H. Taddiken, unpublished results.

20. T. Uenoyama, S. Odanaka, and T. Onuma, Analysis of narrow channel effect in small-size GaAs MESFET. *Conf. Ser. Inst. Phys.* **83** 447–452 (1987).

21. W. R. Curtice, A MESFET model for use in the design of GaAs integrated circuits. *IEEE Trans. Microwave Theory Tech.* **MTT-28**(5), 448–456 (1980).

22. C. T. M. Chang, T. Vrotsos, M. T. Frizzell, and R. Carroll, A subthreshold current model for GaAs MESFETs. *IEEE Electron Device Lett.* **EDL-8**(2), 69–72 (1987).

23. L. W. Nagal, "SPICE2: A Computer Program to Simulate Semiconductor Circuits," Memo ERL-M520. (Electron. Res. Lab., University of California, Berkeley, 1975).

24. T. Takada, K. Yokoyama, M. Ida, and T. Sudo, A MESFET variable-capacitance model for GaAs integrated circuit simulation. *IEEE Trans. Microwave Theory Tech.* **MTT-30**(5), 719–724 (1982).

25. C. T. M. Chang, M. R. Namordi, and W. A. White, The effect of parasitic capacitances on the circuit speed of GaAs MESFET ring oscillators. *IEEE Trans. Electron Devices* **ED-29**(11), 1805–1809 (1982).

26. N. G. Alexopoulos, J. A. Maupin, and P. T. Greiling, Determination of the electrode capacitance matrix for GaAs FETs. *IEEE Trans. Microwave Theory Tech.* **MTT-28** 459–466 (1966).

27. J. D. Ullman, "*Computational Aspects of VLSI.*" Computer Science Press, Rockville, Maryland, 1983.

28. D. E. Ward and K. Doganis, Optimized extraction of MOS model parameters. *IEEE Trans. Comput.-Aided Des. Integr. Circuits Syst.* **CAD-1**(4), 163–168 (1982).

29. P. Yang, and P. K. Chatterjee, SPICE modeling for small geometry MOSFET circuits. *IEEE Trans. Comput.-Aided Des. Integr. Circuits Syst.* **CAD-1**(4), 169–182 (1982).

30. F. S. Lee, G. R. Kaelin, B. M. Welsh, R. Zucca, E. Shen, P. Asbeck, C. P. Lee, C. G. Kirkpatrick, S. I. Long, and R. C. Eden, A high-speed LSI GaAs 8×8 parallel multiplier. *IEEE J. Solid-State Circuits* **SC-17**(4), 638–647 (1982).

31. W. V. McLevige, W. A. White, A. H. Taddiken, and C. T. M. Chang, High-speed GaAs static RAMs for digital RF memory applications. *Dig. Pap. Gov. Microcircuit Appl. Conf.*, *1986* pp. 173–176 (1986).

8 GaAs FET Amplifier and MMIC Design Techniques

THOMAS R. APEL*

AVANTEK, Inc., Santa Clara, California

8.1 INTRODUCTION

The objective of this chapter is to present a systematized approach to GaAs FET amplifier design. The broadband design techniques discussed here are also applicable to narrowband amplifier design as an inclusive subset. A two-stage power amplifier design is included as an example. Finally, MMIC (monolithic microwave integrated circuit) realization of lumped element designs is discussed.

With the exception of distributed amplifiers, microwave amplifiers are usually comprised of several GaAs FET devices interconnected with input, interstage, and output impedance matching networks. This is shown conceptually in Fig. 8.1. The specific amplifier application will usually determine the necessary impedance behavior which must be provided by each network. The important point here is that the methodology by which the networks are obtained remains the *same, regardless of application*. For example, the techniques that we are about to consider are applicable to both low-noise amplifiers and power amplifiers. They are equally applicable to single- or multistage amplifier design requirements. In fact, they also provide an effective means of insuring optimum narrowband design as well.

In order to see that the network problem for all broadband amplifier applications is really the same, several applications will be now considered. Each case will then be reduced to the (same) problem of obtaining an LC network with some desired driving-point impedance behavior.

1. *Low-noise amplifier*. From the optimum noise reflection coefficient ρ_{opt}, the desired matching network driving-point impedance Z_S can be determined

$$Z_S = 50 \frac{1 + \rho_{opt}}{1 - \rho_{opt}}$$

*Present address: Teledyne Monolithic Microwave, Mountainview, California.

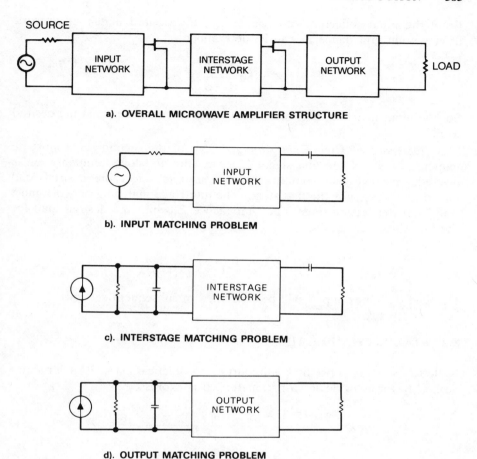

SOURCE

a). **OVERALL MICROWAVE AMPLIFIER STRUCTURE**

b). **INPUT MATCHING PROBLEM**

c). **INTERSTAGE MATCHING PROBLEM**

d). **OUTPUT MATCHING PROBLEM**

Figure 8.1   Three basic matching problems of a multi stage amplifier.

Hence, the input design requirements are in the desired form. Flat amplifier gain can be achieved by controlled mismatch at the output port of the device. Typically, constant gain circles are plotted for this interface. A desired load impedance Z_L is then determined. Therefore, the output matching network is also specified in the desired form.

2. *Power Amplifier.* Either from load line considerations or load–pull data the optimum load impedance $Z_{L\,opt}$ is determined. The design task is to obtain an output matching network that provides this optimum load impedance behavior. Usually, minimum input port reflection is desired. The device input reflection coefficient S'_{11} is determined:

$$S'_{11} = S_{11} + \frac{S_{12}S_{21}\rho_{L\,opt}}{1 - S_{22}\rho_{L\,opt}}$$

From the input reflection coefficient S'_{11}, the desired matching network driving-point impedance Z_S can be determined:

$$Z_S = 50 \frac{1 + S'^*_{11}}{1 - S'^*_{11}}$$

So, the input matching network design requirements are also in the desired form.

3. *Interstage of High Gain Amplifier.* S'_{11} of the second stage must be matched to S'_{22} of the first stage. If mismatch gain slope compensation is desired, constant gain contours can be used to select the desired load impedance $Z_{L\,opt}$, for the first stage. The interstage matching network must transform the second-stage input impedance Z_{IN} into the desired load for the first stage, $Z_{L\,opt}$:

$$Z_{IN} = 50 \frac{1 + S'_{11}}{1 - S'_{11}}$$

Detailed design examples will be presented in subsequent sections.

8.2 STABILITY AND GAIN

Rollette's stability constant k is important to practical GaAs FET amplifier design. In terms of device S-parameters, it is expressed as

$$k = \frac{1 - |S_{11}|^2 - |S_{22}|^2 - |S_{11}S_{22} - S_{12}S_{21}|^2}{2|S_{21}|\,|S_{12}|}$$

Although one-port unilateral models (as seen in Fig. 8.27) are frequently used to represent impedance matching requirements, the complete device representation is nonunilateral. Therefore, at frequencies at which useful gain is available, the potential for oscillations must be examined. Even for power amplifier applications, where the device is being operated nonlinearly, small-signal stability should also be considered.

The significance of Rollette's stability constant is that for $k > 1$, the device is unconditionally stable and no combination of load and source impedances can produce oscillations. For this case, simultaneous complex conjugate matching of the FET input and output ports is possible. When this is done, maximum available gain (MAG) results. The expression for MAG is given by

$$\text{MAG} = \frac{|S_{21}|}{|S_{12}|} (k - \sqrt{k^2 - 1})$$

If, on the other hand, $k < 1$, some load and source impedances can cause oscillations. In such cases, the impedance regions to be avoided can be plotted or represented graphically by circular regions on the Smith chart. This method is adequately described in several references [1, 2]; hence, it will not be repeated here. Often, these difficulties are avoided when lossy negative feedback or lossy branch amplitude equalization techniques are employed. The lossy branch technique will be discussed in the two-stage design example of Section 8.4.4.

8.3   Q-BANDWIDTH LIMITS ON IMPEDANCE MATCH

Before any attempt to design a broadband matching network, the achievable match performance must first be determined. A common pitfall that the inexperienced circuit designer often encounters is an attempt to obtain matching networks blindly by numerical optimization. If the desired performance level is not achievable, considerable computer and engineering time can be wasted. This section addresses the limits imposed on impedance match performance by load behavior.

The relative reactive to resistive (susceptive to conductive) behavior of the load immittance sets the limits on achievable broadband performance. This behavior is sometimes described in terms of a parameter called load-Q. Bode [3] showed that the integral of return loss is bound by a constant. This constant is dependent on the behavior of the reflection function. The load that was initially considered by Bode was a simple parallel RC. In this case, the match performance limit can be described in terms of load-Q and the complex frequency location of matching network reflection function zeros. Several years later, Fano [4] extended Bode's work to address more general cases. For our purposes here, a detailed look at Bode's work will suffice. In addition to the parallel RC case that Bode considered, we will show that the circuit-dual (series RL case) also yields the same results. So, all single reactance absorption lowpass cases are covered. These results can then be extended to the two-element bandpass cases by the well-known lowpass to bandpass transformation.

The two cases that will now be considered are illustrated in Fig. 8.2. An LC matching network that absorbs to complex valued load behavior and provides an impedance-matched filter response to the R_0 source is desired. Matching networks are filter structures. However, a filter that provides a low reflection match between a resistive source and load is not necessarily a matching network. The additional requirement that is imposed on matching networks is reactance absorption at one or both sides of the structure. Typical filter responses have zero flat-loss (offset) due to reflection zeros on the imaginary axis. Since matching networks have additional constraints placed on them, additional degrees of freedom in the realization are required. General placement of reflection zeros allows this freedom. The

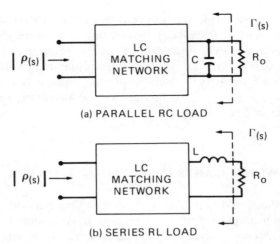

(a) PARALLEL RC LOAD

(b) SERIES RL LOAD

Figure 8.2   Load representations for match limit analysis.

significance of which half of the complex plane the zeros are placed will become clear when Eq. (13) is discussed.

The match limit relationships that we seek are obtained by considering the following contour integral:

$$\oint \log \left| \frac{1}{\rho'} \right| dS$$

where the path of integration is the simple closed contour shown in Fig. 8.3. The reflection function in the above expression is related to the reflection function in Fig. 8.2 by

$$\rho' = \rho \, \frac{(S + S_1)(S + S_2)(---)(S + S_n)}{(S - S_1)(S - S_2)(---)(S - S_n)} \tag{1}$$

Both functions have identical steady-state magnitude, since they differ only by an allpass factor. Equation (1) provides a convenient means of accounting for any right-half plane (RHP) zeros of ρ. Since the contour integral contains the reciprocal of Eq. (1) in the integrand, zeros of ρ' represent poles in the integrand. The allpass factor allows the presence of any RHP zeros of ρ to be removed from ρ' without changing the steady-state magnitude characteristic. For each RHP zero of ρ, S_i, A corresponding RHP pole and LHP zero appear in the allpass function. This integral can be expressed in two parts, by considering the closed contour in two segments, as seen in[†]

$$\oint \log \left| \frac{1}{\rho} \right| dS = j \int_{-\infty}^{\infty} \log \left| \frac{1}{\rho} \right| d\omega + \int \lim_{|S| \to \infty} \log \left| \frac{1}{\rho} \right| dS \tag{2}$$

[†] (Editor's Note): ρ and ρ' have same steady state magnitude.

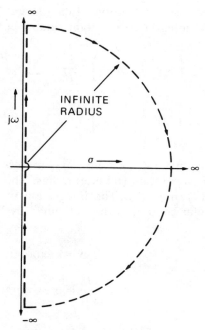

Figure 8.3   Contour integration.

Hence, no RHP poles appear in the integrand. By the Cauchy–Goursat theorem, the integral is zero. This leads to Eq. (3), or in admittance form Eq. (4).

$$-2j \int_0^\infty \log\left|\frac{1}{\rho}\right| d\omega = \int \lim_{|S|\to\infty} \log\left|\frac{1+Z_L/R_0}{1-Z_L/R_0}\right| + \lim_{|S|\to\infty} \sum_i \log\left|\frac{1-S_i/S}{1+S_i/S}\right| dS \quad (3)$$

$$-2j \int_0^\infty \log\left|\frac{1}{\rho}\right| d\omega = \int \lim_{|S|\to\infty} \log\left|\frac{1-Y_L/G_0}{1+Y_L/G_0}\right| + \lim_{|S|\to\infty} \sum_i \log\left|\frac{1-S_i/S}{1+S_i/S}\right| dS \quad (4)$$

The infinite radius limits of the load impedance and admittance for the RC and RL cases are presented in Eqs. (5) and (6), respectively. Similarly, the infinite radius limit of the allpass factors are presented in Eq. (7).

$$\lim_{|S|\to\infty} \log\left|\frac{1+Z_L/R_0}{1-Z_L/R_0}\right| = \frac{2}{SR_0 C} \quad (5)$$

$$\lim_{|S|\to\infty} \log\left|\frac{1-Y_L/G_0}{1+Y_L/G_0}\right| = \frac{2R_0}{SL} \quad (6)$$

$$\lim_{|S|\to\infty} \log\left|\frac{1-S_i/S}{1+S_i/S}\right| = -2\frac{S_i}{S} \quad (7)$$

By making these limit case substitutions, Eqs. (3) and (4) yield Eqs. (8) and (9), respectively. The integral on the right side of both

$$-j \int_0^\infty \log\left|\frac{1}{\rho}\right| d\omega = \int \frac{1}{CR_0} - \sum_i S_i \frac{dS}{S} \tag{8}$$

$$-j \int_0^\infty \log\left|\frac{1}{\rho}\right| d\omega = \int \frac{R_0}{L} - \sum_i S_i \frac{dS}{S} \tag{9}$$

can be evaluated by making a change in variables and performing the integration in theta at a constant (infinite) radius. This change in variables is indicated in Eqs. (10) and (11). The integration in a clockwise direction around the semicircular contour is then obtained by Eq. (12).

$$S = r \exp(j\Theta) \tag{10}$$

$$\frac{dS}{d\Theta} = jr \exp(j\Theta) \tag{11}$$

$$\int_{\pi/2}^{-\pi/2} \frac{jr \exp(j\Theta)}{r \exp(j\Theta)} d\Theta = -j\pi \tag{12}$$

Hence, the results of Eq. (13) are obtained. This is Bode's limit. We can clearly see from this that the presence of any RHP reflection zeros will degrade the match performance. Later, it will be demonstrated that RHP zeros allow greater reactance absorption at the source side of the matching network. This zero-placement trade-off is important in the design of inter-stage matching networks, where reactance absorption at both ports is important.

$$\int_0^\infty \log\left|\frac{1}{\rho}\right| d\omega = \frac{\pi}{CR_0} - \pi \sum_i S_i \tag{13a}$$

$$\int_0^\infty \log\left|\frac{1}{\rho}\right| d\omega = \frac{\pi L}{R_0} - \pi \sum_i S_i \tag{13b}$$

From Eq. (13), it is clear that with LHP zero placement the integral of return loss is equal to a constant that is inversely proportional to load-Q. Consider Fig. 8.4. The area under the curve represents the value of the integral. If load-Q is decreased, the available area is correspondingly increased; so, the same level of match performance is achievable over a wider band. It is also true that if the load-Q is unchanged, a reduction in

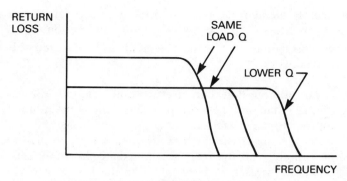

Figure 8.4 *Q*-Bandwidth trade-off.

desired bandwidth permits an improvement in match performance. These qualitative observations are intuitive. Equation (13) is important because it allows a quantitative assessment. Sometimes Bode's limit is expressed with the inequality seen in Eq. (14). This result is easily obtained from both forms of Eq. (13) with LHP zeros and ideal "brick wall" lowpass amplitude response.

$$|\rho| \geq \exp(-\pi/Q_L) \tag{14}$$

Clearly, to make use of Eq. (13) or (14), a lumped model that represents the load impedance behavior is required. Load modeling techniques and the bandpass equivalent of Eqs. (14) will be detailed in Sections 8.4.3 and 8.4.2.1, respectively.

8.4 BROADBAND MATCHING NETWORK DESIGN

The design of lumped *LC* impedance matching networks is now considered. Applicability of the approach to be discussed is wide-ranging, since most practical applications can be reduced to the problem of obtaining some desired impedance behavior at one or both sides of a two-port network.

Direct distributed synthesis techniques, such as those using the Richards transformation and Kuroda identities, will not be considered here. This was decided for several reasons:

1. Discontinuities are not accounted for in distributed synthesis.
2. Lumped designs can be easily converted to an equivalent distributed form.
3. Discontinuities can be folded into the design at the lumped-to-distributed conversion step.

4. Commensurate element designs are virtually always larger and therefore more lossy and less desirable for monolithic integration.
5. Reactance absorption requires unit elements to be inserted only from one side of the network.

It is important to note that there are two fundamentally different ways of addressing the *LC* matching network problem. One is the more classic filter synthesis approach, which requires that a load model be available for absorption into the matching network through the synthesis process. The other operates directly on the interface impedance requirements without a model. The material presented here falls into the first category.

The classic filter synthesis approach to broadband impedance matching network design usually involves four activities:

1. *Approximation*: a functional representation of the desired response.
2. *Realization*: the network synthesis step that satisfies the approximation and absorbs the load model.
3. *Mapping*: a frequency domain transformation to the desired passband.
4. *Load Model*: a one-port formed by an *LC* two-port that is resistively terminated.

These four aspects are discussed in the next three subsections.

8.4.1 The Lowpass Prototype

Although most design requirements are for passbands that do not extend down to dc, the lowpass representation is very useful. Since frequency domain mappings can be used to extend approximations to other bands, it is not necessary to address the approximation problem separately for each type of network to be considered. In some cases, even the completed network realizations can be transformed directly. The lowpass (LP) to bandpass (BP) mapping is one such case.

It is clear that an optimum matching network will only provide good matching inside the desired passband. This is a consequence of the Bode analysis, since the area under the return-loss frequency response is fixed. One of several classic approximations can be applied here to achieve an abrupt transition between passband and stopband. Usually, the phase-transfer characteristic is not a primary consideration. Cases in which phase linearity is important can have additional bandwidth margin designed in or use another approximation, such as the Bessel or Gaussian. Of potential interest here are the Butterworth, Chebyshev, and elliptic responses. Each of these has previously been used for matching network design. However, the Chebyshev (equal-ripple) response is by far the most popular because it offers superior performance to the Butterworth form and is more easily realized than the elliptic form. Consequently, the sensitivity to element variations also falls between the Butterworth and elliptic cases. This section will consider the Chebyshev LP approximation and LP prototype synthesis.

The LP prototype network is frequency- and impedance-normalized. The general Chebyshev amplitude-transfer function is illustrated in Fig. 8.5. Note that in addition to ripple loss, allowance has been made for nonzero offset loss (flat loss). This is represented algebraically in Eq. (15) the expression for insertion loss, where T_N is the Nth Chebyshev polynomial of the first kind:

$$|IL|^2 = 1 + K^2 + \epsilon^2 T_N^2(\Omega) \tag{15}$$

Since insertion loss in LC ladder networks is due to reflection, the reflection function ρ is representable as a function of the desired insertion-loss response. This relationship is expressed in Eq. (16) and from it Eq. (17) follows:

$$|\rho|^2 = \frac{|IL|^2 - 1}{|IL|^2} \tag{16}$$

$$|\rho|^2 = \frac{(K/\epsilon)^2 + T_N^2}{(1 + K^2)/\epsilon^2 + T_N^2} \tag{17}$$

The roots of the numerator and denominator of Eq. (17) allow the reflection zeros and poles to be determined. In order to represent a realizable network, the poles are restricted to the LHP. The only restriction placed on the zeros of reflection is that they must appear in complex conjugate pairs or on the real axis. Note that reflection zeros may appear in RHP or LHP, and the *same* magnitude insertion-loss response may be achieved. The network

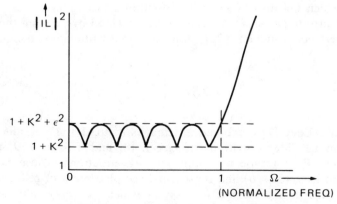

Figure 8.5 Chebyshev LP prototype response.

element values will be affected, since symmetric (with respect to the imaginary axis) reflection zero shifts result in an interchange in the port Q's. The S-plane poles and zeros of reflection are represented in Eqs. (18–21), where i is an integer between 1 and N. The subscript pi or zi is used to indicate the ith pole or zero, respectively.

$$S_{pi} = -\sin\left[\frac{\pi(2i-1)}{2N}\right]\sinh[a] \pm j\cos\left[\frac{\pi(2i-1)}{2N}\right]\cosh[a] \qquad (18)$$

$$S_{zi} = -\sin\left[\frac{\pi(2i-1)}{2N}\right]\sinh[b] \pm j\cos\left[\frac{\pi(2i-1)}{2N}\right]\cosh[b] \qquad (19)$$

$$a = \frac{1}{N}\sinh^{-1}\left(\sqrt{\frac{1+K^2}{\epsilon^2}}\right) \qquad (20)$$

$$b = \frac{1}{N}\sinh^{-1}\left(\frac{K}{\epsilon}\right) \qquad (21)$$

Equations (20) and (21) provide two constants (a and b) called Fano's parameters, which appear in Eqs. (18) and (19). One should not be surprised to find that there is a unique combination of response offset (K^2) and ripple (ϵ^2) that provides a *minimum* mismatch loss in passband. This optimum case is sometimes called the Fano match case. The corresponding Fano parameter values are given in Figs. 8.6 and 8.7 as a function of LP prototype load-Q. Our purpose here is to provide a practical means of designing optimum wideband circuits with a minimum of esoteric complexity. Intentionally, the class of LP loads has been limited to single reactance cases, since Matthaei's [5] "situation 1" can thereby be ensured. The reader should keep in mind that this allows freedom for two-element bandpass models. The bandpass load modeling technique that will be presented in Section 8.4.3 provides a means of ensuring that fourth-order models can be obtained which fall into Matthaei's "situation 1" category.

The LP prototype network can now be obtained by forming the driving-point impedance function (22), followed by continued fraction [6] expansion:

$$Z_{dp}(S) = \frac{1+\rho(S)}{1-\rho(S)} \qquad (22)$$

Significantly, Levy [7] combined Matthaei's "situation 1" results with the closed-form LP filter equations of Green [8] and Takahasi [9] to obtain closed-form LP prototype matching network equations. These expressions are given in Eqs. (23–26) and represent LHP placement of reflection zeros. Equation (25) is a recursion relation in which "i" may take on integer values between 1 and $N-1$, for an Nth-order network.

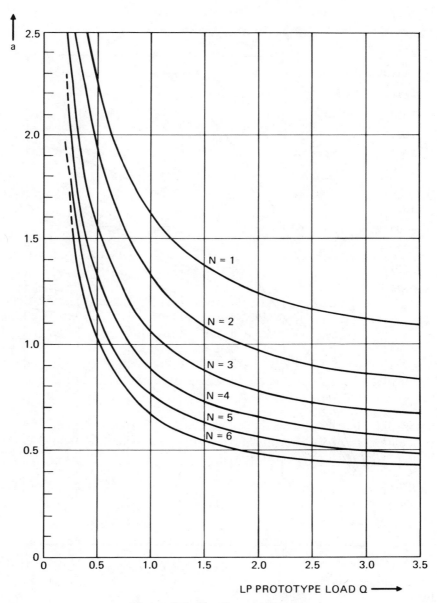

Figure 8.6 Fano's *a* parameter.

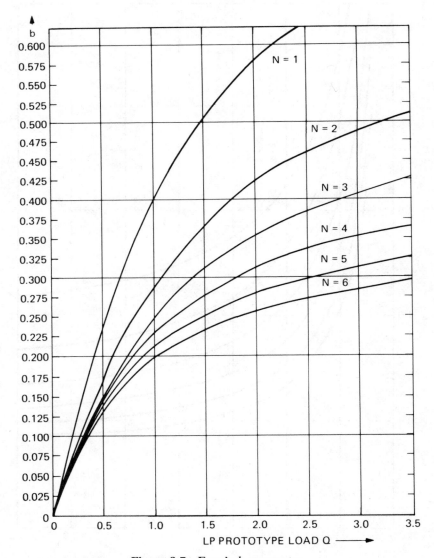

Figure 8.7 Fano's *b* parameter.

$$g_0 = 1 \tag{23}$$

$$g_1 = \frac{2 \sin[\pi/(2N)]}{\sinh[a] - \sinh[b]} \tag{24}$$

$$g_i g_{i+1} = \frac{4 \sin[\pi(2i-1)/(2N)] \sin[\pi(2i+1)/(2N)]}{\sinh^2[a] + \sinh^2[b] + \sin^2[\pi i/N] - 2 \sinh[a] \sinh[b] \cos[\pi i/N]} \tag{25}$$

$$g_N g_{N+1} = \frac{2 \sin[\pi/(2N)]}{\sinh[a] + \sinh[b]} \tag{26}$$

At this point, we note that $g_0 g_1$ and $g_N g_{N+1}$ are load and source Q, respectively. For convenience, they will be denoted Q_L and Q_S. It should be noted that Eqs. (24) and (26) can be solved simultaneously for Fano's a and b when Q_S and Q_L are both specified. This provides a means of designing interstage matching networks, as seen in Eqs. (27) and (28):

$$a = \sinh^{-1}\left\{ \sin\left[\frac{\pi}{2N} \right]\left(\frac{1}{Q_L} + \frac{1}{Q_S} \right) \right\} \tag{27}$$

$$b = \sinh^{-1}\left\{ \sin\left[\frac{\pi}{2N} \right]\left(\frac{1}{Q_S} - \frac{1}{Q_L} \right) \right\} \tag{28}$$

From a substitution of Eqs. (20) and (21) into Eq. (17), we obtain Eq. (29). By evaluating this expression at $T_N = 0$ and $T_N = 1$, expressions for the minimum and maximum reflection coefficients are obtained for this Nth-order case. These match limit results, which are seen in Eqs. (30) and (31), complement the previously discussed Bode limit.

$$|\rho|^2 = \frac{\sinh^2[Nb] + T_N^2}{\sinh^2[Na] + T_N^2} \tag{29}$$

$$|\rho|_{\min} = \frac{\sinh[Nb]}{\sinh[Na]} \tag{30}$$

$$|\rho|_{\max} = \frac{\cosh[Nb]}{\cosh[Na]} \tag{31}$$

A LOWPASS MATCH EXAMPLE

Before moving on to the more important consequences of the preceding LP formulation, an example of a LP matching network is in order.

Suppose we wish to obtain a baseband (video) match from dc to 10 MHz into a parallel load that is comprised of a 1000 pF capacitor and a 31.8 Ω resistor. Use a third-order matching network.

Match Limits: $Q_L = (2\pi 10^7)(10^{-9}\ F)(31.8\ \Omega) = 2.0$

$$\text{Rho}_{\text{Bode}} = \exp\left(\frac{-\pi}{2}\right) = 0.208$$

$$\text{VSWR}_{\text{Bode}} = 1.5{:}1$$

$$\text{VSWR}_{\min} = 1.66{:}1$$

$$\text{VSWR}_{\max} = 1.90{:}1$$

Fano Parameters: $a = 0.783$

$$b = 0.358$$

LP Prototype: $g_0 = 1.0\ \Omega$

$$g_1 = 2.0\ \text{F}$$

$$g_2 = 0.760\ \text{H}$$

$$g_3 = 1.351\ \text{F}$$

$$g_4 = 0.602\ \Omega$$

Denormalized LP Prototype: Frequency scale by $2\pi\ 10^7$

Impedance scale by 31.8

Hence, the LP matching network shown in Fig. 8.8 is obtained:

(From g_0) $R_L = 31.8\ \Omega$

(From g_1) $C_1 = 1000\ \text{pF}$

(From g_2) $L_1 = 0.385\ \mu\text{H}$

(From g_3) $C_2 = 677\ \text{pF}$

(From g_4) $R_s = 19.12\ \Omega$

Note that the desired load model was forced. This occurred through the selection of the Fano parameters and LHP zero placement, which ensured

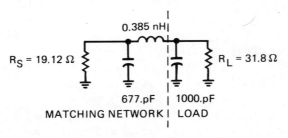

Figure 8.8 LP match example.

optimum performance with this load-Q. Unfortunately, the LP matching procedure does not provide a means of adjusting the source resistance, since transformers (and transformer equivalent circuits) do not function at dc. The networks obtained by the frequency transformation in the next section do not share this limitation.

8.4.2 Mappings

When impedance matching down to dc is not required, as is usually the case, a number of frequency transformations can be applied to the LP formulation. Three mappings are considered in this section: the bandpass BP, the degree doubling quasi-lowpass QLP, and the degree doubling quasi-highpass QHP.

8.4.2.1 Lowpass-to-Bandpass Transformation. Perhaps the most well-known frequency transformation is the LP to BP. It is illustrated in Fig. 8.9, and defined in Eq. (32), where S and p are the LP and BP complex frequencies, respectively. The transform Q, Q_T, is defined in Eq. (33), with passband corner frequencies f_1 and f_2.

$$S = Q_T\left[\frac{\omega_0}{p} + \frac{p}{\omega_0}\right] \tag{32}$$

$$Q_T = \frac{\sqrt{f_2 f_1}}{f_2 - f_1} \tag{33}$$

Reflection functions, pole and zero locations, and transfer functions can all be transformed to BP form with Eq. (32). However, the most surprising feature of this mapping is that it allows network conversions on an element-by-element impedance basis! Consider an inductor of LP complex impedance SL. Two impedance terms result when Eq. (32) is applied: $(Q_T\omega_0 L)/p + (Q_T L/\omega_0)p$. The first term appears to be a capacitor equal to $1/(Q_T\omega_0 L)$, while the second term behaves as an inductor equal to $(Q_T L/\omega_0)$.

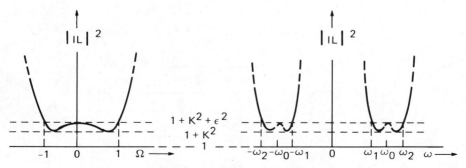

Figure 8.9 LP to BP mapping.

Hence, a series LC branch results from a LP inductor. Similarly, a capacitor of LP complex admittance SC will yield two terms under this mapping. The first will appear as an inductor equal to $1/(Q_T \omega_0 C)$. In parallel with this inductor is a capacitor equal to $(Q_T C/\omega_0)$, contributed by the second susceptance term. So, LP capacitors map into parallel LC branches. The inductors in the series resonators and the capacitors in the parallel resonators are obtained numerically by frequency scaling the corresponding LP prototype elements by the desired bandwidth, in radian frequency. The associated branch element is obtained by setting the resonance at midband (geometric center). Figure 8.10 illustrates the possible LP prototype to BP network relationships.

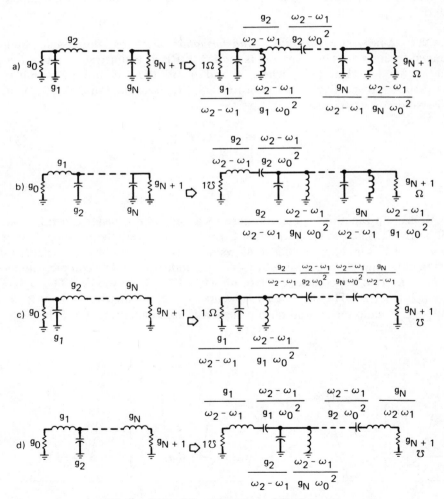

Figure 8.10 LP prototype to BP elements.

Since the requirements (of Q_L and impedance level) present at the load side of the bandpass network necessary to force the desired load model usually do not provide the desired source resistance, additional transformer action is needed. Clearly, it can now be seen that broadband matching is comprised of two distinct actions: reactance absorption and impedance transformation. Usually, both are required for a complete impedance match. The need for transformers is satisfied by Norton [10] subcircuits, discussed next.

The Norton transformer subnetworks are comprised of an ideal transformer cascaded with a pair of series- and shunt-connected inductors or capacitors. Figure 8.11 illustrates the subnetworks that will be considered. They are realized with equivalent "T" or "pi" structures. The "T" equivalents are obtained by coefficient matching Z-parameter representations. Similarly, coefficient matching in Y-parameter form yields the pi structures.

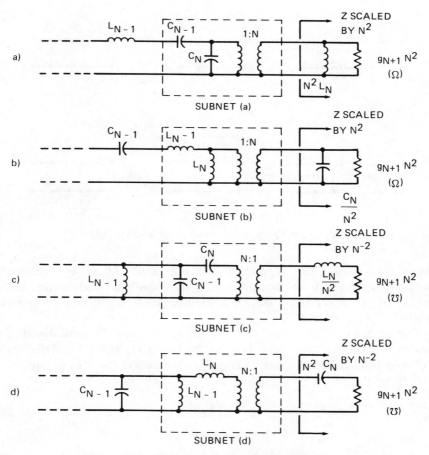

Figure 8.11   Norton transformer subnetworks.

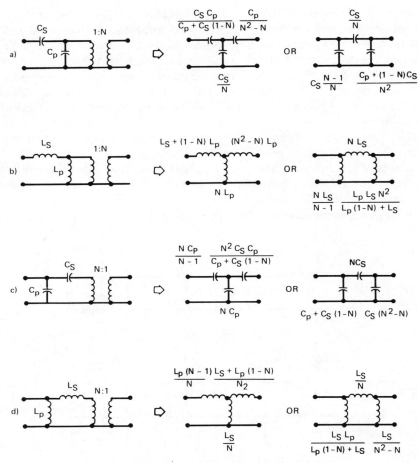

Figure 8.12   Norton transformer equivalents.

These equivalences are represented in Fig. 8.12. These are frequency-invariant representations. So, they are truly broadband equivalents, unlike the "J" or "K" inverters, which are sometimes used in narrow and moderate bandwidth filter designs.

The limit to the available transformer action must be considered. The maximum effective turns ratio is given by Eqs. (34) and (35). Attempts to go beyond this limit will result in negative elements in the T or pi equivalent circuits.

$$N_{\max} = \frac{C_S + C_P}{C_S} \tag{34}$$

$$N_{\max} = \frac{L_S + L_P}{L_P} \tag{35}$$

Since contiguous LP prototype elements yield pairs of inductors and capacitors in the BP network, it is useful to relate $N_{\max}$ to the relevant LP prototype elements ($g_i g_{i+1}$). This is expressed in

$$N_{\max} = 1 + Q_T^2 g_i g_{i+1} \tag{36}$$

If one is willing to sacrifice some match performance in order to minimize the required number of network elements, Eq. (36) can be used. It is possible, by deviating from the Fano solution, to obtain adjacent LP prototype elements ($g_i g_{i+1}$) that simultaneously allow the necessary react-ance absorption (g_1) and provide precisely the required $N_{\max}$ to complete the design. These bandpass forms are called minimum element [11] BP matching networks and are obtained by combining computer-aided synthesis with nonlinear programming.

Several practical BP matching network examples will be presented in Section 8.4.4.

8.4.2.2 Lowpass-to-Quasi-lowpass Transformation.
Since practical mi-crowave designs must behave as BP structures, it is natural also to seek LP or highpass (HP) configurations which behave as "pseudobandpass" struc-tures. The HP and LP network forms of these pseudobandpass structures are often called quasi-highpass (QHP) and quasi-lowpass (QLP) networks. Varieties of frequency-variable transformations that address this need are available in the literature. Some provide the same network order in the transformed variable as in the LP domain, as in the work done by Christian and Eisenmann [12]. Perhaps the most well-known degree doubling QLP transformation was used by Matthaei [13] to form Chebyshev transformer networks. This work was later extended by Cottee and Joines [14] to permit prescribed reactance absorption, by allowing nonzero offset (flat-loss) in the response. The Butterworth QLP response approximation has also employed [15].

The QLP mapping will be considered in this section. It is illustrated in Fig. 8.13, and defined in Eq. (37), where S and p are the LP and QLP

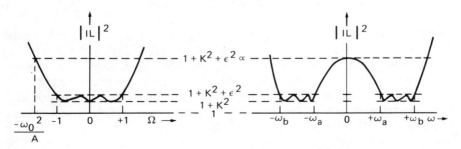

Figure 8.13 LP to QLP mapping.

complex frequencies, respectively. The band corner frequencies are f_1 and f_2.

$$S = \frac{-j(p^2 + \omega_0^2)}{A} \tag{37}$$

where

$$\omega_0^2 = \frac{\omega_a^2 + \omega_b^2}{2}, \qquad A = \frac{\omega_a^2 - \omega_b^2}{2},$$

$$\omega_a = \frac{2f_1}{f_2 + f_1}, \text{ and } \omega_b = \frac{2f_2}{f_2 + f_1}.$$

Reflection functions, pole and zero locations, and transfer functions can all be transformed to QLP form with Eq. (37). However, unlike BP mapping, network conversion on an element-by-element impedance basis is unsuccessful. Synthesis of this type of network can be accomplished by mapping LP reflection poles and zeros into QLP roots. The QLP reflection function can then be formed by expansion of the pole and zero factors. By Eq. (22), the driving-point impedance function can then be formed. Finally, the network is obtained from Eq. (22) by continued fractions. Denormalization is performed with respect to the low impedance port impedance and the arithmetic mean frequency, $(f_2 + f_1)/2$.

In the QLP synthesis procedure, just described, there was no step comparable to the Norton transformer insertion seen in BP design. How is the impedance transformation specified? The answer can be seen in Fig. 8.13, when the dc insertion loss is considered. Since, at dc, the resistive terminations are connected directly, the source-to-load impedance transformation can be controlled by adjusting the mismatch loss there. In so doing, a degree of freedom is lost. The response offset (flat loss) cannot be specified independently of the response ripple. As in Matthaei's case, with zero offset, a unique ripple solution was specified. As with Cottee and Joines, ripple and offset were constrained together uniquely to meet the design requirements. The linking relationship is given in Eq. (38), where r is the desired transformation ratio and N is the desired QLP network order:

$$\epsilon^2 = \left[\frac{(r-1)^2}{4r} - K^2 \right] / \cosh^2 \left\{ 2N \cosh^{-1} \left[\frac{\omega_0^2}{A} \right] \right\} \tag{38}$$

Nonzero offset is a result of reflection zeros that are not on the imaginary axis. This raises the issue of optimum placement (RHP or LHP) of the zeros. With this type of network, LHP zero placement provides maximum reactance at the low impedance (normalized) port. Similarly, RHP zero placement provides maximum reactance absorption at the high impedance port. In order to determine the match performance and necessary offset, Figs. 8.14–8.18 are provided. In each case, as a function of desired transformation ratio, several families of curves that are indexed by network order N and normalized bandwidth W are plotted. The vertical axis indicates

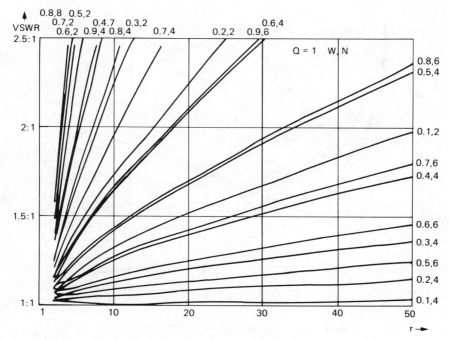

Figure 8.14 QLP reactance absorption, $Q = 1$.

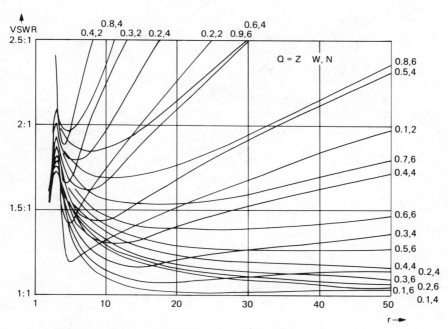

Figure 8.15 QLP reactance absorption, $Q = 2$.

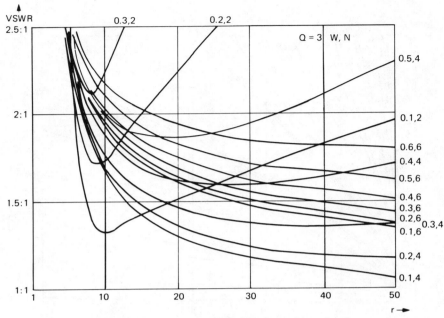

Figure 8.16 QLP reactance absorption, $Q = 3$.

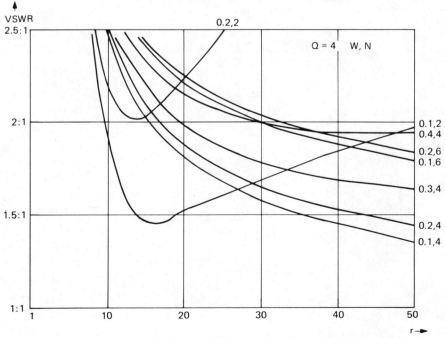

Figure 8.17 QLP reactance absorption, $Q = 4$.

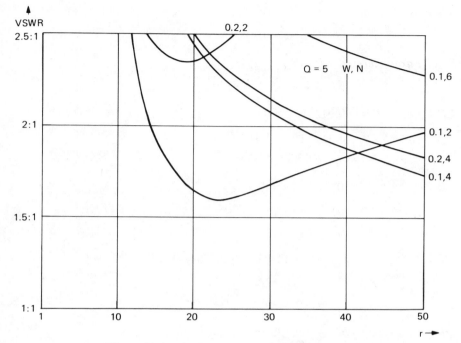

Figure 8.18 QLP reactance absorption, $Q = 5$.

the match performance. The normalized bandwidth is defined by

$$W = \frac{2(f_2 - f_1)}{f_2 + f_1} \tag{39}$$

A QUASI-LOWPASS MATCH EXAMPLE

Before moving on to QHP design, a QLP example is in order.

Suppose we wish to obtain an octave band match from 4 GHz to 8 GHz into a series load that is comprised of a 0.132 nH inductor and a 2.0 Ω resistor. Use a sixth-order matching network. A 50 Ω source is given. So, $r = 25$.

Match Limits: $Q_{\text{QLP}} = \dfrac{(2\pi 6)(10^9)(0.132 \, \text{nH})(10^{-9})}{2.0 \, \Omega} = 2.5$

$Q_{\text{Bode}} = (Q_{\text{QLP}})(W) = (2.5)(0.667) = 1.667$

$\rho_{\text{Bode}} = \exp\!\left(\dfrac{-\pi}{1.667}\right) = 0.152$

$\text{VSWR}_{\text{Bode}} = 1.34{:}1$

$\text{VSWR}_{\text{max}} = 1.7{:}1$ (from Figs. 8.15 and 8.16)

Response Parameters: Offset = 0.19 dB

$\qquad\qquad\qquad\qquad$ Ripple = 0.13 dB

Fano Parameters: $a = 0.818$

$\qquad\qquad\qquad b = 0.339$

LP Poles: $S_{p1} = -0.912815 + j\,0$

$\qquad\qquad S_{p2} = -0.456408 + j\,1.172571$

$\qquad\qquad S_{p3} = -0.456408 - j\,1.172571$

LP Zeros: $S_{z1} = -0.351723 + j\,0$

$\qquad\qquad S_{z2} = -0.172586 + j\,0.916165$

$\qquad\qquad S_{z3} = -0.172586 - j\,0.916165$

QLP Poles: $p_{p1}, p_{p1}^{*} = -0.279045 \pm j\,1.090402$

$\qquad\qquad p_{p2}, p_{p2}^{*} = -0.110227 \pm j\,1.380208$

$\qquad\qquad p_{p3}, p_{p3}^{*} = -0.243953 \pm j\,0.623627$

QLP Zeros: $P_{z1}, p_{z1}^{*} = -0.108579 \pm j\,1.059670$

$\qquad\qquad P_{z2}, p_{z2}^{*} = -0.043817 \pm j\,1.312939$

$\qquad\qquad P_{z3}, p_{z3}^{*} = -0.080805 \pm j\,0.711944$

$$\rho = \frac{p^6 + 0.4664p^5 + 3.4421p^4 + 1.0960p^3 + 3.5131p^2 + 0.5599p + 1.0053}{p^6 + 1.2665p^5 + 4.1353p^4 + 3.3118p^3 + 4.5699p^2 + 1.7900p + 1.0891}$$

$$\frac{Z_{dp}}{R_0} = \frac{2p^6 + 1.7329p^5 + 7.5774p^4 + 4.4079p^3 + 8.0830p^2 + 2.3500p + 2.0944}{0.800p^5 + 0.6932p^4 + 2.2160p^3 + 1.057p^2 + 1.2300p + 0.08378}$$

QLP Prototype: $g_0 = 1.0\ \Omega$

$\qquad\qquad g_1 = 2.5\ \text{H}$

$\qquad\qquad g_2 = 0.3925\ \text{F}$

$\qquad\qquad g_3 = 8.1518\ \text{H}$

$\qquad\qquad g_4 = 0.1486\ \text{F}$

$\qquad\qquad g_5 = 17.3982\ \text{H}$

$\qquad\qquad g_6 = 0.04617\ \text{F}$

$\qquad\qquad g_7 = 25.00\ \Omega$

Denormalized QLP Prototype: Frequency scale by $2\pi 6(10^9)$

$\qquad\qquad\qquad\qquad\qquad\qquad$ Impedance scale by 2.0

REFLECTION ZERO PLACEMENT: LHP, LHP, LHP

BANDWIDTH = 66.66 %
LOWER STOPBAND LOSS = 8.2994 dB
OFFSET = 0.192 dB, RIPPLES = 0.128 dB
PASSBAND MAXIMUM LOSS = 0.3192 dB
PASSBAND VSWR (RIPPLE + OFFSET) = 1.725 : 1

QUASI-LOWPASS MATCH

R = 2	Ohm
L = 0.132	NANO-HENRY
C = 5.206	PICO-FARAD
L = 0.432	NANO-HENRY
C = 1.971	PICO-FARAD
L = 0.923	NANO-HENRY
C = 0.612	PICO-FARAD
R = 50	Ohm

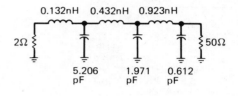

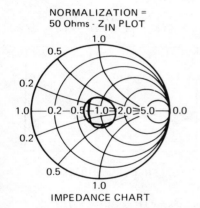

NORMALIZATION =
50 Ohms - Z_{IN} PLOT

IMPEDANCE CHART

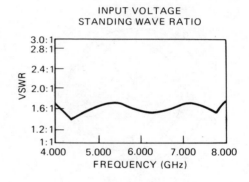

INPUT VOLTAGE
STANDING WAVE RATIO

Figure 8.19   QLP match example: LHP zeros.

Hence, the QLP matching network shown in Fig. 8.19 is obtained:

$$(\text{From } g_0) \qquad R_L = 2.000 \ \Omega$$

$$(\text{From } g_1) \qquad L_1 = 0.132 \ \text{nH}$$

$$(\text{From } g_2) \qquad C_1 = 5.206 \ \text{pF}$$

$$(\text{From } g_3) \qquad L_2 = 0.432 \ \text{nH}$$

$$(\text{From } g_4) \qquad C_2 = 1.971 \ \text{pF}$$

$$(\text{From } g_5) \qquad L_3 = 0.932 \ \text{nH}$$

$$(\text{From } g_6) \qquad C_3 = 0.612 \ \text{pF}$$

$$(\text{From } g_7) \qquad R_s = 50.00 \ \Omega$$

REFLECTION ZERO PLACEMENT: RHP, RHP, RHP

BANDWIDTH = 66.66 %
LOWER STOPBAND LOSS = 8.2994 dB
OFFSET = 0.192 dB, RIPPLES = 0.128 dB
PASSBAND MAXIMUM LOSS = 0.3192 dB
PASSBAND VSWR (RIPPLE + OFFSET) = 1.725 : 1

<u>QUASI-LOWPASS MATCH</u>

R = 2	Ohm
L = 0.061	NANO-HENRY
C = 9.23	PICO-FARAD
L = 0.197	NANO-HENRY
C = 4.324	PICO-FARAD
L = 0.52	NANO-HENRY
C = 1.326	PICO-FARAD
R = 50	Ohm

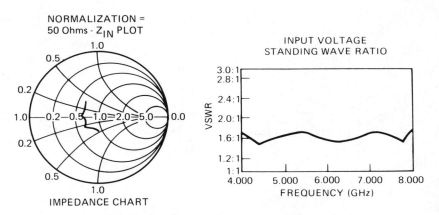

Figure 8.20   QLP match example: RHP zeros.

Note that the desired load model was forced. This occurred through the selection of the Fano parameters and LHP zero placement. If all reflection zeros are symmetrically moved to the RHP, the QLP network in Fig. 8.20 is obtained. Note that the mismatch magnitude performance is identical, but the input port capacitor has doubled and the output port inductor has correspondingly reduced in size! For comparison, consider the Matthaei zero offset QLP network, which results from the same requirements except for prescribed reactance absorption. This case, shown in Fig. 8.21, results when reflection zeros are placed on the imaginary axis. Note that the size of the reactive elements at each port falls between the RHP and LHP cases.

8.4.2.3   Lowpass to Quasi-Highpass Transformation. A pseudobandpass alternative to the QLP is the quasi-highpass (QHP) network. Since realization of QHP designs is quite similar to that of the QLP, this section will be brief.

BANDWIDTH = 66.66%
LOWER STOPBAND LOSS = 8.2994 dB
PASSBAND RIPPLE LOSS = 0.1347 dB
PASSBAND RIPPLE VSWR = 1.423 : 1

QUASI-LOWPASS MATCH

R = 2	Ohm
L = 0.084	NANO-HENRY
C = 7.42	PICO-FARAD
L = 0.29	NANO-HENRY
C = 2.9	PICO-FARAD
L = 0.742	NANO-HENRY
C = 0.849	PICO-FARAD
R = 50	Ohm

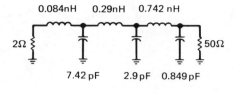

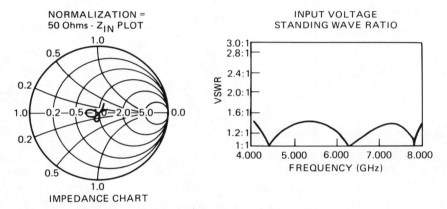

IMPEDANCE CHART

Figure 8.21 QLP match example: no offset.

The QHP mapping to be considered in this section is illustrated in Fig. 22, and defined in Eq. (40), where S and p are the LP and QHP complex frequencies, respectively. The band corner frequencies are f_1 and f_2.

$$S = -j\left[\frac{(\omega_1\omega_2)^2}{Ap^2} + \frac{\omega_0^2}{A}\right] \qquad (40)$$

where $\omega_0^2 = \dfrac{\omega_a^2 + \omega_b^2}{2}$, $A = \dfrac{(\omega_a^2 - \omega_b^2)}{2}$,

$\omega_a = \dfrac{2f_1}{f_2 + f_1}$, and $\omega_b = \dfrac{2f_2}{f_2 + f_1}$.

Like Eq. (37), Eq. (40) does not allow network conversion on an element-by-element impedance basis. However, element-by-element conversion from QLP to QHP can be accomplished with Eq. (41), where P and S are

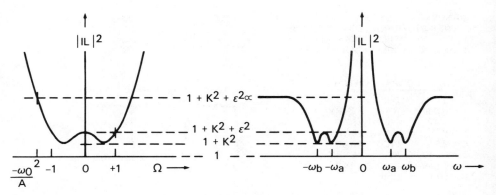

Figure 8.22 LP to QHP mapping.

the QLP and QHP complex frequencies, respectively. Synthesis of this type of network can be accomplished in a manner identical to that of the QLP.

$$P = \frac{\omega_1 \omega_2}{S} \qquad (41)$$

where $\omega_1 = 2\pi f_1$, and $\omega_2 = 2\pi f_2$.

8.4.3 One-port Load Modeling

The unilateral FET model, previously discussed in Section 8.2, represents an embodiment of two simple one-port impedance models. It is well known that intrinsic device impedance behavior can be accurately represented in this manner. However, when feedback or lossy branch gain equalization is employed, a single RC section is an inadequate representation. Sometimes nonunilateral behavior can be strong enough to require a more sophisticated model. Recently, Mellor [16] has derived closed-form expressions for second-order LP and HP one-port impedance models. Second- and fourth-order closed-form BP models have been used successfully by this author [17] for many years. This modeling technique will now be discussed. When the occasional need arises for higher-order models, this method can be easily extended to cover those cases.

In order to provide a form that is compatible with the subsequent synthesis step, all one-port models are required to be two-port LC networks that are resistively terminated. The two topologies that will be considered are illustrated in Fig. 8.23. Each reactance or susceptance block is allowed up to second-order behavior. So, fourth-order models can thereby be realized. To avoid Matthaei's "situation 2," the innermost branch of the model must provide the dominant behavior. The outermost branch provides a perturbational contribution. Therefore, the two forms to be considered can be seen in Fig. 8.24.

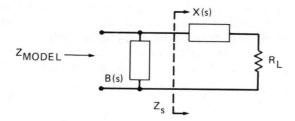

(a) SERIES BEHAVIOR WITH SHUNT PERTURBATION

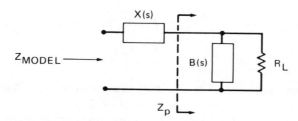

(b) SHUNT BEHAVIOR WITH SERIES PERTURBATION

Figure 8.23   One-port model topologies.

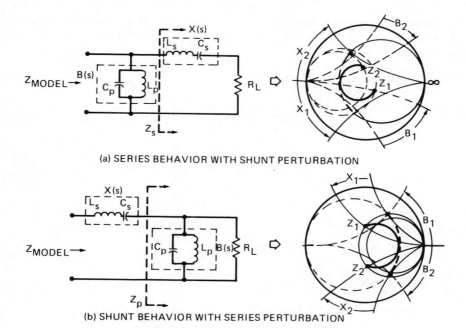

(a) SERIES BEHAVIOR WITH SHUNT PERTURBATION

(b) SHUNT BEHAVIOR WITH SERIES PERTURBATION

Figure 8.24   Fourth-order models.

The *LC* values in the second-order subsections can be determined from the reactance (series case) or susceptance levels at the band corners, f_1 and f_2. These calculations, which result from simultaneously solving two equations in two unknowns, are given in Eqs. (42–45). The reactance and susceptance levels at f_1 and f_2 are X_1, X_2, B_1 and B_2, respectively.

$$L_S = \frac{f_2 X_2 - f_1 X_1}{2\pi(f_2^2 - f_1^2)} \tag{42}$$

$$C_S = \frac{f_2^2 - f_1^2}{2\pi(f_1^2 f_2 X_2 - f_1 f_2^2 X_1)} \tag{43}$$

$$L_P = \frac{f_2^2 - f_1^2}{2\pi(f_1^2 f_2 B_2 - f_1 f_2^2 B_1)} \tag{44}$$

$$C_P = \frac{f_2 B_2 - f_1 B_1}{2\pi(f_2^2 - f_1^2)} \tag{45}$$

The fourth-order models seen in Fig. 8.24 are obtained by a decomposition into second-order subnetworks and solution of Eqs. (42–45). To see this decomposition, consider Fig. 8.25. The closed-form solution for the reactance and susceptance shifts (X_1, X_2, B_1, and B_2) are given in Eqs. (46–53). Since the end points of the locus are fixed, the value of the terminal resistance adjusts the amount of curvature. For series-dominant behavior, R_L is usually set to the lowest value that is encountered in the real part of the data to be modeled. Similarly, shunt-dominant behavior usually requires R_L to be set to the largest value encountered in the real part of the data to be modeled.

Shunt with Series Perturbation

$$B_1 = \frac{A\sqrt{[R_L/R_A - 1]}}{R_L} \tag{46}$$

$$X_1 = X_A + AR_A\sqrt{\frac{R_L}{R_A - 1}} \tag{47}$$

$$B_2 = \frac{B\sqrt{[R_L/R_B - 1]}}{R_L} \tag{48}$$

$$X_2 = X_B + BR_B\sqrt{\frac{R_L}{R_B - 1}} \tag{49}$$

where $Z(f_1) = R_A + jX_A$, $Z(F_2) = R_B + jX_B$,

$A = +1$ if $\text{Imag}\{Z_p(f_1)\} \leq 0$, $A = -1$ if $\text{Imag}\{Z_p(f_1)\} > 0$,

$B = +1$ if $\text{Imag}\{Z_p(f_2)\} \leq 0$, and $B = -1$ if $\text{Imag}\{Z_p(f_2)\} > 0$.

IMPEDANCE COORDINATES

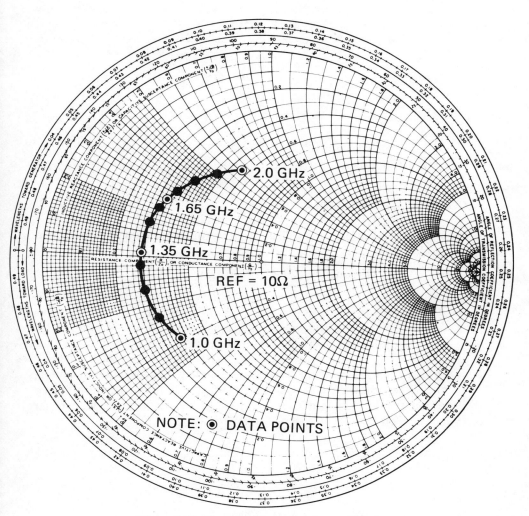

REF = 10Ω

NOTE: ⊙ DATA POINTS

Figure 8.25 Impedance data to be modeled.

Series with Shunt Perturbation

$$X_1 = AR_L \sqrt{\frac{R_A(1 + Q_1^2)}{R_L - 1}} \tag{50}$$

$$B_1 = \frac{A\sqrt{[R_A(1 + Q_1^2)/R_L - 1]}}{R_A(1 + Q_1^2)} - \frac{Q_1^2}{X_A(1 + Q_1^2)} \tag{51}$$

$$X_2 = BR_L \sqrt{\frac{R_B(1 + Q_2^2)}{R_L - 1}} \tag{52}$$

$$B_2 = \frac{B\sqrt{[R_B(1 + Q_2^2)/R_L - 1]}}{R_B(1 + Q_2^2)} - \frac{Q_2^2}{X_B(1 + Q_2^2)} \tag{53}$$

where $Z(f_1) = R_A + jX_A$, $\quad Z(f_2) = R_B + jX_B$,

$$Q_1 = \frac{|X_A|}{R_A}, \qquad Q_2 = \frac{|X_B|}{R_B},$$

$A = -1$ if $\mathrm{Imag}\{Z_S(f_1)\} \le 0$, $\quad A = +1$ if $\mathrm{Imag}\{Z_S(f_1)\} > 0$,

$B = -1$ if $\mathrm{Imag}\{Z_S(f_2)\} \le 0$, and $B = -+1$ if $\mathrm{Imag}\{Z_S(f_2)\} > 0$.

A LOAD MODEL EXAMPLE

As a numerical example, we will model the following data with a fourth-order model:

Frequency (GHz)	R_{Load}	X_{Load}
1.00	4.0	−4.0
1.35	3.0	0.0
1.65	3.5	3.0
2.00	6.0	7.0

As seen in Fig. 8.25, the behavior to be modeled is predominantly series-resonant. So, we choose the series model with the shunt perturbation. Since the minimum value seen in the real part of the data is $3.0\,\Omega$, we set $R_L = 3.0\,\Omega$. From Eqs. (50–53), we obtain

$$X_1 = -3.873$$

$$B_1 = -0.0364$$

$$X_2 = 5.788$$

$$B_2 = 0.0538$$

The *LC* element values within the model are then obtained from Eqs. (42–45). The model and modeled results can be seen in Fig. 8.26.

8.4.4 A Two-stage Power Amplifier Design Example

In this section, we will put to use the circuit design procedures that have been detailed in the previous sections of this chapter. As an illustrative

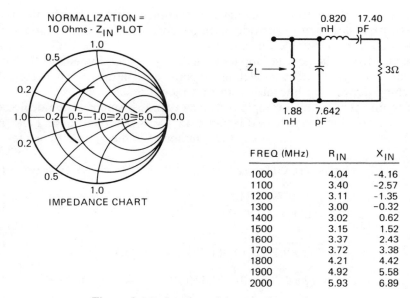

FREQ (MHz)	R_{IN}	X_{IN}
1000	4.04	-4.16
1100	3.40	-2.57
1200	3.11	-1.35
1300	3.00	-0.32
1400	3.02	0.62
1500	3.15	1.52
1600	3.37	2.43
1700	3.72	3.38
1800	4.21	4.42
1900	4.92	5.58
2000	5.93	6.89

Figure 8.26   Load model example results.

vehicle, we will design a two-stage $\frac{1}{2}$ W amplifier for operation between 8 and 12 GHz. Therefore, three matching network designs are required. The FET model that will serve as a basis for this example is the Hughes TRC-4080. This is an X-band geometry with a gate length of 0.8 µm. The large-signal unilateral model for a total gate width of 1 mm is given in Fig. 8.27. Similarly, a 330 µm driver device model is shown as a 3:1 impedance scaling.

The output port impedance behavior of this model represents the complex conjugate of the optimum power load, rather than the FET source impedance strictly. This is because the optimum power match impedance

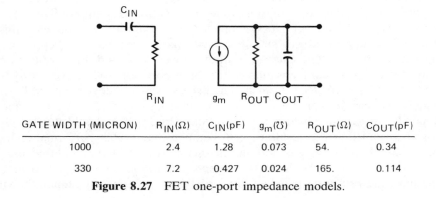

GATE WIDTH (MICRON)	$R_{IN}(\Omega)$	$C_{IN}(pF)$	$g_m(\mho)$	$R_{OUT}(\Omega)$	$C_{OUT}(pF)$
1000	2.4	1.28	0.073	54.	0.34
330	7.2	0.427	0.024	165.	0.114

Figure 8.27   FET one-port impedance models.

requirements differ from small-signal reflection match requirements. When a network is designed to provide an impedance match with this model, that network will also provide an optimum power load for the FET. From this example, it should now be apparent that any desired impedance behavior can be approximated by modeling its complex conjugate and designing a matching network to interface with that model.

The first step in this example is to make an assessment of the device Q's and corresponding match performance in the 8–12 GHz band. Fano networks are used in this example; thus, midband is the geometric center.

$$f_0 = \sqrt{(8\ \text{GHz})(12\ \text{GHz})} \qquad\qquad = 9.798\ \text{GHz} \qquad (64)$$

$$Q_T = \frac{9.798\ \text{GHz}}{12\ \text{GHz} - 8\ \text{GHz}} \qquad\qquad = 2.44949 \qquad (55)$$

$$Q_{In} = \frac{1}{(2\pi 9.798)(10^9)(1.28)(10^{-12})(2.4)} = 5.2877 \qquad (56)$$

$$Q_{Out} = (2\pi 9.798)(10^9)(0.34)(10^{-12})(54) \ = 1.1635 \qquad (57)$$

From these Q's we can calculate the Bode match limits. To do this, the bandpass Q must transformed into its equivalent LP Q with Eq. (58):

$$Q_{LP} = \frac{Q_{BP}}{Q_T} \qquad (58)$$

$$|\rho|_{\text{Bode-in}} = \exp\left(\frac{-\pi 2.4495}{5.2877}\right) = 0.233 \qquad (59)$$

$$|\rho|_{\text{Bode-out}} = \exp\left(\frac{-\pi 2.4495}{1.1635}\right) = 0.0013 \qquad (60)$$

Clearly, the output match performance appears to be quite good. From (59), the input performance does not look attractive. However, since microwave FET devices exhibit $-6\ \text{dB/octave}$ roll-off in $|S_{21}|$ with increasing frequency, some form of amplitude equalization is required. This could be achieved by feedback, lossy branch compensation, or controlled mismatch. For superior power amplifier performance, feedback and controlled mismatch are rejected due to output power loss and high reflection, respectively. Lossy compensation of the input port can most conveniently be applied with a shunt RLC branch. The excess available device gain at lower frequencies can be compensated for by applying the proper LC reactance slope. From gain-bandwidth considerations (Bode), we should not be surprised to find that the resultant gain compensated match performance has improved. Another way of looking at this is as a frequency tailored input "de-Qing." Figure 8.28 shows the compensated input models. When these one ports are reflection-matched, a properly gain-compensated amplifier will

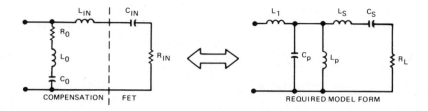

Figure 8.28 Lossy branch compensation input models.

GATE WIDTH	R_{IN}	C_{IN}	L_{IN}	R_0	L_0	C_0	L_1	C_p	L_p	C_S	L_S	R_L
1000 MICRON	2.4Ω	1.28pF	0.137nH	21Ω	0.25nH	3.0pF	−0.0516nH	0.251pF	0.274nH	1.597pF	0.167nH	2.851Ω
330 MICRON	7.2Ω	0.427pF	0.412nH	65Ω	0.75nH	1.0pF	−0.1549nH	0.0843pF	0.822nH	0.523pF	0.501nH	8.55Ω

result. For matching network design by the techniques already discussed, these compensated input ports must be remodeled into a proper form: an *LC* two-port that is terminated with a single resistor. These alternative representations are also illustrated in Fig. 8.28. As a result, the input BP Q is 3.588. The corresponding Bode match reflection limit is now 0.117.

The input, interstage, and output matching networks can now be designed. A sixth-order network (including the model absorbed) will be used at the amplifier input. At the interstage, an eight-order model is needed. And due to the relatively low-output Q, only a fourth-order output matching network is required.

Input Match

$$N = 6$$

$$Q_{\text{Load}} = 3.588$$

$$a = 0.882$$

$$b = 0.313$$

$$\text{VSWR}_{\text{Bode}} = 1.27{:}1$$

$$\text{VSWR}_{\text{max}} = 1.53{:}1$$

$$\text{VSWR}_{\text{min}} = 1.37{:}1$$

$$g_0 = 1.0$$

$$g_1 = 1.465$$

$$g_2 = 0.890$$

$$g_3 = 1.036$$

$$g_4 = 0.733$$

The bandpass transformed input matching network is depicted in Fig. 8.29a. An inductive Norton transformer is then inserted to obtain the 50 Ω input port (Fig. 8.29b). Input match performance is shown in Fig. 8.29c.

Interstage Match

$$N = 8$$

$$Q_{\text{Source}} = 1.164$$

$$Q_{\text{Load}} = 3.588$$

$$a = 0.928$$

$$b = 0.521$$

$$\text{VSWR}_{\text{max}} = 1.50{:}1$$

$$\text{VSWR}_{\text{min}} = 1.48{:}1$$

$$g_0 = 1.0$$

$$g_1 = 1.465$$

$$g_2 = 0.867$$

$$g_3 = 1.617$$

$$g_4 = 0.317$$

$$g_5 = 1.497$$

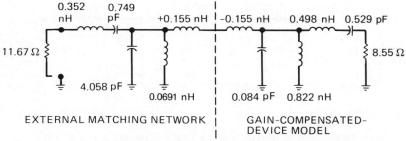

a) BP TRANSFORMED INPUT MATCHING NETWORK/MODEL

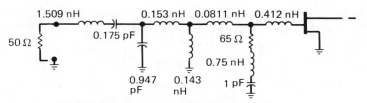

b) COMPLETED INPUT NETWORK, AFTER NORTON
TRANSFORMER INSERTION

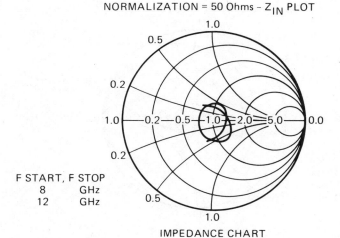

c) INPUT MATCH PERFORMANCE

Figure 8.29 Input matching network.

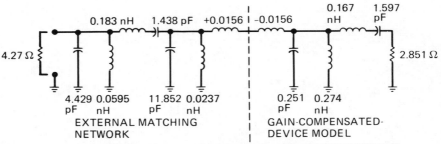

a) BP TRANSFORMED INTERSTAGE MATCHING NETWORK/MODEL

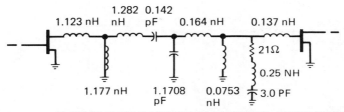

b) COMPLETED INTERSTAGE NETWORK, AFTER NORTON
TRANSFORMER INSERTION

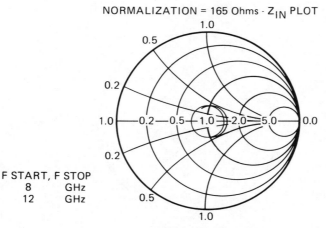

c) INTERSTAGE MATCH PERFORMANCE

Figure 8.30   Interstage matching network.

The bandpass transformed interstage matching network is illustrated in Fig. 8.30a. Two inductive Norton transformers are then inserted to obtain the 165 Ω and 0.114 pF source (FET drain) model (Fig. 8.30b). Interstage match performance is shown in Fig. 8.30c. This represents the power match provided to the driver stage, since a large-signal output model was used.

Output Match

$$N = 4$$

$$Q_{Source} = 1.164$$

$$a = 0.928$$

$$b = 0.521$$

$$VSWR_{Bode} = 1.00:1$$

$$VSWR_{max} = 1.11:1$$

$$VSWR_{min} = 1.03:1$$

$$g_0 = 1.0$$

$$g_1 = 0.475$$

$$g_2 = 0.388$$

$$g_3 = 1.107$$

The bandpass transformed output matching network is illustrated in Fig. 8.31a. Since the topology here is dictated by the necessity of a parallel device model, the only available Norton transformation is downward (with this $N = 4$ case). Upward transformation is desired; so, the order could be increased to $N = 6$. Since the external load is 50 Ω to within a VSWR of 1.03:1, no Norton will be used. The power match provided by this circuit is can be seen in Fig. 8.31b.

The overall two-stage amplifier circuit and response are shown in Fig. 8.32. We have systematically and optimally obtained this design. It is important to note that no numerical optimization was used with this procedure. Monolithic realization of this lumped design is straightforward with quasi-lumped or distributed elements. The conversion of lumped designs into MMIC-compatible form is discussed in the next section.

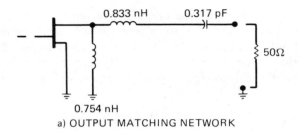

a) OUTPUT MATCHING NETWORK

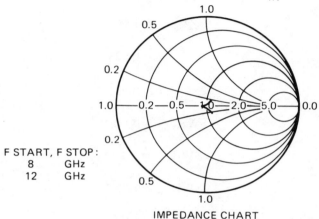

IMPEDANCE CHART

b) OUTPUT MATCH PERFORMANCE

Figure 8.31   Output matching network.

8.5   MMIC CIRCUIT ELEMENTS

This section addresses the realization of passive circuit elements in a monolithically integrated form. We will begin this discussion with a bold statement: *"Ideal lumped elements do not exist!"* Distributed effects are always present. "Lumped" inductors and capacitors will always self-resonate at some frequency. Similarly, reactive effects can always be found in resistors. A key to achieving good equivalent lumped-element performance is small element size compared with the operating wavelength.

8.5.1   Transmission-line Element Approximations

A short segment of transmission line can be used to approximate an inductor or capacitor, depending on its characteristic impedance and how it is used (interconnection topology). For purposes here, we will consider lossless quasi-TEM line representations. Lossy lines can easily be substi-

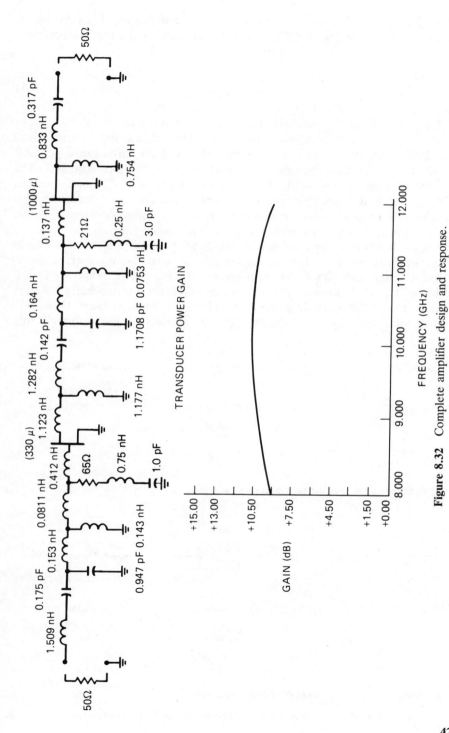

Figure 8.32 Complete amplifier design and response.

423

tuted. The quasi-TEM assumption is good, since microstrip lines are by far the preferred media for MMICs. On an incremental basis, the characteristic impedance Z_0 of the line is given by

$$Z_0 = \sqrt{\frac{L}{C}} \qquad (61)$$

In Eq. (61), L and C are the incremental series inductance and shunt capacitance per unit length, respectively. The higher the characteristic impedance, the more series inductance and less shunt capacitance per unit length are seen. Lumped inductor behavior can be approximated if a sufficiently large Z_0 is used. Alternatively, from low Z_0 lines, lumped approximations for capacitors can be obtained. The conversion relationships are obtained by coefficient matching Y- or Z-parameter representations of transmission-line segments with those of third-order lowpass LC models. The equivalent forms are depicted in Fig. 8.33. Clearly, a series inductor is represented by the pi-section of Fig. 8.33a. In a similar manner, a short-circuited shunt stub can be used to approximate a shunt inductor. The T section of Fig. 8.3b allows a practical shunt capacitor approximation. Series capacitors are a bit more difficult where microstrip lines are used. However, the "lumped"-overlay-type capacitor can be modeled quite well with this

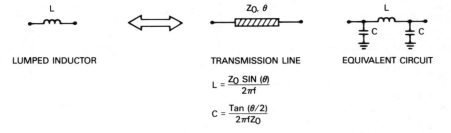

LUMPED INDUCTOR   TRANSMISSION LINE   EQUIVALENT CIRCUIT

$$L = \frac{Z_0 \, \text{SIN} \, (\theta)}{2\pi f}$$

$$C = \frac{\text{Tan} \, (\theta/2)}{2\pi f Z_0}$$

a). TRANSMISSION LINE INDUCTOR APPROXIMATION

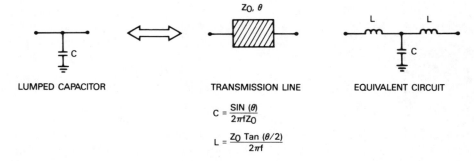

LUMPED CAPACITOR   TRANSMISSION LINE   EQUIVALENT CIRCUIT

$$C = \frac{\text{SIN} \, (\theta)}{2\pi f Z_0}$$

$$L = \frac{Z_0 \, \text{Tan} \, (\theta/2)}{2\pi f}$$

b). TRANSMISSION LINE CAPACITOR APPROXIMATION

Figure 8.33 Transmission-line equivalents for lumped elements.

method, as will be discussed in the next section. Microstrip lines offer characteristic impedances between 15 and 100 Ω on 75–100 μm GaAs substrates. Good lumped equivalent performance can be obtained when transmission-line length is less than 30°.

8.5.2 Lumped Capacitors

Monolithic circuit realization of lumped capacitors is commonly fabricated with several geometries. The most often used form is called the "overlay." To a much lesser extent, the so-called "interdigitated" geometry is employed. Most contemporary designs use the overlay exclusively, since it offers a lower shunt parasitic capacitance, a much wider range of practical realizations, and more compact size. Therefore, our discussion will be directed to this geometry. For more information about interdigital capacitors, see Alley [18] and Esfandiari [19]. When either of the capacitor forms is used as a series of dc blocking elements, a parasitic shunt capacitance is also seen. This is due to the electric field path through the GaAs substrate. Of course, if one side of the capacitor is to be grounded, then this additional capacitance can easily be included into the capacitor design.

Overlay capacitors are fabricated by depositing a thin-film dielectric layer between thin-film metal plates. This is illustrated in Fig. 8.34. Typical dielectric layer thicknesses are between 0.2 and 0.25 μm. Common dielectric materials are silicon nitride (Si_3N_4), silicon dioxide (SiO_2), and tantalum pentoxide (Ta_2O_5). When SiO_2 or $Ta2O_5$ is used, the result is sometimes called a metal–oxide–metal (MOM) capacitor. Dielectric constant properties are as follows:

Material	Dielectric Constant	Temperature Coefficient
Si_3N_4	6–7	25–35
SiO_2	4–5	100–500
Ta_2O_5	20–25	0–200

High-quality MMIC capacitors must exhibit low microwave energy loss, high breakdown field capability, capacitance stability with temperature, and good film integrity (low pinhole density and stability). Clearly, the dielectric film plays a central role in determining the performance in each of these categories.

Conductor losses are also important in determining the microwave Q-factor. This can be seen in the overlay capacitor models of Fig. 8.35.

Two modes must be allowed, since the overlay structure is suspended over a second dielectric layer (the GaAs substrate). The odd-mode characteristic impedance is equivalent to that of a microstrip line of the same width and half the dielectric thickness as the capacitor. The even-mode characteristic impedance is set by the capacitance of the microstrip mode through the GaAs substrate. When the coupled line model of Fig. 8.35a is converted

Figure 8.34   Scanning electron micrograph (SEM) of MMIC overlay capacitor.

Fig. 8.35a is converted into a lumped approximation, the model forms of Fig. 8.35b and c are obtained. The transmission-line representation method which was discussed in Section 8.5.1 is applicable here if lossy lines are used. The applicable equations are

$$C_1 = C_3 - \text{Parallel plate substrate capacitance}$$

$$C_2 = \frac{\sin(\Theta)/(\omega Z_{0o}) - \sin(\Theta)/(\omega Z_{0e})}{2}$$

$$C_3 = \frac{\sin(\Theta)}{\omega Z_{0e}}$$

$$L_S = \frac{Z_{0o} \tan(\Theta/2)}{\omega}$$

$$R_S = \frac{Z_{0o} \tan(\Theta/2)}{Q_{\text{even}}}$$

$$R_P = \frac{2Z_{0o} Q_{\text{odd}}}{\sin(\Theta)}$$

It is interesting to note that accurate modeling of "lumped" capacitors has necessitated a distributed model.

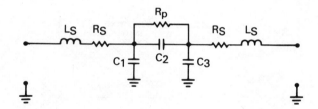

a). **ASSYMETRIC BROADSIDE COUPLED LINE MODEL**

b). **LUMPED EQUIVALENT OF BROADSIDE COUPLED LINE MODEL**

c). **SIMPLIFIED LUMPED EQUIVALENT MODEL**

Figure 8.35 Circuit models for MMIC overlay capacitors.

8.5.3 Lumped Inductors

When relatively large inductance values are required, uncoupled transmission lines of the form described in Section 8.5.1 are not always useful. This limitation is set by the maximum practical realizable microstrip characteristic impedance. The rectangular spiral configuration shown in Fig. 8.36a offers a means of exceeding the uncoupled-line inductance limit. Additionally, more efficient utilization of MMIC surface area results, when compared with meander-line or S-line ("uncoupled") configurations. The notation for the present discussion is also indicated in Fig. 8.36a. The horizontal lines are labeled H_x and the vertical lines labeled V_x.

Two modeling approaches for rectangular spiral inductors will now be discussed. The modified Grover [20] method used by Greenhouse [21] is

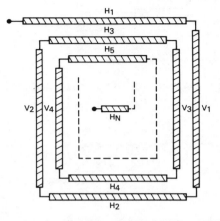

a). RECTANGULAR "SPIRAL" CONFIGURATION/NOTATION

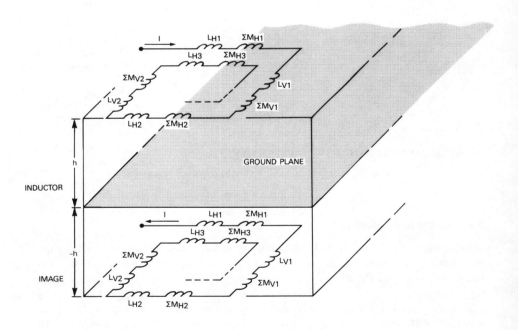

b). GREENHOUSE/GROVER MODEL

Figure 8.36 Rectangular spiral inductor circuit models.

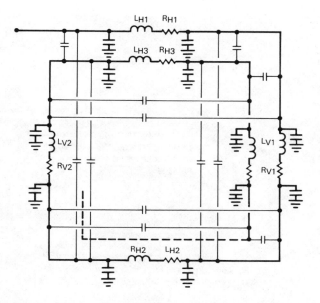

c). **LUMPED EQUIVALENT REPRESENTATION OF TRANSMISSION LINE MODEL FOR RECTANGULAR "SPIRAL"**

Figure 8.36 continued

illustrated in Fig. 8.36b. Only magnetic interactions are considered in this model. Ground-plane effects are equivalently represented by an antiphased image, separated by a distance of twice the ground-plane spacing. The total inductance is equal to the sum of the segment self-inductances added to the sum of the mutual inductive contributions. These effects are shown separately in Fig. 8.36b for each inductor segment. For MMIC applications, the self-inductances are approximately given by

$$L \ (\mu H) = 0.002a \left[\ln\left(\frac{2a}{b}\right) + 0.5005 + \frac{b}{3a} \right] \tag{62}$$

where a = length and b = width of the line segment in centimeters. The mutual inductance between two segments is approximated by

$$M \ (\mu H) = 2a \left[\ln\left(\frac{a}{d} + 1 + \frac{a^2}{d^2}\right) + \left(1 - 1 + \frac{a^2}{d^2}\right) \frac{d}{a} \right] \tag{63}$$

where d = distance between centers. The mutual inductive contribution to each segment is due to the summation over all parallel-line segments. This includes all image-line segment contributions. More details about this method can be found in Greenhouse [21]. Good results can be obtained for moderate-to-small rectangular spirals, since the current in each line segment

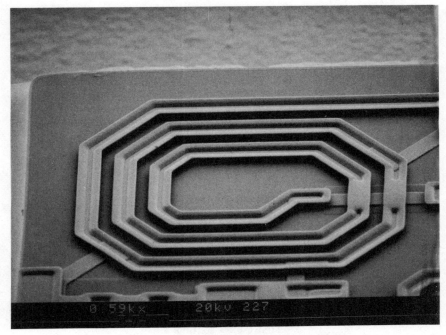

Figure 8.37   SEM of an MMIC spiral inductor.

is approximately the same and near-resonance effects (due to distributed capacitance) are minimal.

For large rectangular spiral inductors, where phase shift between line segments become significant, coupled transmission lines provide a more accurate model. Similar to the application of transmission-line (T) equivalent circuits from Section 8.5.1 to overlay capacitors in Section 8.5.2, pi equivalent circuits can be used to model the coupled-line segments. This approach can be seen in Fig. 8.36c. The adjacent even-mode impedance and line-segment lengths are used to determine the series inductance and even-mode shunt capacitance for each segment. This can be done with the equations of Fig. 8.33a. Losses can be represented by the series resistance in each segment. The coupling capacitances are then determined by calculating half the difference between the odd-mode and even-mode shunt capacitance.

A scanning electron micrograph of an MMIC rectangular spiral inductor is shown in Fig. 8.37.

8.5.4 Lumped Resistors

The thin-film monolithic resistor is realized either by isolating a substrate region that carries a conductive epitaxial layer or by vacuum deposition of a metallic conductive film on the substrate. The isolation process in the semiconductor resistors is achieved by either implantation or mesa etch. Metal films are usually preferred, however, since semiconducting films exhibit nonlinear behavior at high current levels. Also, semiconductor resistors in which a relatively low value of resistance is desired also require that particular attention be given to resistance contributed by the ohmic contact.

The most commonly used thin-film resistor materials are tantalum nitride and nichrome. Resistivities for these materials are 280–300 and 60–600 $\Omega \cdot$ cm, respectively. The first-order calculation for a films resistance is expressed in

$$R = \rho \, \frac{\text{length}}{\text{area}} \tag{64}$$

where ρ is the bulk resistivity (in $\Omega \cdot$ cm). When the thickness of the film is identified, the resistivity is sometimes specified as a sheet resistivity ρ_s (in $\Omega/\square$). In this manner, Eq. (65) is obtained. Clearly, this form is of great practical use, since one must only count the number of squares (length to width ratio) that comprise the resistive path to determine the resistance.

$$R = \rho \, \frac{\text{length}}{\text{area}} = \frac{\rho_s \, \text{length}}{\text{width}} \tag{65}$$

$$\rho_s = \frac{\rho}{\text{thickness}} \tag{66}$$

When corners or steps are encountered, the number of squares in the path must be modified to allow for the discontinuity. For example, a uniform right-angle bend has effectively 0.559 squares of path length. A nonuniform right-angle bend of aspect ratio a has an effective path length given by Eq. (67). Keep in mind that these are dc current crowding effects. We have yet to apply the rf effects.

$$\text{Length} = \frac{1}{a} - \frac{2 \ln[4a/(a^2+1)]}{\pi} + \frac{(a^2-1)\cos^{-1}[(a^2-1)/(a^2+1)]}{a\pi} \tag{67}$$

It should come as no surprise to find that "lumped" thin-film resistors are no more ideal than were the capacitors and inductors of the previous two sections. Since thin-film resistors occupy surface area over the GaAs substrate, distributed capacitance to the ground plane is present. Similarly, the path length in the direction of current flow contributes a series inductance, along with the desired resistance. If the resistor film is laser-trimmed, the modified current flow path must be considered in order to determine its microwave behavior. The inductive and capacitive branches are in the necessary lowpass form for application of the equations from Fig. 8.33 (Section 8.5.1). This approach to modeling thin-film resistors will now be discussed.

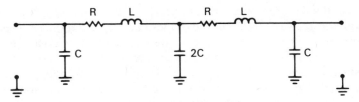

Figure 8.38   Equivalent circuit for thin-film resistors.

When a thin-film resistor is viewed as a lossy transmission line that has conceptually been partitioned into two cascaded sections, the lumped equivalent form illustrated in Fig. 8.38 is obtained. The inductance value L is calculated from the pi model in Fig. 8.33 for half the resistors electrical length. Similarly, C is determined from the same set of equations. The total thin-film resistance is divided by two to obtain R in Fig. 8.38. This two section (fifth-order) approach is valid as long as the resistor path is less than 90° in effective length. For longer structures, the number of sections in the model can be increased; so, 45° per section is not exceeded.

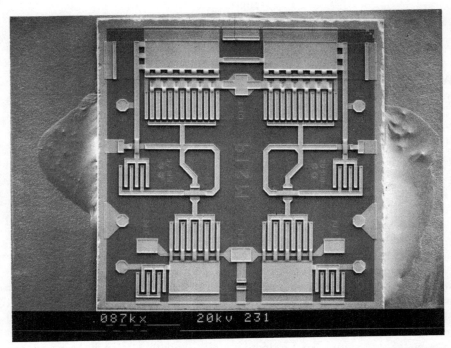

a) **TWO FET STAGES WITH INTERSTAGE MATCHING AND SELF-BIAS RESISTORS.**
   —ILLUSTRATES: TRANSMISSION LINE INDUCTORS,
                    OVERLAY CAPACITORS, AND
                    THIN-FILM RESISTORS.

Figure 8.39   Application of lumped elements to MMICs.

**b) FOUR FET STAGES OF DISTRIBUTED AMPLIFICATION
—ILLUSTRATES: SPIRAL INDUCTORS,
            OVERLAY CAPACITORS,
            AND THIN-FILM RESISTORS.**

Figure 8.39 continued

8.6   SUMMARY

This chapter began with a discussion of similarities in various FET amplifier applications from the standpoint of matching network design constraints. It was shown that the needs of each application could be reduced to the same problem: obtaining a network that provides a desired driving-point impedance behavior.

Much of the material in the chapter was accordingly devoted to presenting a systematic method for obtaining optimum matching networks. An understanding of the limits of achievable match bandwidth performance is essential before a design begins, so a detailed discussion of Bode's limit analysis was included. The methods presented did not require numerical optimization nor other computer support to obtain solutions; thus, sufficient detail was included for application by the reader. To facilitate an understanding of these techniques and their application to practical design problems, a two-stage power amplifier design example was included.

The material that followed the design example was directed to the realization of FET microwave amplifier designs in a monolithic form. The use of transmission lines to approximate inductors and capacitors was first presented. Next, "lumped" inductors, capacitors, and resistors were examined. In each case, it was seen that nonideal distributed effects were needed to model the element adequately. There are no truly lumped elements! The most useful forms were discussed. These included overlay capacitors, rectangular spiral inductors, and metallic thin-film resistors. Figure 8.39 illustrates application of these elements in MMIC realizations. (Photos courtesy of AVANTEK, Inc.)

REFERENCES

1. R. S. Carson, "High Frequency Amplifiers." Wiley, New York, 1975.
2. "S-Parameter Design," HP Appl. Note 154. April 1972. (Hewlett Packard in house publication)
3. H. W. Bode, "Network Analysis and Feedback Amplifier Design." Van Nostrand, New York, 1945.
4. R. M. Fano, "Theoretical Limitations on the Broadband Matching of Arbitrary Impedances," MIT Tech. Rep. No. 41. Massachusetts Institute of Technology, Cambridge, 1948.
5. G. L. Matthaei, Synthesis of Tchebycheff impedance-matching networks, filters, and interstages. *IRE Trans. Circuit Theory* **3**, 163–172 (1956).
6. L. Weinberg, "Network Analysis and Synthesis." McGraw-Hill, New York, 1962.
7. R. Levy, Explicit formulas for Chebyshev impedance matching networks, filters, and interstages. *Proc. Inst. Electr. Eng.* **111**, 1099–1106 (1964).
8. E. Green, "Amplitude-Frequency Characteristics of Ladder Networks." Marconi's Wireless Telegraph Co., Essex, England, 1954.
9. H. Takahasi, On the ladder-type filter network with Tchebyshev response. *J. Inst. Electr. Commun. Eng. Jpn.* **34** No. 2 (1951).
10. A. I. Zverev, "Handbook of Filter Synthesis." Wiley, New York, 1967.
11. T. R. Apel, "Bandpass Matching Networks Can Be Simplified By Maximizing Available Transformation," pp. 105–117. Microwave Systems News, Palo Alto, CA, 1983.
12. E. Christian and E. Eisenmann, Broad-band matching by lowpass transformations. *Allerton Conf. Circuit Syst. Theory 4th, 1966* pp. 155–164 (1966).
13. G. L. Matthaei, Tables of Chebyshev impedance-transforming networks of low-pass filter form. Proc. *IEEE* **52**, 939–963 (1964).
14. R. M. Cottee and W. I. Joines, Synthesis of lumped and distributed networks for impedance matching of complex loads *IEEE Trans. Circuits Syst.* **CAS-26**, 316–329 (1979).
15. E. G. Cristal, Tables of maximally flat impedance transforming networks of low-pass filter form. *IEEE Trans. Microwave Theory Tech.* **MTT-13**, 693–695 (1965).
16. P. J. T. Mellor, Improved computer-aided synthesis tools for the design of matching networks for wide-band microwave amplifiers. *IEEE Trans. Microwave Theory Tech.* **MTT-34**, 1276–1281 (1986).
17. T. R. Apel, "One-Port Impedance Models Prove Useful for Broadband RF

Power Amplifier Design," pp. 96–105. Microwave Systems News, Palo Alto, CA, 1984.

18. G. D. Alley, Interdigital capacitors and their application to lumped element microwave integrated circuits. *IEEE Trans. Microwave Theory Tech.* **MTT-18**, 1028–1032 (1970).

19. R. Esfandiari, P. W. Maki, and M. Siracusa, Design of interdigitated capacitors and their application to gallium arsenide monolithic filters. *IEEE Trans. Microwave Theory Tech.* **MTT-31**, 57–64 (1983).

20. F. W. Grover, "Inductance Calculations: Working Formulas and Tables." Van Nostrand, New York, 1946.

21. H. M. Greenhouse, Design of planar rectangular microelectronic inductors. *IEEE Trans. Parts, Hybrids, Packag.* **10**, 101–109 (1974).

9 GaAs Optoelectronic Device Technology

PAUL KIT-LAI YU
University of California at San Diego, La Jolla, California

PEI-CHUANG CHEN
Ortel Corp., Alhambra, California

9.1 INTRODUCTION

Gallium arsenide (GaAs) has been extensively used as emitters in the infrared region. Common applications have been in handheld remote controls, intrusion alarms, and optoisolators. Many of these applications make use of rather simple and basic technologies such as simple diffusion or epitaxial growth to form a p-n junction. While these simple technology are adequate for many applications, new and more advanced technologies have opened up many exiciting arenas of applications never before envisioned. The new technologies developed have, in turn, improved the performance characteristics of the older-version emitters. For example, by adopting heterojunction structures, which will be described later, brighter LEDs that consume only a fraction of the power of the old versions are now commercially available. Yields have also increased steadily due to the amount of research into better substrate materials, growth and diffusion properties, and processing techniques.

This treatment of GaAs technology will cover the newer ones that made possible commercialization of optoelectronic devices such as high-performance LEDs, lasers and various heterojunction photodetectors, as well as those that enabled improved performance of older generation LEDs.

The invention of semiconductor injection lasers in the early 1960s has been the prime source of motivation for optoelectronic device development in the following two decades. However, the attention in this period has been mostly directed to the improvement of the semiconductor laser, initially more a scientific curiosity than as a practical device. In 1968, homojunction injection lasers were replaced by those using single GaAs/AlGaAs heterostructures for current confinement and were capable of room-temperature pulsed operation with a threshold current density of $10 \, kA/cm^2$. This effort was quickly extended to double heterostructure in 1970 and resulted in room-temperature CW operation. It should be noted that many of these developments coincided with those in achieving low-loss fiber, as a major push behind the scenes had been high-data-rate fiber-optic communication networks. Most of the optoelectronic device development in the 1970s was focused on GaAs/AlGaAs materials [1]. As a consequence, many of today's fiber links are still employing GaAs/$Al_xGa_{1-x}As$ emitters, although the lowest attenuation and dispersion-free windows of fibers near 1.3 and 1.55 μm wavelengths [2] have caused all of the recently installed fiber links for long-haul communication to be based on InP/InGaAsP materials as emitters and detectors.

As it turns out, optoelectronic device applications are not limited to long-haul communication systems. Local area networks (or, equivalently, metropolitan networks), interconnects between computers, processors, and chips, and consumer product applications such as compact audio disc, remote controls (for traffic, for instance), and laser printers all require high-quality optoelectronic devices, especially those made of GaAs/AlGaAs materials. This is mainly due to the fact that operation wavelengths of AlGaAs devices (680–850 nm) are near the visible spectrum and that light diffraction is directly proportional to wavelength, and thus shorter wavelength sources are favored in many applications.

Also, due to the higher electron mobility in GaAs, as compared with that in Si, many recent advances have focused on electronic devices on GaAs substrates. Devices such as GaAs MESFETs, HEMTs/MODFETs, HBT's and other electronic devices have been studied and applied to large-scale integrated circuits (LSIs), as covered in other chapters of this book. These GaAs ICs have become very popular, especially in applications where speeds in the microwave and millimeter-wave frequency ranges are involved.

In view of optical and electronic device development, there has been a strong motivation for achieving optoelectronic integrated circuits (OEICs) on the same substrate for high-speed application. So far, such OEICs are mostly demonstrated on GaAs substrates due the maturity of GaAs/AlGaAs materials and technology.

9.2 GaAs EMITTER TECHNOLOGY

9.2.1 Introduction

Emitter devices such as lasers and LEDs require a strong radiative recombination cross section between holes and electrons in the active region of the device. This necessitates the use of direct bandgap materials for the active region. For efficient carrier injection and carrier confinement, a double heteroinjection (DH) diode concept [3] is used. As shown in Fig. 9.1, electrons are injected under forward bias from the wide-bandgap n-type materials into the active region (smaller bandgap) where they form a high

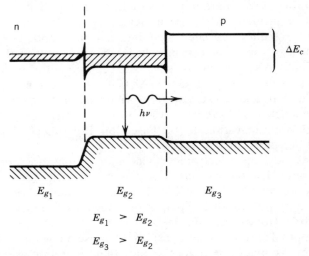

Figure 9.1 Energy band diagram of a double heterostructure: E_{g1} and E_{g3} denote the bandgap energy of the n- and p-type cladding layers, respectively, while E_{g2} is the bandgap energy of the active region.

density electron gas ($n \sim 10^{18}$ cm^{-3}) and are prevented from reaching the adjacent p-type region by the heterobarrier ($\Delta E_c >$ a few kTs). Similar conditions exist for holes injected from the p-type region. Such variations in bandgap can be easily achieved by varying the Al content in the $Al_x Ga_{1-x} As$ materials whose bandgap E_g as a function of the Al content x can be expressed as [1]:

$$E_g \sim 1.424 + 1.247x \quad \text{(eV)} \tag{1}$$

for $x < 0.45$. It should be noted that for values of x greater than 0.45, the resulting AlGaAs materials are of indirect bandgap.

Besides carrier confinement, the design and fabrication of emitters should minimize the effect of other nonradiative recombination processes, such as those caused by bulk and heterointerface defects. Fortunately, GaAs and AlAs are identical in lattice structure and are very close in lattice constant (5.6533 Å and 5.6605 Å, respectively). As a consequence, good-quality, low-defect-density multilayer structure of GaAs/AlGaAs heterojunctions can be readily grown by liquid phase epitaxy (LPE), molecular beam epitaxy (MBE), or metal-organic chemical vapor deposition (MOCVD).

The availability of sufficiently large number of both electrons and holes with large likelihood of recombining radiatively constitutes the source for optical generation. The density of these injected electrons and holes, due to the deviation from their thermal equilibrium values, are respectively designated by using quasi-fermi levels, E_{Fn} and E_{Fp} [4, 5]:

$$N = \int \frac{\rho_c(E_c)dE_c}{\exp(E_c - E_{Fn})/kT + 1} \tag{2a}$$

$$P = \int \frac{\rho_v(E_v)dE_v}{\exp(E_v - E_{Fp})/kT + 1} \tag{2b}$$

where ρ_c and ρ_v denote the conduction and valence band density of states, respectively, E_c the energy of the electron with respect to the bottom of the conduction band, and E_v the energy of the hole with respect to the top of the valence band. For the operation of LEDs, spontaneous recombination of electrons and holes is accompanied by photon emission. The emitted photons bear no fixed-phase relationship with each other and their energy can be quite spread around the bandgap energy. For successful operation of lasers, however, it is required that the injected carriers result in population inversion and thus net stimulated transition, a condition which is often expressed as [5]:

$$E_{Fn} - E_{Fp} > h\nu > E_g \tag{3}$$

where ν is the frequency of the emitted photon.

Lasing threshold is reached when the round-trip total cavity losses are overcome by the optical gain supplied by the injected carriers. The relationship between the threshold gain g_{th} and various losses is derived in Eq. (15).

It should be noted that the DH structure shown in Fig. 9.1 also results in waveguiding in the active layer, which has a higher refractive index than the neighboring layers. The refractive index of $Al_xGa_{1-x}As$ as a function of photon energy with composition as a parameter is shown in Fig. 9.2 [1]. For a photon energy of 1.38 eV, the refractive index of $Al_xGa_{1-x}As$ may be represented by [1]:

$$n(x) = 3.590 - 0.71x + 0.091x^2 \tag{4}$$

The confinement of the optical mode inside the laser active layer depends primarily on the thickness of the layer and the difference of its refractive index with the cladding layers [5].

In designing device parameters for GaAs/AlGaAs lasers and LEDs to meet the radiance and bandwidth requirements for various applications, considerations must be given to effects of self-absorption in GaAs materials, recombination at the GaAs/AlGaAs interfaces, active region doping concentration, dimensions, and carrier confinement. Unlike InGaAsP/InP materials, which are commonly used for the 1.1–1.6 μm wavelength fiber transmission systems, some of the nonradiative recombination processes such as Auger processes and heterobarrier carrier leakage are relatively unimportant in AlGaAs. This is attributed to the larger bandgap (E_g) in

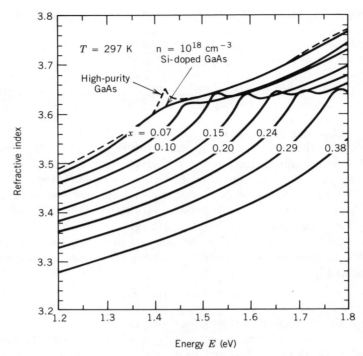

Figure 9.2 Refractive index as a function of photon energy for different composition of $Al_xGa_{1-x}As$; the refractive index of high-purity GaAs and Si-doped GaAs are also given. (From Casey and Panish [1]. Reprinted with permission from Academic Press.)

AlGaAs materials [6]. By optimizing these parameters, high emission efficiency can in principle be achieved in both GaAs/AlGaAs LEDs and lasers [7].

In the past, LPE has been the main epitaxial method for the material growth of the GaAs/AlGaAs emitters. With the advent of MOCVD and MBE, high-quality epilayers with good control of doping and layer thickness uniformity are achieved over a large-diameter (2–3 in.) wafer. At present, MOCVD is becoming economical for the manufacturing of GaAs/AlGaAs emitter devices [8].

9.2.2 GaAs LEDs: Structures and Properties

Many of the low bit rate (<200 Mbit/s) fiber-optic transmission systems presently use LEDs as light sources rather than lasers. This is attributed largely to the LED's higher reliability, the simplicity of coupling LED output to multimode fiber, and the simplicity of drive circuitry. In addition, the low nonlinear distortion (>40 dB) of LEDs is highly desirable for analog video applications [9]. For short-haul multimode fiber networks, GaAs/AlGaAs LEDs are adequate for many applications. In addition, infrared (IR) GaAs LEDs are used extensively in remote controls, and red-emitting AlGaAs LEDs are increasingly used as displays.

There are two basic generic structures for LEDs, namely, the surface-emitting LEDs (SELEDs) and the edge-emitting LEDs (EELEDs). Schematic diagrams of both structures are shown in Fig. 9.3a and b. These

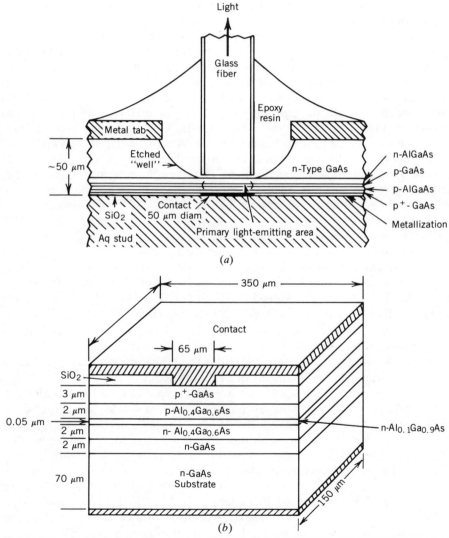

Figure 9.3 Schematic diagram of (a) AlGaAs/GaAs SELEDs (b) EELEDs. Saul et al. [1]. Reprinted with permission from Academic Press.)

two LEDs are quite different in many respects, such as emission pattern, spectral linewidth, temperature dependence, and modulation capacity.

9.2.2.1 Surface-emitting LEDs (SELEDs).

The surface-emitting LED (SELED) structure was first conceived by Burrus and Miller [10] in 1971. In this structure, a double heterostructure (DH) is employed to increase injection efficiency and radiative recombination. The light-emitting (active) layer is sandwiched between higher-bandgap AlGaAs materials to enhance the carrier confinement in the vertical direction. Light can be collected from either the substrate side or the epilayer side. The advantage of collecting light from the substrate side is that heat generated by the device near the surface can be properly conducted away by mounting the hot region near a good heat sink. However, since the active layer consists of either GaAs or $Al_xGa_{1-x}As$ with low Al content, the GaAs substrate must be removed in order to avoid reabsorption (or residual absorption) of the emitted radiation. To achieve this, one can selectively etch a well through the GaAs substrate underneath the light-emitting region as in Burrus' original design, shown in Fig. 9.3a. Also, the whole GaAs substrate can be completely etched away [11, 12] to form planar devices. Alternatively, at the expense of giving up a little efficiency due to less efficient heat dissipation, the SELED can be mounted substrate side down. This is preferable in manufacturing, since it greatly simplifies assembly and reduces cost, as bad devices can be eliminated by testing in wafer form before dicing and final assembly.

Typically, five epilayers are grown on a n-GaAs substrate to form a DH SELED. As shown in Fig. 9.3a, these epilayers are (1) a 2–5 μm thick n-GaAs buffer layer doped with Sn or Te to about $2 \times 10^{18} cm^{-3}$, (2) a 1–2 μm thick n-AlGaAs cladding layer doped with Sn or Te to about $2 \times 10^{18} cm^{-3}$, (3) a 0.5–1.5 μm thick p-GaAs or AlGaAs active layer ($\lambda_g = 0.8$–0.9 μm) doped with Ge (5×10^{17}–$2 \times 10^{18} cm^{-3}$), (4) a 1–2 μm thick p-AlGaAs cladding layer doped with Ge (0.5×10^{18}–$5 \times 10^{18} cm^{-3}$), and (5) a 0.2 μm thick cap layer of heavily p-doped GaAs ($1 \times 10^{19} cm^{-3}$) for contact. The thickness and doping level of the active layer as well as the composition of other layers are related to the power output and modulation speed requirements, as discussed in Section 9.2.2.3.

Self-absorption can be significant in LEDs, especially for those with thick active regions. In direct bandgap materials such as GaAs, the self-absorption coefficient can be as high as 10^3–$10^4 cm^{-1}$ at low injection levels. Under higher bias, the absorption saturates as the injected carrier concentration increases. This effect causes a superlinear increase in light output with current. The peak emisson wavelength depends on the doping level of the active layer and injection level. It tends to shift to higher values in more heavily doped GaAs due to band tailing [13]. In DH LEDs, the p-type active region is normally used, since higher injection efficiency of electrons than holes gives rise to better light conversion efficiency [13].

9.2.2.2 Edge-emitting LEDs (EELEDs).

The edge-emitting LED (EELED) structure shown in Fig. 9.3b was introduced by Ettenberg et al.

[14]. The EELED is similar in structure to the stripe geometry laser diode. However, unlike laser diodes, there is no lasing action in EELEDs due to the lack of optical feedback. This can be achieved by intentionally destroying the facets or by using antireflection-coated facets. Alternatively, the contact to the active stripe can be partially metallized [15], as shown in Fig. 9.4, so that light propagating toward the rear facet as well as the feedback from that facet is heavily attenuated in the unpumped region due to self-absorption.

A typical EELED also consists of five epitaxial layers with similar dopings and thicknesses to those of SELEDs, as shown in Fig. 9.3b, with the exception that the active layer is much thinner (0.1–0.25 μm). Unlike SELEDs, where waveguiding effect is absent, light from EELEDs is guided by the active layer directionally along the stripe length and then emitted through the front facet. This leads to a narrow vertical emission angle of ≈30°. A thinner active layer allows a larger portion of the optical field to spread into the cladding layers, thus minimizing the self-absorption of the emitted light along the typically 150–250 μm long active stripe. Highly efficient edge emitters have been made this way.

Conventional LEDs usually emit at a wavelength corresponding closely to the bandgap of the active region. However, by incorporating a superlattice consisting of many alternating thin (~100 Å) p- and n-type GaAs layers as the active region of the EELED, the emission wavelength of LED can be longer than that accessible by GaAs bulk materials. An emission spectra of such superlattice is shown in Fig. 9.5 [16]. The inserts in Fig. 9.5 show excess electrons and holes recombine in a "quasi-vertical" mode, due to the spatial proximity of quantized carriers, resulting in peak emission wavelength in excess of 900 nm. The superlattice effect can also lead to shorter carrier radiative lifetimes (and hence higher bandwidth) and higher efficiency, as described in Section 9.2.2.4.

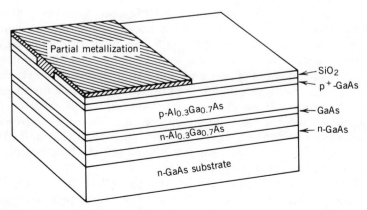

Figure 9.4 Schematic diagram of a partially metallized AlGaAs/GaAs EELED.

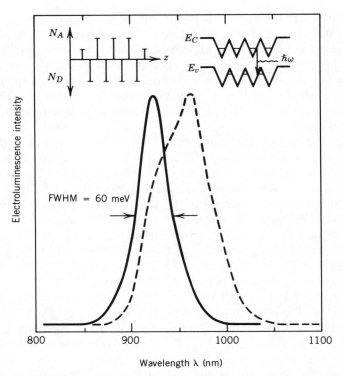

Figure 9.5 Emission spectra of two EELEDs with a sawtoothlike superlattice active region. The inserts show the the doping profile as well as the sawtooth shape of the conduction and valence bands of the active region. (From Schubert et al. [16]. Reprinted with permission from IEE.)

In many applications, it is desirable to have sources whose optical power are comparable to those of laser diodes, but the spatial and/or temporal coherence are either not required or not desired. Such applications include fiber gyroscopes [17], fiber sensors, light-triggered switches [18], optical sources for printers, and so on. In these cases, a superluminescent diode (SLD) or monolithic array of SLDs [19] can be very useful. SLDs are distinguished from both laser diodes and LEDs in that the emitted light consists of amplified spontaneous emission with a spectrum much narrower than that of LEDs but wider than that of the lasers. The first investigation of the GaAs/AlGaAs SLD was carried out by Kurbatov et al. in 1971 [20]. The EELED can be designed to operate in the SLD mode by narrowing the stripe dimensions to achieve such a high injection level that optical gain is achieved. Lasing action must be suppressed and this can be accomplished by antireflection coating of mirror facets.

The SLD can be treated as a limited case of laser which has a much higher threshold current density [21] and very "soft" light-versus-current characteristics, due to its very low facet reflectivity. At low current level, the SLD operates like an EELED; at high injection current, the output power of an SLD increases superlinearly and its spectral width narrows due to the onset of optical gain. As the injection level is increased, the output beam also narrows spatially as a result of the high gain at the center. The output power, at a given current, depends strongly on the length of the pumped section. By reducing the stripe length while keeping the same current, the current density and the output power can be increased [7]. However, increases in power are limited eventually by the increased heating of the shorter pumped stripe section.

9.2.2.3 LED Emission Properties

Spectral Characteristics. Among various characteristics of LEDs, the emission wavelength and the spectral width of the source are the most critical ones for lightwave transmission considerations. Both these properties affect the maximum data rate and distance of transmission due to fiber loss and dispersion at various wavelengths. The peak emission wavelength λ_p of an LED is determined mainly by the bandgap E_g of the active layer, as the spontaneous emission rate $I(h\nu)$ depends on the densities of electrons and holes and transition matrix element across the bandgap [5]. Generally, $I(h\nu)$ takes the form [22]:

$$I(h\nu) \sim \nu^2 (h\nu - E_g)^{1/2} \exp\left[\frac{-(h\nu - E_g)}{kT} \right] \quad (5)$$

However, under pulsed operation, λ_p can be shifted to a shorter wavelength by increasing the injection current, as a result of a band-filling effect [23]. Alternatively, it can be shifted toward a longer wavelength ($\approx 0.6\,\text{nm}/°\text{C}$) by heating [24], which affects the bandgap and thus shifts the gain profile, or by increasing the active layer doping. The latter shifts the gain profile as a result of the formation of band-tail states in highly doped materials [25].

The LED output spectrum is distinguished by its wide, full width at half maximum (FWHM) spectral width $\Delta\lambda$. For an LED with an undoped active layer, the spectral width can vary roughly [26] as λ_p^2. There are, however, significant variations in $\Delta\lambda$ among different LED structures. For GaAs SELEDs, the spectral width is typically about 25–50 nm. For EELEDs, due to self-absorption along the active layer, the spectral width is about $\frac{3}{5}$ that of the SELED and is usually 15–30 nm. Since the SLD operates in the region of amplified spontaneous emission, its spectral width is narrowed to about $\frac{1}{4}$ that of the SELED. The LED spectral width can be broadened by heating or by operating at higher current ($\sim 0.3\,\text{nm}/°\text{C}$), both related to band filling [27]. Alternatively, it can be broadened by increasing the active layer doping due to the formation of the band tail states [28].

Radiation Pattern and Radiance. Radiation patterns of LEDs are highly structure-dependent. The SELED has a large, planar p-n junction at the interface between the active layer and one of the cladding layers. However, in order to reduce the injection current and increase the emission efficiency, the light-emitting area is restricted to a small circular disc, generally 20–50 μm in diameter. This can be achieved by confining current to this region through methods such as proton-bombarding the rest of the area to form highly resistive materials, dielectric (e.g., silicon dioxide) isolation for injection current, mesa formation of the light-emitting region [29], or selective doping by diffusing zinc through a circular opening in the dielectric to reach the light-emitting region [30].

SELEDs usually have a large diameter-to-thickness ratio with light emitting relatively uniformly over the area. As a consequence, the structure shown in Fig. 9.3a has a radiation pattern close to Lambertian [31], as shown in Fig. 9.6, and the external radiant intensity (power per unit area per unit solid angle) $I(\Theta)$, can be expressed as

$$I(\Theta) \approx I_0 \cos(\Theta) \tag{6}$$

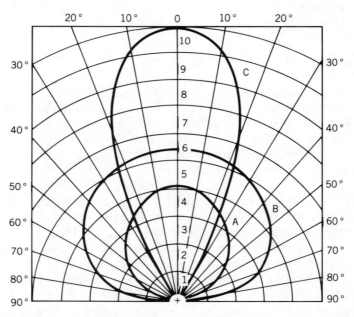

Figure 9.6 Calculated radiation patterns of SELEDs with A, a flat semiconductor–air interface at the emission surface; B, a hemispherical interface; and C, a parabolic interface. (From Galginaitis [31]. Reprinted with permission from the American Institute of Physics.)

where Θ is the angle between the emission direction and the normal to the emitting surface and I_0 is the axial radiance. However, in reality, due to imperfect current confinement near the edge of the active region, SELEDs usually have a slightly nonuniform spatial emission profile that is more closely represented by a "stretched Gaussian" function, with a Gaussian tail near the edge. But for most practical purposes, the above equation suffices.

The axial radiance, I_0 of an SELED depends on the thickness of the radiative recombination region, the current density and the internal quantum efficiency. It can be approximated by

$$I_0 \approx \frac{\eta_{\text{int}} T' JV}{4\pi n^2} \tag{7}$$

where η_{int} is the internal quantum efficiency of the LED, as defined in Eq. (11), J is the current density, V is the junction voltage, n is the refractive index of the semiconductor, and T' is a parameter that takes into account the absorption loss of emitted photons through semiconductor materials and the Fresnel loss at the semiconductor–air interface (it should be noted that photons incident to the interface at angles greater than the critical angle are totally internally reflected).

A high radiance can be obtained if the internal absorption is low, the reflectivity of the facet at the heat sink side is close to unity, and the emission interface is antireflection coated. Radiance as high as 200 $\text{Wcm}^{-2} \cdot \text{sr}^{-1}$ has been achieved [7] for an SELED with an active region 2.5 μm thick and a doping level of 2×10^{17} cm^{-3}. The output beam from SELEDs can be made directional by employing lens of hemispherical or parabolic profiles at the emission surface, with corresponding radiaiton patterns shown in Fig. 9.6 [31].

In EELEDs, lateral current spreading is limited by means of stripe injection, such as in the oxide stripe structure shown in Fig. 9.3b. However, in this case, considerable current spreading can occur in the cladding and active layers underneath the contact stripe. To improve the injection efficiency as well as the lateral waveguiding, EELEDs can be fabricated with a buried-heterostructure (BH) configuration [32] or mesa structure [33]. Due to the waveguiding effect of the stripe geometry, the EELED far-field radiation pattern is asymmetric and is more directional than that of the SELED. The output beam in the direction perpendicular to the junction plane is similar to that of the laser diode, with a far-field angle $\Theta_\perp$ of about 30° (FWHM). The far-field pattern in this direction depends on the active region thickness and the index difference between the active region and the cladding layers. Typically, a strong main lobe can be obtained with an active layer thickness of about 0.12–0.15μm in GaAs/AlGaAs EELEDs.

However, if an EELED and a laser with the same stripe width are compared, it will be seen that the output beam in the junction plane of the EELED will be much wider than that of the laser, due to the lack of optical

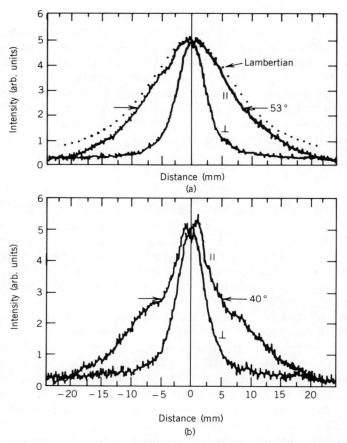

Figure 9.7 Measured far-field patterns in both directions of an AlGaAs/GaAs superluminescent diode at two input currents: (a) 600 mA, (b) 2 A. (From Paoli et al. [19]. Reprinted with permission from the American Institute of Physics.)

feedback. The distribution of the emitted radiation in this plane can be approximated by a cos Θ function. The emission angle can be reduced by using lateral index guiding or strong gain guiding, such as in SLDs. Figure 9.7 shows the far-field patterns of an SLD at two injection levels, the emission angle narrows as the injection level is raised [19]. High radiance values in the $1000 \, \text{Wcm}^2 \cdot \text{sr}^{-1}$ range at a 100 mA current level can be obtained for EELEDs.

9.2.2.4 *LED Bandwidth.* One of the main advantages of using GaAs LEDs is the fact that its light output intensity can be modulated directly with current. The maximum modulation rate of an LED is ultimately limited by the carrier recombination lifetime. Parasitic elements such as space charge capacitance and spreading diffusive capacitance can also cause a delay of carrier injection into the junction and thus result in an increase in the light output rise time [34]. The modulation bandwidth Δf is defined as the frequency at which the detected power becomes half of its low frequency value:

$$\Delta f = \frac{\Delta \omega}{2\pi} = \frac{1}{2\pi\tau} \tag{8}$$

where the carrier lifetime τ includes the contribution from interfacial recombinations as well as those from radiative and nonradiative recombinations in the active region. In the limit when the active layer thickness d is much less than the carrier diffusion length and when the interfacial recombination velocity s is small (s is generally less than 10^3 cm/s for $Al_xGa_{1-x}As/Al_yGa_{1-y}As$ interfaces), τ can be expressed as [7]:

$$\frac{1}{\tau} = \frac{1}{\tau_r} + \frac{1}{\tau_{nr}} + \frac{2s}{d} \tag{9}$$

where τ_r and τ_{nr} are, respectively, bulk carrier radiative and nonradiative lifetimes. The carrier radiative lifetime is given by [35]:

$$\tau_r = \frac{edN}{2J}\left[\sqrt{1 + \frac{4J}{edN^2B}} - 1\right] \tag{10}$$

where B is the radiative recombination coefficient, N the active layer doping, and J the injection current density. Therefore, for cases where τ_r is dominant in Eq. (9), the bandwidth can be enhanced, at low injection level, by increasing the dopant concentration in the active region. Alternatively, the bandwidth can be enhanced by increasing the current density per unit active layer thickness d, since τ_r is proportional to $(d/J)^{1/2}$ at high injection level [36].

Figures 9.8a and b show the effect of injection level and active layer doping, respectively, on modulation bandwidth. In direct bandgap materials, the radiative lifetime is usually much shorter than the nonradiative lifetime. As the doping level is increased, however, impurity-assisted nonradiative processes become dominant and thereby reduce τ_{nr} [6]. This can cause an overall reduction in the carrier lifetime and lead to a larger modulation bandwidth. However, the increase in nonradiative processes also reduces the LED output as the radiation conversion efficiency (i.e., quantum efficiency) η_{int}, which can be expressed as

$$\eta_{int} = \frac{\tau}{\tau_r} \tag{11}$$

becomes smaller. For an SELED with a doping level of approximately $1 \times 10^{17} - 5 \times 10^{17}$ cm^{-3}, a nominal bandwidth of 50–100 MHz is obtained

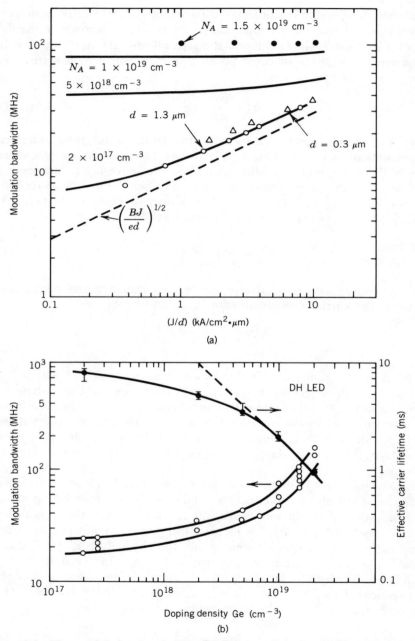

Figure 9.8   The modulation bandwidth of LEDs as a function of (a) the normalized current density (J/d), (b) active layer doping concentration. (From Lee and Dentai [13]. Reprinted with permission, copyright 1978, IEEE).

[37]. When the doping level is increased to 2×10^{18}–4×10^{18} cm^{-3}, the bandwidth increases to more than 150 MHz [38]. When the doping level exceeds 10^{19} cm^{-3}, bandwidths as high as 1.2 GHz are obtained [39].

LED response time can also be strongly affected by extrinsic factors such as *RC* time constant, which is a function of the series resistance of the diode, the space charge capacitance of the junction depletion region, and the diffusion capacitance at the edge of the light-emitting region. The series resistance is dominated by the contact resistance and can be improved by employing a low contact resistance metallization scheme. The depletion region capacitance depends on device parameters such as area and thickness and must be optimized with respect to output power and speed. The diffusion capacitance is mainly due to lateral current spreading. It can be reduced by raising the bias current level (which causes a strong current crowding effect in the emitting region), by a small dc prebias [40], or by employing current spikes [34]. Alternatively, the lateral current spreading can be avoided by employing structural designs with built-in lateral current confinement mechanisms as described earlier. In designing high-speed drive circuits, this LED capacitance should be taken into consideration. When the LED is in the off state, carriers in the charged-up junction capacitor remain in the active region and must be discharged through recombination by τ. This fall time can be improved by designing the circuit with a small forward bias to the LED even in the off state, thus allowing the LED capacitance to discharge itself. Alternatively, an intentional momentary reverse bias which rapidly drains the carriers in the active region can be applied to shorten the fall time [41].

9.2.3 GaAs Laser Technology

9.2.3.1 *GaAs/AlGaAs Lasers*: *General Properties*. Similar to LEDs, lasers are employed as emitters for a wide range of applications. However, unlike LEDs, which emit incoherent light, laser output consists of highly coherent radiation. This difference arises mainly from the fact that lasers have built-in feedback mechanisms. This permits the building up of stimulated emission, which results in laser oscillation as the injection level reaches a threshold value. In this section, some general properties of GaAs/AlGaAs lasers will be summarized; more in-depth discussions of these properties can be found in the literature, especially those in Refs. 1, 4, and 5.

Most of today's GaAs/AlGaAs lasers consist of a double heterostructure (DH) with an active layer of either GaAs or AlGaAs (low Al content) surrounded by higher-bandgap AlGaAs layers. As in the DH EELED structures, conventional DH lasers have a p-n junction at the active region, and injected carriers are confined to the active region by heterobarriers, as shown in Fig. 9.1. Under forward bias and below threshold, lasers behave like other p-n junction diodes, with current–voltage characteristics described by

$$I \approx I_T \left\{ \exp\left[\frac{q(V - IR_s)}{kT} \right] - 1 \right\} \tag{12}$$

where I_T is the saturation current and R_s is the diode's series resistance. As the injection level is increased and the electron and hole quasi-Fermi levels are separated, lasing threshold can be reached when the gain produced by the injected carriers overcomes the cavity loss. Once lasing occurs at threshold, the quasi-Fermi levels are "clamped," as the excess injected carriers are converted into photons. This effect can be observed in the I–V characteristics as a kink in the I–V curve at the onset of lasing. This clamping effect is the result of homogeneous broadening [5]. Since semiconductor laser media are not perfectly homogeneous-broadened, the quasi-Fermi levels are not totally clamped and do change slowly with bias. The threshold gain g_{th} can be derived from the condition of unity round-trip gain, where

$$R_1 R_2 e^{(\Gamma g_{th} - \alpha)2L} = 1 \tag{13}$$

where L is the cavity length, R's are the end facet reflectivity, α is the total cavity loss, given by

$$\alpha = \alpha_{fc}\Gamma + \alpha_{cl}(1 - \Gamma) + \alpha_s \tag{14}$$

α_{fc} and α_{cl} denote respectively the loss in the active region and the cladding layers, α_s is the scattering loss of the waveguide, and Γ is the mode confinement factor which is defined as the fraction of the lasing mode within the active region. Thus the threshold gain is given by [1]:

$$g_{th} = \frac{\alpha}{\Gamma} + \frac{1}{2L\Gamma} \ln\left(\frac{1}{R_1 R_2} \right) \tag{15}$$

As α is attributed to free-carrier absorption and waveguide scattering losses, g_{th} becomes somewhat structure-dependent. Γ also depends greatly on structure and can be tailored by adjusting the active layer thickness and Δn, the refractive index difference between the active and cladding layers. For an AlGaAs material system, α's are typically on the order of 10 cm^{-1}, L is generally around $300 \, \mu\text{m}$, and R's are around 0.3 (for AlGaAs–air interface); with an active region of ~ 0.15–$0.2 \, \mu\text{m}$, a confinement factor Γ of 0.3 is obtained, these values correspond to a threshold gain of $\sim 60 \text{ cm}^{-1}$. On the other hand, the optical gain, $g(h\nu)$, supplied by injected carriers, is related to the cross section for radiative recombination and is highly sensitive to temperature [42]:

$$g(h\nu) = \frac{e^2}{2\epsilon_0 m_0^2 cn\nu} \int_{-\infty}^{\infty} \rho_c(E)\rho_v(E - h\nu)|M(E, E - h\nu)|^2 (f(E - h\nu)$$
$$- f(E))dE \tag{16}$$

where m_0 is the free electron mass, ϵ_0 is permittivity of free space, n is the refractive index, ρ_c and ρ_v are densities of states in the conduction and valence bands, respectively, $M(E_1, E_2)$ denotes the transition matrix element between conduction band state at energy E_1 and valence band state at energy E_2, f's are the Fermi factors [42]. A plot of the spectral dependence of gain at several injected electron densities is shown in Fig. 9.9 [42]. Moreover, since the injected carriers can also decay via nonradiative recombination processes which are temperature-sensitive as well, the injection current density J can be expressed as

$$J = ed\left(\frac{N}{\tau_r} + \frac{N}{\tau_{nr}}\right) \qquad (17)$$

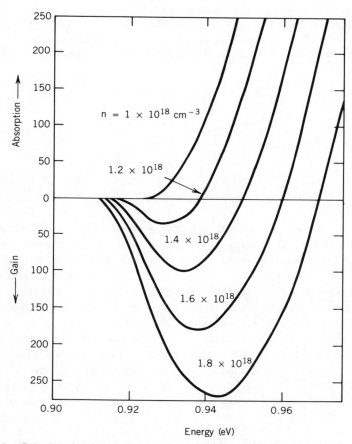

Figure 9.9 Calculated gain spectra at several inject carrier densities. The parameters used are temperature $T = 297$ K, $N_A = N_D = 2 \times 10^{17}$ cm^{-3}, and $m_c/m_0 = 0.059$. (From Dutta [42]. Reprinted with permission from the American Institute of Physics).

where τ_r and τ_{nr} denote the radiative and nonradiative recombination lifetime of carriers, respectively, N is the carrier density in the active region and d the active region thickness.

Threshold current density J_{th} corresponds to a threshold electron density N_{th}, at which the gain is given by Eq. (15). Empirically, J_{th} is found to increase exponentially with temperature T as

$$J_{th}(T) \approx C \exp\left(\frac{T}{T_0}\right) \tag{18}$$

where C is a parameter that depends weakly on temperature and T_0 is a parameter ranging from 120 to 200 K in value for AlGaAs lasers. From Eq. (18), it is seen that ohmic heating, which raises the junction temperature of the diode, can increase the threshold current.

One important laser parameter is the differential quantum efficiency η_{ext}, which is defined as the change in the optical output with respect to the change in the injection current. For injection above threshold, η_{ext} is given by [4]:

$$\eta_{ext} = \eta_i \frac{\frac{1}{2L} \ln\left(\frac{1}{R_1 R_2}\right)}{\alpha + \frac{1}{2L} \ln\left(\frac{1}{R_1 R_2}\right)} \tag{19}$$

where η_i is the internal quantum efficiency, which is close to 0.8 for AlGaAs lasers and is not very temperature-sensitive for well-designed lasers, as described below.

With the use of double heterostructure alone, the threshold current of broad area laser diodes has been drastically reduced and room-temperature CW operation is possible when injection is limited to a narrow stripe region $\sim 10\ \mu m$ wide. The earliest form of these stripe geometry lasers accomplished this by restricting current through a dielectric stripe opening, similar to the EELED structure shown in Fig. 9.3b. However, the laser output versus current characteristics of these lasers are not very linear, and kinks are observed [1]. This is believed to be related to unstable spatial mode at high injection, when the fundamental optical mode causes excessive depletion of carriers in the stripe center through stimulated emission. This effect, called spatial hole burning, creates a dip in the gain profile, lowering net gain of the fundamental mode, and allows a higher-order mode to achieve threshold condition. As a consequence, a higher degree of lateral carrier and optical mode confinement is desired. Examples of these are shown in Fig. 9.10 [43–46]. The buried-heterostructure lasers [43] have an active region surrounded by materials with higher bandgap (E_g) and lower index in both vertical and lateral directions. The channel substrate planar (CSP) stripe lasers [46] have almost flat active regions, and the buried convex

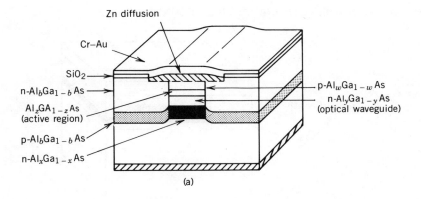

(a)

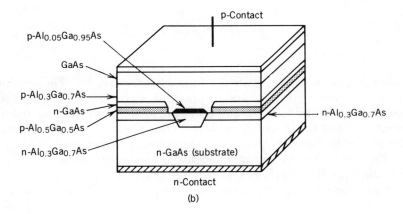

(b)

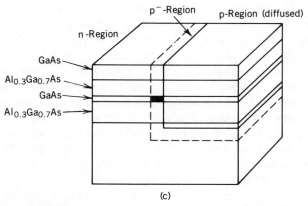

(c)

Figure 9.10 Schematic diagrams of four index-guided AlGaAs/GaAs lasers: (a) buried-heterostructure laser with buried optical waveguide (From Chinone et al. [43]; reprinted with permission from the American Institute of Physics), (b) buried convex waveguide laser (from Shima et al. [44]; reprinted with permission, copyright 1982, IEEE), (c) transverse junction stripe laser (from Kumabe et al. [45]; reprinted with permission, from *Japanese Journal of Applied Physics*), (d) channeled substrate planar laser (Aiki et al. [46]; reprinted with permission, copyright 1978, IEEE).

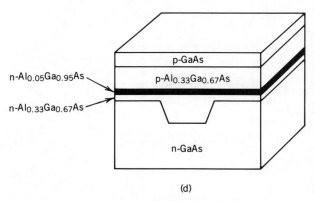

(d)

Figure 9.10 continued

waveguide lasers [44] have gradually tapered active region. The common objectives of these structures are to create a waveguide that favors the propagation of only the fundamental optical mode and to match the overlap of current flow and carrier confinement with this mode. By proper design of the dimension of the active region, these lasers achieve low threshold current (as low as 1 mA) and a stable single transverse mode. The single lobe emission far-field pattern of these lasers have an FWHM of 10–20° in the junction plane and 20–40° in the direction perpendicular to the junction plane.

Another laser structure that merits mentioning is the transverse junction stripe (TJS) laser [45], shown in Fig. 9.10, which consists of a DH and a lateral p-n junction. This structure is realized by diffusing p-type dopants (e.g., Zn) into a AlGaAs/GaAs/AlGaAs DH (all n-type) and by a subsequent drive-in diffusion. In this laser, carriers are injected laterally and are confined to the active region (p^--GaAs region) due to the slightly smaller bandgap of the p^- region. The higher resistivity of AlGaAs cladding layers as well as the higher bandgap contributes to the current confinement in the vertical direction. The lateral mode confinement is aided by the carrier induced index step between the lightly p-doped region and neighboring regions. Typical threshold current of this laser is approximately 20–30 mA, and output power is ~3–5 mW. The simplicity in this structure allows mass production of this laser.

More recent development in lasers for mass production involves the v-channel substrate inner stripe laser (VSIS) on p-GaAs substrate [47], as shown in Fig. 9.11. The transverse mode is stabilized again by a built-in optical waveguide, which is self-aligned with an internal current confinement channel. This laser emits in the visible wavelength (725–790 nm) region and has good reliability. Most of today's lasers used in compact disc players make use of this structure.

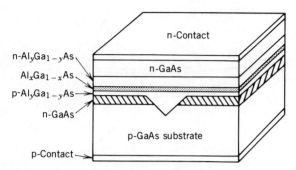

Figure 9.11 Schematic diagram of v-channeled substrate inner stripe laser on p-GaAs. (From Hayakawa et al. [47]. Reprinted with permission from the American Institute of Physics.)

9.2.3.2   Recent Laser Development. Recently, the emphasis of laser diode research and development efforts has been shifted toward optimization to suit various applications. In particular, issues of (1) high-speed modulation, (2) stable single mode and linewidth control, and (3) high emission power have been investigated by many laboratories.

High-frequency Modulation. For present and future optical transmission systems, it is necessary to modulate the optical laser signals at high frequency or at high data rates. Because of its simplicity, direct current modulation has been the most popular method [48]. Under small-signal modulation, the output intensity spectra of laser diodes are relatively flat from low frequency, as shown in Fig. 9.12 [49], until a resonant frequency—the relaxation oscillation frequency—is reached. This resonance is caused by the intrinsic coupling between photons and injected carriers inside the laser cavity. The output response rolls off quickly beyond this frequency. Coincidentally, the intensity noise spectrum of lasers, whose intensity varies as the square root of the injection current above the laser threshold current [50], also peaks at this frequency [51]. From experimental results and from analysis, the existence of this resonance peak often limits the maximum direct modulation bandwidth of laser diodes to the multigigahertz range.

The dynamics of photons and carriers in lasers can be described with the help of the following one-dimensional rate equations [52]:

$$\frac{\partial I^+}{\partial t} + c\,\frac{\partial I^+}{\partial z} = ANI^+ + \beta\,\frac{N}{\tau_s} \tag{20a}$$

$$\frac{\partial I^-}{\partial t} - c\,\frac{\partial I^-}{\partial z} = ANI^- + \beta\,\frac{N}{\tau_s} \tag{20b}$$

$$\frac{dN}{dt} = \frac{J}{ed} - \frac{N}{\tau_s} - AN(I^+ + I^-) - \alpha N \tag{20c}$$

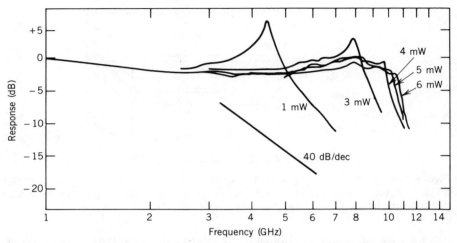

Figure 9.12 A modulation response of a 175 μm AlGaAs/GaAs BH laser at −50°C. (From Lau and Yariv [49]. Reprinted with permission from the American Institute of Physics.)

where I^+ and I^- denote the photon intensities of the traveling waves going in the positive and negative z directions, respectively, along the laser waveguide, N is the carrier density, A is the differential optical gain constant (which is stronly dependent on material and temperature), β is the fraction of spontaneous emission entering the lasing mode, τ_s is the carrier lifetime, d is the active layer thickness, α is the distributed cavity loss, and J is the injected current density.

Using a small-signal analysis with Eqs. (20a − c) and by spatially averaging the photon and carrier densities inside the laser cavity, the relaxation oscillation frequency f_r can be expressed [49] as

$$f_r = \frac{1}{2\pi} \sqrt{\frac{AP_0}{\tau_p}} \qquad (21)$$

where P_0 is the steady-state photon density and τ_p is the photon cavity lifetime (which is related to device parameters such as cavity length, facet reflectivity, and the distributed loss α.)

As can be seen from Eq. (21), the intrinsic laser modulation bandwidth can be enhanced by increasing the optical gain coefficient or the photon density, or by decreasing the photon lifetime. These parameters can be separately adjusted to a certain degree through various means. For example, the gain coefficient A can be increased by about a factor of five by cooling the laser from room temperature down to 77 K [53]. Photon density in the active region can be increased by raising the forward-bias level of the laser, but the optical output intensity at the laser facet must stay within the catastrophic optical damage (COD) limit of mirrors [54]. For the GaAs/ AlGaAs material system, this limits the maximum power density to ≈1 MW/ cm^2 before permanently damaging the mirrors. A way to alleviate this problem is to incorporate a window region [55] near the facets to expand the optical mode volume there or to make window regions out of a higher bandgap (i.e., nonabsorbing) material. With windowed buried-heterostructure (BH) or transverse junction stripe (TJS) lasers, modulation bandwidth in excess of 10 GHz has been demonstrated in the GaAs/AlGaAs system. The modulation response of these lasers shows a reduction and sometimes an absence of the resonance peak; this is attributed to superluminescence damping, which reduces the facet reflectivity [56]. Modulation bandwidth can also be enhanced by reducing the photon lifetime, which can be effected by decreasing the laser cavity length, as shown in Fig. 9.13 [49]. However, in practice, there is a limit to which the cavity length can be shortened. As the cavity length shortens, the current density within the active stripe will rise [57], until a point is reached beyond which long-term reliability is compromised. In addition, the maximum photon density for lasers with shorter cavity may also be smaller due to heating effects.

More sophisticated schemes to enhance the modulation bandwidth have

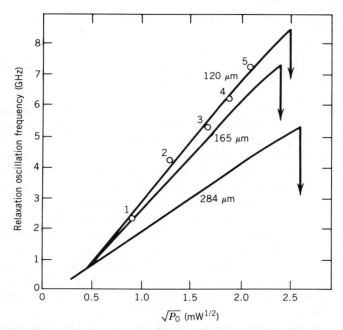

Figure 9.13 Measured relaxation oscillation frequency of lasers of various cavity lengths as a function of $P_0^{1/2}$ (see Eq. [21]). The points of catastrophic damage are indicated by downward arrows. (From Lau and Yariv [49]. Reprinted with permission from the American Institute of Physics.)

been investigated. In one approach, the laser is placed next to an external cavity formed by one of laser facets and an external mirror. By adjusting the length of the external cavity, the cavity can be detuned such that the differential gain is always increasing during modulation [58]. The direct modulation bandwidth thus increases according to Eq. (21). In a proposed approach [59], it is noted, from the rate equation analysis, that the product AP_0 is proportional to the inverse of the carrier lifetime. Therefore, the effect of the resonance peak can be reduced by putting the laser junction near the base of a bipolar transistor. The carrier lifetime in the laser active region can be reduced as a result of transistor action, and this causes a damping action near the resonance frequency.

Since Eq. (21) is based on one-dimensional rate equations, the effect of the lateral profile of the carrier has been neglected. It turns out that the strength of the resonance depends, among other things, on the spontaneous emission factor, lateral carrier diffusion, and the presence of saturable absorber along the active cavity. In particular, the inhomogeneity in the lateral carrier profile can enhance damping of the resonant peak [60, 61]. The rate equation analysis, which includes lateral diffusion and spatially averaging in the cavity length direction, can be expressed as [61]:

$$\frac{\partial N}{\partial t} = D\,\frac{\partial^2 N}{\partial x^2} + \frac{J(x,t)}{ed} - \frac{N}{\tau_s} - g_t P(x) \tag{22a}$$

$$\frac{d}{dt} \int_{-\infty}^{\infty} P(x)dx = \int_{-\infty}^{\infty} \left(\Gamma g_t - \frac{1}{\tau_p}\right) P(x)dx + \beta \int_{-\infty}^{\infty} \frac{N}{\tau_s}\,dx \tag{22b}$$

where g_t denotes the local gain coefficient and $P(x)$ is the optical intensity profile. This effect of lateral diffusion on f_r is not significant in strongly guided lasers such as BH and channel substrate planar (CSP) lasers where both lasers exhibit 90° phase shift between the modulating current and the modulated optical output. However, with gain-guided laser structures such as TJS lasers where lateral carrier confinement is not strong, the inhomogeneous carrier injection in the lateral direction results in a in-phase relation between the current and the modulated output near the resonance frequency [62]. As a consequence, 20 Gbit/s optical multiplexing has been demonstrated in TJS AlGaAs lasers with direct modulation [63].

In addition to intrinsic modulation limitations, laser bandwidth limitations can be caused by parasitic capacitances, which come from the laser structure itself, or from the associated wire-bonding and package. Laser structures with small junction area usually have small capacitance. However, in these lasers, the main parasitic capacitance comes from the area between the top and bottom contact outside the lasing region. For example, the reverse blocking junctions in the BH laser contribute much of the capacitance [64, 65]. Lasers fabricated on semi-insulating substrates have been

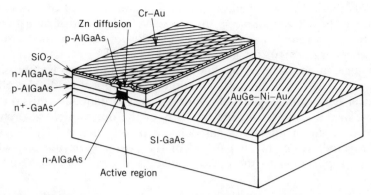

Figure 9.14 Schematic diagram of an AlGaAs/GaAs buried heterostructure fabricated on semi-insulating GaAs. (From Bar-Chaim et al. [66]. Reprinted with permission, copyright 1981, IEE).

shown to have small parasitic capacitance element due to the distributed network nature of the parasitic reactances; an example is shown in Fig. 9.14 [66].

Single Mode and Linewidth Control. Single mode lasers denote those lasers that oscillate with a stable single longitudinal mode and a stable single transverse mode. In contrast to the output of single mode lasers, the optical output of multilongitudinal mode laser diodes usually consist of higher relative intensity noise (RIN) [40], which is mainly caused by the random switching among different modes. This noise can seriously degrade the quality of the transmitted signals. However, for multimode fiber communication systems (for short haul), this multimode property can be an advantage, since the less coherent the laser emission is, the less are the optical feedback induced noise and the fiber modal noise. On the other hand, for single mode fiber communication systems, it is highly desirable to have stable, single mode laser sources with narrrow linewidth (i.e., for long-haul application) at the emitting end.

Besides controlling the transverse mode structure, lateral guiding mechanisms of laser diodes can also affect the longitudinal mode structure. Strongly index-guided lasers usually have few longitudinal modes when operated slightly above the threshold. Gain-guided lasers, on the other hand, usually have many longitudinal modes. Such behaviors can be accounted for by the large spontaneous emission factor β in the lasing modes [67]. As defined in Eq. (20), the spontaneous emission factor β is the ratio of the spontaneous emission power in the lasing mode to the total spontaneous emission power and is given as

$$\beta = K \frac{\lambda^4}{4\pi^2 n^3 V \Delta\lambda} \tag{23}$$

where λ is the lasing wavelength, $\Delta\lambda$ the spectral width of the spontaneous emission, V the active volume, n the refractive index, and K an astigmatism parameter. It is found that K is close to unity for index-guided lasers and can be very large (>10) for gain-guided lasers [68].

For single mode lasers, the linewidth of the mode above threshold is related primarily to the phase fluctuation of the optical field. The phase of the optical field changes abruptly during each spontaneous emission event; however, in the same event, the intensity of the optical field also changes, which can result in an additional delayed phase change [69]. These two effects together result in an FWHM of the power spectrum $\Delta\nu$, which is inversely proportional to the optical power and can be expressed as [70]:

$$\Delta\nu = \frac{R(1 + \alpha_n^2)}{4\pi I} \tag{24}$$

where R is rate of spontaneous emission, I the optical intensity, and α_n is a parameter defined as [70]:

$$\alpha_n = \frac{\Delta n'}{\Delta n''} \tag{25}$$

with $\Delta n'$ and $\Delta n''$ denoting the changes in the real and imaginary part of the refractive index, respectively. This power dependence of linewidth was measured by Welford and Mooradian [71] and agrees with Eq. (24). The value of α_n varies from one laser structure to another, and for BH lasers a value of $\alpha_n \sim 6$ was measured.

It is still not wholly clear why some index-guided lasers can operate in a single mode whereas others operate in multiple longitudinal modes. However, it is found that even index-guided lasers that operate in a single mode at dc can become unstable in emission wavelength when subjected to disturbances such as temperature change, optical feedback, and high carrier fluctuations under modulation. The resultant emission linewidth-broadening and mode-switching can lead to temporal pulse broadening in fiber due to fiber dispersion and degrade the quality of the transmitted signals for long-haul systems. To maintain dynamic single mode operation, extra means are needed to stabilize the laser frequency, including (1) short cavity length, (2) mode selection by means of composite cavity, and (3) optical feedback by means of a grating incorporated inside the laser cavity. Besides stabilizing the lasing frequency, these can also provide means of tuning the emission wavelength over a limited range (from a few Å to a few nm).

Short-cavity lasers are useful for stabilizing single mode operation because, for a gain spectrum with a finite spectral width, the number of longitudinal modes within the positive region is inversely proportional to the length L of the Fabry–Perot cavity. The longitudinal mode spacing $\Delta\lambda$ can be expressed as

$$\Delta\lambda = \frac{\lambda^2}{2L\left(n_{\text{eff}} - \lambda\,\dfrac{dn_{\text{eff}}}{d\lambda}\right)} \tag{26}$$

where n_{eff} is the effective refractive index. Thus, by reducing the cavity length, the Fabry–Perot mode spacing can be increased to a point where only a single longitudinal mode within the laser gain spectrum has a much higher gain than the next mode, so that other modes cannot achieve threshold.

For instance, by means of microcleaving technique [57], as shown in Fig. 9.15, cavity lengths as short as 30 μm have been obtained. Also, by forming the facets with dry or wet chemical etching techniques, cavities as short as 23 μm can be achieved. In both cases, stable single mode operation is observed at CW operation. However, even for these short-cavity lasers, multimode emission still appears under pulsed modulation, as shown in Fig. 9.16 [72]. This is caused by a transient effect on the gain spectrum and the refractive index as the carrier density changes [69].

To stabilize the longitudinal mode structure further, the composite cavity approach can be used; it incorporates additional mode selection to stabilize the oscillation frequency. One such scheme places an external mirror close to one of the laser facets to create a multicavity effect. It is found that when good phase matching is obtained between the optical feedback and the lasing mode in a BH laser, the optical-feedback-induced intensity noise can be suppressed by more than 30 dB, and a single mode oscillation is obtained [73]. In another scheme, with a combination of a short cavity and an external mirror, stable single longitudinal mode operation is obtained even

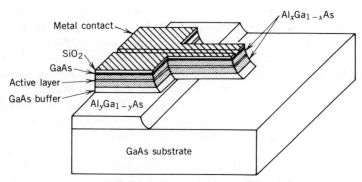

Figure 9.15 Schematic diagram of the cantilever part of an AlGaAs/GaAs laser prior to microcleaving. The cantilever can be ultrasonically cleaved to form a laser mirror. (From Blauvelt et al. [57]. Reprinted with permission from the American Institute of Physics.)

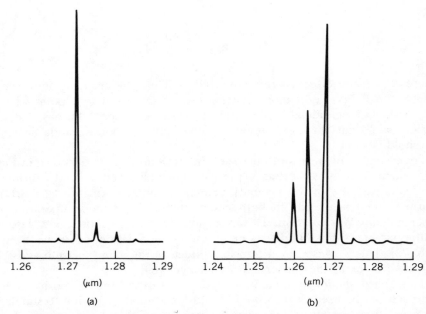

Figure 9.16 Longitudinal mode spectra during (a) CW and (b) pulsed (100 ps) operation of an InGaAs/InP short cavity laser. (From Lin and Lee [72]. Reprinted with permission, copyright 1984, IEEE.)

under pulsed modulation [74]. In the same manner, a single longitudinal mode can be obtained with cleaved coupled cavity lasers [75] and interferometric index-guided lasers [76]. The former method employs two separate laser cavities coupled to each other and the latter approach uses a laser cavity with active and passive sections. All of these lasers show a stable single longitudinal mode with a 7–10°C locking range for temperature variation (within which $\Delta\lambda_p$ changes slowly, ~0.7 Å/°C), beyond which mode hopping occurs. Since λ_p also varies with current, the emission wavelength can be tuned rapidly over a certain range by adjusting the current in the passive section, which makes these lasers attractive for many applications such as wavelength-division multiplexing. Single mode operation under modulation can also be obtained by external injection [77], in which the wavelength of one laser (the slave) is locked in value by light injected from a master laser. Laser linewidth narrowing can be achieved with this scheme.

Single mode laser diode operation with stable frequency can also be achieved with a wavelength selective grating incorporated in the laser. Distributed feedback (DFB) [78], as shown in Fig. 9.17 [79], and distributed Bragg reflector (DBR) lasers are examples using this scheme. The former uses a grating throughout the laser cavity and the latter uses a grating as reflectors outside the pumped region. With the incorporation of a grating

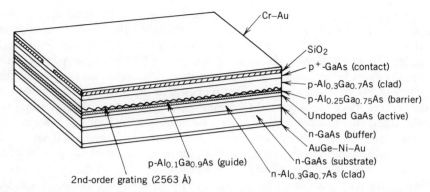

Figure 9.17   Schematic diagram of an AlGaAs DFB laser with oxide stripe current confinement. The laser structure can be grown with MOCVD. (From Kojima et al. [79]. Reprinted with permission, copyright 1986, IEEE.)

structure, the oscillation frequencies for the DFB lasers becomes [80]:

$$\lambda_0 = \lambda_B \pm \frac{(q + \frac{1}{2})\lambda_B^2}{2n_r L} \qquad (27)$$

where q is an integer, n_r is the effective index of the medium, and L is length of the grating region; the parameter λ_B is the Bragg wavelength and is related to the grating period Λ by

$$\lambda_B = \frac{2n_r \Lambda}{m} \qquad (28)$$

where m is an integer.

For ordinary DFB lasers, two degenerate longitudinal modes, corresponding to $q = 0$, have the lowest threshold gain and will oscillate. To obtain single mode oscillation in DFB lasers, one has to split the degeneracy by inducing a difference in threshold gain of the two modes in Eq. (27). This can be done by means of end facet reflectivity discrimination, a chirped grating [77], or the use of an asymmetrical device structure (including those incorporating a $\lambda/4$ phase shifting section in the grating) [81]. Wavelength stability can be maintained over a wide temperature range (0–50°C) in these schemes [79, 82].

In contrast, DBR lasers can only oscillate at the Bragg wavelength λ_B, as the grating is outside the pumped region. DBR lasers, on the other hand, usually have a higher threshold current that DFB lasers due to the weaker coupling between the pumped and grating regions [83]. Although DBR can be maintained in single mode operation, its emission wavelength is susceptible to temperature variation, since the Bragg wavelength has a temperature coefficient of [84]:

$$\frac{d\lambda_B}{dT} = \frac{\lambda_B}{n_{eff}} \frac{\partial n_{eq}}{\partial T} \qquad (29)$$

where n_{eq} and n_{eff} are the equivalent indices of the pumped and grating regions, respectively. When the Bragg wavelength drifts, the adjacent mode ($\Delta m = 1$) can acquire a higher gain and begins to lase, causing mode hopping. In DBR lasers, since the optical coupling between pumped and grating regions is not as strong as that in the DFB lasers, it is expected that the linewidth of the oscillation frequency can be broadened by carrier-induced refractive index variation during modulation. It is observed that the modal linewidth reaches a maximum when modulated near the laser resonant frequency [85]. Fabrication of the gratings in DBR and DFB lasers are achieved through a combination of photosensitive exposure by two interfering laser beams and wet or dry chemical etching. By adjusting the angles between the two incident beams, the grating period Λ can be varied. An alternative way of creating the grating patterns uses electron beams to write on a wafer coated with electron beam sensitive material such as PMMA.

High-power Operation. The study of high-power aspect of laser diodes has recently attracted a great deal of attention for applications such as optical recording systems and laser printers. However, the most widespread use of high-power lasers (< 0.5 W) has been the pumping of YAG lasers. The pumping is most efficient when the lasing wavelength is centered around 0.8 μm, as this overlaps with the absorption line in YAG crystals. Very compact and energy-efficient YAG lasers made this way have found diverse applications in medicine (opthalmalogy, surgery), and instrumentation (e.g., gyroscopes). Also, as mentioned earlier, for individual lasers, higher output power implies higher optical power density inside the laser cavity, which is usually accompanied by a larger modulation bandwidth, according to Eq. (21).

The two major factors limiting the laser output power are the catastrophic failure due to facet damage at high intensity and the junction temperature, which can cause power saturation at high injection. It is generally believed that facet damage is caused by surface recombination of carriers at the facet that makes the region near the facet absorbing. At high power levels, the absorption causes local heating and the region becomes more absorbing, eventually causing thermal runaway and material meltdown.

A facet passivation coating can be applied to increase the threshold power density before catastrophic optical damage (COD) occurs. To increase output from the front facet, a highly reflective multilayer coating is applied to the rear facet while an antireflective coating is applied to the front facet.

The front facet coating alleviates the problem of COD only to a certain degree. Methods to circumvent material limitations are necessary if more power is needed. There are two general approaches to accomplish this. The first is to increase the volume of the waveguiding region while maintaining stable single modal characteristics. The second uses closely spaced multiple stripe lasers which lase coherently to give high combined outputs.

In the first approach, several schemes are studied to increase the optical modal volume. One interesting scheme is to employ a very thin active region so that laser light spreads further into the cladding layers. With the twin-ridge-substrate laser shown in Fig. 9.18 [86], growth characteristics over mesas that are etched into the substrate allow controllable growth of thin active layers by LPE. Resulting laser output as high as 100 mW, which corresponds to an optical flux intensity of 4 MW/cm^2, has been achieved with an active region thickness as small as 0.05 μm [86]. This twin-ridge feature also permits stable single mode operation at high power. However, the thin active region also results in higher threshold current density. To retain low threshold current density, the laser can be designed with the active layer thickness of ~0.1–0.15 μm, which corresponds to minimum threshold current density but thins out near the facets [55], where the optical mode can spread out due a smaller index step in the guiding region. In practice, this is done by different spacings of the two mesas within the laser and the "window" region near the facets.

Alternatively, a layer with index and bandgap energy intermediate between those of the active layer and the cladding layer can be incorporated throughout the length of the laser cavity. This additional guiding layer controls the effective modal volume and at the same time helps confine the injected carriers to the active region. These large optical cavity (LOC) lasers usually can operate at high power with stable single mode and relatively low threshold current characteristics [87, 88]. Highly efficient lasers, which have relatively low junction heating, have been achieved by incorporating the LOC layer in the buried heterostructures [89, 90].

The second approach, which utilizes the combined output from multiple stripe laser diodes, has achieved the highest output power (~4 W at CW operation) from semiconductor lasers [91]. However, besides obtaining large

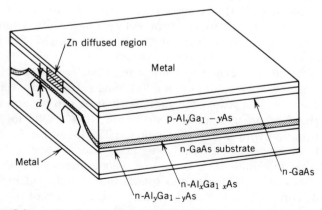

Figure 9.18 Schematic diagram of an AlGaAs/GaAs twin-ridge-substrate laser. (From Wada et al. [86]. Reprinted with permission from IEE.)

output power, the control of the near-field and thus the far-field distributions is very essential for effective usage of the optical power. Therefore, applications of laser array critically rely on the degrees of phase-locking of the output from lasing stripes which are arranged spatially next to each other in an array manner. Analysis of the arrays is similar to antenna array analysis, and permitted modes, so-called supermodes, have been successfully used to describe the operation of these lasers. Some authors [92, 93] have used the coupled mode theory to analyze phase-backed arrays. However, in strongly coupled systems, coupled mode theory cannot be applied.

The typical far-field pattern of supermodes of a phase array consists of peaks with an angular separation inversely proportional to the spatial separation of the elements, and the width of each peak is proportional to the overall width of the array. The phase between two adjacent lasers determines the relative positions of these peaks with respect to the forward direction (i.e., 0°). When all the lasers operate in the same phase, the far field usually has a main lobe in the forward direction, which is useful for most applications. This usually corresponds to the fundamental mode in the supermode system. However, when adjacent lasers operate 180° out of phase with each other, the far field will have the characteristic double lobe pattern, which can be a disadvantage in some applications, such as optical recording. It turns out most of the arrays favor double lobe operation at high output level (>1 W). This is largely due to the lack of gain (or presence of loss) in the regions between stripes, thus the 180° out-of-phase mode which has less optical field in this region can achieve higher net threshold gain compared with the fundamental mode. As a consequence, many schemes are devised to control the mutual interaction between the array elements such that desirable far-field distribution can be achieved [93]. In principle, for fundamental (single lobe) mode operation, the loss (gain) in the interstripe region should be reduced (increased). A more successful means to force all stripes to lase in the 0° phase mode takes the approach of placing a Y junction [94], as shown in Fig. 9.19, to mix the internal oscillating photon flux and to discourage 180° phase-shifted modes, since these modes combine destructively at the Y junction. This has resulted in high CW power (>1 W) of single lobe output.

Another scheme is designed to control the injection of each laser stripe separately, instead of equal pumping at each stripe. By tailoring the pumping current at each stripe, the gain distribution can be closely matched to that of the fundamental mode distribution; this results in single lobe far-field distribution with a divergence angle (FWHM of the far-field angle) as small as 1.28° for a four-element gain-guided laser array [95]. However, it is observed that the single lobe beam divergence increases as the pumping current increases. This is due to the spatial hole burning at high injection and the mode competition among the possible eigenmodes of the supermode system. Mode competition becomes more significant as the number element N in an array increases, since then the threshold gain of the higher-order

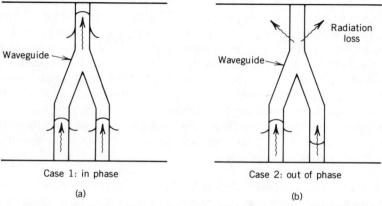

Case 1: in phase   Case 2: out of phase

(a)   (b)

Figure 9.19   To illustrate the principle of a Y-junction coupler: when the incident modal amplitudes at the input waveguides are in phase, they add to excite a mode in the output waveguide; the out-of-phase excitation produces destructive interference, which results in radiation loss of all the incident power. (From Streifer et al. [94]. Reprinted with permission, copyright 1987, IEEE.)

modes approaches that of the fundamental mode. Further stabilization can be achieved by means of index guiding (e.g., ridge-guided laser) [96, 97], which minimizes the effects of gain distortion at high injection, and by means of variable spacing between stripes [98] such that the pumping regions match well to a gain distribution favoring the fundamental mode operation, as illustrated in Fig. 9.20. Both schemes have resulted in phase-locked behaviors for CW outputs as high as 100 mW.

So far, laser arrays fabricated with LPE show variations in far-field behaviors possible due to the nonuniformity in layer thickness across the

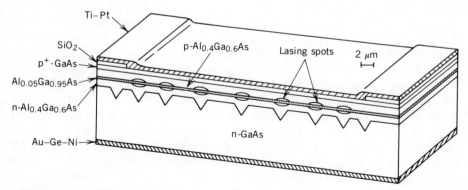

Figure 9.20   Schematic diagram of a phase-locked CSP laser array with variable spacing between stripe. (From Ackley et al. [98]. Reprinted with permission, copyright 1986, IEEE.)

width of the array. More uniform behavior, as well as narrower far-field angle, are obtained from arrays made with MBE or MOCVD [91]. This also brings forth a back-to-basics approach to achieve high power. In the early days of laser research, large area lasers (widths up to 100 μm) would lase in separate filaments due to inhomogeneity in the layers incorporated during the growth process. With the advent of MOCVD and MBE, extremely uniform layers can be grown and a 100 μm wide stripe can be electrically pumped to lasing without filamentation and achieve high efficiency and high power. Since the entire stripe lases coherently, the output far-field pattern is single lobed and narrow. This approach has already produced commercially available high-power (>1 W) lasers [99].

9.2.3.3 *Advances in Laser Research.* As seen from Section 9.2.3.2, the modal, modulation, and emission power characteristics of laser diodes have been studied and controlled. Recent research on laser diodes has taken on two main routes, namely, the study of quantum well lasers and surface-emitting lasers. The motivation behind the former is the advance in the understanding of two-dimensional phenomena in solid-state physics and advances in the fabrication technology that enable the design and growth of ultrathin layers. The motivation behind surface-emitting lasers lies in the desirability of realizing large-scale OEICs with a laser technology that can be implemented with the usual IC processing technology, as well as their possible applications as sources in optical interconnect and optical computing technologies.

Quantum Well Lasers. A quantum well (QW) laser is a DH laser whose active layer incorporates a QW structure, in contrast to conventional lasers whose active layer is so thick that no quantum size effect is expected. The family of QW lasers consists of multiple quantum well (MQW) lasers, single quantum well (SQW) lasers, graded-index waveguide separate confinement heterostructure (GRIN-SCH), single quantum well [100] lasers, etc. Since the electronic and waveguiding properties of QW structures are covered in detail elsewhere in this book, only main features of laser structures incorporating QWs will be summarized here [101].

SQW lasers can have an active region (the well) as thin as 60 Å. Consequently, the energy levels at the active region are quantized in the direction perpendicular to the quantum well layer. The separation of these levels and the band edge is determined from the properties of both the well layer and barrier layers, which are located next to the well layer. MQW lasers have an active region consisting of many well layers separated by thicker barrier layers. Compared with conventional DH lasers, QW lasers can exhibit a smaller J_{th} [100]. This is primarily due to the staircase density of state in QWs, shown in Fig. 9.21 [101], which gives an optical gain with a higher peak value and a narrower spectral width as compared with the parabolic density of state in conventional DH lasers. With MQW lasers, this

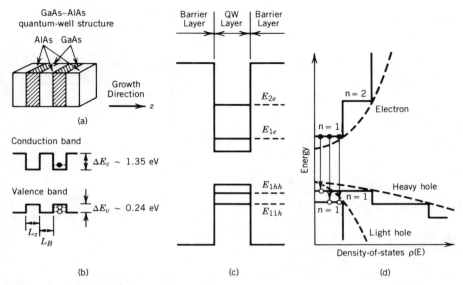

Figure 9.21 AlGaAs/GaAs quantum well structure: (a) layer structure with z direction as the growth direction, (b) energy band structure, (c) quantized energy levels in a quantum well, (d) density of states as a function of energy for a QW structure (solid line) and for a bulk crystal (broken line). (From Okamoto [101]. Reprinted, with permission, from *Japanese Journal of Applied Physics*.)

reduction in threshold current can easily be observed [100]. However, in SQW lasers, due to the insufficient carrier and optical mode confinement of the thin active region, low threshold cannot be readily obtained. This is solved by SCH structures where each of two cladding layers consisting of a thin $Al_xGa_{1-x}As$ layer with low Al content ($x \approx 0.2$) is followed by a AlGaAs layer of higher Al content ($x \approx 0.4$) to increase carrier confinement efficiency and optical confinement Γ [102]. Alternatively, graded-index (GRIN) regions consisting of many $Al_xGa_{1-x}As$ layers with increasing (or decreasing) x values can be inserted between outer confinement AlGaAs layers and the active region, as shown in Fig. 9.22 [103].

QW lasers are expected to have higher T_0 values than conventional DH lasers [104]. This can again be explained by the staircase density of state, which makes the energy distribution of carriers less temperature-sensitive [105]. The nonradiative recombination lifetime in QW lasers is also believed to be longer than that in conventional DH lasers due to the density of states and the decrease in the number of configurations that can satisfy the requirement of both momentum and energy conservation simultaneously. This feature becomes significant in long wavelength lasers, since conventional DH InGaAsP/InP lasers have a small T_0. In addition, as optical gain spectra of QW lasers are narrower than that of the conventional DH lasers, QW lasers are easier to oscillate in a stable single longitudinal mode than

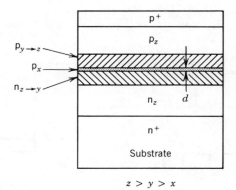

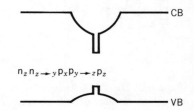

Figure 9.22 Schematic layered structure of a graded-index waveguide separate confinement heterostructure laser with the corresponding energy band diagram. The x, y, and z the Al content of the AlGaAs layers. (From Tsang [103]. Reprinted with permission from the American Institute of Physics.)

conventional lasers, even under current modulation at high frequency [106].

Experimentally, MQW lasers have exhibited a factor-of-three reduction in chirping (i.e., wavelength shift under modulation) when compared with conventional lasers. This can have significant consequences for future coherent communication systems [107]. The output from QW lasers exhibit stable TE polarization as compared with conventional DH lasers [108]. This is due to the fact that the TE polarized optical wave whose electric vector lies in the plane of QW layers can couple both with the conduction band (whose angular momentum in the direction perpendicular to the well layers is $J_z = \pm\frac{1}{2}$) to heavy hole band ($J_z = \pm\frac{3}{2}$) transition and with the conduction band to light hole band ($J_z = \pm\frac{1}{2}$) transition, whereas the TM polarized optical wave can only couple with the latter transition, [108]. Consequently, the optical gain is anisotropic and favors TE mode oscillation.

Since the energy levels of the QW shifts upward as the thickness of the well layer decreases, AlGaAs/GaAs QW lasers can be tailored to emit in the visible region by decreasing the well width and/or by increasing the Al content (x) in the well layer [109].

One interesting phenomenon that may have significant applications has been observed in thin QW multilayers. When a dopant such as zinc or silicon is introduced into thin multiple layers of AlAs and GaAs, it is discovered that the crystal composition becomes mixed such that the region becomes a homogeneous $Al_xGa_{1-x}As$ material, as illustrated in Fig. 9.23 [110]. When this lattice disordering is done selectively on the wafer, regions can be created that are of different bandgaps and refractive indices and thus

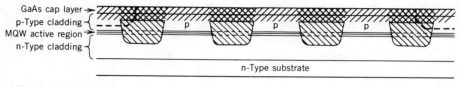

GaAs cap layer
p-Type cladding
MQW active region
n-Type cladding

n-Type substrate

`///` n-Type Si diffused
`\\\` p-Type Zn compensated
`===` Disordered active region
`——` Proton bombardment

Figure 9.23 A buried-heterostructure laser array fabricated by silicon-impurity-induced disordering. (From Thornton et al. [110]. Reprinted with permission from the American Institute of Physics.)

have different electrical and optical properties. Index-guided lasers have been demonstrated using zinc diffusion or ion implantation to produce this lattice disordering [110].

Surface-Emitting Lasers. Surface-emitting (SE) lasers are characterized by the fact that light is emitted vertically from the surface of a chip. There are many advantages of surface-emitting lasers. Most importantly, since the laser facets can be formed with conventional processing techniques without applying mirror cleavage, monolithic integration of lasers and other IC components can be readily achieved. Also, from a manufacturing point of view, vertical emission allows wafer scale probing and testing, thus simplifying the selection process. Furthermore, spatial arrangement of these SE lasers is very flexible, and this means two-dimensional array of coherent emitters may be feasible

There are many schemes to achieve SE lasers. Iga and co-workers at Tokyo Institute of Technology have focused on realizing lasers with an ultrashort vertical cavity enclosed by two facets that are parallel to the surface [111]. Due to the limited active layer thickness ($1–3\ \mu m$), the round-trip gain is rather small, and thus high reflectivity facets (R's $> 95\%$) are needed, as can be seen from Eq. (15). This can be achieved by depositing dielectric multilayers of SiO_2/TiO_2. Alternatively, AlGaAs/GaAs multiple mirror layers can also be used, although the index difference between AlGaAs and GaAs is smaller than that of SiO_2/TiO_2 and therefore many layers are needed to achieve the same reflectivity. The second problem associated with this scheme is the high threshold current density that is needed even with high reflectivity mirrors. Consequently, very good lateral current confinement and heat sink are required. By placing the junction side down to the heat sink, efficient heat sinking can be achieved, as in the case of SELEDs, although this defeats one of the proposed attractiveness of SE lasers—that of integration with other OEICs. For a small emission area, the lateral leakage can be very severe. Iga et al. have used the circular buried heterostructure to confine current and carriers, as

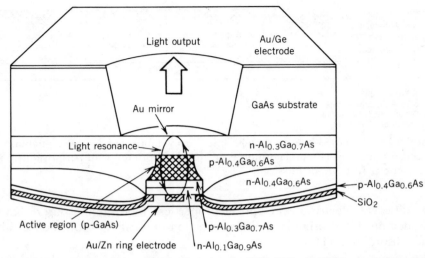

Figure 9.24 Schematic structure of a circular BH AlGaAs/GaAs surface-emitting laser. (From Kinishita et al. [112]. Reprinted, with permission, from *Japanese Journal of Applied Physics*.)

shown in Fig. 9.24, where a circular mesa is etched in the conventional DH layers, followed by a regrowth of the reverse-bias junction layers [112]. Since the emission is from the substrate side and the GaAs substrate is absorptive of 0.8 μm radiation, a groove has to be etched in the back side beneath the lasing region. The facets in this case consist of Au coating and Au/SiO$_2$ (rear facet), and pulsed operation at room temperature is achieved with a nominal threshold current density (normalized with respect to the active region thickness) of 20 A/cm$^2 \cdot$ μm.

Another scheme to achieve SE lasers is to employ a cavity consisted of a conventional horizontal DH waveguide with slanted facets to deflect the laser beam into the vertical direction. High reflectance dielectric coating (at the facets) and semiconductor multilayer reflector (at the substrate side) are used for optical feedback. This scheme has been demonstrated in the InGaAsP/InP system, as shown in Fig. 9.25, and threshold current as low as 65 mA is achieved [113].

As in conventional lasers, the distributed feedback (DFB) approach, which has the property of wavelength selectivity, has been investigated in SE laser studies. In one approach, the (horizontal) output from an MQW AlGaAs/GaAs laser is coupled to a passive DFB grating region. The resultant radiation angle ϕ with respect to the direction to normal to the substrate, however, is not always vertical, and can be expressed as [114]:

$$\phi \sim n_{\text{ext}} \frac{\lambda - \lambda_{\text{B}}}{\lambda_{\text{B}}} \tag{30}$$

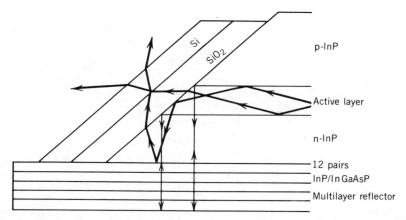

Figure 9.25 A diagram to illustrate the beam reflection and emission at the slant edge of laser; both the SiO_2/Si dielectrics and InP/InGaAsP multilayer reflectors are for feedback to the active layer. (From Ohshima et al. [113]. Reprinted with permission from the Institute of Physics.)

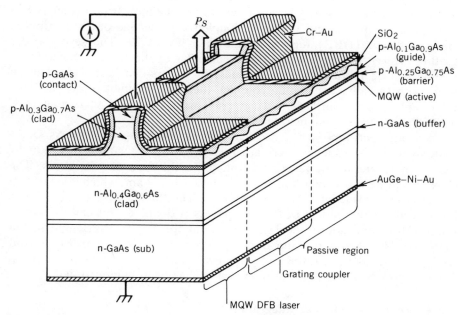

Figure 9.26 Schematic diagram of an integrated device consisting of an AlGaAs/ GaAs MQW DFB laser and a surface-emitting grating coupler. (From Noda et al. [114]. Reprinted with permission from the American Institute of Physics.)

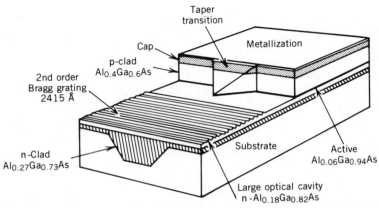

Figure 9.27 Schematic diagram of a large optical cavity surface-emitting CSP laser. (From Evan et al. [115]. Reprinted with permission from the American Institute of Physics.)

where n_{ext} is the effective index of the passive region, λ and λ_B are the lasing wavelength and the Bragg wavelength of the passive region, respectively. By changing the injection current, the radiation angle can be changed. Very narrow far-field angle ($\sim 0.22°$), as shown in Fig. 9.26, and low threshold current (~ 40 mA) are obtained with this approach.

In a similar approach, a large optical cavity channel stripe planar (CSP) laser with a tapered end region is coupled to a second-order distributed Bragg reflector (DBR) region, as shown in Fig. 9.27. The second-order grating diverts the beam to emit perpendicular to the surface. Since DBR grating can be fabricated on both ends, the lasers can be arranged in an array with the adjacent grating coupling the phase of light from individual laser. Five-element surface-emitting arrays with a far-field angle as narrow as $0.25°$ have been demonstrated [115].

In a different approach, a vertical grating consisting of AlGaAs/GaAs multilayers is used and carriers are injected from the lateral direction by means of a lateral p-n junction [116], as in the case of TJS lasers. The thickness of each pair of AlGaAs/GaAs layers is 124 nm, which is set to half of the lasing wavelength in these materials. The thickness of a single GaAs layer at about the middle of the multilayers is doubled; this can work as a phase shifter and allows the laser to lase at the Bragg wavelength in the material. The lateral p-n junction consists of selectivity regrown p- and n-type AlGaAs layers, respectively, on each side of the vertical multilayers. Continuous operation at room temperature has been demonstrated.

9.2.4 Emitter – Amplifier Integration

Monolithic integration of high-speed emitters and other electronics components will remain desirable as long as there is need to increase the data capacity and speed requirements of systems. Early stages of optoelectronic integration circuits (OEICs), such as integration of a laser diode and an FET driver, have demonstrated the feasibility of optoelectronic integration [117, 118]. In the next stages, the electronic part of the circuit will evolve toward large-scale integration (LSI) in order to accommodate more complex functions such as signal processing besides simple amplification. To be successful at those stages, it is imperative that the laser diode fabrication will not impose any limitation such as dimension and layout to the fabrication of other OEIC components. In the following, the key technologies as well as the milestones achieved so far will be summarized.

As is the case for high-speed ICs, OEICs should be operated with as low current levels as possible to minimize heat dissipation. However, this is aggravated by the fact that in planar integrated circuit configuration, where the lasers are usually fabricated on the topside of the chip, lasing junctions are far away from heat sinks on the substrate side and the resultant local heating can degrade the laser performance. For instance, at an injection level of 50 mA, the junction temperature of a well-designed laser diode having a series resistance of 3 Ω will operate at more than 10°C higher than the heat sink. Consequently, lasers with low threshold current and small parasitic resistances are required for OEICs. On the other hand, the laser fabrication should be kept simple to ensure the success of high-yield integration. This limits the laser structure at the present to simple ones such as channel substrate planar (CSP), transverse junction stripe (TJS), and terrace substrate (TS) lasers [119] and rules out complicated ones such as buried heterostructure (BH), which also has inevitable surface unevenness (as a result of the regrowth step) that is undesirable for high-density integration.

Conventional low-threshold laser diodes are typically of the Fabry–Perot type with two cleaved mirror facets separated by 200–300 μm. Although cleaving can provide the best facets as far as quality and reliability are concerned, it also restricts the width of OEIC chips and limits the freedom in layout. Three different ways are studied to remove these restrictions. The first one is the microcleaving technique [57], mentioned in Section 9.2.3.2. This is usually achieved by preferentially etching off the unnecessary portion around the laser active layer to form a cantilever beam. Since this portion is then cleaved off by mechanical shock methods such as ultrasonic vibration, the properties of the rest of the OEICs may be affected. The second method involves the fabrication of at least one laser facet by either wet or dry etching techniques. For the case of wet chemical etching, the etchant, the etching temperature, and masking materials, usually determine the etched morphology [120]. For instance, for GaAs/AlGaAs multilayers, the 1:1:3 solution of $H_2O_2:H_3PO_4:HOCH_2CH_2OH$ at room temperature usually gives good etched profiles with positive photoresist as etch mask. In dry processes, reactive ion etching (RIE) with Cl_2 has been reported to give sufficiently high-quality vertical facets [121]. The third approach is to use facetless laser diodes on OEICs, such as DFB and DBR lasers, as mentioned in Section 9.2.3.2. As the optical feedback in these lasers is accomplished by microcorrugations etched on the substrate, no mirror facet is needed, enhancing the freedom in layout. These lasers have the additional advantages of mode stability and temperature insensitivity. The threshold current as well as yield of these lasers is being continuously improved and commercialization may not be too many years away.

Since both optical and electronic components are to be incorporated adjacent to one another, cross-talk between them may become significant and should not be neglected. It is already known that a MESFET's operation can be significantly changed when light from a laser is absorbed in the layer under the gate. Should this pose a problem, it can be solved by proper layout. Otherwise, compositions of the electronic devices may be made to be of higher bandgap than the laser components to prevent absorption at the lasing wavelength.

The introduction and control of p-type dopants become critical issues in OEICs, since p-type dopants are used in junction devices and possibly as current-conducting paths. Diffusion of zinc can be employed in OEICs; however, this is complicated by the fact that the predominant zinc diffusion mechanism may switch from substitutional to interstitial as the Al content x is increased, thus making the process difficult to control. An alternative technique is to use ion implantation where Be, Mg, or Zn can be selectively implanted [122] into the semiconductor. However, ion implantation alone imposes limitations such as shallow implanted depth and low percentage of ionization. Therefore, a subsequent high-temperature annealing is required, and this would subject the p-type dopants to diffusion.

Another consideration is the substrate flatness throughout the OEIC fabrication. Optical components such as laser diodes at least occupy several microns in vertical space. It is essential for precision in photolithography and in the subsequent patterning that this height be closely matched to that of electronic components without leaving any significant surface steps. One solution to minimize surface steps is vertical integration [120], where the electronics lay on top of the laser diodes. To perform horizontal integration, a counter step can be etched on an original substrate to compensate for the height of the laser [120], as shown in Fig. 9.28. In the case of GaAs (100) substrate, LPE produces an almost flat surface due to the different in growth speeds at the sidewalls of the counter step and at the flat surface. In view of this, a combination of different growth techniques may give more design freedom. For instance, MOCVD may be used to grow conductive layers with uniform thickness along sharp structures etched on the substrate, whereas LPE can be used to grow multilayers on the top to even the surface.

Although semi-insulating (SI) GaAs materials have been readily produced for laser and electronic IC applications, it is still difficult to obtain SI GaAs substrates with low dislocation density. Since dislocations will mani-

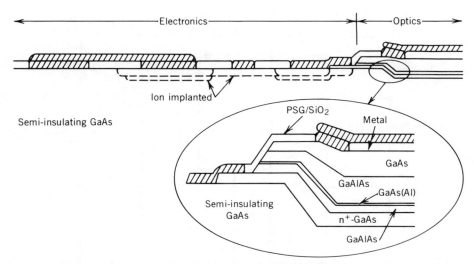

Figure 9.28   An example of horizontal integrated circuit using a growth step. (From Matsueda et al. [120]. Reprinted with permission, copyright 1983, IEEE.)

fest under lasing conditions as dark-line defects and cause degradation, it is anticipated that in future large-scale OEICs, higher-quality (with large defect-free areas) SI GaAs substrates would be in demand.

As an illustration of the OEIC concept, a schematic of an OEIC incorporating laser diodes, photomonitors, and laser-driving and monitoring circuits on semi-insulating GaAs is shown in Fig. 9.29 [123]. Horizontal integration is employed in this structure, with two pairs of laser diodes (LD_1 and LD_2) and photomonitors (PD_1 and PD_2) integrated in the middle of the chip. The lasers consist of an GaAs/AlGaAs MQW active layer, and laser

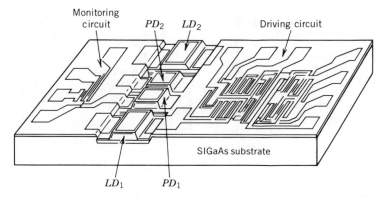

Figure 9.29   A top view of an OEIC incorporating laser diodes, photomonitors, and laser-driving and monitoring circuits on semi-insulating GaAs. (From Nakano et al. [123]. Reprinted with permission, copyright 1986, IEEE.)

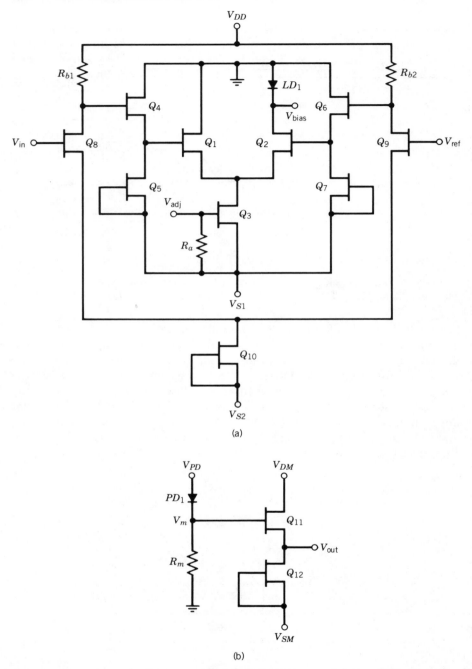

Figure 9.30   Schematic diagrams of the (a) laser-driving and (b) monitoring circuits of Fig. 9.29. (From Nakano et al. [123]. Reprinted with permission, copyright, 1986, IEEE.)

facets are formed by the Cl_2 reactive ion beam etching. The laser-driving circuit consists of a differential current switch and a differential buffer amplifier, as shown in Fig. 9.30a [123]. The output power of the laser can be stabilized by feeding the monitor output signal back to the laser biasing circuit with the monitoring circuit shown in Fig. 9.30b [123]. This OEIC has been operated up to 2 Gbit/s.

Recently, there has been a growing interest in heteroepitaxy among different material systems. Of particular interest to OEICs is the one involving the growth of GaAs/AlGaAs multilayers on Si. This is motivated by the needs to interconnect high-speed Si VLSI subsystems. Employing molecular beam epitaxy (MBE), good-quality GaAs/AlGaAs films have been deposited on Si wafers oriented 3° off (100) toward (111), and electronics devices such as GaAs MESFETs and Si MOSFETs have been made on these wafers with performance similar to that on separate GaAs and Si wafers. With SiO_2 as a mask for selective GaAs/AlGaAs growth, islands of $200 \times 400 \mu m^2$ DH layers are formed. LEDs made from these islands have demonstrated modulation rate up to 27 Mbit/s with signals applied through Si MOSFET gates [124]. The output power of the LEDs, however, is rather low ($\sim 7 \mu W$) due to defects propagating from the Si–GaAs interface.

9.3   GaAs PHOTODETECTOR TECHNOLOGY

For photodetection applications near $0.8 \mu m$ wavelength and for data rates less than few hundred Mbit/s, silicon photodetector technologies, especially those of Si avalanche photodiodes, are well established and have achieved noise and sensitivity characteristics superior to those of GaAs/AlGaAs photodetectors. However, for applications where higher speed or higher data rate is required, the electron velocity in silicon will eventually become a limitation. In these cases, GaAs/AlGaAs photodetectors can offer higher speed, as electrons in GaAs have a higher saturation velocity v_s. The availability of high-quality semi-insulating GaAs facilitates the realization of high-speed photodetectors. Furthermore, GaAs has a large band-to-band absorption coefficient ($\sim 10^4$ cm^{-1}) and thereby a thinner absorption region can be used in photodetectors. In addition, by incorporating DH structures with AlGaAs window layers of higher bandgap, the absorption region within GaAs (or AlGaAs) layer can be vertically defined, which virtually eliminates the effect of diffusion [125].

Another potentially important application of AlGaAs/GaAs photodetectors is in OEICs where other high-speed GaAs electronic devices are incorporated on the same chip; an example was given in Section 9.2.4. GaAs/AlGaAs photodetectors developed on the same substrate as other ICs have the advantages of smaller parasitic capacitance and of reducing the cost of packaging.

In examing the performance of photodetectors as part of a receiver, one always asks, "What is the lowest optical signal level that can be detected without introducing too many errors?" To answer this question, the noise in the entire receiver, especially that contributed by the photodetector and the amplifier circuitry, should be examined and minimized. However, for many applications, there are other parameters that need to be considered, such as speed, quantum efficiency, ease of fabrication, and reliability. The discussion of receiver noise of various photodetectors can be very involved. In what follows, mainly the properties of discrete photodetector components will be emphasized, and the noise properties of receiver can be found in other texts, such as Ref. 126.

9.3.1 Photodiodes and Avalanche Photodiodes

9.3.1.1 Photodiodes: Schottky and PIN Structures. Photodiodes usually have an absorption region where a strong electric field is applied such that photogenerated carriers can be swept out and generate current or voltage signals in the external circuitry. Schottky-barrier and PIN photodiodes are typical examples of photodiodes in which there is no internal photogain. They are similar in that both are potential candidates for high-speed photodetection systems and depend on the reverse-biased, high-field junction region for the photogeneration of carriers.

GaAs Schottky photodiodes consist of an undoped semiconductor layer (usually n-type GaAs) on bulk substrate and a metal layer deposited on top to form a Schottky barrier. As a strong electric field is developed near the metal–semiconductor interface due to the surface barrier, photogenerated electrons and holes are separated and disappear via recombination at the metal contact and in the bulk semiconductor material, respectively. GaAs PIN photodiodes consist of an n^+-AlGaAs, nonabsorbing layer on an n-type GaAs substrate, an undoped GaAs layer (n^- with doping level less than 10^{16} cm^{-3}), and a thin p^+-GaAs (or AlGaAs) layer on top, followed by ohmic contact. Since the GaAs substrate is absorptive of the $0.8\,\mu$m radiation, both types of photodiodes are usually top-illuminated. By applying a reverse bias to these diodes, the depletion region inside the undoped layer can be enlarged and thus both the quantum efficiency and the peak electric field can be increased. The complete absorption of photosignals within the high-field region ensures high-speed performance, since absorption outside the high-field region generates minority carriers which can result in slow diffusion current. However, the undoped layer cannot be made too thick, as the detector speed can then become transit-time-limited. Furthermore, for very high speed operation, the junction capacitance and parasitic capacitance must be reduced to minimize the effect of *RC* time constant. The former can be decreased by reverse-biasing of photodiodes, which increases the depletion width, and by reducing the area of photodiodes [127, 128].

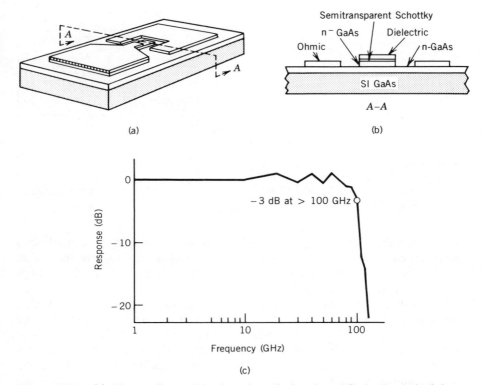

Figure 9.31   (a) Planar GaAs Schotky photodiode: the n-GaAs layer is $0.3\,\mu m$ thick, the n^{+}-GaAs is $0.4\,\mu m$ thick and the Schottky contact consists of 1000 Å of Pt. (b) Cross section along A–A. (c) Fourier transform of an impulse response of the photodiode with FWHM of 5.4 ps exhibiting an bandwidth over 100 GHz. (From Wang and Bloom [128]. Reprinted with permission from IEE).

In comparison with PIN photodiodes, Schottky photodiodes are simpler in structure and Schottky contacts (e.g., Al on GaAs) can be readily made on GaAs materials. As shown in Fig. 9.31, GaAs Schottky photodiodes with a 3 dB bandwidth larger than 100 GHz and operating at less than 4 V reverse bias have been reported by Wang and Bloom [128]. The high speed is accomplished by restricting the photosensitive area to a very small mesa ($5 \times 5\,\mu m^{2}$) and by using a semitransparent Pt Schottky contact [129] for achieving a sizable quantum efficiency with topside illumination.

On the other hand, GaAs PIN photodiodes can achieve higher breakdown voltages and lower dark currents than Schottky diodes and they are not subjected to degradation of the Schottky contact. Planar, low-leakage AlGaAs/GaAs PIN diodes with a zinc diffused p region have been reported with MOCVD [130]. A schematic of the structure is shown in Fig. 9.32; however, its response speed has been limited to 1.2 GHz due to the large

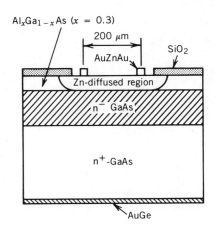

Figure 9.32 Schematic diagram of AlGaAs/ GaAs PIN photodiode grown by MOCVD. (From Ito et al. [130]. Reprinted with permission from IEE.)

area required to achieve high quantum efficiency. By employing a small-area mesa structure with proton bombarding the region surrounding the mesa, GaAs PIN photodiode has been shown to have an impulse response of 19 ps full width at half maximum [131].

For both Schottky and PIN photodiodes, the relative absorption at longer wavelengths (~900 nm) increases with bias voltage. This is believed to be caused by the creation of hot holes in the presence of the electric field enhancement within the n^--GaAs region [129, 130]. At present, both photodiodes are being considered for photodetector–amplifier integrated circuits, as discussed in Section 9.3.4.

9.3.1.2 Avalanche Photodiodes. Avalanche photodiodes (APDs) are based on the impact-ionization effect in semiconductors, where free carriers created by photoabsorption are accelerated by a strong electric field until they gain sufficient energy to promote more electrons from the valence band to the conduction band, thus giving rise to internal photogain. Due to their photogain, APDs are considered alternatives to PINFET or Schottky diode–FET integrations.

The electric field required for impact ionization depends strongly on the bandgap of the material. The minimum energy needed for impact ionization is known as the ionization threshold energy E_i [32], which strongly influences the ionization rates (or coefficients) for electrons (α) and holes (β). These quantities are defined as the reciprocal of the average distance traveled by an electron or a hole, measured along the direction of the electric field, to create an electron-hole pair. Avalanche multiplication is the result of many consecutive occurrences of this process. The multiplication depends on a three-body collision process and is consequently statistical in nature. As a result, it contributes a statistical noise component in addition to the shot noise already present in the diode. This excess noise has been studied by McIntyre for the case of an arbitrary α/β ratio for both uniform and arbitrary electric field profiles [133, 134]. For the uniform electric field case, the excess noise factors for electron and holes are given as

$$F_n = M_n \left[1 - (1 - k)\left(\frac{M_n - 1}{M_n}\right)^2 \right] \qquad (31a)$$

and

$$F_p = M_p \left[1 - \left(1 - \frac{1}{k}\right)\left(\frac{M_p - 1}{M_p}\right)^2 \right] \qquad (31b)$$

where $k = \beta/\alpha$ and M_n and M_p are multiplication factors for pure electron and pure hole injection, respectively. For the case where k is not constant [135], F_n and F_p are given as

$$F_n = k_{\text{eff}} M_n + \left(2 - \frac{1}{M_n}\right)(1 - k_{\text{eff}}) \qquad (32a)$$

and

$$F_p = k'_{\text{eff}} M_p + \left(2 - \frac{1}{M_p}\right)(1 - k'_{\text{eff}}) \qquad (32b)$$

where k_{eff} and k'_{eff} depend on the spatial averages of α and β [135]. Thus, for low-noise APDs, the ionization rate for one type of carrier must be much greater than that of the other, and the carrier with the larger ionization coefficient should initiate the avalanche process.

Ionization coefficients in GaAs have been studied extensively. However, results obtained so far are still controversial, as various methods reported in the literature all involve some measurement uncertainties [136]. It is generally agreed β is less than α for $\langle 100 \rangle$ substrates and α/β ratio decreases from 2.5 to 1.3 as the electric field is increased from 2.2 to 6.25×10^5 Vcm^{-1}. Also, both coefficients decrease with increasing temperature. Figure 9.33 shows the experimental results for these coefficients as a function of 1/(electric field) [137].

Several kinds of GaAs APDs have been demonstrated; earlier work included a Schottky-barrier APD reported by Lindley et al. at Lincoln Laboratory [138]. A schematic diagram is shown in Fig. 9.34 [136], where the proton-bombarded region serves as a guard ring. These diodes exhibit a gain–bandwidth product greater than 50 GHz and an enhanced signal-to-noise ratio greater than 30 dB. Another one involves an GaAsSb/GaAs heterojunction with an inverted mesa structure designed for 1.064 μm detection, as shown in Fig. 9.35 [136], where absorption and multiplication processes take place in the GaAsSb p-n junction [139].

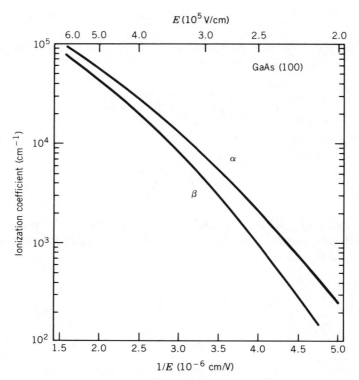

Figure 9.33 Measured ionization rates for both electrons (α) and holes (β) as a function of the inverse electric field for $\langle 100 \rangle$ GaAs. (From Bulman et al. [137]. Reprinted with permission, copyright 1983, IEEE.)

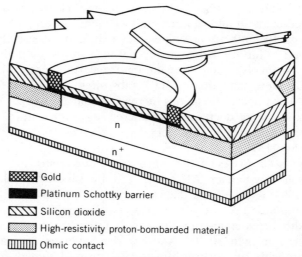

Figure 9.34 Schematic diagram of an APD with proton-bombarded guard-ring APD. The substrate is GaAs. (From Stillman et al. [136]. Reprinted with permission, copyright 1984, IEEE.)

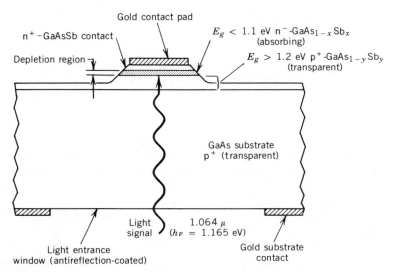

Gold contact pad

n$^+$-GaAsSb contact

$E_g < 1.1$ eV n$^-$-GaAs$_{1-x}$Sb$_x$
(absorbing)

Depletion region

$E_g > 1.2$ eV p$^+$-GaAs$_{1-y}$Sb$_y$
(transparent)

GaAs substrate
p$^+$ (transparent)

Light
signal

1.064 μ
($h\nu = 1.165$ eV)

Light entrance
window (antireflection-coated)

Gold substrate
contact

Figure 9.35 Schematic diagram of an inverted heterojunction mesa APD. (From Stillman et al. [136]. Reprinted with permission, copyright 1984, IEEE.)

9.3.2 Photoconductors and Phototransistors

9.3.2.1 *Photoconductors.* As alternatives to photodiodes, photoconductors can be used in which photogain can be obtained. There are two basic generic types of photoconductors, extrinsic and intrinsic. In intrinsic photoconductors, the conduction is enhanced by absorption of light, which creates electron-hole pairs via band-to-band transitions across the bandgap. For GaAs, the absorption coefficient α is very large due to the large number of available electron states associated with the conduction and valence bands, and α is typically on the order of 10^4 cm^{-1} for photons near the bandgap energy. In extrinsic photoconductors, photons are absorbed at impurity levels, and consequently free electrons are created in n-type semiconductors and free holes are created in p-type semiconductors. The extrinsic photoconductivity is characterized by a small absorption coefficient because of the small number of available impurity levels.

When a steady optical signal with power P_i and wavelength λ is incident on the photoconductor, the optical generation rate of the electron-hole pairs as a function of distance x into the photoconductive sample is given by

$$g(x)dx = \frac{P_i}{h\nu}\,\alpha e^{-\alpha x}dx \tag{33}$$

where α is the absorption coefficient at the energy $h\nu$. If the sample has a finite thickness D, part of the incident photons will be transmitted through without being absorbed, and this decreases the quantum efficiency. For intrinsic photoconductors, the change in average steady-state electron and hole concentration can be written as

$$\Delta n = \Delta p = G\tau = \frac{(1-R)}{WLD} \int_0^D g(x)\tau dx$$

$$= \frac{P_i}{WLDh\nu} (1-R)(1 - e^{-\alpha D})\tau \qquad (34)$$

where G is electron (or hole) generation rate per unit volume; τ is the excess carrier lifetime; W, L, and D are, respectively, the width, length, and thickness of the sample; and R is the reflectivity at the semiconductor–air interface. The Δn and Δp change the conductivity of the sample and give rise to an dc photoconductive gain M, defined as the number of electron–hole pairs collected per incident photon [140] (see Eq. (36) with $\omega = 0$).

GaAs photoconductors have also been proposed for photo-switching [141] and multiplexing [142] applications due to their simplicity and internal photogain. For 0.8 μm applications, several kinds of extrinsic GaAs photoconductive devices have been demonstrated, and they are distinguished from each other by the nature of electrical contacts (ohmic versus Schottky) and the semiconductor layer for photoconduction [140]. Typically, GaAs photoconductors are formed by depositing two ohmic contact electrodes on a layer of conductive GaAs formed over an SI GaAs substrate. The resulting devices usually have large dark current and a fast rise time (10–50 ps) response, as shown in Fig. 9.36. However, the fall time tends to be orders of magnitude longer than the rise time, which may be related to the out-diffusion of deep-level impurities (which are traps for holes) from the SI GaAs substrate [143]. The symmetric interdigitated structure shown in Fig. 9.37 can overcome these problems [144]. The undoped GaAs layer still serves as photoconductive layer, while depletion regions of Schottky barriers extend laterally across the GaAs layer to minimize the dark current. An AlGaAs layer is grown to isolate the photoconductive GaAs layer from the impurities in the substrate and thus shortens the fall time [145]. Alternatively, by incorporating 0.3% of indium in the n-GaAs layer, the influence of impurities from the SI GaAs substrate can be reduced [146]. By subsequently etching a mesa around the photodetective region to reduce parasitic elements, Schottky-barrier photoconductors have demonstrated low dark current (~1 nA) at −10 V [146].

Recently, a high-speed (rise time ~ 100 ps) and moderate gain ($M \sim 3$) GaAs optical FET (OPFET) with a Schottky gate has been demonstrated by Gammel and Ballantyne [147], and it operates essentially in the photoconductive mode. The optimal gain–bandwidth product in these devices is found to be proportional to the peak velocity of the electrons and inversely proportional to the length of the high-field region in the channel [148], and thus shorter-channel devices are prefered.

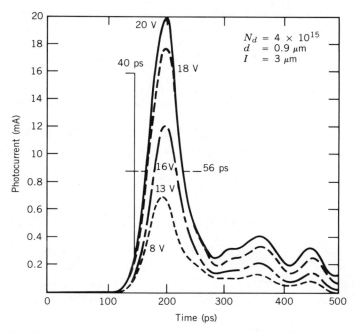

Figure 9.36 Photoresponse of a GaAs photoconductor at different voltages as a function of time. (From Wei et al. [145]. Reprinted with permission from IEE.)

As in APDs, noise considerations are important in many applications of photoconductors. Aside from the dark current noise due to the conductive channel, the generation–recombination current noise i^2_{g-r}, becomes dominant when the device is under strong illumination. This noise depends on the numbers as well as the lifetime τ of the photogenerated electron-hole pairs

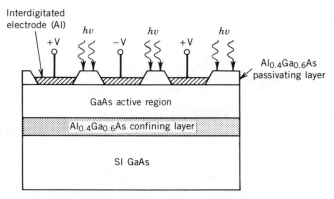

Figure 9.37 Schematic diagram of a GaAs interdigitated photoconductor. The electric field at the substrate is reduced by the AlGaAs confinement layer. (From Figueroa and Slayman [144]. Reprinted with permission, copyright 1981, IEEE.)

and can be expressed as [149]:

$$\langle i^2_{g-r} \rangle = 4e \frac{I_{ph}}{M(0)} \left(\frac{\tau}{\tau_t} \right)^2 \frac{B}{1 + \tau^2 \omega^2} \tag{35}$$

where I_{ph} is the primary photocurrent, $M(0)$ the dc gain, τ_t the carrier transit time across the device, and ω the angular frequency. For a 15 μm channel length GaAs photoconductor, the measured noise at various illumination levels and at different signal frequencies is shown in Fig. 9.38. At ac, the dynamic gain $M(\omega)$ can be expressed as [149]:

$$M(\omega) = \frac{\tau}{\tau_t} \frac{1}{\sqrt{1 + \tau^2 \omega^2}} \tag{36}$$

It can be seen from Eq. (36) and Fig. 9.38 that $M(\omega)$ decreases with increasing signal frequency [150].

Most GaAs photoconductor work in the past was performed employing LPE; however, both MBE and MOCVD have recently achieved layers of lower dopant concentration and improved thickness uniformity. AlGaAs/GaAs photoconductors with modulation doped interface have been reported

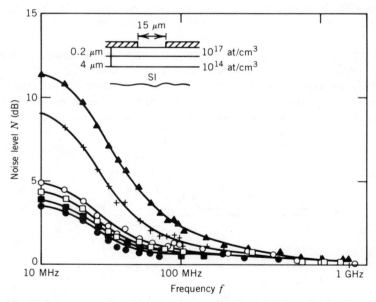

Figure 9.38 Noise power of a photoconductor (structure shown in the insert) as a function of frequency, measured at various optical intensities: the triangle corresponds to 5 μW illumination, the + to 1.5 μW, the open circle to 600 nW, the open square to 60 nW, the filled square to 3.8 nW, and the filled circle to no input. (From Vilcot et al. [149]. Reprinted with permission from IEE.)

and, at 1.5 GHz, a responsivity of 2 A/W which corresponds to a gain of three has been obtained [151, 152]. This device has a built-in electric field at the (undoped) p^--GaAs/n^+-AlGaAs junction and can operate at high speed even in the absence of external bias [152]. These modulation doped photoconductors are usually less noisy than their bulk counterparts due to the higher electron confinement. Also, GaAs photoconductors with a 3 μm thick absorption layer and with a rise times less than 10 ps have been made by laser-stimulated MOCVD [153].

9.3.2.2 Phototransistors.

Although most of the previous work on phototransistors has been based on silicon and germanium [154, 155], recent interest has been focused on III–V phototransistors in view of the good crystal quality of heterojunction devices and their higher bipolar transistor gain. By employing a heterojunction, a wide-bandgap-emitter-configurated phototransistor which greatly improves the emitter injection effeciency becomes available [156]. The heterojunction relieves the restriction on relative dopant concentrations on both sides of the emitter–base junction [157], and a high gain–bandwidth product is in principle possible.

The operation of an n-p-n phototransistor can be briefly described as follows. A bias is applied between the emitter and collector terminals such that the emitter and collector junctions are forward- and reverse-biased, respectively. As incident photons pass through the wide-bandgap emitter, which serves as a transparent window, they are absorbed in the base and collector regions, where electron-hole pairs are created. The photosensitive region of the AlGaAs/GaAs phototransistor has been determined from the EBIC measurement [158], and it has been found that most of the induced current signal originates from the collector depletion region due to the presence of the large electric field there. The holes generated at the base, at the collector depletion region, and within a diffusion length from the edge of the junction in the collector eventually accumulate in the base. These holes change the potential at the emitter–base junction, causing electron injection from the emitter into the base. Current gain is obtained when the lifetime of injected electrons in the base is longer than their transit time across the base. As photosignals act as base bias, the phototransistor usually operates in the floating base configuration. This has the merit of low capacitance due to the absence of base contact.

For a n-p-n phototransistor, the emitter injection efficiency η_e is given by [159]:

$$\eta_e = \frac{\Gamma \cosh(w_b/L_{nb})}{1 + \Gamma \cosh(w_b/L_{nb})} \tag{37}$$

where Γ is proportional to a parameter γ which is expressed as [159]:

$$\gamma = \frac{D_{nb}L_{pe}n_e}{D_{pe}L_{nb}p_b} \left(\frac{m_{nb}^* m_{pb}^*}{m_{ne}^* m_{pe}^*}\right)^{3/2} \exp\frac{\Delta E_v}{kT} \tag{38}$$

In Eqs. (37) and (38), w_b stands for the width of the base region; D's and L's are diffusion coefficients and diffusion lengths, respectively; subscripts e and b refer to emitter and base regions, respectively; n and p stand for electrons and holes, respectively; m^*'s stand for effective masses; and ΔE_v is the valence band discontinuity at the emitter and base junction. In Eq. (38), an abrupt heterojunction is assumed and it is seen that the exponential dependence of γ on ΔE_v can offset the effect of the relative doping levels and that the emitter injection efficiency can be designed very close to unity. Also, the frequency response of phototransistors can be improved by employing designs with a lower emitter capacitance (lightly doped emitter) and a reduced base resistance (heavily doped base).

For phototransistors with a large valence band discontinuity such that Γ is much larger than unity and a base region width much smaller than the minority carrier diffusion length at the base, the optical gain G can be related to the emitter efficiency η_e through

$$G \sim \eta \, \frac{\eta_e \eta_b}{1 - \eta_e \eta_b} \tag{39}$$

where η is the phototransistor quantum efficiency and η_b is the base transport efficiency which is close to unity. By reducing the base width and doping level, very high gain can be obtained. The optical gain, however, has a pronounced collector voltage dependence due to the Early effect [158]. In an actual device, the phototransistor optical gain at low current is limited by the presence of defect current due to recombination centers in the vicinity of the emitter heterojunction [160]. By placing the emitter p-n junction in the wide-bandgap material and at a small distance from the heterojunction interface such that space regions of both overlap, the defect current can be reduced [161] while a high injection efficiency is maintained.

For GaAs/AlGaAs material systems, phototransistors typically consist of an n^--GaAs collector, a p-GaAs base, and an n-$Al_x Ga_{1-x}$As wide-bandgap emitter. The short wavelength cutoff near 0.65 μm is caused by the emitter absorption and the long-length cutoff near 0.88 μm is due to the absorption edge of GaAs. For floating base configurations, current gain of 5000 has been achieved [158], whereas high optical gains occurs only at high signal levels ($\geqslant 1$ μm), due to the defect current at the emitter junction. In a diffused base configuration, as shown in Fig. 9.39, the emitter–base capacitance is minimized, and a current gain of more than 10^4 has been observed [162]. MBE has also been employed in the fabrication of phototransistors, in which a maximum current gain of 300 and a 250 ps rise time has been observed [163].

Due to the dependence of the emitter–base potential on the base charge, the large-signal response of phototransistors depends primarily on the charging times of the junction capacitance through the emitter resistance and that this charging is achieved through the primary photocurrent and the

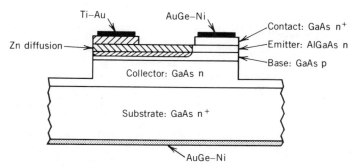

Figure 9.39 Schematic diagram of an AlGaAs/GaAs phototransistor with diffused base contact region. (From Scavennec et al. [162]. Reprinted with permission from IEE.)

dark current. This causes the rise time to be proportional to the emitter–base and collector–base capacitances, and inversely proportional to the collector current, which depends on the gain as well as the sum of the primary photocurrent and dark current.

The phototransistor can also operate in the avalanche mode by biasing the collector–base junction into the avalanche regime. This leads to an increase in the current gain; however, it also causes a softening of the transistor breakdown characteristics. Moreover, a complex bias circuitry is required to control the temperature and bias voltage precisely.

Since phototransistors use relatively high current (~500 mA) output, they can be used directly as a LED driver with optical input. AlGaAs/GaAs phototransistors have been vertically integrated with AlGaAs/GaAs LEDs in an array arrangement [163]. The inherent high gain can be used in image converters with optical amplification. Furthermore, the phototransistor can be used to drive a laser in a repeater. As demonstrated by Bar-Chaim et al. [164], regrowth layers of BH GaAs/AlGaAs lasers can be utilized for phototransistors, as shown in Fig. 9.40. In this case, high responsivity of 75 A/W is obtained for phototransistors, as an additional absorption occurs in the GaAs collector depletion region.

9.3.3 Recent Advances in Photodetection

The capability of MBE and MOCVD for growing ultrathin epitaxial layers and their added flexibility in precise control over doping profiles have led to new ideas on photodetection. Recent efforts on photodetector research, though, tend to emphasize more novel approaches to achieve solid-state, low-noise photodetection and multiplication. Many schemes have been proposed to achieve low-noise avalanche photodiodes [165], most of which rely on the so-called "bandgap engineering" approach, that is, one utilizing the characteristics of heterojunctions, MQWs, and superlattices, in contrast

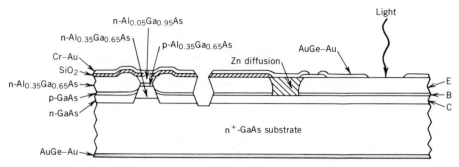

Figure 9.40 Schematic cross section of the AlGaAs/GaAs BH laser integrated with a bipolar phototransistor. (From Bar-Chaim et al. [164]. Reprinted with permission from the American Institute of Physics).

to conventional devices based on bulk material properties. These devices are still under development and may become practical when the control and fabrication of techniques are perfected.

To improve the photodetective device based on avalanche multiplication, it is necessary to reduce the noise generated in the avalanche multiplication process. This can be achieved if the ratio of the ionization coefficients, α and β, can be enhanced. However, in most bulk III–V semiconductor materials, α's and β's are quite close in value. The following describes some novel schemes that make use of the band properties of heterojunctions and quantum wells to enhance the difference of threshold energy for electron and hole impact ionization.

9.3.3.1 Graded Bandgap Avalanche Photodiode. The operation of this device is based on the observation that, inside a graded bandgap region, the extent to which the conduction band is graded can be different from that of the valence band, as a result of different band discontinuities in conduction and valence bands from one material to another. As shown in Fig. 9.41, if the conduction band has a larger grading than the valence band, electrons can be accelerated to a higher energy than holes even when the same electric field (from external bias) is applied on both. Consequently, the α/β ratio can be enlarged, provided that the accelerating electrons travel across the different bandgap region without collision with phonons and reach the threshold energy for impact ionization. In practice, an ionization rate ratio k (which equals $(M_p - 1)/(M_n - 1)$) less than 10 has been obtained for a p-n junction with a 0.4 μm graded bandgap $Al_x Ga_{1-x}As$ (x ranges from 0 to 0.45) layer [166].

9.3.3.2 Avalanche Photodiode Incorporating Superlattice. As in the case of graded bandgap APDs, the disparity in the bandgap discontinuities between two adjacent layers with large bandgap difference can be used to enhance

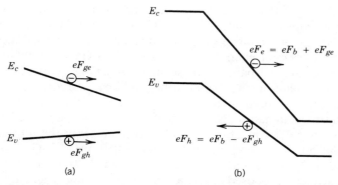

Figure 9.41 (a) Effect of quasi-electric field in a graded gap material, (b) combined effect of applied and quasi-electric field. (From Capasso [165]. Reprinted with permission of Academic Press.)

the α/β ratio [167]. Consider electrons and holes coming from the larger bandgap layer (barrier) to the smaller bandgap layer (well), as shown in Fig. 9.42. These carriers experience a different drop in potential energy due to the different band discontinuities ΔE_c and ΔE_v. Provided that the heterojunction is abrupt enough that the transition region is much shorter than a free path for phonon scattering, the potential energies gained can be added to those gained through the electric field to reach the threshold energy.

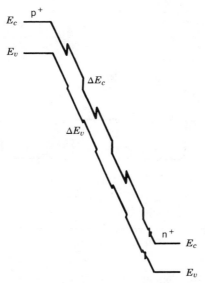

Figure 9.42 Energy band diagram of a superlattice APD with unequal ΔE_c and ΔE_v. (From Capasso et al. [169]. Reprinted with permission from the American Institute of Physics.)

Since ΔE_c is larger than ΔE_v at GaAs/AlGaAs heterojunctions, and GaAs has a lower threshold energy than AlGaAs (due to the dependence of threshold energy on bandgap in the band-to-band impact ionization [168]), electrons have a larger chance to impact-ionize in GaAs. With superlattice structure consisting of pairs of these heterojunctions, two effects can be obtained. The photon can be absorbed in any one of the wells, thus absorption is not limited to the surface. Secondly, the impact ionization, which is statistical in nature, is more likely to occur in one of the "falls." For the case of an APD consisting of a PIN structure with the intrinsic region consisting of 50 alternating layers of GaAs($\sim$450 Å)/Al$_{0.45}$Ga$_{0.55}$As super-lattice, and α/β ratio of eight has been achieved [169]. Other variations of this approach are proposed, such as those employing a superlattice with graded gap section [170] or with periodic doping profile [171]. In the graded gap approach, the wells are graded in composition to reach the bandgap of the barrier layer so that electrons are less likely to be trapped in the well. In the latter approach, as shown in Fig. 9.43, the main feature is that, by

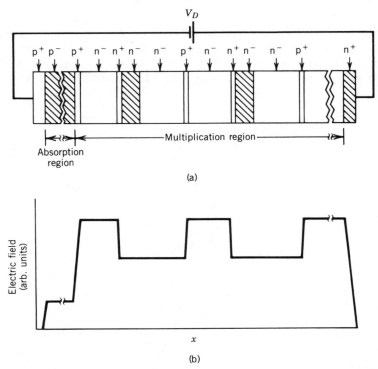

(a)

(b)

Figure 9.43 (a) Schematic diagram of superlattice APD structure with periodic doping profile: material A (shaded region); material B (clear region). (b) Electric field in each layer of the structure. (From Blauvelt et al. [171]. Reprinted with permission from IEE.)

optimizing the doping profile, electrons can be designed to go from a high electric field region to a small bandgap region while holes enter the same region from a low electric field region. This ensures that the fraction of injected electrons with energy above the ionization threshold is significantly larger than that of holes.

9.3.3.3 Staircase Solid-state Photomultiplier. This proposed device is a multistage graded-gap structure, as shown in Fig. 9.44. As in the case of superlattice APDs where impact ionizations are assisted by band discontinuity effects, the electron ionization energy in this structure is entirely provided by conduction band steps [172]. Each stage in this photomultiplier structure is linearly graded in composition from a low bandgap to a high bandgap, with an abrupt transition back to the low bandgap material. The energy step should be equal to or greater than the electron ionization energy. Since the conduction band discontinuity in GaAs/AlGaAs materials accounts for most of the bandgap difference, the corresponding grading and step for the valence band are considerably smaller.

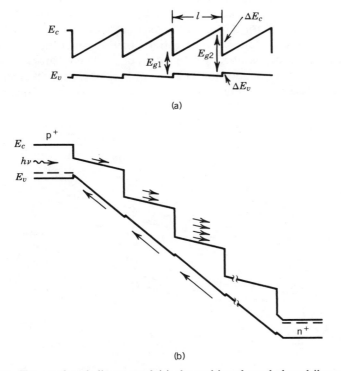

Figure 9.44 Energy band diagram of (a) the unbiased graded multilayer region of the staircase detector and (b) the biased detector. (From Williams et al. [172]. Reprinted with permission, copyright 1983, IEEE.)

In this staircase structure, a photogenerated electron at the p^+ contact will drift, under the net effect of the bias field and grading field ($\Delta E_c/e$), toward the first conductive band step without ionization. However, as it falls down the step, the energy gained can cause impact ionization, and the same process repeats at each stage. For the hole, ionization is caused by the applied electric field and the small grading of the valence band only, since the valence band step is of the wrong sign to assist ionization.

9.3.3.4 Channeling APD. This device is based on a novel interdigitated p-n junction, as shown in Fig. 9.45, where electrons and holes are spatially separated by a transverse electric field and confined in layers of different bandgap. A lateral electric field extending from the p^+ to n^+ regions subsequently drifts them along these layers, where they undergo ionization. Since electrons and holes impact ionize in layers of different bandgap, the ionization rate ratio can be made extremely large by choosing the proper bandgap difference while maintaining a high gain [166]. To maintain a high ionization rate ratio, however, the thickness of the narrow-gap region must be smaller than $1/\beta$ so that the photogenerated or impact-generated holes can be swept out to the higher bandgap layer before they undergo ionization collision.

Another direction for GaAs/AlGaAs photodetector research is in the far-infrared ($\sim 4\ \mu$m) solid-state detection. In this case, the intraband free-carrier transition of heterojunctions can be utilized for photodetection. Although free-carrier absorption effect is not significant in bulk materials, it can be enhanced in modulated doped structures, where a higher electron

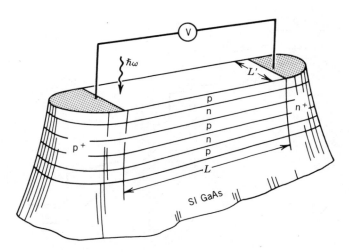

Figure 9.45 Schematic diagram of a channeling APD with the p layer having a wider bandgap than the n layers. (From Capasso [166]. Reprinted with permission, copyright 1982, IEEE.)

density can be achieved. Both phonon-assisted and phononless free-carrier absorption has been analyzed [173–174]. For phonon-assisted free-carrier absorption in GaAs/Al$_{0.3}$Ga$_{0.7}$As system, the theory predicts that the absorption coefficient is considerably larger in the quasi-two-dimensional system, where modulated doped GaAs detection layers are used [173]. An order-of-magnitude improvement in absorption coefficient can be achieved in the case of phononless free-carrier absorption, as depicted in Fig. 9.46, where strong coupling can be obtained when the incident radiation is polarized in the direction perpendicular to the well layers [174]. An important distinction between the two cases is that the excited electron in the phononless case continues to travel in the direction perpendicular to the layers. When an external electric field is applied, a significant fraction of these electrons can be pulled out of the well region and accelerated in the same fashions as the APDs described above. However, in the phonon-assisted case, the electron momentum will be randomized by the phonon emission and absorption processes.

A proposed superlattice far-infrared detector is schematically shown in Fig. 9.47 [174]. Adjacent to the GaAs well region is a layer of Al$_x$Ga$_{1-x}$As followed by a graded bandgap Al$_y$Ga$_{1-y}$As with $y > x$ to provide a thin tunneling barrier for electrons in band E_f (see Fig. 9.47). The graded-gap region has a mobility that can be enhanced when an external bias is applied. However, to minimize the dark current, the lowest lying energy level (E_i) in

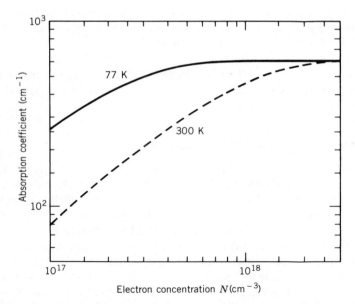

Figure 9.46 Calculated phononless free-electron absorption coefficient as a function of electron concentration at 77 and 300 K for GaAs/Al$_{0.3}$Ga$_{0.7}$As modulated doped material [174]. (Reprinted with permission from Dr. L. C. Chiu.)

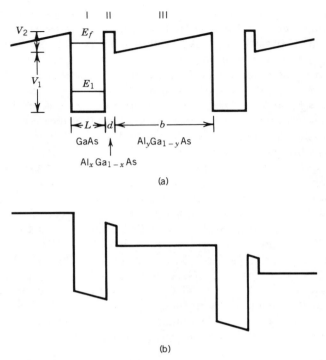

Figure 9.47 Schematic diagram of a proposed superlattice far-infrared detector utilizing tunneling effect (a) at thermal equilibrium, (b) under an externally applied bias [174]. Reprinted with permission from Dr. L. C. Chiu.)

the well region should be effectively decoupled from neighboring regions to prevent tunneling. Devices based on resonant tunneling are discussed further in another chapter of this book.

9.3.4   Photodetector – Amplifier Integration

As for emitters, OEICs for receivers have attracted much interest due to the potential advantages of speed and system simplicity. For applications such as high-speed fiber optic links and recovery of the low power level, high-speed electrical signals generated in the photodetector have become a major challenge. Besides requirements of high speed and low-noise amplification, noise coupled inductively from nearby circuits, the preamplifier, and the photodetector must be minimized. The monolithic integration of the photo-detector with its accompaning amplifying circuits on the same substrate can greatly reduce these problems primarily due to the reduction of parasitic reactances.

As discussed in detail by Smith and Personick [126], in optical receiver designs, two basic approaches are used for the preamplifier: the transimpedance amplifier and the high-impedance amplifier approaches. The former approach can achieve good dynamic range, but is difficult to implement at high frequency; the latter approach can achieve good sensitivity. A typical circuit for the amplifier approach is shown in Fig. 9.48, where i_s denotes the photocurrent, i_d denotes the dark current, i_n is the noise current, e_n is the noise voltage, R_L is the bias resistor, and C_d, C_s, C_i are the capacitances associated with the diode, parasitics, and the amplifier (e.g., FET gate-to-channel capacitance). For the front end, consisting of an FET that is driven by a photodiode, a figure of merit F can be defined to specify the optimal condition for low-noise operation [126]:

$$F = \frac{g_m}{C_T^2} \tag{40}$$

where g_m is the channel transconductance and C_T is equal to the sum of the capacitances. It turns out that g_m/C_i is related to the gain–bandwidth product, which depends only on material properties and the channel dimensions [126]; thus, F can be maximized when C_i equals the sum of C_s and C_d. Consequently, for achieving high-frequency and low-noise integrated circuits for the front end, the parasitic and photodiode capacitances must be reduced as much as possible.

As a large load resistance R_L is used in either high-impedance or transimpedance approaches, it appears that the frequency performance of the front end will be limited by the RC time constant. However, the signal integrated by the large time constant at the front end can be restored by differentiation at the equalizer [159]. An integrated two-stage amplifer receiver employing this concept has been demonstrated by Hamuguichi et al. [175]. A circuit diagram of the receiver is shown in Fig. 9.49. The first stage of the receiver consists of a photodiode and a transimpedance-configurated amplifier (Q_1–Q_6 FETs); the second stage is another amplifier circuit configurated from Q_7–Q_{12} FETs. In a dc potential design, the zero-signal dc input potential at the gate of Q_2 is equalized to the output potential at the drain of Q_4 by connecting the Q_2 gate to the level-shift output stage through a feedback resistance R_f. The optimum dc potential for maximizing the gain of the input inverter can then be obtained by adjusting

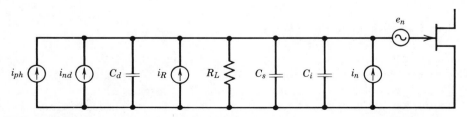

Figure 9.48 A typical circuit for a photodiode driving an input FET amplifier. (From Margalit and Yariv [159]. Reprinted with permission from Academic Press.)

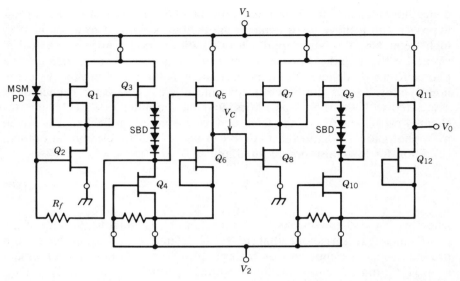

Figure 9.49 Circuit diagram of an optoelectronic integrated receiver involving a two-stage amplifier. (From Hamaguchi et al. [175]. Reprinted with permission, copyright 1987, IEEE.)

the number of level-shift diodes. For a design with a photodiode capacitance of 0.1 pF, gate width of 150 μm for Q_2 and Q_8, 300 μm for those of Q_5 and Q_6, and an overall gain of 5, a response of 2 Gbit/s nonreturn-to-zero has been achieved [175].

In spite of the recent progress in photodetector research, only the PIN photodiode [176] and metal–semiconductor–metal (MSM) photodiode [177] have been seriously considered in integrated receiver designs. Although PIN diodes can be designed to operate at high frequency and at high sensitivity, due to the vertical nature in most of the high-speed design, some structural features such as surface polishing [178] and bridge interconnection [176] have to be introduced for useful monolithic integration. In this respect, the interdigitated MSM diodes are more readily integrated with FETs since both employ Schottky contacts.

9.4 FUTURE PERSPECTIVES

GaAs optoelectronics has come a long way since 1970. We have already witnessed mass commercialization of many devices. The future depends on the successes of current research directions. The impetus will come from two major fronts: commercial exploitation of the unique properties of GaAs and research into more basic properties which may give rise to more exotic devices with unique characteristics.

In the very near future, it is expected that OEICs will be readily available. Some simple functional components such as transmitters, receivers, and repeaters are either in development at major companies or are already available in sample quantities.

Further optimization of lasers will allow them to operate with sub-milliamp thresholds and operation points at a few milliamps. This enables lasers to be compatible with standard logic interfaces for digital and computer applications.

Lasers and electronic devices using epitaxial material grown on silicon substrates have already been demonstrated in laboratories. The big difference in crystal lattice constants usually produces inferior materials with many defects. However, by introducing multiple thin layers of GaAs/ AlGaAs as a buffer layer between the silicon substrate and layers which make up the devices, it is found that defects originating from the substrate may be stopped by the buffer layer, thereby permitting growth of extremely high-quality layers from which even lasers can be fabricated. This is very exciting because, when perfected, the technology permits intermaterial marriage between GaAs optoelectronics and silicon ICs, achieving both higher complexity of chip structure and lower costs, since silicon substrates are much cheaper than GaAs.

In discrete devices, the demand for higher optical output and larger bandwidth seems insatiable. New applications can always be found when a device with even more superior performance becomes available. It is not obvious what the theoretical limits for these characteristics are, but power output will probably be limited by packaging and reliability concerns to several watts, and parasitics and inherent capacitances will probably limit devices to 20–30 GHz operation.

Compared with silicon, GaAs growth and processing techniques are still in their infancies. This is complicated by the fact that many devices require even better controls in epilayer structure, thickness, and composition than silicon, and therefore it is necessary to develop new techniques to enable fabrication of devices dreamed up by physicists and engineers. Advanced techniques such as RIBE (reactive ion beam etching), focused ion beam writing, and laser-assisted CVD are actively being developed. It is hoped that, once perfected, these techniques will pave the way for much more complex three-dimensional structures required by optoelectronics than currently possible.

More exotic devices have been designed on paper, and some have even been demonstrated to some extent. Although many of these devices may not have immediate significant impacts on practical or commercial applications, they help extend the understanding of basic underlying physics of solid-state matter. For example, quantum effects in two-dimensional electron gas are still not completely characterized and understood. Some measured values still cannot be fully explained theoretically, and it is not clear if some effects are fabrication-induced. For example, when the first quantum well lasers were discovered, experimental results seemed to indicate that optical transitions take place with the aid of phonons. However, these results were not reflected in devices fabricated later, using more sophisticated techniques. Despite this, research groups have gone another step further, to quantum wire, that is, one dimensional devices. Presently, confinement in one additional dimension may be accomplished through a strong externally applied magnetic field.

Modeling of quantum well devices using a simple square well potential decoupled from the periodic potentials of the other directions that form the solid crystalline matter can yield rather good first-order agreements with experimental data. But, as device fabrication methods get more sophisticated and device performance reflects the real underlying physics more accurately, quantum size effects cannot be satisfactorily explained quantitatively by the simple model. Proper modeling is now a topic of intense study by many physicists.

In other related areas, devices based on AlInGaP epilayers lattice-matched to GaAs are actively pursued at many companies, notably those in Japan, and lasers with emission wavelengths as short as 6400 Å have already been demonstrated [179]. This material system has an inherent advantage over the AlGaAs/GaAs in the visible spectrum because lattice mismatch of AlGaAs increases with Al content, making lasers with wavelengths less than 7500 Å much less reliable, whereas perfect lattice matching of AlInGaP with GaAs is possible while permitting wavelength tuning by adjusting the composition. It should not be long before truely visible semiconductor lasers become commercially available.

In conclusion, the ferocious appetite for higher-performance devices and new applications in GaAs based optoelectronics will propel future developments at a tremendous pace, and large-scale commercialization will play a major role in the much heralded information age.

ACKNOWLEDGMENTS

The authors are in debt to Dr. L. C. Chiu of PCO Inc. for his valuable suggestions and for proofreading the photodetector section. The authors would also like to thank Prof. J. Meyer of Cornell University for his comments on OEICs.

REFERENCES

1. H. C. Casey, Jr., and M. B. Panish, "Heterostructure Lasers." Academic Press, New York, 1978.
2. S. E. Miller and A. G. Chynoweth, eds., "Optical Fiber Telecommunication." Academic Press, New York, 1979.
3. H. Kroemer, *Proc. IEEE* **51** 1782 (1963).

4. H. K. Kressel and J. K. Butler, "Semiconductor Lasers and Heterojunction LEDs." Academic Press, New York, 1977.

5. G. H. B. Thompson, "Physics of Semiconductor Laser Devices." Wiley, New York, 1980.

6. L. C. Chiu, P. C. Chen, and A. Yariv, *IEEE J. Quantum Electron.* **QE-18**, 938 (1982).

7. R. H. Saul, T. P. Lee, and C. A. Burrus, *in* "Semiconductors and Semimetals" (W. T. Tsang, ed.), Vol. 22, Part C. Academic Press, Orlando, Florida, p. 193, 1985.

8. K. Nitta, T. Komatsubara, H. Kinoshita, Y. Hzuka, M. Nakamura, and T. Beppu, *Electron. Lett.* **21**, 208 (1985).

9. K. Yamashita, Y. Ono, and K. Itoh, *IEEE Electron Device Lett.* **EDL-7**, 81 (1986).

10. C. A. Burrus, and B. I. Miller, *Opt. Commun.* **4** 307 (1971).

11. G. W. Berkstresser, V. G. Keramidas, and C. L. Zipfel, *Bell Syst. Tech. J.* **59**, 1549 (1980).

12. M. Abe, I. Umedu, O. Hasegawa, S. Yamakoshi, T. Yamaoka, T. Kotami, H. Okada, and H. Takanahsi, *IEEE Trans. Electron Devices* **ED-24**, 990 (1977).

13. T. P. Lee, and A. G. Dentai, *IEEE J. Quantum Electron.* **QE-14**, 150 (1978).

14. M. Ettenberg, H. Kressel, and J. P. Wittke, *IEEE J. Quantum Electron.* **QE-12**, 360 (1976).

15. A. C. Carter, J. Ure, M. Harding, and R. C. Goodfellow, *3rd International Conference on Integrated Optics and Optical Fiber Communication*, Tokyo, 1983.

16. E. F. Schubert, A. Fischer, and K. Ploog, *Electron Lett.* **21**, 411 (1985).

17. C. S. Wang, W. H. Cheng, C. J. Hwang, W. K. Burns, and R. P. Moeller, *Appl. Phys. Lett.* **41**, 587 (1982).

18. E. S. Shlegal, and D. J. Pege, *IEEE Trans. Electron. Devices* **ED-27**, 583 (1980).

19. T. L. Paoli, R. L. Thornton, R. D. Burnham, and D. L. Smith, *Appl. Phys. Lett.* **47**, 450 (1985).

20. L. N. Kurbatov, S. S. Shakhidzhanov, L. V. Bystrova, V. V. Krapukhin, and S. J. Kolonenkov, *Sov. Phys—Semicond.* (*Engl. Transl.*) **4**, 1739 (1971).

21. R. Schimpe, and J. Boeck, *Electron Lett.* **17**, 715 (1981).

22. A. Mooradian, and H. Y. Fan, *Phys. Rev.* **148**, 873 (1966).

23. D. F. Nelson, M. Gershenzon, A. Ashkin, L. A. D'Asaro, and J. C. Saraca, *Appl. Phys. Lett.* **2**, 182 (1963).

24. H. Temkin, C. L. Zipfel, M. A. DiGiuseppe, A. K. Chin, V. G. Keramidas, and R. H. Saul, *Bell Syst. Tech.* **62**, 1. (1983).

25. H. C. Casey, Jr. and N. Stern, *J. Appl. Phys.* **47**, 631 (1976).

26. T. Fukui, and Y. Horikoshi, *Jpn. J. Appl. Phys.* **18**, 961 (1979).

27. J. S. Escher, H. M. Berg, G. L. Lewis, C. D. Moyer, T. V. Robertson, and H. A. Wey, *IEEE Trans. Electron Devices* **ED-29**, 1463 (1982).

28. P. D. Wright, Y. G. Chai, and G. A. Antypas, *IEEE Trans. Electron Devices* **ED-26**, 1220 (1975).

29. T. Uji, and J. Hayashi, *Electron. Lett.* **21**, 419 (1985).

30. A. Suzuki, Y. Inomoto, J. Hayashi, Y. Isoda, T. Uji, and H. Nomura, *Electron. Lett.* **20**, 274 (1984).

31. S. V. Galginaitis, *J. Appl. Phys.* **36**, 460 (1965).

32. J. Heinen, *Electron. Lett.* **18**, 23 (1982).

33. J. Hayashi, S. Fujita, Y. Isoda, T. Uji, M. Shikada, and K. Kobayashi, *Postdeadline Pap. Conference on Optical Fiber Communication 1987*, (1987)

34. T. P. Lee, *Bell Syst. Tech. J.* **54**, 53 (1975).

35. H. Namizaki, H. Kan, M. Ishii, and A. Ito, *Appl. Phys. Lett.* **24**, 486 (1974).

36. O. Wada, S. Yamakoshi, M. Abe, Y. Yishitoni, and T. Sakwai, *IEEE J. Quantum Electron.* **QE-17**, 174 (1981).

37. A. G. Dentai, T. P. Lee, and C. A. Burrus, *Electron. Lett.* **13** 484 (1977).

38. D. Gloge, A. Albanese, C. A. Burrus, E. L. Chinnock, J. A. Copeland, A. G. Dentai, T. P. Lee, T. Li, and K. Ogawa, *Bell Syst. Tech. J.* **59**, 1365 (1980).

39. H. Grothe, W. Proebster, and W. Harth, *Electron. Lett.* **12**, 553 (1976).

40. I. Hino, and K. Iwamoto, *IEEE Trans. Electron Devices* **ED-26**, 1238 (1979).

41. R. W. Dawson, *IEEE J. Quantum Electron.* **QE-16,** 697 (1980).

42. N. K. Dutta, *J. Appl. Phys.* **51**, 6095 (1980).

43. N. Chinone, K. Saito, R. Ito, and K. Aiki, *Appl. Phys. Lett.* **35**, 513 (1979).

44. K. Shima, K. Hanamitsu, and M. Takusagawa, *IEEE J. Quantum Electron.* **QE-18**, 1688 (1982).

45. H. Kumabe, T. Tanaka, H. Namigaki, M. Takamiya, M. Ishii, and W. Susaki, *Jpn. J. Appl. Phys.* **18-1**, Suppl., 371 (1979).

46. K. Aiki, M. Nakumura, T. Kuroda, J. Umeda, R. Ito, N. Chinone, and M. Maeda, *IEEE J. Quantum Electron.* **QE-14**, 89 (1978).

47. T. Hayakawa, N. Miyauchi, S. Yamamoto, H. Hayashi, S. Yano, and T. Hijikata, *J. Appl. Phys.* **53**, 7224 (1982).

48. J. M. Osterwalder, and B. J. Rickett, *IEEE J. Quantum Electron.* **QE-16**, 250 (1980).

49. K. Y. Lau, and A. Yariv, *IEEE J. Quantum Electron.* **21**, 121 (1985).

50. L. Figueroa, C. Slayman, and H. W. Yen, *IEEE J. Quantum Electron.* **18**, 1718 (1982).

51. H. Jackel, (1980). Ph.D. Thesis, BTH, Zurich, Switzerland.

52. H. Statz and G. DeMars, *in* "Quantum Electronics" (C. H. Towns, ed.), p. 530. Columbia Univ. Press, New York, 1960.

53. K. W. Wakao, N. Taagi, K. Shima, K. Hanamitsu, K. Hori, and M. Takusagawa, *Appl. Phys. Lett.* **41** 1113 (1982).

54. F. Stern, *J. Appl. Phys.* **47**, 5382 (1976).

55. H. Blauvelt, S. Margalit, and A. Yariv, *Appl. Phys. Lett.* **40**, 1029 (1982).

56. K. Y. Lau, and A. Yariv, *Appl. Phys. Lett.* **40**, 452 (1982).

57. H. Blauvelt, N. Bar-Chaim, D. Fekete, S. Margalit, and A. Yariv, *Appl. Phys. Lett.* **40**, 289 (1982).

58. K. Vahala, L. C. Chiu, S. Margalit, and A. Yariv, *Appl. Phys. Lett.* **42**, 211 (1983).

59. J. Katz, S. Margalit, and A. Yariv, *IEEE Electron Devices Lett.* **EDL-3**, 333 (1982).

60. D. P. Wilt, K. Y. Lau, and A. Yariv, *J. Appl. Phys.* **52**, 4790 (1981).

61. N. Chinone, K. Aiki, M. Nakamura, and R. Ito, *IEEE J. Quantum Electron.* **QE-14**, 625 (1978).

62. K. Kikuchi, T. Fukushima, and T. Okoshi, *Electron. Lett.* **21**, 1088 (1985).

63. A. Alping, T. Andersson, R. Tell, and S. T. Eng, *Electron. Lett.* **18**, 422 (1982).

64. M. Maeda, K. Nagano, M. Tanaka, and K. Chiba, *IEEE Trans. Commun.* **COM-26**, 1076 (1978).

65. R. S. Tucker, *IEEE J. Lightwave Technol.* **LT-3**, 1181 (1985).

66. N. Bar-Chaim, J. Katz, I. Ury, and A. Yariv, *Electron. Lett.* **17**, 108 (1981).

67. K. Petermann, *IEEE J. Quantum Electron.* **QE-15**, 566 (1979).

68. W. Streifer, D. R. Scifres, and R. D. Burnham, *Electron. Lett.* **17**, 933 (1981).

69. C. H. Henry, *in* "Semiconductors and Semimetals" (W. T. Tsang, ed.), Vol. 22, Part B., p. 153, Academic Press, Orlando, Florida, 1985.

70. C. H. Henry, *IEEE J. Quantum Electron.* **QE-18**, 259 (1982).

71. D. Welford, and A. Mooradian, *Appl. Phys. Lett.* **40**, 865 (1982).

72. P. Lin, and T. P. Lee, *IEEE J. Lightwave Technol.* **LT-2**, 44 (1984).

73. H. Sato, T. Fujita, and K. Fujito, *IEEE J. Quantum Electron.* **QE-21**, 46 (1984).

74. C. Lin, and A. Burrus, *Proc. Top. Meet. Opt. Fiber Commun.*, *1983* Paper PD5-1 (1983).

75. W. T. Tsang, R. A. Olsson, and R. A. Logan, *Appl. Phys. Lett.* **42**, 650 (1983).

76. H. K. Choi, and S. Wang, *Electron. Lett.* **19**, 394 (1983).

77. S. Kobayahsi, J. Yamada, S. Machida, and T. Kimura, *Electron. Lett.* **16**, 746 (1980); see also T. E. Bell, *IEEE Spectrum* **20** Dec., p. 38 (1983).

78. H. Kogelnik and C. V. Shank, *J. Appl. Phys.* **43**, 2327 (1972).

79. K. Kojima, S. Noda, K. Mitsunaga, K. Kyuma, and T. Nakayama, *IEEE J. Lightwave Technol.* **LT-4**, 507 (1986).

80. N. Chinone and M. Nakamura, *in* "Semiconductors and Semimetals" (W. T. Tsang, ed.), Vol. 22, Part C., p. 61., Academic Press, Orlando, Florida, 1985.

81. K. Sekartedjo, N. Eda, K. Furuya, Y. Suematsu, F. Koyama, and T. Tanbun-Ek, *Electron Lett.* **20** 80 (1984).

82. W. Streifer, R. D. Burnham, and D. R. Scifres, *IEEE J. Quantum Electron.* **QE-11**, 154 (1975).

83. K. Utaka, S. Akiba, K. Sakai, and Y. Matsushima, *Electron. Lett.* **18**, 863 (1981).

84. Y. Suematsu, K. Kishino, S. Arai, and F. Koyama, *in* "Semiconductors and Semimetals" (W. T. Tsang, ed.), Vol. 22, Part B, p. 205, Academic Press, Orlando, Florida, 1985.

85. K. Abe, K. Koshino, T. Tanbun-Ek, S. Arai, F. Koyama, K. Matsumoto, T. Watanabe, and Y. Suematsu, *Electron Lett.* **18**, 410 (1982).

86. M. Wada, K. Hamada, H. Shimizu, M. Kume, F. Tairi, K. Itoh, and G. Kano, *IEE Proc.* **132**, Pt. J, 3 (1985).

87. J. A. Shimer, W. R. Holbrook, C. L. Reynolds, Jr., C. W. Thompson, N. A. Olsson, and H. Temkin, *J. Appl. Phys.* **57**, 727 (1985).

88. J. K. Butler, and D. Botez, *IEEE J. Quantum Electron.* **QE-20**, 879 (1984).

89. K. Takahashi, K. Ikeda, J. Ohsawa, and W. Susaki, *IEEE J. Quantum Electron.* **QE-19**, 1002 (1983).

90. K. Saito, and R. Ito, *IEEE J. Quantum Electron.* **QE-16**, 205 (1980).

91. D. R. Scifres, C. Lindstrom, R. D. Burnham, W. Streifer, and T. L. Paoli, *IEEE J. Quantum Electron.* **QE-19**, 169. (1983).

92. J. K. Butler, D. E. Ackley, and D. Botez, *Appl. Phys. Lett.* **44**, 293 (1984).

93. E. Kapon, J. Katz, and A. Yariv, *Opt. Lett.* **10**, 125 (1984).

94. W. Streifer, D. F. Welch, P. S. Cross, and D. R. Scifres, *IEEE J. Quantum Electron.* **QE-23**, 744 (1987).

95. J. Katz, E. Kapon, C. Lindsey, S. Margalit, and A. Yariv, *Electron. Lett.* **19**, 661 (1983).

96. Y. Twu, K. L. Chen, A. Dienes, S. Wang, and J. R. Whinnery, *Electron. Lett.* **21**, 224 (1985).

97. N. Kaneno, T. Kadowaki, J. Ohsawa, T. Aoyagi, S. Hinata, K. Ikeda, and W. Susaki, *Electron. Lett.* **21**, 781 (1985).

98. D. E. Ackley, J. K. Butler, and M. Ettenberg, *Electron Lett.* **22**, 2204 (1986).

99. G. Forrest, *Laser Focus Electro-Opt. Mag.* **23**, p. 78 (1987).

100. W. T. Tsang, *Appl. Phys. Lett.* **39**, 786 (1981).

101. H. Okamoto, *Jpn. J. Appl. Phys.* **26**, 315 (1987).

102. S. D. Hersee, M. Baldy, P. Assenat, B. De Cremoux, and J. P. Duchemin, *Electron Lett.* **18**, 870 (1982).

103. W. T. Tsang, *Appl. Phys. Lett.* **39**, 134 (1981).

104. R. Chin, Holonyak, and B. A. Vojak, *Appl. Phys. Lett.* **36**, 19 (1980).

105. N. K. Dutta, *Electron. Lett.* **18**, 451 (1982).

106. K. Uomi, N. Chinone, T. Ohtoshi, and T. Kajimura, *Conf. Ser—Inst. Phys.* **79**, 703 (1986).

107. N. K. Dutta, N. A. Olsson, and W. T. Tsang, *Appl. Phys. Lett.* **45**, 836 (1984).

108. M. Yamada, S. Ogita, M. Yamagishi, K. Tabata, N. Nakaya, M. Asada, and Y. Suematsu, *Appl. Phys. Lett.* **45**, 324 (1984); see also C. Weisbuch, (1987). *in* "Semiconductor and Semimetals (R. Dingle, ed.), Vol. 24, p. 1, Academic Press, San Diego, California, 1987.

109. H. Iwamura, T. Saku, Y. Hirayama, Y. Suzuki, and H. Okamoto, *Jpn. J. Appl. Phys.* **24**, L911 (1985).

110. R. L. Thornton, R. D. Burnham, T. L. Paoli, N. Holonyak, Jr., and D. G. Deppe, *Appl. Phys. Lett.* **48**, 7 (1986).

111. K. Iga, S. Ishikawa, S. Ohkoushi, and T. Nishimura, *IEEE J. Quantum Electron.* **QE-21**, 38 (1984).

112. S. Kinishita, T. Odagawa, and K. Iga, *Jpn. J. Appl. Phys.* **25**, 1264 (1986).

113. M. Ohshima, N. Takenaka, N. Hirayama, Y. Toyoda, and N. Hase, *Conf. Ser.—Inst. Phys.*, No. 83, p. 343, (1986).

114. S. Noda, K. Kojima, K. Mitsunaga, K. Kyuma, K. Hamanaka, and T. Nakayama, *Appl. Phys. Lett.* **51**, 1200 (1987).

115. G. A. Evans, J. M. Hammer, N. W. Carlson, F. R. Elia, E. A. James and J. B. Kirk, *Appl. Phys. Lett.* **49**, 314 (1986).

116. M. Ogura, and S. Mukai, *Electron. Lett.* **23**, 759 (1987).

117. N. Bar-Chaim, S. Margalit, A. Yariv, and I. Ury, *IEEE Trans. Electron Devices* **ED-29**, 1372 (1982).

118. S. Yamakoshi, T. Sanada, O. Wada, T. Fujii, and T. Sakurai, *Electron. Lett.* **19**, 1020 (1983).

119. T. Sugino, M. Wada, H. Shimizu, K. Itoh, and I. Teramoto, *Appl. Phys. Lett.* **34**, 270 (1979).

120. H. Matsueda, S. Sasaki, and M. Nakamura, *J. Lightwave Technol.* **LT-1**, 261 (1983).

121. L. A. Coldren, K. Iga, B. I. Miller, and J. A. Rentschler, *Appl. Phys. Lett.* **37**, 681 (1980).

122. D. Wilt, N. Bar-Chaim, S. Margalit, I. Ury, M. Yust, and A. Yariv, *IEEE J. Quantum Electron.* **QE-16**, 390 (1980).

123. H. Nakano, S. Yamashita, T. P. Tanaka, M. Hirao, and M. Maeda, *IEEE J. Lightwave Technol.* **LT-4**, 574 (1986).

124. H. Choi, G. W. Turner, T. H. Windhorn, and B. Y. Tsaur, *IEEE Electron Device Lett.* **EDL-7**, 500 (1986).

125. N. Bar-Chaim, K. Y. Lau, I. Ury, and A. Yariv, *Appl. Phys. Lett.* **43**, 261 (1981).

126. R. S. Smith, and S. D. Personick, *in* "Semiconductor Device" (H. Kressel, ed.), 2nd ed., Chapter 4. Springer-Verlag, Berlin and New York, 1982.

127. Z. Rav-Nov, C. Harder, U. Schreter, S. Margalit, and A. Yariv, *Electron. lett.* **19**, 753 (1983).

128. S. Y. Wang, and D. M. Bloom, *Electron. Lett.* **19**, 554 (1983).

129. W. T. Lindley, R. J. Phelan, C. M. Wolfe, and A. G. Foyt, *Appl. Phys. Lett.* **14**, 204 (1969).

130. M. Ito, O. Wada, S. Miura, K. Nakai, and T. Sakurai, *Electron. Lett.* **19**, 523 (1983).

131. W. Lenth, A. Chu, L. J. Mahoney, R. W. McClelland, and R. W. Mountain, *Appl. Phys. Lett.* **46**, 191 (1985).

132. F. Capasso, *in* "Semiconductors and Semimetals" (W. T. Tsang, ed.), Vol. 22, Part D., P. 2., Academic Press, Orlando, Florida, 1985.

133. R. J. McIntyre, *IEEE Trans. Electron Devices* **ED-13**, 164 (1966).

134. R. J. McIntyre, *IEEE Trans. Electron Devices* **ED-19**, 703 (1972).

135. P. P. Webb, R. J. McIntyre, and J. Conrade, *RCA Rev.* **35**, 234 (1974).

136. G. E. Stillman, V. M. Robbins, and N. Tabatabaie, *IEEE Trans. Electron Devices* **ED-31**, 1643 (1984).

137. G. E. Bulman, V. M. Robbins, K. F. Brennan, K. Hess, and G. E. Stillman, *IEEE Electron Device Lett.* **EDL-4**, 181 (1983).

138. W. T. Lindley, R. J. Phelan, C. M. Wolfe, and A. G. Foyt, *Appl. Phys. Lett.* **14**, 204 (1969).

139. R. C. Eden, *Proc. IEEE* **63**, 32 (1975).

140. D. Long, *in* "Optical and Infrared Detectors" (R. J. Keyes, ed.), 2nd ed., p. 101, Springer-Verlag, Berlin and New York, 1980.

141. D. K. Lam, and R. I. MacDonald, *IEEE Electron. Device Lett.* **EDL-5**, 1 (1984).

142. M. Makiuchi, H. Hamaguchi, M. Kumai, M. Ito, O. Wada, and T. Sukarai, *IEEE Electron. Device Lett.* **EDL-6**, 634 (1985).

143. H. Hasagawa, T. Kitagawa, T. Sawada, and H. Ohno, *Electron. Lett.* **20**, 562 (1984).

144. L. Figueroa, and C. W. Slayman, *IEEE Electron Device Lett.* **EDL-2**, 208 (1981).

145. C. J. Wei, H.-J. Klein, and H. Beneking, *Electron. Lett.* **17**, 688 (1981).

146. H. Schumacher, P. Narozny, C. Werres, and H. Beneking, *IEEE Electron. Device Lett.* **EDL-7**, 26 (1986).

147. J. C. Gammel, and J. M. Ballantyne, *Tech. Dig.—Int. Electron Devices Meet.* p. 120 (1978).

148. H. Beneking, *IEEE Trans. Electron Devices* **ED-29**, 1420 (1982).

149. J. P. Vilcot, D. Decoster, and L. Raczy, *Electron. Lett.* **20**, 275 (1984).

150. J. C. Gammel, H. Ohno, and J. M. Ballantyne, *IEEE J. Quantum Electron.* **QE-17**, 269 (1981).

151. Y. M. Pang, C. Y. Chen, and P. A. Garbinski, *Electron. Lett.* **19**, 716 (1983).

152. C. Y. Chen, C. G. Bethea, A. Y. Cho, and P. A. Garbinski, *IEE Electron. Devices* **18**, 890 (1982).

153. W. Roth, H. Schumacher, and H. Beneking, *Electron. Lett.* **19**, 142 (1983).

154. M. A. Schuster, and G. Strull, *IEEE Trans. Electron Devices* **ED-13**, 907 (1966).

155. J. N. Shive, *Phys. Rev.* **76**, 575 (1949).

156. Z. I. Alferov, F. Akhmedov, V. Korol'kov, and V. Niketin, *Sov. Phys.— Semicon. (Engl. Transl.)* **7**, 780 (1973).

157. W. Shockley, U.S. Patent 2, 569, 347 (1951).

158. M. N. Svilans, N. Crote, and H. Beneking, *IEEE Electron Device Lett.* **EDL-1**, 247 (1980).

159. S. Margalit, and A. Yariv, in "Semiconductors and Semimetals" (W. T. Tsang, ed.), Vol. 22, Part E, p. 203, Academic Press, Orlando, Florida, 1985.

160. H. J. Hovel, and A. G. Milnes, *IEEE Trans. Electron Devices* **ED-16**, 766 (1969).

161. S. C. Lee, and G. L. Pearson, *J. Appl. Phys.* **52**, 275 (1981).

162. A. Scavennec, D. Ankri, C. Besombres, J. Riou, and F. Heliot, *Electron. Lett.* **19**, 394 (1983).

163. D. Ankri, W. J. Schaff, J. Barnard, L. Lunardi, and L. F. Eastman, *Electron. Lett.* **19**, 278 (1983).

164. N. Bar-Chaim, C. Harder, J. Katz, S. Margalit, and A. Yariv, *Appl. Phys. Lett.* **40**, 556 (1982).

165. F. Capasso, W. T. Tsang, A. L. Hutchinson, and P. W. Foy, *Conf. Ser.—Inst. Phys.* **63**, 473 (1982).

166. F. Capasso, *IEEE Trans. Electron Devices* **ED-29**, 1388 (1982).

167. R. Chin, N. Holonyak, Jr., G. E. Stillman, J. T. Tang, and K. Hess, *Electron. Lett.* **16**, 467 (1980).

168. D. J. Robbins, *Phys. Status Solidi B* **97**, Part I, 9; Part II, 387; Part III, 403 (1980).

169. F. Capasso, W. T. Tsang, A. L. Hutchinson, and G. F. Williams, *Appl. Phys. Lett.* **40**, 38 (1982).

170. F. Capasso, W. T. Tsang, and G. F. Williams, *IEEE Trans. Electron Devices* **ED-30**, 381 (1983).

171. H. Blauvelt, S. Margalit, and A. Yariv, *Electron Lett.* **18**, 375 (1982).

172. G. F. Williams, F. Capasso, and W. T. Tsang, *IEEE Electron Device Lett.* **EDL-3**, 71 (1982).

173. J. S. Smith, L. C. Chiu, S. Margalit, A. Yariv, and A. Y. Cho, *J. Vac. Sci. Technol., B* [2] **1**(2), 376 (1983).

174. L. C. Chiu, (1983). Ph.D. Thesis, California Institute of Technology, Pasadena.

175. H. Hamaguchi, M. Makiuchi, T. Kumai, and O. Wada, *IEEE Electron Device Lett.* **EDL-8**, 39 (1987).

176. S. Miura, O. Wada, H. Hamaguchi, M. Ito, M. Makiuchi, K. Nakai, and T. Sakurai, *IEEE Electron Device Lett.* **EDL-4**, 375 (1983).

177. M. Ito, T. Kumai, H. Hamaguchi, M. Makiuchi, K. Nakai, O. Wada, and T. Sakurai, *Appl. Phys. lett.* **47**, 1129 (1985).

178. R. M. Kolbas, J. Abrokwah, J. K. Carney, D. H. Bradshaw, B. R. Elmer, and J. R. Biard, *Appl. Phys. Lett.* **43**, 821 (1983).

179. S. Kawata, H. Fujii, K. Kobayashi, A. Gomyo, I. Hino, and T. Suzuki, *Electron. Lett.* **23**, 1327 (1987).

10 Electro-optical Properties of Hetreojunction and Quantum Well Structures of III–V Compound Semiconductors and Their Applications in Guided-wave Devices

WILLIAM S. C. CHANG
Department of Electrical and Computer Engineering
University of California at San Diego, La Jolla, California

10.1 INTRODUCTION

Optical guided-wave technology based on III–V compound semiconductor materials is important for a number of optical signal processing, fiber communication, and sensor applications. Devices such as lasers, modulators, switches, and detectors have already been developed. They can also potentially be integrated with other electronic or optical devices on the same substrate. The availability of new heterojunction and quantum well materials grown by LPE, MBE, and OMVPE techniques creates many exciting possibilities for optimizing the performance of discrete devices and integrated systems and implementing novel electronic and electro-optic concepts. This report is concerned with the electro-optic properties of heterojunctions and quantum wells made of these materials and their applications in guided-wave electro-optic devices.

10.1.1 The Heterojunction Structure

One of the more important aspects of III–V compound semiconductors is the availability of bulk crystalline substrates of GaAs and InP, which can be obtained from commercial sources in the form of deliberately n-doped, p-doped, or semi-insulating (100)-oriented substrates. Binary, ternary, or quaternary III–V alloy layers can be grown on these substrates using a

512

variety of epitaxial deposition techniques such as liquid phase epitaxy (LPE), molecular beam epitaxy (MBE), organo-metallic vapor phase epitaxy (OMVPE), or organo-metallic molecular beam epitaxy (MOMBE).

The large variety of materials available by the use of ternary and quaternary alloys which are lattice-matched to their GaAs or InP substrates provides a number of important options for electro-optic device and circuit designers. A principal feature of such heterojunctions is the choice of a particular fundamental bandgap E_g and consequently an optical absorption edge by the deliberate choice of the material parameters that constitute the heterojunction. When the E_g is tuned to a value slightly larger than the photon energy of the radiation, electroabsorption and electrorefraction are produced by means of the Franz–Keldysh effect [1]. At the heterojunction, the conduction band and the valence band have discontinuities, while the Fermi level must coincide on both sides of the junction. Because of such discontinuities, the energy bands near the heterojunction interfaces are bent and local electric fields are created on each side of the junction. Band discontinuities represent important parameters that have been utilized for electronic device design such as modulation doping for high-mobility electronic devices and carrier confinement for semiconductor lasers [2]. But, to a large extent, they have not yet being utilized for electro-optic devices. Change in the bandgap energy implies the ability to change the absorption edge energy as well as the refractive index. For example, one can obtain a planar waveguide in $Al_y Ga_{1-y} As$ by using a larger Al content for the cladding layers and a smaller Al content for the guiding layer. For a ternary or a quaternary layer, one can tailor the absorption edge energy to be just slightly larger than the photon energy of the laser radiation transmitted through a waveguide, so that the Franz–Keldysh effect will yield appropriate electroabsorption or electrorefraction modulation.

10.1.2 The Quantum Well Structure

A single quantum well (SQW) structure is a special form of hetreostructure, it consists of a very thin layer of a III–V compound semiconductor (in the z direction perpendicular to the epitaxial layer), called the well, which has a smaller E_g, sandwiched between the two layers, called the barriers, that have a larger E_g, as illustrated in Fig. 10.1. A multiple quantum well (MQW) structure consists of alternating layers of such wells and barriers. To obtain a quantum size effect, the thickness of the well layer L_z must be reduced to less than the electron mean free path, typically on the order of 3–20 nm. As the number of the alternating layers becomes very large and the barriers also become sufficiently thin so that the quantum mechanical wave functions in adjacent wells overlap each other, a superlattice is obtained. Three of the important differences between a QW structure and an ordinary heterostructure are the following [3–6].

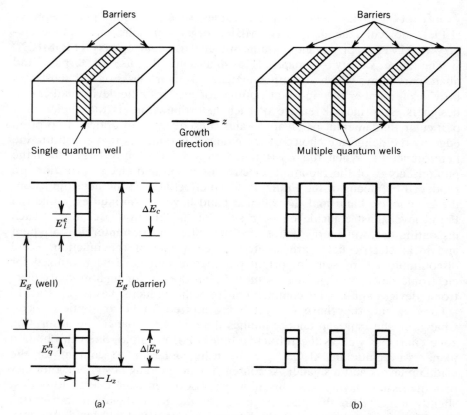

Figure 10.1 (a) Single and (b) multiple quantum well structure.

1. Due to the narrow potential well in the z direction created by the band edge discontinuities between the conduction and valence bands, the energy levels of the electrons and the holes are now discrete energy levels. In other words, the energy for an induced transition between a hole energy level and an electron energy level (corresponding to band edge absorption) is now $E_g + E_l^e + E_q^h$, as shown in Fig. 10.1.

2. The relative motion of the electron and the hole pair in the x–y plane forms an exciton which may be considered as a two-dimensional hydrogen atom. Its discrete resonant energy levels produce discrete excitonic line spectra at $E_g + E_l^e + E_q^h - |E^B|$, in addition to the discrete steps in absorption corresponding to the energies $E_g + E_l^e + E_q^h$. Here, $|E^B|$ is the binding energy of the exciton. Exciton spectra also exist in the bulk material, but they are observed only at liquid-helium temperatures because the E^B for the exciton in the bulk is small (i.e., the exciton line is close to the band edge) and the electron–phonon interaction broadens the linewidth too much for it to be observable at room temperature. Due to the confinement of excitons

in a layer whose thickness is smaller than the excitonic Bohr diameter, both the binding energy and the oscillator strength of the exciton are relatively large in a quantum well. Consequently, excitonic states of QW structures become stable even at room temperature.

3. The density of states as a function of energy, that is, $\rho(E)$, has a discontinuous staircaselike shape, as illustrated in Fig. 10.2, compared with the conventional parabolic $\rho(E)$ curve of the bulk material. The staircase $\rho(E)$ has a finite nonzero value even at the minimum energy. The staircase curve of $\rho(E)$ causes the spectra of the band edge absorption to have a staircase shape also. Figure 10.3 illustrates the total experimentally observed optical absorption spectra of a GaAs/Al$_y$Ga$_{1-y}$As QW structure at 300 K. Note the sharp exciton peaks near the band edges in addition to the staircase-shaped band edge absorption for $L_z \leq 192$ Å. This striking feature of the excitonic absorption leads to a number of important electro-optic properties such as quantum confined electro absorption (EA), electrorefraction (ER), and large nonlinear optical coefficients in QW structures. They have been utilized to realize EA modulators, QW nonlinear devices, tunable detectors, etc. Since the $E_g + E_l^e + E_q^h$ value depends on the thickness L_z of the well, there now exists a second mechanism in the QW structure for tuning (or modulating) the absorption spectra of a given material to fit a specific application in addition to the change of E_g of the material obtainable

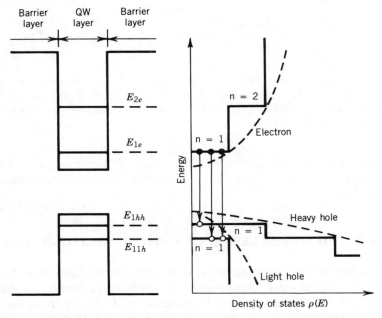

Figure 10.2 The density of states in a quantum well structure. The meanings of heavy hole and light hole are discussed in Section 10.2.

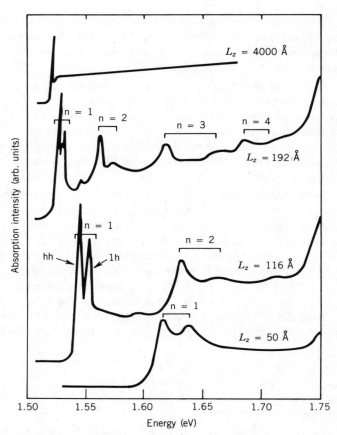

Figure 10.3 Optical absorption for a 4000 Å GaAs film and multiple quantum well structures with GaAs layer thicknesses; $L_z \sim 192$ Å, 116 Å, 50 Å, and $Al_{0.25}Ga_{0.75}As$ layer thicknesses $L_b \sim 200$ Å. (From Chemla [4]. Reprinted with permission from Birkhäuser Verlag AG.)

by varying the alloy composition. The shape of the staircase $\rho(E)$ also affects the saturation (i.e., nonlinear properties) of optical transition near the band edge.

10.2   A DETAILED DISCUSSION OF OPTICAL EXCITONIC SPECTRA

A more detailed understanding of excitonic absorption is important in order to use it for electro-optic components. It has been customary to describe excitonic and band edge absorption quantum mechanically using an effective-mass approximation. In this approximation, the Hamiltonian of the electron-hole pair (including the effect of an applied electronic field in the direction z normal to the layer interface) can be written as [7–9]:

$$H = H_{ez} + H_{hz} + H_{eh} \tag{1}$$

where

$$H_{ez} = \frac{-\hbar^2}{2m_{e\perp}^*} \frac{\partial}{\partial z_e^2} + V_e(z_e) + eFz_e$$

$$H_{hz} = \frac{-\hbar^2}{2m_{h\perp}^*} \frac{\partial}{\partial z_h^2} + V_h(z_h) - eFz_h$$

$$H_{eh} = H^{2D}(r) + H_{int}$$

$$H^{2D}(r) = \frac{-\hbar^2}{2\mu} \frac{1}{r} \frac{\partial}{\partial r}\left(r \frac{\partial}{\partial r}\right) - \frac{e^2}{Kr}$$

$$H_{int} = -\frac{e^2}{K}\left[\frac{1}{\sqrt{r^2 + (z_e - z_h)^2}} - \frac{1}{r}\right]$$

Here, z_e (z_h) is the z coordinate of the electron (hole); $m_{e\perp}^*$ ($m_{h\perp}^*$) is the effective mass of the electron (hole) in the z direction perpendicular to the layer; r is the relative position of electron and hole in the $(x-y)$ plane parallel to the layer; $\mu = m_{e\parallel}^* m_{h\parallel}^*/(m_{e\parallel}^* + m_{h\parallel}^*)$ is the reduced effective mass in the $(x-y)$ plane; K is the dielectric constant; F is the electric field parallel to z; $V_e(z_e)(V_h(z_h))$ is the built-in quantum well potential for electrons (holes) due to the band discontinuity in the conduction (valence) band; and H^{2D} is the cylindrically symmetric Hamiltonian of an hydrogen atom in two dimensions x and y.

The envelope function is the solution ψ of this Hamiltonian, where

$$H\psi = E\psi \tag{2}$$

When H_{int} is neglected, H_{ez}, H_{hz}, and H^{2D} are separable, thus

$$\psi(z_e, z_h, r) = \psi_l^e(z_e)\psi_q^h(z_h)\psi_s^{2D}(r)$$

$$E = E_l^e + E_q^h + E_s^{2D}$$

where

$$H_{ez}\psi_l^e = E_l^e\psi_l^e$$

$$H_{hz}\psi_q^h = E_q^h\psi_q^h$$

$$H_{eh}\psi_s^{eh} = H^{2D}\psi_s^{2D} = E_s^{2D}\psi_s^{2D}$$

ψ_s^{2D} is a two-dimensional hydrogen wave function which is the eigen solution of just $H^{2D}(r)$. q, l, and s are the orders (i.e., quantum numbers) of the respective eigen solutions. When H_{int} is included as a perturbation, we can express

$$\psi_n(z_e, z_h, r) = \sum_{l,q,s} a_n^{l,q,s} \psi_l^e(z_e) \psi_q^h(z_h) \psi_s^{2D}(r)$$

and the effect on the energy is a perturbation term given by $\langle \psi_n | H_{int} | \psi_n \rangle$. Therefore, the excitonic spectra would consist of absorption peaks at the energy

$$E_l^e + E_q^h - |E_s^B| + E_g \tag{3}$$

where E_s^B is the binding energy of the exciton, it has a negative value. E_s^B includes the energy perturbation due to H_{int} on E_s^{2D}. At small L_z, $E_s^B \sim E_s^{2D}$. So far, only the $s = 1$ spectral line has been observed experimentally. The optical absorption coefficient $\alpha(\hbar\omega)$ for the excitonic transition has been shown to be [7]:

$$\alpha(\hbar\omega) = B \sum_{l,q,s} |\langle \psi_l^e | \psi_q^h \rangle|^2 |a_n^{l,q,s} \psi_s^{2D}(0)|^2 \delta(E_l^e + E_q^h - |E_s^B| + E_g - \hbar\omega) \tag{4}$$

where

$$B = \frac{\pi e^2 |P|^2}{3 m_0 c \epsilon_0 \omega n V}$$

P is the momentum matrix element between the conduction and valence unit cell wave functions; P is proportional to the amplitude of the unpolarized optical field. m_0 is the free electron mass, V is the volume of the crystal, and n is the refractive index. $|P|^2/3$ should be replaced by $|P_i|^2$ for polarized radiation along the direction i. A similar expression for α of the continuous states of the H_{eh} representing the band edge absorption can also be obtained.

The details of theoretical solutions are not discussed here. Instead, we will discuss several conclusions that can be drawn from Eqs. (1–4) without actually solving the equations.

1. The quantum size effect in the z direction and the spin–orbit coupling split the degenerate valence band into the spin $\pm\frac{3}{2}$ and $\pm\frac{1}{2}$ combinations with different effective masses (e.g., $m_{h\perp} \sim 0.45 m_0$ and $m_{l\perp} \sim 0.08 m_0$ in GaAs; $m_{h\perp} \sim 0.38 m_0$ and $m_{l\perp} \sim 0.052 m_0$ in InGaAs) [3, 4, 9, 10]. Henceforth, there are two effective masses and two ψ's of the potential wells in the z direction, referred to as the effective mass and the energy states of the heavyhole (hh) and lighthole (lh). The hh and lh transitions are marked in Fig. 10.3 [4]. This means that in some quantum wells, such as in

$Al_yGa_{1-y}As/GaAs$ and in $In_{0.53}Ga_{0.47}As/InP$, there will be distinct electron to hh transitions and electron to lh transitions near to each other in energy. One needs to differentiate the hh and the lh transitions as well as the quantum numbers l and q involved in the transitions in order to identify the various spectra reported in the literature [3, 4, 11].

Moreover, for optical polarization parallel to the plane of the layers, both the hh and the eh exciton peaks are observed. When the polarization is perpendicular to the plane of quantum well layers such as for the TM modes, the hh exciton peak disappears and its strength is transferred to the remaining lh exciton peak [7, 12, 13]. Such an anisotropic behavior is very important to some guided-wave devices where the difference between the TE and TM modes is used to achieve certain polarization sensitive functions such as filtering or selective detection.

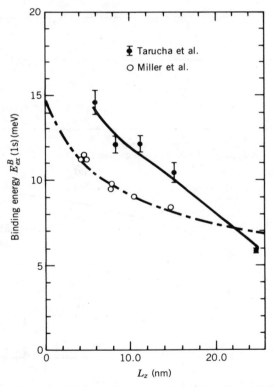

Figure 10.4   Binding energy of the 1st state of $n = 1$ heavy-hole exciton as a function of QW layer thickness L_z. Solid circles are measured by Tarucha et al. Solid line is calculated by assuming the reduced mass $\mu = 0.058m_0$. Dotted line is the result obtained by Miller et al. in photoluminescence excitation spectrum measurement. (From Okamoto [6]. Reprinted, with permission, from *Japanese Journal of Applied Physics*.)

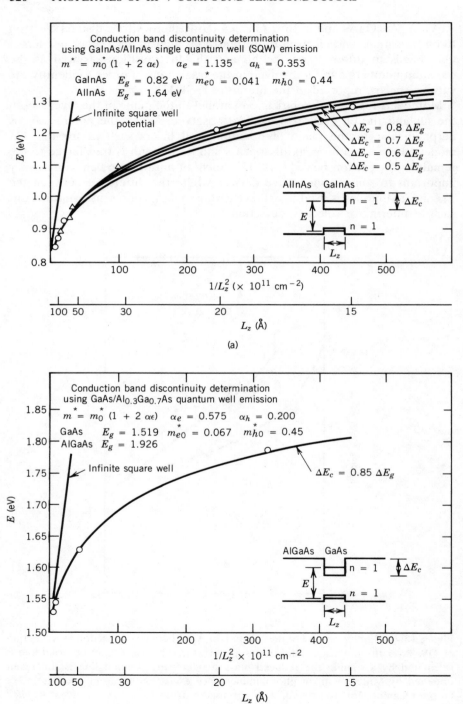

(a)

(b)

2. $|E_s^B|$ is small, typically less than 15 meV. Thus, the absorption edge and the exciton absorption peak are determined to a large extent by E_l^e and E_q^h, which are simple eigenvalue solutions of the one-dimensional potential well. For this reason, a precise knowledge of the discontinuities of the conduction band (ΔE_c) and that of the valence band (ΔE_v) is extremely important for predicting the excitonic and band edge absorption wavelengths. Figure 10.4 shows the binding energy of the 1st state of the $n = 1$ heavy-hole exciton of GaAs as a function of QW layer thickness L_z [6]. Figure 10.5 shows the calculated value of the absorption edge for $Al_yGa_{1-y}As/GaAs$ quantum well structure and for the $In_{0.53}Ga_{0.47}As/In_{0.52}Al_{0.48}As$ quantum well structure based on the rectangular potential well model [14]. The tunability of the exciton absorption wavelength as a function of L_z and of the alloy composition is one of the most important features for optical device design.

3. The oscillator strength is proportional to the value of α/δ in Eq. (4). For practical materials and at temperatures significantly higher than 0 K, the normalized δ function will be broadened to become a normalized line-shape function with a finite linewidth. Thus, the integrated intensity of the exciton absorption line (i.e., the area under a given absorption line) will be proportional to the oscillator strength, whereas the peak absorption coefficient will be directly proportional to the oscillator strength and inversely proportional to the linewidth. Already in 1966, Shimada and Sugano [15] predicted that in the limit of $L_z \to 0$, the oscillator strength of the two-dimensional exciton is eight times that of the three-dimensional exciton in the bulk. Figure 10.6 shows the absorption area of the $n = 1$ hh exciton in the GaAs/AlAs quantum well structure as a function of L_z [6]. For most device applications, one clearly wants to maximize the oscillator strength and to minimize the linewidth. An interesting engineering problem concerns that of obtaining a large oscillator strength and exciton absorption at the desired photon energy by varying the alloy composition as well as L_z.

4. The broadening mechanism of the excitonic absorption linewidth at liquid-helium temperature is not yet precisely known. It has been attributed to a number of causes; these include inhomogeneous broadening due to a

---

Figure 10.5 (a) Plot of emission energy for single quantum wells versus $1/L_z^2$. L_z is the thickness of the quantum well. The solid lines are the theoretically derived emission energies for GaInAs ($E_g = 0.820$ eV) and AlInAs ($E_g = 1.64$ eV). Each line represents a different percentage of the bandgap discontinuity in the conduction band. The experimental points for two separate growth runs are superimposed on the theoretical plots. This gives a $\Delta E_c \sim 70\% \; \Delta E_g$. (b) Plot of emission energy of GaAs quantum wells versus $1/L_z^2$. The data points all lie within experimental error of the calculated curve for a conduction band discontinuity of $\Delta E_c = 0.85 \Delta E_g$. Nonparabolisity is taken into consideration to first order in the equation $m^* = M_0^* (1 + 2a\epsilon)$. (From Welch et al. [14]. Reprinted with permission from the American Institute of Physics.)

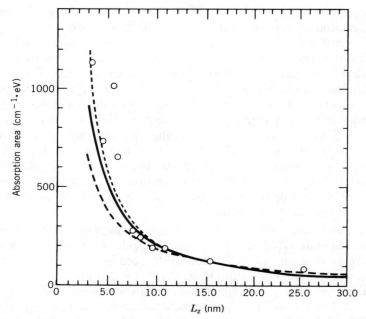

Figure 10.6 Absorption area, which is proportional to the oscillator strength, of $n = 1$ heavy-hole exciton in GaAs/AlAs quantum well structure as a function of the QW layer thickness L_z. The broken line shows $1/L_z$ dependence, reflecting the density of states for the $n = 1$ level. Solid and dotted lines are calculated curves by variational method by Matsumoto et al (From Okamoto [6]. Reprinted, with permission, from *Japanese Journal of Applied Physics*.)

small variation in the thickness L_z or to clustering of the material of the layer, to an inhomogeneous local electric field in the layers, broadening due to the ionization of the exciton, and carrier tunneling. However, for quantum wells with "good" uniformity, the linewidth at low temperatures is much narrower than the linewidth at room temperature; the broadening mechanisms for the large room-temperature linewidth has been attributed primarily to LO-phonon broadening. The temperature dependence of the full-width half-maximum Γ has been given as [3]:

$$\Gamma = \frac{\Gamma_0 + \Gamma_{ph}}{\exp(\hbar\Omega_{LO}/kT) - 1}$$

where, for GaAs, $\hbar\Omega_{LO} = 36$ meV, $\Gamma_0 = 2$ meV, and $\Gamma_{ph} = 5.5$ meV are given by Chemla et al. [3] to fit the experimental data. Linewidth is extremely important in practical applications, since the maximum height of the absorption peak of a normalized line shape will vary inversely with linewidth. Unfortunately, no mechanism has yet been found that can reduce the phonon broadening except lowering the temperature.

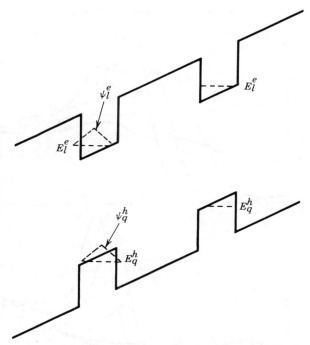

Figure 10.7 Illustration of quantum well potential and wave function with applied electric field perpendicular to the layer.

5. As the electric field is applied perpendicular to the QW layers, the potential well is tilted by the potential eFz, as illustrated in Fig. 10.7 [7, 12, 16]. As a consequence, the E_l^e and E_q^h are changed due to the tilt of the potential well. The net result is a shift of the exciton absorption peak at $E_g + E_l^e + E_q^h - |E_s^B|$ to a lower energy, known as the quantum confined stark effect (QSCE) [7, 8, 13, 17–20]. Moreover, as the exciton absorption peak is shifted, the oscillator strength also decreases, because, as the electric field is increased, ψ_l^e and ψ_q^h are more confined to the two regions of z at the two opposite sides of the quantum well, thereby reducing the value of $|\langle \psi_l^e | \psi_q^h \rangle|$ in Eq. (4). Hence, a second effect of the separation of electron and holes in the z direction is the reduction of the Coulomb binding energy. Eventually, the electric field will ionize the exciton, and there will no longer be exciton absorption. However, this does not happen until at high fields in the z direction (e.g., $f > 10^5$ V/cm) because of the quantum mechanical confinement effect of the thin potential well.

Figure 10.8 shows the absorption spectra of a GaAs/AlGaAs MQW waveguide at various electric fields for optical polarization parallel and perpendicular to the layer. Notice the absence of the hh exciton peak for the perpendicular polarization and the preservation of this polarization dependence even at high electric field values [7, 13].

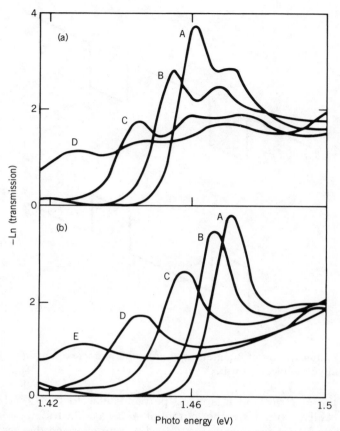

Figure 10.8 Absorption spectra of a quantum well waveguide as a function of electric field applied perpendicular to the layers: (a) incident polarization parallel to the plane of the layers for fields of A, 1.6×10^4 V/cm, B, 10^5 V/cm, C, 1.3×10^5 V/cm, and D, 1.8×10^5 V/cm; (b) incident polarization perpendicular to the plane of the layers for fields of A, 1.6×10^4 V/cm, B, 10^5 V/cm, C, 1.4×10^5 V/cm, D, 1.8×10^5 V/cm, and E, 2.2×10^5 V/cm. The fields were calculated from I–V measurements. (From Miller et al. [7]. Reprinted with permission copyright 1986, IEEE.)

For an electric field applied parallel to the layers (i.e, parallel to the x–y plane), there is no potential well confinement due to the band discontinuities [12]. Therefore, the dominant effect is a broadening of the absorption features with applied field, as shown in Fig. 10.9.

Electroabsorption associated with the change of absorption as a function of electric field is extremely important to the intensity modulation of optical radiation. If the photon energy of a radiation is originally located at the center of the exciton absorption peak, the QCSE can be used to reduce its

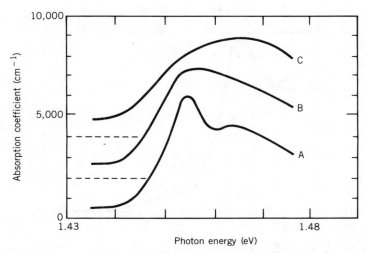

Figure 10.9 Absorption spectra at various electric fields for the parallel-field sample: A, 0 V/cm; B, 1.6×10^4 V/cm; C, 4.8×10^4 V/cm. The MQW region consisted of 60 layers, 95 Å thick, of GaAs separated by 98 Å $Al_{0.32}Ga_{0.68}As$ layers. (From Miller et al. [12]. Reprinted with permission from the American Institute of Physics.)

absorption. If the photon energy of a radiation is at a wavelength slightly longer than the center of the exciton absorption peak, the QSCE can be used to increase its absorption [21].

6. Associated with any variation of absorption as a function of photon energy is always a variation of the refractive index [22–56]. The two are related to each other by the Kramers–Kronig relationship as follows [27];

$$\Delta n(E) = \frac{ch}{2\pi^2} \, PV \int\limits_{0}^{\infty} \frac{\Delta\alpha(E')}{E'^2 - E^2} \, dE' \tag{5}$$

where PV means the principal value of the integral. The Δn associated with the QCSE is much larger than the Δn in the bulk. Figure 10.10 shows the measured electroabsorption, the calculated electrorefraction, and the measured electrorefraction as a function of photon energy for a MQW structure of $In_{0.11}Ga_{0.89}As/GaAs$ under an applied electric field perpendicular to the layers [22]. Wood et al. [28], Glick et al. [29], and Weiner et al. [30] have reported a quadratic dependence of the index on the applied electric field perpendicular to the layer near the exciton absorption line in AlGaAs/GaAs QW structure. Let us express the index as

$$n = n_e - \tfrac{1}{2}n_e^3 rF - \tfrac{1}{2}n_e^3 sF^2$$

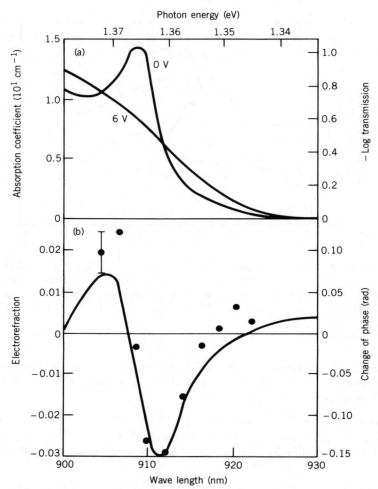

Figure 10.10 Electroabsorption and electrorefraction in $In_{0.11}Ga_{0.89}As$ quantum wells. The absorption spectra for 0 V and 6 V applied to the sample are shown in (a). The solid curve in (b) is the Δn calculated from the measured $\Delta \alpha$ shown in (a). The solid dots in (b) are the measured Δn. The left-hand scale of electrorefraction is the change of refractive index. The sample consists of 60 layers of $In_{0.11}Ga_{0.89}As$(120 Å)/GaAs(140 Å).

where n_e is the zero-field effective refractive index and r and s are the linear and quadratic electrooptic coefficients, respectively. s will be dependent on the photon energy relative to the band edge energy. At 12 meV below the bandgap, Wood et al. obtained a measured value of $|s| = 4.6 \times 10^{-13}$ cm^2/V^2. At 21 meV from the bandgap, they observed $|s| = 2.0 \times 10^{-13}$ cm^2/V^2 for TE polarization, whereas at 16 meV from the bandgap, they had only observed $|s| = 1.4 \times 10^{-13}$ cm^2/V^2 for TM polarization [28].

Moreover, even at photon energy significantly different than the bandgap or the exciton peak absorption energy, the refractive index of a superlattice is affected not only by the composition but also by the well thickness L_z and the barrier thickness L_B, provided L_B is sufficiently small. Figure 10.11 shows the refractive index spectra for four different GaAs/AlAs superlattice structures (solid lines) that have different values of L_z and L_B. The averaged Al content is maintained at 0.55, and the refractive index spectra of GaAs and $Al_{0.53}Ga_{0.47}As$ are shown as broken and dotted curves [6].

The coexistence of electrorefraction and electroabsorption has implications for both phase and amplitude modulation. Electrorefraction in an intensity modulator causes frequency chirping [31], which may be undesirable for applications such as coherent optical fiber communication. For a

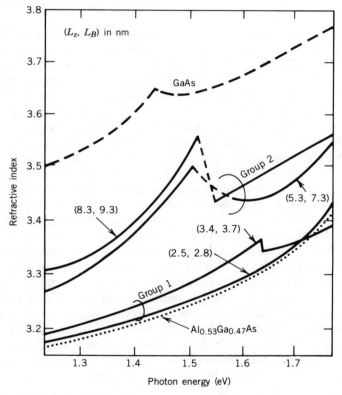

Figure 10.11 Refractive index spectra for four different GaAs/AlAs superlattice structures (solid lines) having different values of QW layer thickness L_z and barrier layer thickness L_B (shown in parentheses) but the same average aluminum content $\bar{x} = L_B/(L_z + L_B) = 0.55 \pm 0.03$. For comparison, refractive index spectra for bulk GaAs and for random alloy $Al_{0.53}Ga_{0.47}As$ are shown by broken and dotted lines. (From Okamoto [6]. Reprinted, with permission, from *Japanese Journal of Applied Physics*.)

phase modulator, electroabsorption causes unwanted intensity modulation. At longer wavelength, both Δn and $\Delta\alpha$ decay with decreasing photon energy. However, $\Delta\alpha$ is expected to decay faster than Δn. Thus, far from the exciton peak, there may be sufficient amount of Δn with an acceptably low value of $\Delta\alpha$. Nagai et al. has also shown experimentally that $\Delta n/n$ takes on the null value at the wavelength where $\Delta\alpha$ is at the peak [23]. These characteristics may be employed to reduce chirping in some modulators.

7. Saturation of the exciton absorption occurs at high light intensity. It originates both from the exhaustion of the number of available states and from the Coulomb interaction screening by the electron-hole plasma [32, 33]. In the small-signal region

$$\alpha(I) \sim \frac{\alpha_0}{1 + I/I_s} \tag{6}$$

where I_s is the saturation intensity and I is the intensity of the radiation. I_s in AlGaAs/GaAs QW samples has been measured to have values ranging from 190 to 580 W/cm^2. The change in the refractive index Δn is again related to $\Delta\alpha$ through the Kramers–Kronig relationship given in Eq. (5). The maximum values of nonlinearities are very large; Chemla et al. reported a $\alpha_2 \sim 39$ cm/W and a $n_2 \sim 2 \times 10^4$ cm^2/W, corresponding to $|\chi^{(3)}| \sim 6 \times 10^{-2}$ ESU [3]. More recently, comparable nonlinear saturation data on InGaAs/InP QW have been reported by Fox et al. [34] and by Tai et al. [35], with saturation intensity ranging from 70 to 200 W/cm^2. Extremely low-intensity optical nonlinearity was reported by Little et al. [36] in asymmetric coupled quantum wells. Nonlinear effects in GaAs/InGaAs strained-layer superlattice was reported by Das et al. [37]. Such large nonlinear coefficients have stimulated interest in using MQW structures for nonlinear guided-wave logic devices. Nonlinear saturation of exciton absorption in quantum wells was utilized to realize the self-electro-optic effect device (SEED) by Miller et al. [38]. A waveguide version of SEED has also been reported [17]. On the other hand, the low I_s values may be detrimental for applications in which linear modulation is desired. For such applications, the Frank–Keldysh effect in the bulk material may be a more desirable mechanism to use rather than the QCSE in MQW structures. The measured response time of the dynamic behavior of the exciton is in the subpicosecond range [17]. However, the relaxation time of the sizable nonlinearities due to saturation is much slower, in the nanosecond range.

In summary, the characteristics of the QW structures important to guided-wave electro-optic devices are

1. Tailoring of band edge absorption wavelength, the band edge discontinuities, the exciton absorption wavelength, and the refractive index for specific applications.

2. QW can be designed and synthesized to obtain the strongest and sharpest absorption line.

3. Utilization of the polarization dependence for differentiating TE and TM modes.

4. Utilization and optimization of electroabsorption and electrorefraction associated with the QCSE for modulation and switching functions.

5. Utilization of the large nonlinear coefficient for a variety of optical signal processing devices.

10.3 MATERIAL CONSIDERATIONS

The two single crystalline bulk III–V compound semiconductor substrates that can be used for epitaxial growth are GaAs and InP. GaAs substrates currently have a crystalline quality and purity much higher than that of InP substrates. Figure 10.12 shows the E_g versus lattice constants diagram of different III–V binary, ternary, and quaternary compound semiconductors. Clearly $Al_yGa_{1-y}As$ is approximately lattice-matched to the GaAs substrates for all values of y. Reasonably deep potential wells can be obtained

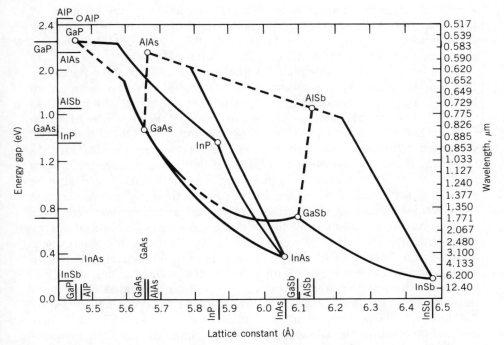

Figure 10.12 Energy gap versus lattice constants for various III–V compound semiconductors.

for GaAs quantum wells with $Al_yGa_{1-y}As$ barriers within a wide range of Al content (i.e., y). $Al_yGa_{1-y}As$ QW structures can be grown by MBE, MOMBE, and OMVPE methods, whereas heterostructures of $Al_yGa_{1-y}As$ and GaAs can be grown by the LPE as well as the MBE and OMVPE methods. Most of the published work on QW structures is based on this material system. However, the use of GaAs as the well in QW structures limits the longest wavelength of exciton absorption to approximately 0.85 μm. Yet, at the wavelength of 0.85 μm or shorter, the substrate is absorbing. Thus, for applications such as spatial light modulation which require transmission of light perpendicular to the layers, the GaAs substrate must be etched away after the epitaxial growth to avoid substrate absorption. Alternatively, if the reflected light can be used, then a multilayer AlGaAs/GaAs mirror may be grown epitaxially on the GaAs substrates to provide the reflection. Even for waveguide applications, a buffer layer with higher Al concentration must be grown to isolate the evanescent tail of the waveguide mode from the substrate. Al may also form deep traps that may slow down the speed of devices, and $Al_xGa_{1-x}As$ for $x > 0.2$ exhibits a persistent photoconductivity.

For operation at the 1.3 μm and 1.5 μm wavelength layers of the lattice-matched heterostructures, $In_xGa_{1-x}As_zP_{1-z}$ is grown epitaxially on InP by LPE, OMVPE, or MOMBE. The InP substrate is transparent in this wavelength range. Lattice-matched QW structures may consist of $In_{0.53}Ga_{0.47}As$ wells and $In_{0.52}Al_{0.48}As$ barriers grown by MBE on InP substrates or $In_{0.53}Ga_{0.47}As$ (or $In_xGa_{1-x}As_zP_{1-z}$) wells and InP barriers grown by OMVPE or MOMBE on InP substrates [39–49].

To avoid substrate absorption at the exciton absorption wavelength, $In_xGa_{1-x}As$ ($x \leq 0.15$) quantum wells and GaAs barriers have been grown by MBE on GaAs substrates for operation at wavelengths longer than 0.9 μm [20, 37, 50, 51]. $In_xGa_{1-x}As/In_xAl_{1-x}As$ MQW grown by MBE on GaAs substrates and AlGaAs/GaAs MQW grown by OMVPE on GaP substrates have also been reported [52, 53]. However, the $In_xGa_{1-x}As$ ($x \leq 0.15$) is not lattice-matched to GaAs. If they are pseudomorphic, such layers are strained. Strained layers of InGaAs and InAlAs can also be grown on InP when their compositions deviate from the lattice-matched composition. Significant modifications of optical and electronic properties are observed in pseudomorphic layers due to strain modification of energy band structure [50, 54]. For example, Fig. 10.13 shows the energy gap increase as a function of In concentration x (due to the biaxial compressive strain). Strained layers may be subject to interfacial dislocations which can propagate through the layers. Dislocations will be generated if the total thickness of all the layers exceeds a certain critical thickness determined by the limit of elastic strain containment in the layer [55–57]. However, strained layers have the advantages of (1) obtaining exciton lines at wavelengths inaccessible by the lattice-matched alloy compositions, (2) the avoidance of elements such as Al in ternary alloys, which are undesirable

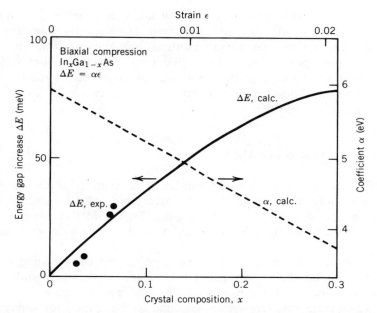

Figure 10.13 Strain-induced bandgap shift for $In_xGa_{1-x}As$ layers biaxially compressed to lattice-match GaAs (no quantum-size effects present). The data points are experimental results. These results are in reasonable agreement with a simple calculation incorporating the deformation potentials and elastic properties on $In_xGa_{1-x}As$ (solid curve). (From Anderson et al. [54]. Reprinted with permission from the American Institute of Physics.)

because deep trapping may create serious problems, and (3) strain can be used to create artifically induced anisotropy or nonlinearity.

A potential solution for increasing the range of available ternary alloy compositions used for QW structures is to relieve the strain partially, and to increase the total critical thickness limitation by first growing a lattice-mismatched thick buffer layer or superlattice buffer layer on the substrate, which prevents the propagation of misfit dislocations. There are encouraging indications that massive dislocation will occur but will be confined near the boundary between the buffer layer and the substrate, thereby partially relaxing the average strain of the MQW. This may also provide an extension of the total thickness of the MQW, which can be grown without exceeding the limit for elastic deformation of the layers. Material research is being conducted to investigate other new or improved approaches of material synthesis. Most notable are the initial success in the growth of GaAs on Si and the growth of GaAs and GaP on fluoride insulators such as CaF_2 [58, 59]. Many other opportunities for material synthesis have yet to be explored. For example, $Ga_{0.49}In_{0.51}P$ can be latticed-matched to GaAs, and it has a wide bandgap with $E_g = 1.9$ eV. GaInAsP or GaInAlP may serve as quantum wells on InP substrates that will allow us to have E_g or exciton absorption within a wide range of wavelength. The selection of the appropriate alloy composition, the growth method, and the substrate (or buffer composition) for specific applications will depend on many factors, including interdiffusion of elements at the boundary, homogeneity of the layer that can be grown, thickness uniformity, crystalline quality, and defects or traps created during growth, etc.

10.4   OPTICAL WAVEGUIDES

In optical waveguides, the refractive index of the material in the waveguide region is slightly higher than the refractive index of the material in the surrounding cladding (or substrate) region. There are three requirements that must be satisfied in order to obtain a good single mode waveguide:

1. The index difference between the guiding and the cladding (or substrate) region and the cross-sectional dimension of the waveguide region must be coordinated so that only a single mode can propagate. If the index difference is large, the cross-sectional dimension required for single mode propagation may be very small (i.e., in the submicron range). Submicron layer thickness is easy to achieve, but waveguides with lateral dimensions less than 2 or 3 μm are very difficult to fabricate. Waveguides with very small cross-sectional dimensions are also difficult to excite with optical laser radiation and to couple efficiently to optical fibers. There will be a serious radiation noise problem in inefficiently excited waveguide, since there will be strong radiation intensity in the substrate (or cladding). Consequently, tailoring of the index variation to suit specific dimension requirements would be important in obtaining the high-quality waveguides. Thus far, two methods have been used to obtain index control. In the first one, the alloy compositions of the layers in the guiding and the cladding region have been used to control the indices. Such a method is applicable to both the heterojunction and the quantum well structures. In the second method, the well thickness L_z and the barrier thickness L_b of QW structures may be used to control the average index of the guiding region, since the averaged index is $(L_z n_w + L_b n_b)/(L_z + L_b)$, where n_w and n_b are the indices of the well and the barrier respectively. As long as the alloy composition or the L_z is chosen so that the wavelength of the absorption edge is far away from the radiation wavelength, there will be little attenuation. Since the barrier thickness L_b can be controlled very precisely, the latter method can control the index much more precisely, yielding a reproducible and small index difference between the guiding and the cladding region desired for single mode operation in a waveguide that has reasonably large cross-sectional dimensions. Recently, an additional method, the impurity-induced disorder, has been investigated to form low-index cladding regions in the transverse

direction to obtain buried-channel waveguides [60–62]. Of particular interest is the fact that the impurity disordered material will have a larger bandgap than the original layered structure and hence lower index of refraction. Impurity doping can be precisely defined in the lateral direction with the use of masks, and the disordering process on a layered planar waveguide structure will result into a buried-channel waveguide.

2. The doping of the semiconductor near (or in) the guiding region must be low (e.g., less than 10^{16} cm^{-3}) to avoid large attenuation due to free-carrier absorption. Figure 10.14 shows the effect of the carrier concentration on the loss of GaAs ridge waveguide [63].

3. The interface between the guiding and the surrounding regions must be optically smooth to avoid attenuation due to surface scattering. Surface scattering is very large when the abrupt boundary between the guiding and the surrounding region has a large index difference. Typically, a ridge channel waveguide is fabricated by etching a mesa on the cladding layer. Since the index difference between semiconductor and air is large and the etched surface is not optically smooth, an etching depth less than or equal to the cladding layer thickness is usually used to avoid large scattering loss [64]. The disadvantage of a ridge waveguide is that the difference of the equivalent index between the channel region and the adjacent region in the lateral direction is small, so that it is weakly confined in the lateral direction. Often, there is also a planar guided mode propagating in the lateral region outside the channel. For this reason, buried-channel single mode waveguides are expected eventually to have much lower scattering loss and

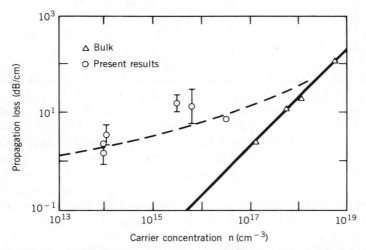

Figure 10.14 Propagation loss at 1.3 μm wavelength as a function of carrier concentration: the dashed curve is for the GaAs ridge waveguide and the solid line is for the GaAs bulk crystal. (From Hiruma et al. [63]. Reprinted with permission from the American Institute of Physics.)

better guiding than the etched channel waveguides. Buried-heterostructure (BH) GaInAsP waveguides grown by LPE, using an etch and regrowth technique, have been reported by Johnson et al. at the 1.3 μm wavelength [65]. However, their attenuation rate is not very low because the regrown layers had a carrier concentration in the low 10^{17} cm^{-3} range. Thus far, very low-loss buried-channel waveguides with attenuation rates 1 dB/cm or less have not yet been reported.

Low-loss GaAs waveguides have been made by both LPE and OMVPE growth of GaAs and AlGaAs on GaAs substrates at the 1.3 μm and 1.5 μm wavelengths. The channel waveguides were formed by etching the GaAs layer to provide a channel region that has a larger n_{eff} value than the lateral region outside of the channel. The lowest loss reported by Hiruma et al. [63] is 0.2 dB/cm for OMVPE-grown waveguide with a 1 μm thick $Al_yGa_{1-y}As$ $(0.01 < y < 0.1)$ cladding layer, a 3 μm thick GaAs waveguide layer, an etching depth of 2 μm, and a channel ridge width larger than 4 μm. However, single mode propagation was not demonstrated by Hiruma et al. Similar structures with thinner GaAs layer (1.54 μm thick) and channel ridge width varying from 2 to 5.5 μm have also been fabricated by Deri et al. They have obtained single mode operation at $\lambda = 1.52$ μm with a 0.2 dB/cm attenuation rate [64]. The GaAs waveguide itself does not have significant electroabsorption or electrorefraction effect at the 1.3 μm and the 1.5 μm wavelengths. However, at the near infrared, where the electroabsorption (or electrorefraction) modulation may be effective, the waveguide attenuation at the near infrared is expected to be considerably higher than 0.2 dB/cm because there will be residual absorption from the GaAs band edge. Optical waveguides at 1.3 μm wavelength in $In_{0.52}Al_{0.48}As$ grown by InP by MBE have been reported by Ritchie et al. [66]. There are, however, uncertainties concerning the refractive index of their $In_{0.52}Al_{0.48}As$, which may very well depend on the growth conditions. InGaP/InP waveguides grown by OMVPE with a 1.25 dB/cm attenuation rate at $\lambda = 1.3$ μm have been reported by Joyner et al. [67]. The most interesting result is the low optical waveguides made from QW structures. A low-loss (0.8 dB/cm) waveguide in the 1.46–1.52 μm wavelength range was reported by Koren et al. using multiple quantum wells of InGaAs/InP grown by OMVPE on InP [68]. In this case, the $In_{0.53}Ga_{0.47}As$ QW absorption edge of the guiding layer is at the 1.18 μm wavelength corresponding to 10 Å L_z and 90 Å L_b. The channel waveguide was formed chemically by etching a mesa, 0.5 μm deep and 5 μm wide. Although it is not clear whether single mode operation was achieved, clearly the L_z and L_b could be adjusted to obtain single mode operation using this approach, and QW with a different L_z can be used to obtain electrorefraction or electroabsorption modulation. Kapon and Bhat reported loss as low as 0.15 dB/cm for single mode waveguides at 1.52 μm wavelength made from 2 μm thick GaAs guiding layer sandwiched between GaAs(30 Å thick)/$Al_{0.1}Ga_{0.9}As$(50 Å thick) QW structures as cladding

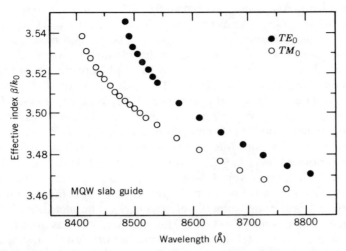

Figure 10.15 Dispersion of the fundamental TE and TM modes in a GaAs/AlGaAs asymmetric MQW slab waveguide. The sample consists of 25 periods of GaAs $(80 \, \text{Å})/\text{Al}_{0.26}\text{Ga}_{0.74}\text{As}(80 \, \text{Å})$ MQW, cladded above and below by $0.1 \, \mu\text{m}$ and $1.5 \, \mu\text{m}$ thick $\text{Al}_{0.26}\text{Ga}_{0.74}\text{As}$ layers.

layers at a channel ridge width varying from 2.7 to 6 μm [69]. In summary, existing results demonstrated that waveguide attenuation due to material absorption can be kept low provided the band edge (or the exciton) wavelength is far shorter than the radiation wavelength; the scattering loss for etched ridge waveguide several microns wide is tolerable. However, single mode operation will require a refined control of refractive index variation and lateral dimension. The scattering loss will increase as the width of the etched mesa is reduced.

Finally, dispersion of the effective index of the TE and the TM modes in $\text{Al}_{0.3}\text{Ga}_{0.7}\text{As}/\text{GaAs}$ MQW waveguide with $L_z = 70 \, \text{Å}$ and $L_b = 75 \, \text{Å}$ has been measured by Sonek et al. [70]. Their results are shown in Fig. 10.15. The $n_{\text{TE}} - n_{\text{TM}}$ increases from 0.011 at $\lambda = 0.8765 \, \mu$m to 0.039 at $\lambda = 0.8480 \, \mu$m.

10.5   MODULATION AND SWITCHING IN HETEROJUNCTION WAVEGUIDES

Electroabsorption modulation in InGaAsP/InP double heterostructure (DH) waveguides was reported by Dutta and Olsson as early as 1984 [71]. High-extinction (30 dB at 10 V) electroabsorption at $\lambda = 0.79 \, \mu$m using the Franz–Keldysh effect in AlGaAs waveguide was reported by Gee et al. [72]. The modulator structure is a ridge channel waveguide which is also a reverse-biased pin diode; it consists of a 350 μm long double heterostructure of 3.2 μm thick n-$Al_{0.19}Ga_{0.81}As$ grown by LPE as the cladding, followed by 3 μm thick n-$Al_{0.15}Ga_{0.85}As$ as the guide, 1 μm thick p-$Al_{0.19}Ga_{0.81}As$ as the top cladding, and 1.3 μm thick p-$Al_{0.28}Ga_{0.72}As$. Etched mesa (10 μm wide and 2.3 μm deep) was used to create the channel waveguide. The 3 dB bandwidth is 3 GHz, limited primarily by the 1 pF capacitance of the device. The total insertion loss is 12 dB for TE and 15 dB for TM polarization. Similarly, Noda et al. reported a high-speed electroabsorption modulator at $\lambda = 1.55$ with 23 dB extinction ratio at 4.5 V using the Franz–Keldysh effect in GaInAsP waveguide [73]. It consists of an undoped InP layer 0.35–1 μm thick grown by VPE on InP substrate as the cladding, followed by a quaternary undoped layer 0.3–1 μm thick with $\lambda_g \sim 1.45$ μm as the guide, an undoped InP cladding layer 0.4 μm thick, and a p^+-InP layer. A mesa 4–8 μm wide was formed by etching the top two InP layers down to the quaternary and preserved immediately by Si_3N_4. The measured 3 dB bandwidth of this device is also 3–4 GHz. The insertion loss was about 10–14 dB, contributed primarily by the band-to-band absorption in the quaternary layer, the end reflection loss, and the coupling loss. Buried-channel InGaAsP quaternary waveguide electroabsorption modulator at $\lambda = 1.3$ μm has also been reported by Lin et al. that has 30 dB extinction at 10 V for a 650 μm long device [74]. In this case, the buried channel was obtained by etching a groove approximately 5 μm wide in the undoped InP substrate, followed by the LPE growth of a very thin undoped InP buffer layer, a quaternary InGaAsP with $\lambda_g \sim 1.25$ μm, and a InP cladding layer. The quaternary layer is about 2 μm thick in the groove with $n \sim 7 \times 10^{15}$ cm^2. Selective Zn diffusion is used to obtain a p-n junction at a position about 0.7 μm from the top of the waveguide.

In short, DH waveguide EA modulators reported so far in the literature seem to behave in a similar manner and have a high extinction ratio at reasonably low applied voltages, large insertion loss, and approximately 1 pF capacitance. No data on the linearity, temperature sensitivity, dynamic range, frequency chirping, or saturation characteristics of such devices have been reported as yet. However, data collected from the existing devices have already demonstrated that there are two major problems to be solved:

1. The insertion loss consists typically of 5 dB reflection loss, 4–6 dB coupling loss, and several dB of propagation loss. Reflection loss and coupling loss can be reduced by antireflection coating and better coupling. Thus, the only real problem is the propagation loss due to the absorption of the band tail. It increases as the length of the device increases and as the separation of the λ_g from the radiation wavelength λ decreases.

2. The capacitance limits the bandwidth due to the RC time constant. In the traveling wave version, large capacitance implies low electrode impedance that is difficult to match. The total capacitance consists of electrode capacitance and parasitic capacitance. The electrode capacitance is large primarily because, electrically, the modulator is just a reverse-biased diode.

Its depletion layer thickness is small, the device is quite long (a few hundred μm), and the width is limited by processing techniques to 2 or 3 μm or more. The lowering of the doping density, new fabrication methods to reduce the width of the channel (i.e. the p-n junction), and shortening of the electrode length will all reduce the electrode capacitance. Unfortunately, reduction of the electrode length and the increase of the separation between λ and λ_g will also reduce the extinction coefficient and increase the voltage needed for obtaining a given modulation depth. Consequently, computer analysis involving the trade-offs of the design would be quite valuable in determining the optimum performance that can be obtained for the DH waveguide EA modulators. Parasitic capacitance can be reduced when the device is fabricated on semi-insulating substrates.

GaAs and other III–V compound semiconductors also have a respectably large conventional electro-optic coefficient. Electroreflection effect in bulk materials such as GaAs and InGaAsP has also been reported. Therefore, phase modulation can be achieved in addition to amplitude modulation. Nevertheless, efficient pure phase modulation (or switching) is difficult to obtain because of the associated electroabsorption when λ_g is close to the radiation wavelength λ. However, as λ becomes sufficiently longer than the λ_g, the electroabsorption effect may be acceptably weak to be useful. A single mode phase modulator that has over 20 GHz of bandwidth has been reported by Wang et al. in a ridge $Al_{0.032}Ga_{0.968}As/GaAs$ waveguide with 8 mm long coplanar electrode at $\lambda = 1.3$ μm, using just the conventional electro-optic effect [75, 76]. The capacitance per unit length is kept small (i.e., the impedance of the electrode transmission line is kept high) by means of both the very low doping density $(3 \times 10^{14}\,cm^{-3})$ achieved in the OMVPE grown epitaxial layer and the large separation $(4.5\,\mu m)$ of the electrodes on top of the waveguide. The propagation loss is small because λ is far away from λ_g. The bandwidth is limited by the mismatch of the microwave and optical phase velocities. The price paid for obtaining the large bandwidth and low insertion loss is the small phase shift per unit voltage per unit length, $\Delta\phi \sim 1.18°/V\text{-mm}$.

Guided-wave phase modulation using the change of index caused by the depletion of the carrier in GaAs has been reported at $\lambda = 1.06$ μm by Alping et al. [77]; a large phase shift $(\Delta\phi \sim 56°/V\text{-mm})$ was obtained for the TE mode at 4 V for a device 4 μm wide and 800 μm long. To obtain efficient modulation, the Alping device had an undepleted free-carrier concentration of $2 \times 10^{17}\,cm^{-3}$, which yields a small depletion depth at the operating voltage range. The estimated achievable capacitance is 2 pF, which will limit the operating frequency to less than 2 GHz. Phase-shift efficiency can be increased further through optimization of the device parameters, including both the linear electro-optic effect and the quadratic electro-optic effects near the band edge in addition to the carrier effect [78, 79]. Carrier-injection-induced refractive index change has also been used by Ishida et al.

to obtain switching in a multimode crossing channel waveguide switch with 20 ns switching time at 90 mA of switching current [80]. P. Rogers et al. have shown that $\Delta\phi$ as large as 8°/V-mm at $\lambda = 1.15\ \mu m$ and 6°/V-mm at $\lambda = 1.3\ \mu m$ can be obtained through the electrorefraction effect near the band edge in an InGaAsP ridge waveguide that has $\lambda_g = 1.07\ \mu m$ and 0.18 μm waveguide layer thickness [81]. Data on both the capacitance and the residual amplitude modulation on this device are unknown.

In summary, large bandwidth phase modulation has clearly been achieved when large drive voltage can be provided to obtain the necessary depth of modulation. Large phase shift $\Delta\phi$, in deg/V-mm, can also be obtained provided the voltage is applied to a thin active layer, thereby producing a large capacitance and limiting the bandwidth to low GHz range. Low doping density is a necessity for large depletion depth and low capacitance. ER will increase $\Delta\phi$ provided the quadratic dependence on voltage and the associated amplitude modulation due to EA are acceptable.

10.6 MODULATION AND SWITCHING IN WAVEGUIDES CONTAINING QUANTUM WELLS

The excitonic absorption line and the absorption edge of the QW structures have a wavelength variation much sharper than the wavelength dependence of the absorption edge in the bulk material. Thus, the propagation loss of the optical radiation at λ due to residual absorption of the band edge at $\lambda_g (\lambda > \lambda_g)$ is expected to be much smaller in QW structures than in the bulk material for the same $\lambda - \lambda_g$. In other words, the λ_g can be placed closer to λ in QW structures without creating excessive propagation loss, and the EA due to the QCSE is expected to be more sensitive to the applied electric field than the EA effect of the Franz–Keldysh effect.

A 100 ps pulse at $\lambda = 0.851\ \mu m$ generated from a MQW waveguide EA modulator with 10/1 on/off ratio at 15 V was reported by Wood et al. [21]. They used a leaky capillary waveguide of GaAs/AlGaAs superlattice (SL) grown by MBE. The waveguide was 3.6 μm thick, 40 μm wide, and 150 μm long. The center section of the waveguide was a 1.1 μm intrinsic region containing two 94 Å thick GaAs QWs. The top section of the waveguide was p-doped, while the bottom section was n-doped. The device has a capacitance of 0.94 pF and an insertion loss of 7 dB consisting primarily of the radiation leakage of the capillary mode and the reflection loss. A capillary leaky waveguide was used for experimental convenience to assure the rapid decay of the higher-order leaky modes excited by the incident radiation. The polarization dependence of the exciton absorption in this device has already been shown in Fig. 10.8 [13, 19]. EA modulator with 9 dB on/off ratio at a driving voltage of 5 V and an operating wavelength of 1.55 μm has also been reported by Wakita et al. using InGaAs(70 Å)/InP(150 Å) MQW structures grown by MOMBE [82]. The device was fabricated with a ridge width of

80 μm and length 162 μm. The device capacitance was 2.2 pF at a reverse bias of 5 V. The 15 dB total insertion loss was attributed primarily to the coupling and the reflection loss. Koren et al. reported an electroabsorption waveguide modulator in an InGaAs(80 Å)/InP(420 Å) structure at 1.67 μm wavelength with a 47/1 on/off ratio, low insertion loss (2.9 dB), 3 GHz bandwidth, and 0.1 mW optical input power [83].

Five conclusions can be drawn from the characteristics of QW electroabsorption devices:

1. Low insertion loss can be achieved in a MQW waveguide modulator [69, 83].

2. On the other hand, the modulation efficiency is not much different from that of the reported modulation of heterostructure waveguide modulators. This is probably caused by the small overlap integral of the QWs with the optical guided mode profile. Much stronger EA effects could probably be achieved if more QWs were used.

3. Note that the capacitance remains (i.e., the bandwidth is low) high. The total device capacitance is the sum of the electrode capacitance and the parasitic capacitance. Koren et al. have shown that the parasitic capacitance can be reduced significantly by burying the sidewalls of the waveguide with semi-insulating material grown by OMVPE [83]. Since the QCSE occurs only when the electric field is applied in the z direction (perpendicular to the QW layers), the electrode capacitance will be inversely proportional to the depletion layer thickness. Reduction of the doping density will reduce the capacitance until the entire undoped layer is fully depleted, but it will also reduce the electric field available for creating the QCSE at a given applied voltage. The width of the electrode ridge can be reduced to reduce the capacitance, but it will eventually be limited by lithographic resolution. Reduction of the modulator length will reduce both the capacitance and the modulation effiency. If the QW structures can be designed such that sufficient modulation depth can be obtained within a relative short length, such as 10 μm or less, then the electrode capacitance can be significantly reduced. In other words, similar to the case of DH modulators, a trade-off study in terms of capacitance and modulation efficiency is necessary in order to optimize the performance of QW EA waveguide modulators.

4. However, as opposed to the heterostructure waveguide EA modulators, the EA properties of the material structure can be optimized and modified by design of the L_z and L_b of the QWs. This is an additional dimension of design optimization available only to the QW structures. For example, Islam et al [84] reported a 14/1 on/off ratio modulator when the QW structure in this capillary leaky waveguide consisted of two 46 Å GaAs wells separated by 11.5 Å $Al_{.295}Ga_{.705}As$ barrier; the application of the electric field could now easily skew the wave function of the electron to one side of the coupled QW structure and the wave function of the hole to the other side of the coupled QW structure. Figure 10.16 shows the measured

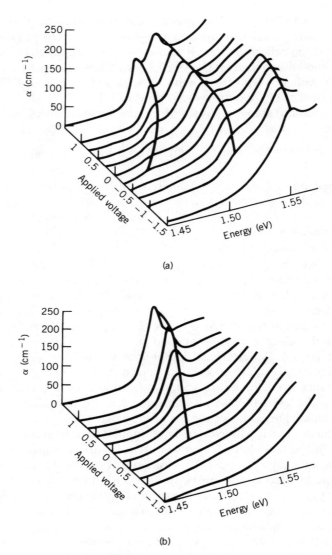

(a)

(b)

Figure 10.16 Measured absorption spectra as a function of energy for different voltages applied perpendicular to the quantum wells. Electric field for the optical wave is polarized (a) parallel to and (b) perpendicular to the CQW layers. The lines connecting the peaks are drawn as a guide to the eye. The adjusted field is calculated by using the estimated built-in field and allowing for a "dead field" of 5×10^4 V/cm $[E \text{ (V/cm)} \cong -77.76 \times 10^3 \text{ V} + 8 \times 10^4]$. (From Islam et al. [84]. Reprinted with permission from the American Institute of Physics.)

absorption spectra for different applied voltage and for TE and TM polarizations. Note the small voltage at which the electroabsorption takes place. Le et al. have also shown that optical spectra near the band edges of $Al_xGa_{1-x}As/GaAs$ coupled QW structures exhibit a rich structure [85], whereas Nishi and Hiroshima proposed the use of graded-gap QW to enhance the QCSE [86]. Enhancement in excitonic absorption due to overlap in heavy-hole and light-hole excitons has also been reported by Kothiyal et al. in a strained InGaAlAs/GaAs MQW structure [87]. These exotic QW structures may reduce significantly the rf drive power needed for the modulators or the device length needed for obtaining a given depth of modulation (or extinction ratio) at a given applied voltage, thereby reducing the electrode capacitance and increasing the bandwidth.

5. The fact that the QW structures have a much larger nonlinear coefficient than the heterojunction structures implies also that the QW EA modulators may saturate at a much lower optical power density than the heterojunction EA modulators. This may mean that the linear dynamic range of the QW EA modulators may be limited. No detailed experimental data have yet been reported on the saturation properties of QW EA modulators.

An InGaAs/InP MQW waveguide phase modulator has also been reported by Koren et al. [88]. In this case 60 layers of InGaAs(30 Å)/InP(230 Å) MQW with λ_g of the wells at 1.4 μm were used as the guiding layer. The undoped guiding layer grown on the InP substrate and a 0.6 μm thick undoped InP top cladding layer. The doping concentration for the InGaAs wells is $n = 3 \times 10^{15}$ cm^{-3}, and the doping concentration for the InP barriers is $n = 10^{16}$ cm^{-3}. The 5 μm wide ridge waveguide is formed by selective chemical etching of the top InP cladding layer. Propagation loss of the TE mode at $\lambda = 1.52$ μm is 9.7 dB/cm, caused primarily by the heavy doping of the p-cladding. Phase modulation $\Delta\phi$ as large as 12°/V-mm is obtained at low voltage at $\lambda = 1.52$ μm. More interesting than the $\Delta\phi$ data is that EA data has also been collected on this device. Voltage swings of -15 V lead to nearly 100% extinction at $\lambda = 1.48$ μm for a 1 mm long device. Even at $\lambda = 1.52$ μm, substantial intensity modulation is measured for applied voltage larger than 10 V. The estimated $\Delta n'/\Delta n''$ is 40 at $\lambda = 1.52$ μm and at low voltage where $\Delta n'$ and $\Delta n''$ are the change of the real part and imaginary part of the index due to the applied voltage. In comparision $\Delta n'/\Delta n''$ is estimated to be 3 at $\lambda = 1.48$ μm. The large $\Delta\phi$ at $\lambda = 1.52$ μm has a significant contribution from ER. There is also intensity modulation. Intensity modulation can probably be reduced if a longer λ or a shorter λ_g is used, but it will be accompanied by a reduction of $\Delta\phi$. However, the $\Delta n'/\Delta n''$ ratio will increase as $\lambda - \lambda_g$ is increased. The propagation loss can be reduced by lowering the p concentration of the bottom cladding layer. Unfortunately, there is no capacitance data on the device reported by Koren et al., although one would expect it to be in the same range as other EA devices. No data on nonlinear saturation is available. Clearly, the five comments made about the electroabsorption modulators are also applicable to phase modulators. An additional design consideration that must be taken into account is the factor $\Delta n'/\Delta n''$.

QW waveguide modulators have shown a potential for superior performance in insertion loss and control of guided-wave optical mode profile and depth of modulation for a given applied voltage, as compared with the heterojunction waveguide modulators. What is not clear is whether they would also have a superior performance in reducing the capacitance and in increasing the linear dynamic range. Small capacitance requires low doping density, large depletion depth, or short device length, since width is limited by lithography. It is difficult to obtain doping density in the 10^{14} cm^{-3} range for some ternary and quaternary QWs, as reported by Wang et al. for the heterostructures. The use of planar electrode also does not meet the requirement of applying the electric field in the z direction to produce the QCSE. Perhaps the electrode length can be shortened if the QW structures can be designed to yield very large $\Delta n'$ or $\Delta n''$ at reasonable applied voltage.

10.7 NOVEL DEVICES

Research results on complex channel waveguide devices such as Y-branches, directional coupler switches, etc., have begun to appear in the literature. Initial results concerning the investigations of two or more optical or electronic-optic devices integrated on the same chip have also begun to be reported. However, it is difficult to review the significance of these results at the present time, as there are so few of them. Instead, we shall only discuss here the result of a proposed type of high-speed low-device-power gate-controlled FET optical modulator which has the long-term significance as an example of a coupled optical-electronic device [89, 90]. The material structure for Kastalsky's proposed device [89] consists of a 3 μm thick n$^-$-Al$_{0.45}$Ga$_{0.55}$As grown on GaAs, serving as the cladding layer for the waveguide, followed by 3000 Å of n$^-$-Al$_{0.27}$Ga$_{0.73}$As grown as the waveguiding layer. Within the guiding layer, there is a single undoped GaAs QW, 100 Å thick, located 1000 Å below the top surface. There is also a 200 Å thick n$^+$ layer grown on the top surface. Because of this n$^+$ layer, the energy band of the Al$_{0.27}$Ga$_{0.73}$As near the top surface, above the QW, is bent as shown in Fig. 10.17b. The solid curve represents the case where there is no voltage applied. A FET is fabricated on this structure using the single QW as the two-dimensional conducting channel of an undoped heterojunction transistor. The Schottky barrier gate is 5 μm long and 125 μm wide. As a voltage V_G is applied to the gate, the energy band diagram is modified as illustrated by the dashed curves in Fig. 10.17b, creating carriers in the QW channel. Unlike a conventional FET, the gate area is also etched down 500 Å deep into the shape of a mesa as shown in

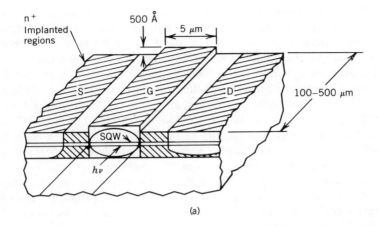

(a)

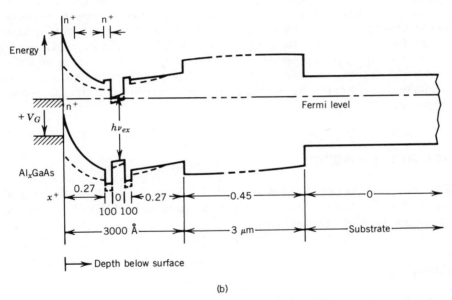

(b)

Figure 10.17 (a) Structure of the FET optical modulator (FETOM). Optical signals are guided under a gate self-aligned to a 5 μm wide, 100–500 μm long, 500 Å high ridge. The normally-off FET channel is formed by the SQW optical absorber. Ion-implanted n$^+$ regions provide electrical contact to the active region (S, source; G, gate; D, drain). (b). Epilayer and band structure sketch of an Al$_x$Ga$_{1-x}$As FETOM. For the gate–source bias $V_g = 0$ (solid line), the SQW conduction band is depleted of carriers; excitonic optical absorption is strongest. With a positive V_g (dotted line), the SQW conduction band is sufficiently filled to quench the excitonic absorption. The 3 μm Al$_{0.45}$Ga$_{0.55}$As layer provides optical confinement. (From Kastalsky et al. [89]. Reprinted with permission from the American Institute of Physics.)

Fig. 10.17a. The ridge structure under the gate thus serves simultaneously as a single mode ridge channel waveguide. The direction of propagation of guided waves is parallel to the width of the FET gate. At $V_G = 0$, there is electroabsorption in the single QW for the channel waveguide mode. The carrier produced by the gate voltage will induce transparency in the single QW due to the nonlinear saturation effect [89, 90], thereby modulating the optical guided-wave intensity, while the effect of the gate voltage on the source–drain current will be similar to that of a conventional FET. The physical mechanism for the quenching of optical absorption was speculated by Kastalsky. A similar device was reported by Chemla et al. [91] in which they identified the physical mechanism of the quenching of absorption due to carrier injection to be the filling of the phase space by the electrons as they are swept in the conducting QW channel. This mechanism was identified experimentally and the quenching of the absorption was observed by Bar-Joseph et al. [92]. Sakaki et al. have also observed the carrier concentration dependence of absorption spectra in modulation doped AlGaAs/ GaAs QWs [93]. The attractiveness of this type of device is that a high-speed FET is coupled effectively to an optical modulator, providing an effective coupling of high-speed electronic and optical devices. The apparent drawback is that the gate capacitance of a 125 μm wide gate is large (estimated at 0.75 pF by the authors), which will limit the frequency response.

ACKNOWLEDGMENTS

The author would like to thank Prof. H. H. Wieder, Prof. P. K. L. Yu, and Dr. T. Van Eck for many interesting discussions and helpful suggestions in the preparation of the manuscript.

REFERENCES

1. G. E. Stillman, C. M. Wolfe, C. O. Bozler, and J. A. Rossi, Electroabsorption in GaAs and its application to waveguide detectors and modulators. *Appl. Phys. Lett.* **28** 544 (1976).

2. H. Kroemer, Heterostructure device physics: Band discontinuities as device design parameters. *in* "VLSI Electronics" (N. G. Einspruch and W. R. Wisseman, eds.), Vol. 10, Chapter 4. Academic Press, Orlando, Florida, 1985.

3. D. S. Chemla, D. A. B. Miller, P. W. Smith, A. C. Gossard, and W. Wiegmann, Room temperature excitonic non-linear absorption and refraction in GaAs/ AlGaAs multiple quantum well structures. *IEEE J. Quantum Electron.* **QE-20**, 265 (1984).

4. D. S. Chemla, Quasi-two-dimensional excitons in GaAs/Al$_x$Ga$_{1-x}$As semiconductor multiple quantum well structures. *Helv. Phys. Acta* **56**, 607 (1983).

5. R. Dingle, Confined carrier quantum states in ultra thin semiconductor hetero-structures. *Festkoerperprobleme* **15** 21 (1975).

6. H. Okamoto, Semiconductor quantum-well structures for optoelectronics—recent advances and future prospects. *Jpn. J. Appl. Phys.* **26**, 315 (1987).

7. D. A. B. Miller, J. S. Weiner, and D. S. Chemla, Electric-field dependence of linear optical properties in quantum well structures: Waveguide electroabsorption and sum rules. *IEEE J. Quantum Electron.* **QE-22**, 1816 (1986).

8. D. A. B. Miller, D. S. Chemla, T. C. Damen, A. C. Gossard, W. Wiegmann, T. H. Wood, and C. A. Burrus, Relation between electroabsorption in bulk semiconductors and in quantum wells: The quantum confined Franz-Keldysh effect. *Phys. Rev. B. Condens. Matter* [3] **33**, 6976 (1986).

9. D. Coffey, Photoabsorption due to excitons for strained superlattice quantum wells in the presence of an applied electric field. *J. Appl. Phys.* **63**, 4626 (1988).

10. I. Bar-Joseph, C. Klinshirn, D. A. B. Miller, D. S. Chemla, U. Koren, and B. I. Miller, Quantum confine stark effect in InGaAs/InP quantum wells grown by organometallic vapor phase epitaxy. *Appl. Phys. Lett.* **50**, 1010 (1987).

11. Y. Kawamura, K. Wakita, and H. Asahi, Observation of heavy-hole and light-hole excitons in InGaAs/InAlAs MQW structures at room temperature. *Electron. Lett.* **21**, 371 (1985).

12. D. A. B. Miller, D. S. Chemla, T. C. Damen, A. G. Gossard, W. Wiegmann, T. H. Wood, and C. A. Burrus, Electric-field dependence of optical absorption near the bandgap of quantum-well structures. *Phys. Rev. B: Condens. Matter* [3] **32**, 1043 (1985).

13. J. S. Weiner, D. A. B. Miller, D. S. Chemla, T. C. Damen, C. A. Burrus, T. H. Wood, A. C. Gossard, and W. Wiegmann, Strong polarization—sensitive electroabsorption in GaAs/AlGaAs quantum well waveguides. *Appl. Phys. Lett.* **47**, 1148 (1985).

14. D. F. Welch, G. W. Wicks, and L. F. Eastman, Calculation of the conduction band discontinuity for $Ga_{.47}In_{.53}As/Al_{.48}In_{.52}As$ heterojunction. *J. Appl. Phys.* **55**, 3176 (1984).

15. M. Shimada and S. Sugano, Interband optical transitions in extremely aniso-tropic semiconductors. *J. Phys. Soc. Jpn.* **21**, 1936 (1966).

16. D. A. B. Miller, D. S. Chemla, T. C. Damen, A. C. Gossard, W. Wiegmann, T. H. Wood, and C. A. Burrus, Band-edge electroabsorption in quantum well structures, quantum confined stark effect. *Phys. Rev. Lett.* **53**, 2173 (1984).

17. W. H. Knox, D. A. B. Miller, T. C. Damen, D. S. Chemla, C. V. Shank, and A. C. Gossard, Subpicosecond excitonic electroabsorption in room-temperature quantum wells. *Appl. Phys. Lett.* **48**, 864 (1986).

18. K. Wakita, Y. Kawamura, Y. Yoshikuni, and H. Asahi, High temperature excitons and enhanced electroabsorption in InGaAs/InAlAs multiple quantum wells. *Electron. Lett.* **21**, 574 (1985).

19. T. H. Wood, Direct measurement of the electric-field-dependent absorption coefficient in GaAs/AlGaAs multiple quantum wells. *Appl. Phys. Lett.* **48**, 1413 (1986).

20. T. E. Van Eck, P. Chu, W. S. C. Chang, and H. H. Weider, Electroabsorption in a InGaAs/GaAs strained-layer multiple quantum wells structure. *Appl. Phys. Lett.* **49**, 135 (1986).

21. T. H. Wood, C. A. Burrus, R. S. Tucker, J. W. Weiner, D. A. B. Miller, D. S. Chemla, T. C. Damen, A. C. Gossard, and W. Wiegmann, 100 ps waveguide multiple quantum well (MQW) optical modulator with 10:1 on/off ratio. *Electron. Lett.* **21**, 693 (1985).

22. T. E. Van Eck and W. S. C. Chang, Electrorefraction and electroabsorption in InGaAs/GaAs multiple-quantum-well structures. *Tech. Dig. Conf. Lasers Electro Opt. 1987*, p. 338 (1987).

23. H. Nagai, M. Yamanishi, Y. Kan, and I. Suemune, Field-Induced modulation of refractive index and absorption coefficient in a GaAs/AlGaAs quantum well structure. *Electron. Lett.* **22**, 888 (1986).

24. H. Nagai, Y. Kan, M. Yamanishi, and I. Suemuene, Electrorefluctance spectra and field-induced variation in refractive index of a GaAs/AlAs quantum well structure at room temperature. *Jpn. J. Appl. Phys.* **25**, L640 (1986).

25. T. Hiroshima, Electric field induced refractive index changes in GaAs-$Al_xGa_{1-x}As$ quantum wells. *Appl. Phys. Lett.* **50**, 968 (1987).

26. K. B. Kahen and J. P. Leburton, Exciton effects in the index of refraction of multiple quantum wells and superlattices. *Appl. Phys. Lett.* **49**, 734 (1986).

27. F. Stern, Dispersion of the index of refraction near the absorption edge of semiconductors. *Phys. Rev. A.* **133**, 1653 (1964).

28. T. H. Wood, R. W. Tkach, and A. R. Chraplyvy, Observation of large quadratic electro-optic effect in GaAs/AlGaAs multiple quantum wells. *Appl. Phys. Lett.* **50**, 798 (1987).

29. M. Glick, F. K. Reinhart, G. Weimann, and W. Schlapp, Quadratic electro-optic light modulation in a GaAs/AlGaAs multiple quantum well heterostructure near the excitonic gap. *Appl. Phys. Lett.* **48**, 989 (1986).

30. J. S. Weiner, D. A. B. Miller, and D. S. Chemla, Quadratic electro-optic effect due to quantum confined Start effect in quantum wells. *Appl. Phys. Lett.* **50**, 842 (1987).

31. F. Koyama and K. Iga, Frequency chirping of extenal modulation and its reduction. *Electron. Lett.* **21**, 1065 (1985).

32. S. Schmitt-Rink, D. S. Chemla, and D. A. B. Miller, Theory of transient excitonic optical nonlinearities in semiconductor quantum-well structures. *Phys. Rev. B: Condens. Matter.* [3] **32**, 6601 (1985).

33. D. S. Chemla and D. A. B. Miller, Room-temperature excitonic nonlinear-optical effects in semiconductor quantum well structures. *J. Opt. Soc. Am. B: Opt. Phys.* **2**, 1155 (1985).

34. A. M. Fox, A. C. Maciel, M. G. Shorthose, J. F. Ryan, M. D. Scott, J. I. Davies, and J. R. Riffat, Nonlinear excitonic optical absorption in GaInAs/InP quantum wells. *Appl. Phys. Lett.* **51**, 30 (1987).

35. K. Tai, J. Hegarty, and W. T. Tsang, Nonlinear spectroscopy in $In_{0.53}Ga_{0.47}As$/InP multiple quantum wells. *Appl. Phys. Lett.* **51**, 86 (1987).

36. J. W. Little, J. K. Whisnaut, R. P. Leavitt, and R. A. Wilson, Extremely low-intensity optical nonlinearity in asymmetric coupled quantum wells. *Appl. Phys. Lett.* **51**, 1786 (1987).

37. V. Das, Y. Chen, and P. Bhattacharya, Nonlinear effects in coplanar GaAs/InGaAs strained-layer superlattice directional couplers. *Appl. Phys. Lett.* **51**, 1679 (1987).

38. D. A. B. Miller, D. S. Chemla, T. C. Damen, T. H. Wood, C. A. Burrus, A. C. Gossard, and W. Wiegmann, The quantum well self-electro-optic effect devices: Optoelectronic bistability and oscillation, and self-linearized modulation. *IEEE J. Quantum Electron.* **QE-21**, 1462 (1985).

39. H. Temkin, M. B. Panish, P. M. Petroff, R. A. Hamm, J. M. Vandenberg, and S. Sumski, GaInAs(P)/InP quantum well structures grown by gas source molecular beam epitaxy. *Appl. Phys. Lett.* **47**, 394 (1985).

40. J. S. Weiner, D. S. Chemla, D. A. B. Miller, T. H. Wood, D. Sivco, and A. Y. Cho, Room temperature excitons in 1.6 μm band-gap GaInAs/AlInAs quantum wells. *Appl. Phys. Lett.* **46**, 619 (1985).

41. M. B. Panish, H. Temkin, R. A. Hamm, and S. N. G. Chu, Optical properties of very thin GaInAs(P)/InP quantum wells grown by gas source molecular beam epitaxy. *Appl. Phys. Lett.* **49**, 164 (1986).

42. M. S. Skolnick, L. L. Taylor, S. J. Bass, A. D. Pitt, D. J. Mowbray, A. G. Cullis, and N. G. Chew, InGaAs-InP quantum wells grown by atmospheric pressure metalorganic chemical vapor deposition. *Appl. Phys. Lett.* **51**, 24 (1987).

43. H. Temkin, D. Gershoni, and M. B. Panish, InGaAsP/InP quantum well modulators grown by gas source molecular beam epitaxy. *Appl. Phys. Lett.* **50**, 1776 (1987).

44. Y. Kawaguchi and H. Asahi, High temperature observation of heavy and light-hole excitons in InGaAs/InP multiple quantum well structures grown by metalorganic molecular beam epitaxy. *Appl. Phys. Lett.* **50**, 1243 (1987).

45. M. Razeghi, J. Nagle, P. Maurel, F. Omnes, and J. P. Pocholle, Room-temperature excitons in $Ga_{0.47}In_{.53}As$-InP superlattices grown by low-pressure metalorganic chemical vapor deposition. *Appl. Phys. Lett.* **49**, 1110 (1986).

46. B. I. Miller, E. F. Schubert, U. Koren, A. Ourmazd, A. H. Dayem, and R. J. Capik, High quality narrow GaInAs/InP quantum wells grown by atmospheric organometallic vapor phase epitaxy. *Appl. Phys. Lett.* **49**, 1384 (1986).

47. D. J. Westland, A. M. Fox, A. C. Maciel, J. F. Ryan, M. D. Scott, J. I. Davies, and J. R. Riffat, Optical studies of excitons in $Ga_{0.47}In_{0.53}As$/InP multiple quantum wells. *Appl. Phys. Lett.* **50**, 839 (1987).

48. K. W. Carey, R. Hall, J. E. Fouquet, F. G. Kellert, and G. R. Trott, Structural and photoluminescent properties of GaInAs quantum wells with InP barriers by organometallic vapor phase epitaxy. *Appl. Phys. Lett.* **51**, 910 (1987).

49. W. T. Tsang and E. F. Schubert, Extremely high quality $Ga_{0.47}In_{0.53}As$/InP quantum wells grown by chemical beam epitaxy. *Appl. Phys. Lett.* **49**, 220 (1986).

50. David A. Dahl, L. J. Dries, F. A. Junga, W. G. Opyd, and P. Chu, Photo-luminescence in strained InGaAs/GaAs superlattices. *J. Appl. Phys.* **61**, 2079 (1987).

51. G. Burnes, C. R. Wie, F. H. Dacol, G. D. Pettit, and J. M. Woodall, Phonon shifts and strains in strained-layered $(Ga_{1-x}In_x)As$. *Appl. Phys. Lett.* **51**, 1919 (1987).

52. H. C. Lee, K. M. Dzurko, P. D. Dapkus, and E. Garmire, Electro-absorption in AlGaAs/GaAs multiple quantum well structures grown on a GaP transparent substrate. *Appl. Phys. Lett.* **51**, 1582 (1987).

53. K. H. Chang, P. R. Berger, J. Singh, and P. K. Bhattacharya, Molecular beam epitaxial growth and luminescence of $In_xGa_{1-x}As/In_xAl_{1-x}As$ multi-quantum wells on GaAs. *Appl. Phys. Lett.* **51**, 262 (1987).

54. N. G. Anderson, W. D. Laidig, R. M. Kolbas, and Y. C. Lo, Optical characterization of pseudomorphic $In_xGa_{1-x}As$-GaAs single quantum well structures. *J. Appl. Phys.* **60**, 2361 (1986).

55. K. Oe, Y. Shinoda, and K. Sugiyama, Lattice deformations and misfit dislocations in GaInAsP/InP double-heterostructure layers. *Appl. Phys. Lett.* **33**, 962 (1978).

56. R. Hall, J. C. Bean, F. Cerdeira, A. T. Fiory, and J. M. Gibson, Stability of semiconductor strained-layer superlattices. *Appl. Phys. Lett.* **48**, 56 (1986).

57. T. G. Anderson, Z. G. Chen, V. D. Kulakovskii, A. Uddin, and J. T. Vallin, Variation of the critical layer thickness with In content in strained $In_xGa_{1-x}As$-GaAs quantum wells grown by molecular beam epitaxy. *Appl. Phys. Lett.* **51**, 752 (1987).

58. P. W. Sullivan, J. E. Bower, and G. M. Metze, Growth of semiconductor/insulator structures-GaAs/fluoride/GaAs(001). *J. Vac. Sci. Technol. B* [2] **3**, 500 (1985).

59. K. Tsutsui, H. Ishiwara, and S. Furukawa, Lattice matching at elevated substrate temperature for growth of GaAs films with good electrical properties on $Ga_xSr_{1-x}F_2$/GaAs (100) structures. *Appl. Phys. Lett.* **48**, 587 (1986).

60. W. D. Laidig, N. Holonyak, Jr., M. D. Camras, K. Hess, J. J. Coleman, P. D. Dapkus, and J. Bardeen, Disorder of an AlAs-GaAs superlattice by impurity diffusion. *Appl. Phys. Lett.* **38**, 776 (1981).

61. J. E. Eplear, R. D. Burnham, R. L. Thornton, and T. L. Paoli, Low threshold buried heterostructure quantum well diode lasers by laser-assisted disordering. *Appl. Phys. Lett.* **50**, 1637 (1987).

62. F. Julien, P. D. Swanson, M. A. Emanuel, D. G. Deppe, T. A. Detemple, J. J. Coleman, and N. Holonyak, Jr., Impurity-induced disorder-delineated optical waveguide in GaAs-AlGaAs superlattices. *Appl. Phys. Lett.* **50**, 866 (1987).

63. K. Hiruma, H. Inoue, K. Ishida, and H. Matsumura, Low loss GaAs optical waveguide grown by the metalorganic chemical vapor deposition method. *Appl. Phys. Lett.* **47**, 186 (1985).

64. R. J. Deri, E. Kapon, and L. M. Schiavone, Scattering in low-loss GaAs/AlGaAs rib waveguides. *Appl. Phys. Lett.* **51**, 788 (1987).

65. L. M. Johnson, Z. L. Liau, and S. H. Groves, Low-loss GaInAsP buried-heterostructure optical waveguide branches and bends. *Appl. Phys. Lett.* **44**, 278 (1984).

66. S. Ritchie, E. G. Scott, and P. M. Rodgers, Optical waveguides in $In_{0.52}Al_{0.48}As$ grown on InP by MBE. *Electron. Lett.* **22**, 1066 (1986).

67. C. H. Joyner, A. G. Dentai, R. C. Alferness, L. L. Buhl, M. D. Divino, and W. C. Dautremont-Smith, InGaP/InP waveguides. *Appl. Phys. Lett.* **50**, 1509 (1987).

68. U. Koren, B. I. Miller, T. L. Koch, G. D. Boyd, R. J. Capik, and C. E. Soccolich, Low-loss InGaAs/InP multiple quantum well waveguides. *Appl. Phys. Lett.* **49**, 1602 (1986).

69. E. Kapon and R. Bhat, Low-loss single-mode GaAs/AlGaAs optical waveguides grown by organometallic vapor phase epitaxy. *Appl. Phys. Lett.* **50**, 1628 (1987).

70. G. J. Sonek, J. M. Ballantyne, Y. J. Chen, G. M. Carter, S. W. Brown, E. S. Koteles, and J. P. Salerno, Dielectric properties of GaAs/AlGaAs multiple quantum well waveguides. *IEEE J. Quantum Electron.* **QE-22**, 1015 (1986).

71. N. K. Dutta and N. A. Olsson, Electro-absorption in InGaAsP-InP double heterostructures. *Electron. Lett.* **20**, 634 (1984).

72. C. M. Gee, H. Blanvelt, G. D. Thurmond, and H. W. Yen, High extinction AlGaAs/GaAs waveguide modulator. *Tech. Dig. Top. Meet. Integr. Guided-Wave Opt. 1986*, p. 22 (1986).

73. Y. Noda, M. Suzuki, Y. Kushiro, and S. Akiba, High-speed electro-absorption modulator with strip-loaded GaInAsP planar waveguide. *IEEE J. Lightwave Technol.* **LT-4**, 1445 (1986).

74. S. C. Lin, X. L. Jing, M. K. Chin, L. M. Walpita, P. K. L. Yu, and W. S. C. Chang, GaInAsP buried channel EA modulator for 1.3 μm fiber links. *Electron. Lett.* **23**, 1257 (1987).

75. S. Y. Wang, S. H. Lin, and Y. M. Houng, GaAs traveling-wave polarization electro-optic waveguide modulator with bandwidth in excess of 20 GHz at 1.3 μm *Appl. Phys. Lett.* **51**, 83 (1987).

76. S. H. Lin and S. Y. Wang, High-throughput GaAs PIN electrooptic modulator with a 3-dB bandwidth of 9.6 GHz at 1.3 μm. *Appl. Opt.* **26**, 1696 (1987).

77. A. Alping, X. S. Wu, T. R. Hansken, and L. A. Coldrin, Highly efficient waveguide phase modulator for integrated opto-electronics. *Appl. Phys. Lett.* **48**, 1243 (1986).

78. A. Alping and L. A. Coldrin, Electrorefraction in GaAs and InGaAsP and its application to phase modulators. *J. Appl. Phys.* **61**, 2430 (1987).

79. L. A. Coldrin, J. G. Mendoza-Alvarez, and R. H. Yan, Design of optimized high-speed depletion-edge-translation optical waveguide modulators in III–V semiconductors. *Appl. Phys. Lett.* **51**, 792 (1987).

80. K. Ishida, H. Nakamura, H. Matsumura, T. Kadoi, and H. Inoue, InGaAsP/InP optical switches using carrier induced refractive index change *Appl. Phys. Lett.* **50**, 141 (1987).

81. P. M. Rogers, M. J. Robertson, A. K. Chaterjee, and S. Y. Wong, High-performance InGaAsP/InP phase modulators for integrated optics. *Tech. Dig. Top. Meet. Integr. Guided-Wave Opt. 1986*, p. 22 (1986).

82. K. Wakita, S. Nojima, K. Nakashima, and Y. Kawaguchi, GaInAs/InP waveguide multiple-quantum-well optical modulator with 9 dB on/off ratio. *Electron. Lett.* **23**, 1067 (1987).

83. U. Koren, B. I. Miller, T. L. Koch, G. Eisenstein, R. S. Tucker, I. Bar-Joseph, and D. S. Chemla, Low-loss InGaAs/InP multiple quantum well optical electroabsorption waveguide modulator. *Appl. Phys. Lett.* **51**, 1132 (1987).

84. M. N. Islam, R. L. Hillman, D. A. B. Miller, D. S. Chemla, A. C. Gossard, and J. H. English, Electroabsorption in GaAs/AlGaAs coupled quantum well waveguides. *Appl. Phys. Lett.* **50**, 1098 (1987).

85. H. Q. Le, J. J. Zayhowski, and W. D. Goodhue, Stark effect in $Al_xGa_{1-x}As$/GaAs coupled quantum wells. *Appl. Phys. Lett.* **50**, 1518 (1987).

86. K. Nishi and T. Hiroshima, Enhancement of quantum confined Stark effect in a graded gap quantum well. *Appl. Phys. Lett.* **51**, 320 (1987).

87. G. P. Kothiyal, S. Hong, N. Dabbar, P. K. Battacharya, and J. Singh, Enhancement in excitonic absorption due to overlap in heavy-hole and light-hole excitons in GaAs/InAlGaAs quantum well structures. *Appl. Phys. Lett.* **51**, 1091 (1987).

88. U. Koren, T. L. Koch, H. Presting, and B. I. Miller, InGaAs/InP multiple quantum well waveguide phase modulator. *Appl. Phys. Lett.* **50**, 368 (1987).

89. A. Kastalsky, J. H. Abeles, and R. F. Leheny, Novel optoelectronic single quantum well devices based on electron bleaching of exciton absorption. *Appl. Phys. Lett.* **50**, 708 (1987).

90. C. Delalande, J. Orgonasi, J. A. Brum, G. Bastard, M. Voss, G. Weimann, and W. Schlapp, Optical studies of a GaAs quantum well based field-effect transistor. *Appl. Phys. Lett.* **51**, 1346 (1987).

91. D. S. Chemla, I. Bar-Joseph, C. Klingshirn, D. A. B. Miller, J. M. Kuo, and T. Y. Chang, Optical reading of field-effect transistors by phase-space absorption quenching in a single InGaAs quantum well conducting channel. *Appl. Phys. Lett.* **50**, 585 (1987).

92. I. Bar-Joseph, J. M. Kuo, C. Klingshirn, G. Livescu, T. Y. Chang, D. A. B. Miller, and D. S. Chemla, Absorption spectroscopy of the continuous transition from low to high electron density in a single modulation-doped InGaAs quantum well. *Phys. Rev. Lett.* **59**, 1357 (1987).

93. H. Sakaki, H. Yoshimura, and T. Matsusue, Carrier concentration dependent absorption of modulation doped n-AlGaAs/GaAs quantum wells and performance analysis of optical modulators and switches using carrier induced bleaching (CIB) and refractive index change (CIRIC). *Jpn. J. Appl. Phys.* **26**, L1104 (1987).

11 Quantum Mechanical Corrections to Classical Transport in Submicron/ Ultrasubmicron Dimensions

MICHAEL A. STROSCIO*
Electronics Division, U.S. Army Research Office
Research Triangle Park, North Carolina

11.1 INTRODUCTION

Anticipated data processing requirements in the coming decades are such that future electronic chips must have device densities greatly in excess of the current value of approximately 1 million devices per chip. To achieve substantially greater densities, device features must be of the order of 1000 Å or less. For such small spatial scales, quantum transport techniques must be employed to model accurately a wide variety of phenomena underlying device operation. These small device scales also lead to unusually large electric field gradients which drive a number of nonequilibrium processes. Furthermore, device–device and device–environment couplings are complicated significantly by the fact that carrier de Broglie wavelengths may be approximately equal to device spatial scales; many such devices may constitute open systems.

To model the quantum phenomena associated with such ultrasmall electronic structures, it is necessary to formulate quantum transport and many-body theories. Such fully quantum mechanical theories may be based on the density matrix or on equivalent approaches such as the Wigner distribution function. Before discussing the theoretical framework needed to model quantum-based electronic devices, some of the issues associated with microstructure technology needed to realize envisioned quantum-based electronic devices will be briefly reviewed.

A key element of microstructure technology is the epitaxial growth of

*The author is also affiliated with the Departments of Electrical Engineering and Physics at Duke University, Durham, North Carolina, and with the Department of Electrical and Computer Engineering at North Carolina State University, Raleigh, North Carolina.

ultrasmall electronic structures. These growth techniques provide the means to produce ultrasmall compound semiconductor structures [1] with compositional changes that vary on a scale small compared with the de Broglie wavelength of charge carriers [2]. In these structures, thermal energies are frequently comparable to differences between allowed energy states [3]. Under such conditions, carrier transport cannot be predicted accurately without including quantum mechanical effects [2, 4]. Indeed, quantum phenomena, such as tunneling, provide the basis for ultrasmall electronic devices with minimum feature sizes well below the 0.25 μm limit [5] projected for current integrated-circuit technologies. In addition, these phenomena lead to devices with intrinsic response times less than 0.1 ps [6, 7]. As basic concepts for quantum-based electronic devices emerge, it is becoming clear that the full capabilities of envisioned devices can be realized only after significant advances are made in key supporting elements of the technology underlying quantum-based electronic devices.

Theoretical approaches for formulating adequate descriptions of device performance in the quantum regime are based on the use of the Wigner function [8], moment equations [9], and the density matrix [10, 11]. These approaches are based on formulations suitable for incorporating a variety of important phenomena, including many-body interactions, dissipation in quantum systems [12], Pauli exclusion, finite-temperature effects [12], interactions between open systems [11], and self-consistent time-dependent potential variations [11]. The complexity of these phenomena has precluded accurate predictions of device performance from *ab initio* calculations.

The Wigner function and density matrix techniques have been applied recently [10] to describe the dynamical evolution of many-body quantum mechanical systems in terms of a closed self-consistent set of moment equations. Recently, this approach has been extended [13] to include dissipative quantum phenomena [12] associated with the interaction of a charge carrier with a heat bath comprised of an ensemble of harmonic oscillators. As will be demonstrated in this chapter, the role of quantum corrections to the classical dynamics of quantum mechanical systems may be indicated clearly by the use of moment equations [13].

Numerous experimental results bearing on envisioned quantum-based electronic devices are available. In this discussion, several of these results are reviewed, since they illustrate currently available or conspicuously absent elements of the full technology of quantum-based electronic devices. Other examples demonstrate existing capabilities for fabricating ultrasmall electronic structures.

The well-known tunneling experiments of Sollner et al. [6, 7] demonstrate negative differential resistance (NDR) associated with quantum tunneling via a quantum level in a 50 Å wide GaAs quantum well. These experiments also illustrate the importance of size quantization, finite-temperature effects, and Fermi degeneracy. There is currently some uncertainty regarding the dominant physical mechanisms responsible for the observed NDR; it has been suggested recently [14] that it is not due solely to the resonant Fabry–Perot effect, but instead arises, to a large extent, because the number of electrons available for tunneling into the quantum well depends on the magnitude of the time-dependent bias potential.

Although the total length of the active region of the tunneling structure and AlGaAs barriers is only 150 Å, the entire device is much larger as a result of the requisite substrate and contact structures. This situation is typical of the current state of the art and illustrates clearly that technologies underlying contacts, interconnects, and packaging of ultrasmall electronic structures limit the practical performance levels achievable from epitaxially grown quantum-based electronic devices.

Some classes of envisioned quantum-based electronic devices depend on the confinement of carriers in two or three dimensions; in such devices the one-dimensional confinement of conventional epitaxially grown superlattices is inadequate. For example, one often-discussed architecture for quantum-based electronic systems relies on coupling vertical superlattices by lateral tunneling between adjacent, localized superstructures. At temperatures of a few degrees Kelvin, lateral confinement is achieved in semiconductor structures with dimensions as large as $\frac{1}{4}$ μm. Electron beam lithography [3], ion-induced disordering [15], and temperature-induced layer interdiffusion [16] have been employed to modify fundamental compound–semiconductor properties. As an example, the three-dimensional confinement of carriers offers an explanation for dramatic changes in quantum branching ratios involving light-hole and heavy-hole luminescence experiments [3]. Likewise, ion-induced disordering provides a technique for selectively disordering GaAs/AlGaAs superlattices into homogeneous crystals of AlGaAs [15]. Temperature-induced layer interdiffusion produces changes in energy level splittings in compound-semiconductor quantum wells. These techniques provide the means to modify the quantum properties of compound semiconductors and are essential elements of the full technology underlying quantum-based electronic devices.

A class of quantum-based electronic devices based on wave function engineering [17] has been pursued actively by Sakaki and co-workers at the University of Tokyo. As an example, experiments with dual-gate modulation doped field effect transistors (MODFETs) demonstrate that minor potential-induced distortions of the wave function of the two-dimensional electron gas (2DEG) significantly change the mobility of the 2DEG; the change in the mobility is determined primarily by the overlap of the electron gas wave function with ionized impurities near the MODFET heterojunction.

Still another class of quantum devices [18] is based on quantum interference phenomena.

The quantum analog of Hess real-space transfer has been demonstrated by the important results of Kirchoefer et al. [19]. In these experiments, resonant tunneling through a barrier separating two quantum wells is observed to produce NDR during room-temperature operation. As in the Sollner experiments, the contact layers, cladding layers, ohmic contacts, and substrate structures are much larger than the active tunneling structure.

The recent observation [20] of NDR in a system of two strongly coupled superlattices provides evidence of tunneling between minibands of two superlattices separated by a tunnel barrier. Quantum-based electronic devices that use tunneling between minibands may not be subject to complications associated with intervalley transfer, since these devices will operate at lower potential energies.

Device concepts continue to emerge as a wide variety of experimental observations demonstrate the utility of basing device operation on quantum phenomena. Many of these results are encouraging, but they also serve to highlight the need to formulate theoretical models for small, ultrahigh-speed quantum-based electronic devices. This chapter presents the theoretical techniques available to describe quantum-based devices and to model the quantum mechanical corrections to classical transport in submicron/ ultrasubmicron dimensions.

11.2   BOLTZMANN TRANSPORT EQUATION

Charge transport in the bulk crystalline solids of conventional electronic devices may be modeled accurately by computing quantities such as current from the probability distribution function $f(\mathbf{r}, \mathbf{p}, t)$. At time t, the probability of finding an electron of momentum $\mathbf{p} = \hbar\mathbf{k}$ in a volume $d\mathbf{r}$ centered at $\mathbf{r}$ is $f(\mathbf{r}, \mathbf{p}, t)d\mathbf{r}$. For conventional electronic structures, the probability distribution function satisfies the Boltzmann transport equation:

$$\frac{\partial f}{\partial t} + \mathbf{v} \cdot \nabla_{\mathbf{r}} f + \mathbf{F} \cdot \nabla_{\mathbf{p}} f = \left(\frac{\partial f}{\partial t}\right)_{\text{collisions}} \tag{1}$$

where $(\partial f/\partial t)_{\text{collisions}}$ is the rate of change of f due to collisions, $\mathbf{F}$ is the force on the charge carrier, and the velocity $\mathbf{v}$ is given in terms of the energy momentum dispersion relation $\varepsilon(\mathbf{p})$ by

$$\mathbf{v} = \nabla_{\mathbf{p}} \varepsilon(\mathbf{p}) \tag{2}$$

The Boltzmann transport equation describes a wide variety of classical transport phenomena; however, for a number of reasons, the Boltzmann transport equation fails to yield accurate predictions for systems with size features less than 1000 Å. Since the distribution function $f(\mathbf{r}, \mathbf{p}, t)$ depends on both position and momentum, it is unclear how $f(\mathbf{r}, \mathbf{p}, t)$ should be interpreted for dimensional scales of the order of the de Broglie wavelength of the particles in the distribution. For example, the de Broglie wavelength of an electron in GaAs at room temperature is about 250 Å; thus, if wave packets are formed with dimensions of approximately Δl, then the spread in momentum is of the order of $\hbar/\Delta l$. The finite extent of such a wave packet is incompatible with the Boltzmann description of transport for several reasons, including

1. Boltzmann transport is based on the assumption of localized collisions.
2. **F** may vary greatly over the width of a wave packet.
3. Many-particle quantum effects may occur in regions of size Δl for some physical parameters.

To overcome these shortcomings, the techniques of quantum transport must be used to formulate alternatives to the Boltzmann transport description. These quantum transport theories may be formulated in terms of several fully quantum mechanical theories; this treatment will emphasize the use of the density matrix formalism, the Feynman path integral, and the Wigner distribution function. Finally, this treatment will consider selected applications of the quantum transport theories to illustrate quantum mechanical corrections to classical transport in systems with ultrasubmicron dimensions.

11.3   DENSITY MATRIX FORMULATION

The density matrix description of quantum mechanics is well suited for describing transport in ultrasmall electronic devices. The density matrix formalism provides a means of modeling complex dynamical systems composed of mixed states as well as systems interacting with thermal environments.

The density operator $\hat{\rho}$ for a quantum mechanical system described by m independently prepared state vectors, $|\psi_m\rangle$, is defined by

$$\hat{\rho} = \sum_m W_m |\psi_m\rangle\langle\psi_m| \tag{3}$$

where W_m is the statistical probability for the system to be in state $|\psi_m\rangle$.

As an example of the statistical weight W_m, consider a system of N spin-$\frac{1}{2}$ particles with N_1 particles prepared in a state $|\psi_1\rangle = |+\frac{1}{2}\rangle$ and with $N_2 = (N - N_1)$ particles prepared in a state $|\psi_2\rangle = |-\frac{1}{2}\rangle$; then W_1 and W_2 are given by N_1/N and N_2/N, respectively. The density operator is given by

$$\hat{\rho} = W_1|+\tfrac{1}{2}\rangle\langle +\tfrac{1}{2}| + W_2|-\tfrac{1}{2}\rangle\langle -\tfrac{1}{2}| \tag{4}$$

and the density matrix in diagonal representation is

$$\langle\psi_i|\hat{\rho}|\psi_j\rangle = W_i\delta_{ij} \tag{5}$$

Another example of the statistical weight W_m is provided by the canonical ensemble of quantum statistical mechanics where

$$\rho_{mn} = \delta_{mn} W_m = \delta_{mn} \frac{e^{-\beta E_m}}{Tr\, e^{-\beta H}}; \qquad \beta = 1/kT \tag{6}$$

For this case

$$\hat{\rho} = \sum_m |\psi_m\rangle \frac{e^{-\beta E_m}}{Tr\, e^{-\beta H}} \langle \psi_m| \tag{7}$$

Thus, for a system in thermal equilibrium with a larger system of temperature T, the probability for the smaller system to have an energy E_m is $W_m = e^{-\beta E_m}/Tr\, e^{-\beta H}$

The density matrix ρ_{ij} is obtained from the density operator through the relationship

$$\rho_{ij} = \langle \phi_i|\hat{\rho}|\phi_j\rangle \tag{8}$$

where $|\phi_n\rangle$ is a complete orthonormal basis set for $|\psi_m\rangle$:

$$|\psi_m\rangle = \sum_m a_n^{(m)}|\phi_n\rangle \tag{9a}$$

and

$$\langle \psi_m| = \sum_{n'} a_{n'}^{(m)*} \langle \phi_{n'}| \tag{9b}$$

With these equations, it follows that

$$\begin{aligned}
\rho_{ij} &= \langle \phi_i|\hat{\rho}|\phi_j\rangle \\
&= \langle \phi_i| \sum_m W_m \sum_n \sum_{n'} a_n^{(m)} a_{n'}^{(m)*} |\phi_n\rangle\langle \phi_{n'}|\phi_j\rangle \\
&= \sum_m W_m a_i^{(m)} a_j^{(m)*}
\end{aligned} \tag{10}$$

Thus, the density matrix in the $|\phi_n\rangle$ representation is given in terms of W_m and the probability amplitudes $a_i^{(m)}$.

The equation of motion for the density matrix may be derived by considering the time derivative of a pure state $|\psi\rangle$:

$$|\psi\rangle = \sum_n a_n(t)|\phi_n\rangle \tag{11}$$

That is

$$i\hbar \sum_n \frac{\partial a_n}{\partial t} |\phi_n\rangle = \sum_n a_n H|\phi_n\rangle \tag{12}$$

Operating on this equation from the left with $\langle \phi_p |$

$$i\hbar \frac{\partial a_p}{\partial t} = \sum_n H_{pn} a_n \tag{13a}$$

$$-i\hbar \frac{\partial a_p^*}{\partial t} = \sum_n H_{pn}^* a_n^* \tag{13b}$$

where

$$H_{pn} = \langle \phi_p | H | \phi_n \rangle$$

For a pure state

$$\begin{aligned}
i\hbar \frac{\partial \rho_{np}}{\partial t} &= i\hbar \frac{\partial (a_p^* a_n)}{\partial t} \\
&= i\hbar \left(\frac{\partial a_p^*}{\partial t} a_n + a_p^* \frac{\partial a_n}{\partial t} \right) \\
&= \sum_j (H_{nj} \rho_{jp} - \rho_{nj} H_{jp})
\end{aligned} \tag{14}$$

where $H_{jp} = H_{pj}^*$; this result may be expressed in terms of the density operator as,

$$i\hbar \frac{\partial \hat{\rho}}{\partial t} = [\hat{H}, \hat{\rho}] \tag{15}$$

This equation for the time evolution of the density operator is known as the quantum Liouville equation.

Alternatively, the dynamical equation for the density matrix may be derived by writing,

$$i\hbar \frac{\partial}{\partial t} |\psi(\mathbf{r}, t)\rangle = \hat{H} |\psi(\mathbf{r}, t)\rangle \tag{16}$$

as,

$$\frac{\partial}{\partial t} \left[\exp\left(\frac{it\hat{H}}{\hbar} \right) |\psi(\mathbf{r}, t)\rangle \right] = 0 \tag{17}$$

Integrating this last result from 0 to t gives

$$|\psi(\mathbf{r}, t)\rangle = \exp\left(\frac{-it\hat{H}}{\hbar} \right) |\psi(\mathbf{r}, 0)\rangle = \hat{U} |\psi(\mathbf{r}, 0)\rangle \tag{18}$$

Hence, the density operator may be written as

$$\begin{aligned}
\hat{\rho}(t) &= \sum_m W_m \hat{U}(t) |\psi(\mathbf{r}, 0)\rangle \langle \psi(\mathbf{r}, 0)| \hat{U}(t)^+ \\
&= \hat{U}(t) \hat{\rho}(0) \hat{U}(t)^+
\end{aligned} \tag{19}$$

Thus

$$
\begin{aligned}
i\hbar \, \frac{\partial \hat{\rho}(t)}{\partial t} &= \hat{H}(t)\hat{U}(t)\hat{\rho}(0)\hat{U}(t)^{+} \\
&\quad - \hat{U}(t)\hat{\rho}(0)\hat{U}(t)^{+}\hat{H}(t) \\
&= [\hat{H}(t), \hat{\rho}(t)]
\end{aligned}
\tag{20}
$$

where use has been made of the relations

$$
i\hbar \, \frac{\partial U(t)}{\partial t} = \hat{H}(t)U(t)
\tag{21a}
$$

$$
-i\hbar \, \frac{\partial U(t)^{+}}{\partial t} = U(t)^{+}\hat{H}(t)
\tag{21b}
$$

The expectation value of $\langle a(x, p)\rangle$, of an observable, $a(x, p)$, may be constructed on the $|\phi_k\rangle$ basis as follows:

$$
\begin{aligned}
\langle a(x, p)\rangle &= \langle \psi | a(x, p) | \psi \rangle \\
&= \sum_{l,k} \langle \psi | \phi_l \rangle \langle \phi_l | a(x, p) | \phi_k \rangle \langle \phi_k | \psi \rangle \\
&= \sum_{l,k} \langle \phi_k | \psi \rangle \langle \psi | \phi_l \rangle \langle \phi_l | a(x, p) | \phi_k \rangle \\
&= \sum_{l,k} \rho_{k,l} a(x, p)_{l,k} \\
&= \text{Trace}[\hat{\rho}(x, p)]
\end{aligned}
\tag{22}
$$

The density matrix, therefore, contains all of the physical information needed to calculate the expectation value of an observable. This result is of special significance in quantum transport theory where one of major aims is to predict measurable values of observables such as current, charge density, etc.

Frensley has applied the density matrix approach to formulate a model [11] for carrier transport through a quantum well bound state in a double-barrier compound-semiconductor structure. In this formulation, dissipative interactions are included by augmenting the quantum Liouville equation with a term containing a phenomenological relaxation time τ, as well as an equilibrium value of the density matrix ρ_0; that is

$$
\frac{\partial \rho}{\partial t} = -\frac{i}{\hbar} [H, \rho] - \frac{1}{\tau} (\rho - \rho_0)
\tag{23}
$$

We shall return to the topic of dissipative interactions in quantum systems in subsequent sections on the path-integral approach and the moment equation approach.

The density matrix formalism provides a convenient means of describing N-body quantum mechanical systems that may include complex interactions such as the interaction of a charge carrier with a thermal environment. Indeed, a major tool in quantum statistical mechanics is the density matrix. As an example, the value of W_m computed for a canonical ensemble predicts the probability of finding an energy eigenvalue E_m for a small system in contact with a much larger thermal reservoir. Consider a single charge carrier of coordinate, x, interacting with a thermal environment represented by an ensemble of interaction centers of coordinates, $\mathbf{R}$. $\mathbf{R}$ is an n-dimensional vector with one component for each of the n particles in the ensemble representing the thermal environment. For such a system, a general element of the density matrix in the coordinate-space representation is $\langle x, \mathbf{R} | \rho | x', \mathbf{R}' \rangle$. Frequently, it is convenient to consider the one-particle reduced density matrix. In modeling the transport of a charge carrier in a quantum-based electronic device, it is convenient to retain only the coordinate of the charge carrier and to integrate over the coordinates of the n particles in the thermal ensemble. Such an integration is accomplished by tracing over the ensemble coordinates of the density matrix to give the one-particle reduced density matrix $\tilde{\rho}(x, x', t)$:

$$\tilde{\rho}(x, x', t) \equiv \int d\mathbf{R} \langle x\mathbf{R} | \hat{\rho}(t) | x'\mathbf{R} \rangle \ . \tag{24}$$

The one-particle reduced density matrix will also be discussed in the subsequent sections on the path-integral and moment equation approaches to quantum transport in ultrasmall electronic devices.

11.4 FEYNMAN PATH-INTEGRAL FORMULATION

The discussion thus far of theories suitable for modeling quantum-based electronic devices has included a discussion of quantum transport formalisms based on the density matrix theory. While this theory is extremely useful in treating quantum mechanical systems, it is convenient to introduce path-integral techniques to formulate models for

1. Quantum dissipation.
2. Quantum Monte Carlo transport.
3. Strong electron–LO-phonon coupling in polar semiconductors.

These important phenomena are essential to the full description of transport in quantum-based electronic devices. Very convenient methods for describing these phenomena were formulated by Richard Feynman in 1948 and by Feynman and Vernon in 1963. In 1948, Feynman published a classic paper [21] on the path-integral formulation of quantum mechanics; in 1963,

Feynman and Vernon published a key work [22] that established the influence-functional approach to describing interaction in complex systems. The influence-functional approach is well suited to describing interacting systems such as a small superlattice structure in contact with a large wafer. As discussed in connection with the density matrix, the canonical ensemble may be used to derive the statistical coefficient W_m for the small superlattice structure. Before the path-integral formulation of quantum mechanics is presented, it is advantageous to review briefly the principle of least action in the context of classical mechanics as well as the propagator method of quantum mechanics. These techniques will permit detailed examination of the three phenomena mentioned earlier as regards recent major papers.

The principle of least action may be used to describe the behavior of conservative classical systems. For such systems, the action S along a path $x(t)$ is defined by

$$S[x(t)] = \int_{t_1}^{t_2} [T(t) - V(t)] \, dt \tag{25}$$

where $T(t)$ and $V(t)$ are the kinetic and potential energies, respectively. The action $S[x(t)]$ is a functional of the path $x(t)$; that is, the numerical value of the action depends on the path $x(t)$. We shall use brackets to indicate that we are dealing with a functional. According to the principle of least action, the classical (or "true") path is the path $\underline{x}(t)$ for which the action is a minimum.

In practice, the path $\underline{x}(t)$ is determined by exploiting the facts that:

1. First-order variations in x about the minimum value of a function $f(x)$ cause only second-order variations in $f(x)$.
2. First-order variations in x about positions away from the minimum value of a function $f(x)$ cause first-order variations in $f(x)$.

To determine $\underline{x}(t)$, these facts may be applied formally through use of the calculus of variations [23] or in a unique manner for a given action.

As an example of the application of the principle of least action, consider

$$S[x(t)] = \int_{t_1}^{t_2} \left[\frac{m}{2} \left(\frac{dx(t)}{dt} \right)^2 - V(x) \right] dt \tag{26}$$

and take

$$x(t) = \underline{x}(t) + \varepsilon(t) \tag{27}$$

Then

$$S = \int_{t_1}^{t_2} \left[\frac{m}{2} \left(\frac{dx}{dt} + \frac{d\varepsilon}{dt} \right)^2 - V(\underline{x} + \varepsilon) \right] dt \qquad (28)$$

$$= \int_{t_1}^{t_2} \left[\frac{m}{2} \left(\frac{d\underline{x}}{dt} \right)^2 - V(\underline{x}) + m \frac{d\underline{x}}{dt} \frac{d\varepsilon}{dt} - \varepsilon V'(\underline{x}) + \text{higher order terms} \right] dt$$

where

$$V'(\underline{x}) = \frac{\partial V(\underline{x})}{\partial \underline{x}} \qquad (29)$$

Thus

$$S = \bar{S} + \delta S^{(1)} \qquad (30)$$

where

$$\bar{S} = \int_{t_1}^{t_2} \left[\frac{m}{2} \left(\frac{d\underline{x}}{dt} \right)^2 - V(\underline{x}) \right] dt , \qquad (31)$$

and

$$\delta S^{(1)} = \int_{t_1}^{t_2} \left[m \frac{d\underline{x}}{dt} \frac{d\varepsilon}{dt} - \varepsilon V'(\underline{x}) \right] dt \qquad (32)$$

$\delta S^{(1)}$ denotes the first-order variation in the action. The integral $\delta S^{(1)}$ must be zero when evaluated along the classical path $\underline{x}(t)$, since the variation in δS along $\underline{x}(t)$ is second order or higher. In particular, δS_1 equals zero when evaluated along $\underline{x}(t)$; this must be true independently of the value of $\varepsilon(t)$. Upon integrating by parts and using the fact that all paths must begin at $x_1 = x(t_1) = \underline{x}(t_1)$ and end at $x_2 = x(t_2) = \underline{x}(t_2)$

$$\delta S^{(1)} = m \frac{d\underline{x}(t)}{dt} \varepsilon(t) \Big|_{t_1}^{t_2} - \int_{t_1}^{t_2} \left[m \frac{d^2\underline{x}}{dt^2} + V'(\underline{x}) \right] \varepsilon(t) \, dt$$

$$= \int_{t_1}^{t_2} \left[-m \frac{d^2\underline{x}}{dt^2} - V'(\underline{x}) \right] \varepsilon(t) \, dt = 0 \qquad (33)$$

Since this integral must vanish independent of $\varepsilon(t)$, it follows that

$$m \frac{d^2\underline{x}}{dt^2} = -\frac{\partial V(\underline{x})}{\partial \underline{x}} = F(\underline{x}) \qquad (34)$$

Therefore, for a conservative system, the principle of least action selects the path $\underline{x}(t)$ for which Newton's law holds.

In 1948, Feynman published a classic paper [21] on the path-integral formulation of quantum mechanics. In Feynman's 1948 theory, the basic quantum mechanical propagator, $K(x_2, t_2; x_1, t_1)$ is given by

$$K(x_2, t_2; x_1, t_1) = \sum_{\text{all paths from } x_1,\, t_1 \text{ to } x_2,\, t_2} \phi[x(t)] \tag{35}$$

where the functional $\phi[x(t)]$ is defined by

$$\phi[x(t)] = \text{constant } \exp\left\{\frac{i}{\hbar} S[x(t)]\right\} \tag{36}$$

In the classical limit, the action $S[x(t)]$ is large compared to $\hbar$ and $\phi[x(t)]$ varies very rapidly as a function of the phase $S[x(t)]/\hbar$. In this classical limit, the terms in the sum over $\phi[x(t)]$ are nearly all canceled by rapid phase variations in adjacent paths; however, variations about the path $\underline{x}(t)$ produce no first-order changes in the phase $S[x(t)]/\hbar$. For arbitrary values of $S[x(t)]$, contributing paths are, in general, not restricted to those near $\underline{x}(t)$. In the quantum mechanical limit, many paths contribute to $K(x_2, t_2; x_1, t_1)$.

From elementary quantum mechanics, a complete formal solution for the wave function of a particle in an external field is provided by

$$\psi(x, t) = \int \psi(x', t') K(x', t'; x, t)\, dx' \tag{37}$$

where

$$K(x', t'; x, t) = \sum_n \psi_{E_n}^*(x') \psi_{E_n}(x)\, e^{-iE_n(t-t')/\hbar} . \tag{38}$$

In this description, the arbitrary state function $\psi(x, t)$ is expressed as a superposition:

$$\psi(x, t) = \sum_n a_{E_n}(t) \psi_{E_n}(x) \tag{39}$$

where $\psi_{E_n}(x)$ are eigenfunctions of E and $|a_{E_n}(t)|^2$ is the probability that the particle will be found in state n at time t. Equations (35) and (36) are frequently written in the form [23]:

$$K(x_2, t_2; x_1, t_1) = \int_{t_1}^{t_2} \exp\left\{\frac{i}{\hbar} S[x(t)]\right\} \mathcal{D}x(t) \tag{40}$$

These expressions are completely equivalent. $\mathcal{D}x(t)$ indicates that the integral is taken over all paths $x(t)$ from t_1 to t_2.

As described in connection with the density matrix formalism, it is convenient to consider a charge carrier of coordinate, x, interacting with an ensemble of n interaction centers located at the positions defined by the n components of the vector $\mathbf{R}$. The quantum mechanical propagator for such a system is defined by $K(x_2, \mathbf{R}_2, t_2; x_1, \mathbf{R}_1, t_1)$; at time t_1, the system coordinates are $(x_1, \mathbf{R}_1)$ and at time t_2 they are $(x_2, \mathbf{R}_2)$. As discussed previously, the propagator $K(x_2, \mathbf{R}_2, t_2; x_1, \mathbf{R}_1, t_1)$ may be written in terms of a Feynman path integral as follows:

$$K(x_2, \mathbf{R}_2, t_2; x_1, \mathbf{R}_1, t_1) = \int_{t_1}^{t_2} \int_{t_1}^{t_2} \mathscr{D}x(t)\mathscr{D}\mathbf{R}(t)\, \exp\left\{ \frac{i}{\hbar}\, S[x, \mathbf{R}] \right\} \qquad (41)$$

The time-dependent density operator for such a system is represented by

$$\rho(t) = \exp\left(\frac{-iHt}{\hbar} \right) \rho(0) \exp\left(\frac{iHt}{\hbar} \right) \qquad (42)$$

and the density matrix in coordinate representation may be written as [12]:

$$\langle x, \mathbf{R}|\hat{\rho}(t)| y, \mathbf{Q} \rangle = \int dx'\,dy'\,d\mathbf{R}'\,d\mathbf{Q}' \Big\langle x, \mathbf{R}|\exp\left(\frac{-iHt}{\hbar} \right)|x', \mathbf{R}' \Big\rangle$$

$$\times \langle x', \mathbf{R}'|\hat{\rho}(0)| y', \mathbf{Q}' \rangle \Big\langle y', \mathbf{Q}'|\exp\left(\frac{iHt}{\hbar} \right)|y, \mathbf{Q} \Big\rangle$$

$$(43)$$

where

$$\Big\langle x, \mathbf{R}|\exp\left(\frac{-iHt}{\hbar} \right)|x', \mathbf{R}' \Big\rangle = K(x, \mathbf{R}, t; x', \mathbf{R}', 0)$$

$$= \int \int \mathscr{D}x\mathscr{D}\mathbf{R}\, \exp\left\{ \frac{i}{\hbar}\, S[x, \mathbf{R}] \right\}$$

$$(44a)$$

and

$$\Big\langle y', \mathbf{Q}'|\exp\left(\frac{iHt}{\hbar} \right)|y, \mathbf{Q} \Big\rangle = K^*(y, \mathbf{Q}, t; y', \mathbf{Q}', 0)$$

$$= \int \int \mathscr{D}y\mathscr{D}\mathbf{Q}\, \exp\left\{ \frac{-i}{\hbar}\, S[y, \mathbf{Q}] \right\}$$

$$(44b)$$

The endpoints of these path integrals are $x = x(t)$, $x' = x(0)$, $y = y(t)$, $y' = y(0)$, $\mathbf{R} = \mathbf{R}(t)$, $\mathbf{R}' = \mathbf{R}(0)$, $\mathbf{Q} = \mathbf{Q}(t)$ and $\mathbf{Q}' = \mathbf{Q}(0)$.

11.5 WIGNER DISTRIBUTION FUNCTION

One of the earliest formulations of quantum transport was given by E. Wigner in 1932 [8]. For an n-particle system with a pure-state wave function $\psi(x_1, \ldots x_n)$, Wigner made the assertion that the probability distribution function of the simultaneous values of $x_1, \ldots, x_n$ for the coordinates, and $p_1, \ldots, p_n$ for the momenta, is

$$P(x_1, \ldots, x_n; p_1, \ldots, p_n) = \left(\frac{1}{2\pi\hbar}\right)^n \int_{-\infty}^{\infty} \cdots \int_{-\infty}^{\infty} dy_1, \ldots, dy_n$$

$$\times \psi^*\left(x_1 + \frac{y_1}{2}, \ldots, x_n + \frac{y_n}{2}\right)$$

$$\times \psi\left(x_1 - \frac{y_1}{2}, \ldots, x_n - \frac{y_n}{2}\right)$$

$$\times e^{i(p_1 y_1 + \cdots + p_n y_n)/\hbar} \tag{45}$$

This distribution function was assumed in spite of the fact that knowledge of both position and momentum variables overspecifies a quantum mechanical system. One consequence of this overspecification is that the distribution function P may take on negative values! However, as is to be argued, the Wigner distribution function $P(x, p)$ may be used to compute physically meaningful values of observable quantities. Thus, while $P(x_1, \ldots, x_n; p_1, \ldots, p_n)$ cannot be rigorously interpreted as a probability distribution function, meaningful probabilities can be extracted by taking appropriate moments.

For a single coordinate and momentum, the Wigner distribution function reduces to

$$P(x, p) = \frac{1}{2\pi\hbar} \int_{-\infty}^{\infty} dy \psi^*\left(x + \frac{y}{2}\right)\psi\left(x - \frac{y}{2}\right)e^{ipy/\hbar} \tag{46}$$

where $\psi(x)$ is the coordinate representation of the wave function. Wigner determined the dynamical behavior of $P(x, p)$ through evaluation of the expression

$$E\psi(x, t) = \left[\frac{p^2}{2m} + V(x)\right]\psi(x, t) \tag{47a}$$

or

$$i\hbar \frac{\partial}{\partial t} \psi(x, t) = \left[-\frac{\hbar^2}{2m} \frac{\partial^2}{\partial x^2} + V(x)\right]\psi(x, t) \tag{47b}$$

Through straightforward mathematical manipulations, it is evident that

$$\frac{\partial P(x, p)}{\partial t} = \frac{1}{(2\pi\hbar)}$$

$$\times \int_{-\infty}^{\infty} dy\, e^{ipy/\hbar} \left\{ \frac{i\hbar}{2m} \left[-\frac{\partial^2 \psi^*(x + y/2)}{\partial x^2} \psi\left(x - \frac{y}{2}\right) \right. \right.$$

$$\left. + \psi^*\left(x + \frac{y}{2}\right) \frac{\partial^2 \psi(x - y/2)}{\partial x^2} \right] \tag{48}$$

$$+ \frac{i}{\hbar} \left[V\left(x + \frac{y}{2}\right) - V\left(x - \frac{y}{2}\right) \right]$$

$$\left. \times \psi^*\left(x + \frac{y}{2}\right) \psi\left(x - \frac{y}{2}\right) \right\}$$

Since

$$\frac{\partial P(x, p)}{\partial x} = \frac{1}{2\pi\hbar} \int_{-\infty}^{\infty} dy \left[\frac{\partial \psi^*(x + y/2)}{\partial x} \psi\left(x - \frac{y}{2}\right) \right.$$

$$\left. + \psi^*\left(x + \frac{y}{2}\right) \frac{\partial \psi(x - y/2)}{\partial x} \right] e^{ipy/\hbar} \tag{49}$$

it follows after one partial integration that

$$\frac{\partial P(x, p)}{\partial t} = -\frac{p}{m} \frac{\partial P(x, p)}{\partial x}$$

$$+ \frac{1}{(2\pi\hbar)} \int_{-\infty}^{\infty} dy\, e^{ipy/\hbar} \frac{i}{\hbar} \left[V\left(x + \frac{y}{2}\right) \right.$$

$$\left. - V\left(x - \frac{y}{2}\right) \right] \psi^*\left(x + \frac{y}{2}\right) \psi\left(x - \frac{y}{2}\right) \tag{50a}$$

or

$$\frac{\partial P(x, p)}{\partial t} + \frac{p}{m} \frac{\partial P(x, p)}{\partial x} = \frac{i}{\pi\hbar^2} \int_{-\infty}^{\infty} dy [V(x + y) - V(x - y)]$$

$$\times \psi^*(x + y)\psi(x - y)e^{2ipy/\hbar} \tag{50b}$$

Expanding $V(x + y)$ in a Taylor series about x

$$V(x + y) = \sum_{\lambda=0}^{\infty} \frac{y^\lambda}{\lambda!} \frac{\partial^\lambda V(x)}{\partial x^\lambda} = \sum_{\lambda=0}^{\infty} \frac{y^\lambda}{\lambda!} V^{(\lambda)}(x) \tag{51}$$

it follows that

$$V(x + y) - V(x - y) = 2 \sum_{\substack{\lambda = 1 \\ (\text{odd})}}^{\infty} \frac{y^{\lambda}}{\lambda!} V^{(\lambda)}(x) \tag{52}$$

The dynamical behavior of $P(x, p)$ is

$$\frac{\partial P(x, p)}{\partial t} + \frac{p}{m} \frac{\partial P(x, p)}{\partial x} =$$

$$\frac{2i}{\pi \hbar^2} \int_{-\infty}^{\infty} dy \sum_{\lambda = 0}^{\infty} \frac{y^{\lambda}}{\lambda!} V^{(\lambda)}(x) \psi^*(x + y) \psi(x - y) e^{2ipy/\hbar}$$

$$= \sum_{\substack{\lambda = 1 \\ (\text{odd})}}^{\infty} \frac{1}{\lambda!} \left(\frac{\hbar}{2i} \right)^{\lambda - 1} \frac{\partial^{\lambda} V(x)}{\partial x^{\lambda}} \frac{\partial^{\lambda} P(x, p)}{\partial p^{\lambda}} \tag{53}$$

This equation of Wigner's is a Boltzmann-like transport equation. Through order λ^3, the Wigner distribution function $P(x, p)$ satisfies

$$\frac{\partial P(x, p)}{\partial t} + \frac{p}{m} \frac{\partial P(x, p)}{\partial x} - \frac{\partial V(x)}{\partial x} \frac{\partial P(x, p)}{\partial p}$$

$$= \frac{1}{3!} \left(\frac{\hbar}{2i} \right)^2 \frac{\partial^3 V}{\partial x^3} \frac{\partial^3 P(x, p)}{\partial p^3} + \mathcal{O}(\hbar^4) \tag{54}$$

The left-hand side of this equation is of the form of the collisionless Boltzmann equation, also known as the Vlasov equation. On the right-hand side are the lowest-order quantum mechanical corrections to the Wigner–Boltzmann-like transport equation.

For the multidimensional case of an n particle, n-dimensional configuration space, Wigner demonstrated that

$$\frac{\partial P(x_1, \ldots, x_n; p_1, \ldots, p_n)}{\partial t} + \sum_{k=1}^{n} \frac{p_k}{m_k} \frac{\partial P(x_1, \ldots, x_n; p_1, \ldots, p_n)}{\partial x_k}$$

$$= \sum \frac{\partial^{\lambda_1 + \cdots + \lambda_n} V}{\partial x_1^{\lambda_1}, \ldots, \partial x_n^{\lambda_n}} \frac{(\hbar/2i)^{\lambda_1 + \cdots + \lambda_n - 1}}{\lambda_1!, \ldots, \lambda_n!}$$

$$\times \frac{\partial^{\lambda_1 + \cdots + \lambda_n} P(x_1, \ldots, x_n; p_1, \ldots, p_n)}{\partial_{p_1}^{\lambda_1}, \ldots, \partial_{p_n}^{\lambda_n}} \tag{55}$$

where summation on the right-hand side is extended over all positive integer values of $\lambda_1, \ldots, \lambda_n$ for which the sum $\lambda_1 + \lambda_2 + \cdots + \lambda_n$ is odd.

The Wigner distribution function may be generalized from the pure-state case to the mixed-state case by summing over the m components of the mixed state; that is

$$P(x, p) = \sum_m W_m \left[\frac{1}{2\pi\hbar} \int\limits_{-\infty}^{\infty} dy \, \psi_m^*\left(x + \frac{y}{2}\right) \psi_m\left(x - \frac{y}{2}\right) e^{ipy/\hbar} \right] \tag{56}$$

where W_m is the statistical probability for the system to be in state $|\psi_m\rangle$. From the previously defined density operator for a mixed-state system

$$\hat{\rho} = \sum_m W_m |\psi_m\rangle\langle\psi_m| \tag{57}$$

it follows that the density matrix $\langle x - y/2|\hat{\rho}|x + y/2\rangle$ is related to the mixed-state Wigner distribution function through the relation

$$P(x, p) = \frac{1}{2\pi\hbar} \int\limits_{-\infty}^{\infty} dy \left\langle x - \frac{y}{2} \,|\hat{\rho}|x + \frac{y}{2}\right\rangle e^{ipy/\hbar}$$

$$= \frac{1}{2\pi\hbar} \int\limits_{-\infty}^{\infty} dy \sum_m W_m \psi_m^*\left(x + \frac{y}{2}\right) \psi_m\left(x - \frac{y}{2}\right) e^{ipy/\hbar} \tag{58}$$

where the coordinate-space projections of the state vector define the wave functions

$$\psi_m\left(x - \frac{y}{2}\right) = \left\langle x - \frac{y}{2} \,|\psi_m\right\rangle \tag{59a}$$

$$\psi_m^*\left(x + \frac{y}{2}\right) = \left\langle \psi_m|x + \frac{y}{2}\right\rangle \tag{59b}$$

As an example of the statistical weight W_m, consider a system of N spin-$\frac{1}{2}$ particles with N_1 particles prepared in a state $|\psi_1\rangle = |+\frac{1}{2}\rangle$ and with $N_2 = (N - N_1)$ particles prepared in a state $|\psi_2\rangle = |-\frac{1}{2}\rangle$; then, W_1 and W_2 are given by N_1/N and N_2/N, respectively. The density operator is given by Eq. (4):

$$\hat{\rho} = W_1|+\tfrac{1}{2}\rangle\langle+\tfrac{1}{2}| + W_2|-\tfrac{1}{2}\rangle\langle-\tfrac{1}{2}| \tag{4}$$

and density matrix in diagonal representation is given by Eq. (5):

$$\langle \psi_i|\hat{\rho}|\psi_j\rangle = W_i \delta_{ij} \tag{5}$$

Another example of the statistical weight W_m is provided by the canonical ensemble of quantum statistical mechanics where

$$\rho_{mn} = \delta_{mn} W_m = \delta_{mn} \frac{e^{-\beta E_m}}{Tr \, e^{-\beta H}} \tag{60}$$

with $\beta = 1/kT$. For this case

$$\hat{\rho} = \sum_m |\psi_m\rangle \frac{e^{-\beta E_m}}{Tr\, e^{-\beta H}} \langle \psi_m| \tag{61}$$

Thus, for a system in thermal equilibrium with a larger system of temperature T, the probability for the smaller system to have an energy E_m is $W_m = e^{-\beta E_m}/Tr\, e^{-\beta H}$.

The Wigner distribution function derives much of its utility from the fact that the integral of the distribution over all momenta equals the quantum mechanical probability density in coordinate space; for the pure-state case

$$\int_{-\infty}^{\infty} P(x, p)dp = \int_{-\infty}^{\infty} dp \int_{-\infty}^{\infty} \frac{dy}{2\pi\hbar} \psi^*\left(x + \frac{y}{2}\right)\psi\left(x - \frac{y}{2}\right)e^{ipy/\hbar}$$

$$= \int_{-\infty}^{\infty} \frac{dy}{2\pi\hbar} \psi^*\left(x + \frac{y}{2}\right)\psi\left(x - \frac{y}{2}\right)2\pi\hbar\delta(y - 0)$$

$$= \psi^*(x)\psi(x) \tag{62}$$

where use has been made of the identity

$$\int_{-\infty}^{\infty} e^{ipy/\hbar}\, dp = 2\pi\hbar\delta(y - 0) \tag{63}$$

Taking the momentum-space wave function to be

$$\phi(p) = \frac{1}{\sqrt{2\pi\hbar}} \int_{-\infty}^{\infty} e^{-ipx/\hbar}\psi(x)\, dx \tag{64}$$

a similar but longer derivation leads to

$$\int_{-\infty}^{\infty} P(x, p)dx = \phi^*(p)\phi(p) \tag{65}$$

Consider an observable $a(x, p)$, which depends on the position operator alone, on the momentum operator alone, or on an additive combination thereof. From the expressions for the coordinate-space and momentum-space integrals over the Wigner distribution function, it is evident that the expectation value $\langle A \rangle$ of the observable, $a(x, p)$, is

$$\langle A \rangle = \int_{-\infty}^{\infty} \int_{-\infty}^{\infty} a(x, p)\, P(x, p)\, dx\, dp \tag{66}$$

Thus, notwithstanding the facts that the Wigner distribution function is nonunique and may take on negative values, it is possible to treat some aspects of quantum transport theory in terms of the classical theory; to achieve such a parallelism, the classical distribution function must be replaced by the Wigner distribution function. This parallelism will be exploited when deriving the moment equations that exhibit quantum corrections to the usual classical terms.

11.6 MOMENT EQUATION APPROACH

The moment equation approach provides a straightforward means of illustrating the basic quantum mechanical corrections to classical transport phenomena. In particular, such corrections are evident in the moment equations obtained through the density matrix, path-integral, and Wigner function techniques. This section illustrates the moment equation approach for two important cases: (1) the general moment equation expansion of the Wigner–Boltzmann equation; and (2) the expansion of the dissipative quantum Liouville equation in terms of the difference between the quantum mechanical and classical momenta.

11.6.1 Moment Expansion

The use of moment equations to describe quantum transport in ultrasubmicron devices has been discussed previously by Iafrate et al. [2]. Herein, their major findings will be summarized. As demonstrated in Section 11.5, the Wigner distribution function obeys the equation

$$\frac{\partial P(x, p)}{\partial t} + \frac{p}{m} \frac{\partial P(x, p)}{\partial x} - \sum_{\substack{\lambda=1 \\ (\text{odd})}}^{\infty} \frac{1}{\lambda!} \left(\frac{\hbar}{2i}\right)^{\lambda-1} \frac{\partial^\lambda V(x)}{\partial x^\lambda} \frac{\partial^\lambda P(x, p)}{\partial p^\lambda} = 0 \tag{67a}$$

or

$$\frac{\partial P(x, p)}{\partial t} + \frac{p}{m} \frac{\partial P(x, p)}{\partial x} - \frac{2}{\hbar} \sum_{n=0}^{\infty} (-1)^n \frac{(\hbar/2)^{2n+1}}{(2n+1)!} \frac{\partial^{2n+1} V(x)}{\partial x^{2n+1}} \frac{\partial^{2n+1} P(x, p)}{\partial p^{2n+1}}$$
$$= 0 \tag{67b}$$

This equation may be written

$$\frac{\partial P(x, p)}{\partial t} + \frac{p}{m} \frac{\partial P(x, p)}{\partial x} + \Theta \cdot P(x, p) = 0 \tag{68}$$

where

$$\Theta \cdot P(x, p) = -\frac{2}{\hbar} \left[\sin \frac{\hbar}{2} \left\{ \frac{\partial^{(V)}}{\partial x} \frac{\partial^{(P)}}{\partial_p} \right\} \right] V(x) P(x, p) \tag{69}$$

and it is understood that $\partial^{(V)}/\partial x$ and $\partial^{(P)}/\partial_p$ operate only on the potential $V(x)$ and the Wigner distribution function $P(x, p)$, respectively. By including a term to describe the time rate-of-change due to collisions of $P(x, p)$, the Wigner–Boltzmann equation may be written as

$$\frac{\partial P(x, p)}{\partial t} + \frac{p}{m} \frac{\partial P(x, p)}{\partial x} + \Theta \cdot P(x, p) = \left(\frac{\partial P(x, p)}{\partial t} \right)_{\text{collisions}} \tag{70}$$

Following Ref. 2, Eq. (70) is multiplied by a function of momentum $\chi(p)$ and integrated over all values of momentum; the resulting expression is

$$\frac{\partial \langle x \rangle}{\partial t} + \frac{1}{m} \frac{\partial}{\partial x} \langle xp \rangle$$

$$- \sum_{n=0}^{\infty} \left(\frac{\hbar}{2i} \right)^{2n} \frac{1}{(2n+1)!} \frac{\partial^{2n+1}}{\partial x^{2n+1}} V(x) \int \chi(p) \frac{\partial^{2n+1}}{\partial p^{2n+1}} P(x, p) \, dp$$

$$= \left\langle \chi \left(\frac{\partial p(x, p)}{\partial t} \right)_{\text{collisions}} \right\rangle \tag{71}$$

with the momentum integration denoted by $\langle \ \rangle$. Upon integrating by parts as described in Ref. 2, Eq. (71) takes the form

$$\frac{\partial \langle \chi \rangle}{\partial t} + \frac{1}{m} \frac{\partial}{\partial x} \langle xp \rangle + \sum_{n=0}^{\infty} \left(\frac{\hbar}{2i} \right)^{2n} \frac{1}{(2n+1)!} \left(\frac{\partial^{2n+1}}{\partial x^{2n+1}} V(x) \right) \left\langle \frac{\partial^{2n+1} \chi}{\partial p^{2n+1}} \right\rangle$$

$$= \left\langle \chi \left(\frac{\partial P}{\partial t} \right)_{\text{collisions}} \right\rangle \tag{72}$$

For $\chi = 1$, Eq. (72) yields the continuity equation

$$\frac{\partial \rho}{\partial t} + \frac{1}{m} \frac{\partial}{\partial x} \langle p \rangle = 0 \tag{73a}$$

where $\rho = \psi^* \psi$. Taking $\chi = p$, Eq. (72) reduces to the momentum equation

$$\frac{\partial \langle p \rangle}{\partial t} + \frac{1}{m} \frac{\partial}{\partial x} \langle p^2 \rangle + \rho \frac{\partial V}{\partial x} = \left\langle p \left(\frac{\partial P(x, p)}{\partial t} \right)_{\text{collisions}} \right\rangle \tag{73b}$$

With $\chi = p^2/2m$, Eq. (72) simplifies to an energy conservation equation of the form

$$\frac{1}{2m} \frac{\partial}{\partial t} \langle p^2 \rangle + \frac{1}{2m^2} \frac{\partial}{\partial x} \langle p^3 \rangle + \frac{\langle p \rangle}{m} \frac{\partial V}{\partial x} = \left\langle \frac{p^2}{2m} \left(\frac{\partial P(x, p)}{\partial t} \right)_{\text{collisions}} \right\rangle \tag{73c}$$

As in Ref. [2], these results may be reduced to

$$\langle p^n \rangle = \left(\frac{\hbar}{2i} \right)^n \sum_{j=0}^{n} (-1)^j \frac{n!}{(n-j)!j!} \frac{\partial^j \psi^*(x)}{\partial x^j} \frac{\partial^{n-j} \psi(x)}{\partial x^{n-j}} \tag{74}$$

and a wave function of the form

$$\psi(x, t) = A(x, t)e^{iS(x, t)/\hbar} \tag{75}$$

may be invoked to demonstrate that

$$\langle p^0 \rangle = \rho(x, t) = \rho \tag{76a}$$

$$\langle p^1 \rangle = mv\rho \tag{76b}$$

$$\langle p^2 \rangle = (mv)^2\rho - \frac{\hbar}{4} \rho \frac{\partial^2}{\partial x^2} \ln \rho \tag{76c}$$

$$\langle p^3 \rangle = (mv)^3\rho - \frac{\hbar^2}{4} \rho\left\{3mv \frac{\partial^2}{\partial x^2} \ln \rho + \frac{\partial^2}{\partial x^2} (mv)\right\} \tag{76d}$$

In deriving these results, $A^2(x, t)$ and $S(x, t)$ have been related to the probability density and ensemble velocity through the relations

$$A^2(x, t) = \psi^*(x, t)\psi(x, t) = \rho(x, t) \tag{77a}$$

$$v(x, t) = \frac{1}{m} \frac{\partial S(x, t)}{\partial x} \tag{77b}$$

The expressions for both $\langle p^2 \rangle$ and $\langle p^3 \rangle$ contain quantum mechanical corrections to moments of the momentum. As could have been anticipated from the absence of first-order terms in $\hbar$ in the dynamical equation for the Wigner distribution, the lowest-order quantum corrections in $\langle p^n \rangle$ are of the order $\hbar^2$.

11.6.2   Quantum Corrections to Classical Terms in Moment Equations

To illustrate the quantum mechanical corrections for the time evolution of an N-body system interacting dissipatively with its environment, it is convenient to derive a set of moment equations from the dissipative quantum Liouville equation. In this section, such a moment equation expansion shall be derived.

The dynamical evolution of N-body quantum mechanical systems is of central importance in a variety of physical phenomena that range from the fluid behavior of nuclear collisions [24] to the transport of carriers in ultrasmall electronic structures [25]. Both the Wigner phase-space representation of quantum mechanics [8] and the quantum Liouville equation [26] have been applied extensively to the study of N-body quantum mechanical systems.

Significant recent developments in the description of N-body quantum mechanical systems deal with the separation of classical and quantum contributions to N-body dynamics as well as with the inclusion of dissipative interactions in N-body systems. In the first of these developments, phase-space distribution functions were used to derive a closed set of self-consistent moment equations [10]. These moment equations are structured so that classical and quantum terms are readily distinguishable; the lowest-order solution is classical and the quantum corrections occur in higher orders. In the second development, dissipative quantum phenomena have been modeled by considering the interaction of a particle with a heat bath comprised of an ensemble of harmonic oscillators [12]. In the high-temperature limit, it is found that the system is described by a Wigner–Boltzmann equation with Fokker–Planck collision terms.

In this work, the moment equation representation of the quantum Liouville equation is applied to derive two new sets of moment equations for both dissipative and nondissipative quantum systems. The present treatment provides an analysis of dissipative quantum effects in the high-temperature limit of Ref. 12.

Consider the Wigner–Boltzmann equation for a quantum system described by one- and two-body potential energy terms [10]:

$$\frac{\partial f}{\partial t} + \frac{p_j}{m}\frac{\partial f}{\partial x_j} - \frac{2}{\hbar}\sin\left\{\frac{\hbar}{2}\frac{\partial^{(V)}}{\partial x_j}\frac{\partial^{(f)}}{\partial p_j}\right\}V(x_j)f$$

$$= -\frac{2}{\hbar}\sum_{i=2}^{n}\int d^3x_i\rho_0(x_i)\sin\left\{\frac{\hbar}{2}\frac{\partial^{(v)}}{\partial x_{ij}}\frac{\partial^{(f)}}{\partial p_j}\right\}v(r_{ij})f \tag{78}$$

where, for such a combination of potential energy terms, the one-body Wigner distribution is denoted by $f(x_j, p_j, t)$, and is related to the one-body density matrix $\rho(x_j - \frac{1}{2}y_j; x_j + \frac{1}{2}y_j; t)$ by

$$f = f(x_j, p_j, t) = (2\pi\hbar)^{-3}\int d^3y_j e^{ip_j y_j/\hbar}\rho\left(x_j - \frac{y_j}{2}; x_j + \frac{y_j}{2}; t\right) \tag{79}$$

In Eq. (78), the position and momentum for the particle j are represented by x_j and p_j, respectively. $V(x_j)$ is the one-body mean field for particle j; $v(r_{ij})$ is the two-body potential operator with $r_{ij} = r_i - r_j$; $\rho_0(x_i)$ is the diagonal one-body density matrix for particle i. In the case of a one-dimensional N-body system, with the approximation

$$\rho(x, x'; t) = \int e^{-ip(x'-x)/\hbar}f\left(\frac{x+x'}{2}, p; t\right)dp$$

$$= \rho_0\left(\frac{x+x'}{2}; t\right)e^{-i(x'-x)\bar{p}/\hbar}\langle e^{-i(x'-x)(p-\bar{p})/\hbar}\rangle$$

$$= \rho_0\left(\frac{x+x'}{2}; t\right)e^{-i(x'-x)\bar{p}/\hbar}(1 - i\hbar^{-1}(x'-x)\langle(p-\bar{p})\rangle$$

$$\quad - \hbar^{-2}(x'-x)^2\langle(p-\bar{p})^2\rangle + \cdots) \tag{80}$$

the moments of Eq. (78) are given by [10, 27]:

$$\frac{\partial}{\partial t}(\rho_0\langle p^n\rangle) = -\frac{1}{m}\frac{\partial}{\partial x}(\rho_0\langle p^{n+1}\rangle) + \frac{4\rho_0}{\hbar^2}\sum_{k=1}^{n}\left(\frac{-\hbar}{2i}\right)^{k+1}\frac{n!}{k!(n-k)!}$$

$$\times\langle p^{n-k}\rangle\frac{\partial^k V_{\text{eff}}}{\partial x^k} \tag{81}$$

where the moment functions $\langle p^n\rangle$ are defined by [28]:

$$\rho_0(x,t)\langle p^n(x,t)\rangle \equiv \int f(x,p)p^n\,dp \tag{82}$$

and where V_{eff} is the sum of $V(x_j)$ and the effective one-body field that results from the right-hand side of Eq. (78) in the one-dimensional approximation. Equation (80) has been applied previously to construct the first three moment equations and a constraining condition on the spatial derivative of $\rho_0\langle(p-\bar{p})^2\rangle$. A more detailed calculation of the first six moment equations reveals that

$$\frac{\partial}{\partial t}\rho_0 = -\frac{1}{m}\frac{\partial}{\partial x}(\rho_0\bar{p}) \tag{83a}$$

$$\frac{\partial}{\partial t}\bar{p} = -\frac{\partial}{\partial x}H_{\text{CL}} - \frac{1}{m\rho_0}\frac{\partial}{\partial x}(\rho_0\langle(p-\bar{p})^2\rangle) \tag{83b}$$

$$\rho_0\frac{\partial}{\partial t}\langle(p-\bar{p})^2\rangle = -\frac{1}{m}\frac{\partial}{\partial x}(\rho_0\langle(p-\bar{p})^3\rangle) - \frac{1}{m}\bar{p}\frac{\partial}{\partial x}(\rho_0\langle(p-\bar{p})^2\rangle)$$

$$-\frac{2\rho_0}{m}\langle(p-\bar{p})^2\rangle\frac{\partial\bar{p}}{\partial x} + \frac{\bar{p}\langle(p-\bar{p})^2\rangle}{m}\frac{\partial\rho_0}{\partial x} \tag{83c}$$

$$\rho_0\frac{\partial}{\partial t}\langle(p-\bar{p})^3\rangle = -\frac{1}{m}\frac{\partial}{\partial x}(\rho_0\langle(p-\bar{p})^4\rangle) - \frac{\rho_0\bar{p}}{m}\frac{\partial}{\partial x}\langle(p-\bar{p})^3\rangle$$

$$-\frac{3}{m}\frac{\partial}{\partial x}(\rho_0\bar{p}\langle(p-\bar{p})^3\rangle)$$

$$+\frac{3\langle(p-\bar{p})^2\rangle}{m}\frac{\partial}{\partial x}(\rho_0\langle(p-\bar{p})^2\rangle) + \frac{\hbar^2}{4}\rho_0\frac{\partial^3 V_{\text{eff}}}{\partial x^3} \tag{83d}$$

$$\rho_0\frac{\partial}{\partial t}\langle(p-\bar{p})^4\rangle = -\frac{1}{m}\frac{\partial}{\partial x}(\rho_0\langle(p-\bar{p})^5\rangle) - \frac{\rho_0\bar{p}}{m}\frac{\partial}{\partial x}\langle(p-\bar{p})^4\rangle$$

$$-\frac{4\rho_0\langle(p-\bar{p})^4\rangle}{m}\frac{\partial\bar{p}}{\partial x} + \frac{12\bar{p}^2}{m}\frac{\partial}{\partial x}(\rho_0\langle(p-\bar{p})^3\rangle)$$

$$+\frac{4\langle(p-\bar{p})^3\rangle}{m}\frac{\partial}{\partial x}(\rho_0\langle(p-\bar{p})^2\rangle) \tag{83e}$$

$$\rho_0 \frac{\partial}{\partial t} \langle (p - \bar{p})^5 \rangle = -\frac{1}{m} \frac{\partial}{\partial x} (\rho_0 \langle (p - \bar{p})^6 \rangle) - \frac{\rho_0 \bar{p}}{m} \frac{\partial}{\partial x} \langle (p - \bar{p})^5 \rangle$$

$$-\frac{5\rho_0 \langle (p - \bar{p})^5 \rangle}{m} \frac{\partial \bar{p}}{\partial x}$$

$$+\frac{10}{m} \frac{\partial}{\partial x} (\rho_0 \bar{p}^3 \langle (p - \bar{p})^3 \rangle)$$

$$-\frac{40\bar{p}^3}{m} \frac{\partial}{\partial x} (\rho_0 \langle (p - \bar{p})^3 \rangle)$$

$$+\frac{5\langle (p - \bar{p})^4 \rangle}{m} \frac{\partial}{\partial x} (\rho_0 \langle (p - \bar{p})^2 \rangle)$$

$$-\frac{5\bar{p}^4}{m} \frac{\partial}{\partial x} (\rho_0 \langle (p - \bar{p})^2 \rangle)$$

$$+\frac{5}{m} \frac{\partial}{\partial x} (\rho_0 \bar{p}^4 \langle (p - \bar{p})^2 \rangle) + \frac{5}{2} \rho_0 \langle (p - \bar{p})^2 \rangle \hbar^2$$

$$\times \frac{\partial^3 V_{\text{eff}}}{\partial x^3} - \frac{1}{16} \rho_0 \hbar^4 \frac{\partial^5 V_{\text{eff}}}{\partial x^5} \tag{83f}$$

where

$$H_{\text{CL}} = \frac{\bar{p}^2}{2m} + V_{\text{eff}} \tag{83g}$$

is the classical Hamiltonian with $\bar{p} = \langle p^1 \rangle$ and with a one-body effective potential V_{eff}. The results of Ref. 10 are recovered from Eq. (83) by setting

$$\langle (p - \bar{p})^3 \rangle = \langle (p - \bar{p})^4 \rangle = \langle (p - \bar{p})^5 \rangle = \langle (p - \bar{p})^6 \rangle = 0 \tag{84}$$

Upon taking $\langle (p - \bar{p})^5 \rangle = \langle (p - \bar{p}))^6 \rangle = 0$ in eq. (83), a closed self-consistent set of equations is obtained for the fourth-order approximation to the quantum Liouville equation; Eq. (83f) reduces to an equation constraining the solution through the values of $\partial^3 V_{\text{eff}}/\partial x^3$ and $\partial^5 V_{\text{eff}}/\partial x^5$.

The generalization of Eq. (81) to the case where dissipative interactions occur through contact with a heat bath is accomplished by replacing the two-body interaction term in Eq. (78) with the Fokker–Planck collision operator [12]; that is

$$\frac{\partial f}{\partial t} + \frac{p}{m} \frac{\partial f}{\partial x} - \frac{2}{\hbar} \sin\left\{ \frac{\hbar}{2} \frac{\partial^{(V)}}{\partial x} \frac{\partial^{(f)}}{\partial p} \right\} Vf = 2\gamma \frac{\partial}{\partial p} (pf) + D \frac{\partial^2 f}{\partial p^2} \tag{85}$$

where γ is the effective relaxation constant of the ensemble of harmonic oscillators and $D = \eta kT$; η is the damping constant and kT is the heat-bath temperature.

Applying Eqs. (79–82) to Eq. (85) results in a moment equation representation:

$$\frac{\partial}{\partial t}\left(\rho_0\langle p^n\rangle\right) + \frac{1}{m}\frac{\partial}{\partial x}\left(\rho_0\langle p^{n+1}\rangle\right)$$

$$-\frac{4\rho_0}{\hbar^2}\sum_{k=1}^{n}\left[\frac{-\hbar}{2i}\right]^{k+1}\frac{n!}{k!(n-k)!}\langle p^{n-k}\rangle\frac{\partial^k V}{\partial x^k}$$

$$= -2\gamma n\rho_0\langle p^n\rangle + Dn(n-1)\rho_0\langle p^{n-2}\rangle \qquad (86)$$

The first four moment equations in the hierarchy represented by Eq. (86) are

$$\frac{\partial}{\partial t}\rho_0 = -\frac{1}{m}\frac{\partial}{\partial x}\left(\rho_0\bar{p}\right) \qquad (87a)$$

$$\frac{\partial\bar{p}}{\partial t} + 2\gamma\bar{p} = -\frac{\partial}{\partial x}H_{\text{CL}} - \frac{1}{m\rho_0}\frac{\partial}{\partial x}\left(\rho_0\langle(p-\bar{p})^2\rangle\right) \qquad (87b)$$

$$\rho_0\frac{\partial}{\partial t}\langle(p-\bar{p})^2\rangle + 4\gamma\rho_0\langle(p-\bar{p})^2\rangle = 2D\rho_0 - \frac{1}{m}\frac{\partial}{\partial x}\left(\rho_0\langle(p-\bar{p})^3\rangle\right)$$

$$-\frac{\bar{p}}{m}\frac{\partial}{\partial x}\left(\rho_0\langle(p-\bar{p})^2\rangle\right)$$

$$-\frac{2\rho_0}{m}\langle(p-\bar{p})^2\rangle\frac{\partial\bar{p}}{\partial x}$$

$$+\frac{\bar{p}\langle(p-\bar{p})^2\rangle}{m}\frac{\partial\rho_0}{\partial x} \qquad (87c)$$

$$\rho_0\frac{\partial}{\partial t}\langle(p-\bar{p})^3\rangle + 6\rho_0\gamma(\langle(p-\bar{p})^3\rangle + 3\bar{p}\langle(p-\bar{p})^2\rangle + \bar{p}^3)$$

$$= 6D\rho_0\bar{p} - \frac{1}{m}\frac{\partial}{\partial x}\left(\rho_0\langle(p-\bar{p})^4\rangle\right) - \frac{\rho_0\bar{p}}{m}\frac{\partial}{\partial x}\langle(p-\bar{p})^3\rangle$$

$$-\frac{3}{m}\frac{\partial}{\partial x}\left(\rho_0\bar{p}\langle(p-\bar{p})^3\rangle\right) + \frac{3\langle(p-\bar{p})^2\rangle}{m}\frac{\partial}{\partial x}\left(\rho_0\langle(p-\bar{p})^2\rangle\right)$$

$$+\frac{\hbar^2}{4}\rho_0\frac{\partial^3 V_{\text{eff}}}{\partial x^3} \qquad (87e)$$

where

$$H_{\text{CL}} = \frac{\bar{p}^2}{2m} + V_{\text{eff}}.$$

Upon taking $\langle(p-\bar{p})^3\rangle = \langle(p-\bar{p})^4\rangle$ as well as $\gamma = D = 0$, Eq. (87) reduces to the previously published [10] closed self-consistent set of equations.

The importance of the dissipative interaction in Eq. (87) is illustrated by setting $\langle (p - \bar{p})^3 \rangle = \langle (p - \bar{p})^4 \rangle = 0$ in Eq. (87d):

$$6\rho_0 \gamma (3\bar{p} \langle (p - \bar{p})^2 \rangle + \bar{p}^3) = 6D\rho_0 \bar{p} + \frac{3\langle (p - \bar{p})^2 \rangle}{m} \frac{\partial}{\partial x} (\rho_0 \langle (p - \bar{p})^2 \rangle)$$

$$+ \frac{\hbar^2}{4} \rho_0 \frac{\partial^3 V_{\text{eff}}}{\partial x^3} \tag{88}$$

or

$$\rho_0 \left\{ 2\gamma (3\bar{p} \langle (p - \bar{p})^2 \rangle + \bar{p}^3) - 2D\bar{p} - \frac{\hbar^2}{12} \frac{\partial^3 V_{\text{eff}}}{\partial x^3} \right\}$$

$$= \frac{\langle (p - \bar{p})^2 \rangle}{m} \frac{\partial}{\partial x} (\rho_0 \langle (p - \bar{p})^2 \rangle)$$

Equation (88) provides a constraint on the solution for $\langle (p - \bar{p})^2 \rangle$ as defined by Eq. (87a–87c) when $\langle (p - \bar{p})^3 \rangle = \langle (p - \bar{p})^4 \rangle = 0$. In the nondissipative case, the constraining relation is determined by $\partial^3 V_{\text{eff}} / \partial^3 x$; however, when dissipative interactions are included, the relation depends on γ and D as well as on $\partial^3 V_{\text{eff}} / \partial x^3$. The relative importance of these terms is represented by the ratio

$$\Gamma = \frac{2\gamma (3\bar{p} \langle (p - \bar{p})^2 \rangle + \bar{p}^3) - 2D\bar{p}}{-\dfrac{\hbar^2}{12} \dfrac{\partial^3 V_{\text{eff}}}{\partial x^3}} \tag{89}$$

Upon approximating $\bar{p}$ and $\langle (p - \bar{p})^2 \rangle$ in terms of the expressions for $\langle p \rangle$ and $\langle p^2 \rangle$ given Ref. 2, Eq. (89) reduces to

$$\Gamma = \frac{2mv\gamma \left(\dfrac{3}{4} \hbar^2 \dfrac{\partial^2 \ln \rho_0}{\partial x^2} - (mv)^2 \right) + 2Dmv}{\dfrac{\hbar^2}{12} \dfrac{\partial^3 V_{\text{eff}}}{\partial x^3}} \tag{90}$$

Equation (90) reveals that the role of dissipative effects depends on the diagonal component of the density matrix as well as the spatial gradients of the potential.

The principal application of these results in the field of microstructures will be in the theory of quantum transport of charge carriers in ultrasmall electronic structures. Recent experimental results [29, 30] indicate clearly that many-body interactions, dissipative effects, and finite-temperature effects must be included in models appropriate for predicting the performance of a variety of quantum-based electronic devices. The model presented in this chapter provides a framework for including all of these effects in future device models.

The role of dissipative interactions in N-body quantum systems is identified in this chapter by unifying and extending recent results on quantum dissipation [12] and on the separation of classical and quantum contributions to N-body dynamics [2]. The general formalism established here provides a quantitative estimate of the importance of dissipative interactions and has potential applications to a variety of significant technological problems [29].

11.7 APPLICATIONS OF QUANTUM TRANSPORT THEORY

The quantum mechanical theories outlined in previous sections are well suited for describing a wide variety of complex quantum transport phenomena. In this section, applications of these previously developed quantum mechanical theories will deal with phenomena that include

1. Quantum dissipation.
2. High-field transport in polar semiconductors.
3. Monte Carlo simulations of quantum transport.
4. Resonant tunneling in quantum well structures.

11.7.1 Quantum Dissipation

In order to describe dissipative interactions in quantum mechanical systems, it is convenient to consider a system A interacting with a reservoir R. The Feynman–Vernon theory [22] treats such a composite system with the corresponding Hamiltonian [12]:

$$H = H_A + H_I + H_R \tag{91}$$

where

$$H_A = -\frac{\hbar^2}{2M}\frac{\partial^2}{\partial x^2} + v(x) \tag{92}$$

describes a system of a particle of mass M in potential $v(x)$ and

$$H_R = -\frac{\hbar^2}{2m}\frac{\partial^2}{\partial \mathbf{R}^2} + \frac{1}{2}\sum_{i\neq j} v_R(R_i, R_j) \tag{93}$$

is the Hamiltonian of the N-particle reservoir of particles with mass M; $\mathbf{R}$ is an N-component vector with one component for each of the N particles in the reservoir. The interaction Hamiltonian H_I is defined by

$$H_I = \sum_i v_i(x, R_i) \tag{94}$$

As discussed in Section 11.4, the coordinate-space density matrix of the density operator

$$\hat{\rho}(t) = \exp\left(\frac{-iHt}{\hbar}\right)\hat{\rho}(0)\exp\left(\frac{iHt}{\hbar}\right) \tag{95}$$

may be represented as

$$\langle x, \mathbf{R}|\hat{\rho}(t)|y, \mathbf{Q}\rangle = \int dx'\, dy'\, d\mathbf{R}'\, d\mathbf{Q}'$$

$$\times \left\langle x, \mathbf{R}|\exp\left(\frac{-iHt}{\hbar}\right)|x', \mathbf{R}'\right\rangle\langle x', \mathbf{R}'|\hat{\rho}(0)|y', \mathbf{Q}'\rangle$$

$$\times \left\langle y', \mathbf{Q}'|\exp\left(\frac{iHt}{\hbar}\right)|y, \mathbf{Q}\right\rangle \tag{96}$$

where

$$\left\langle x, \mathbf{R}|\exp\left(\frac{-iHt}{\hbar}\right)|x', \mathbf{R}'\right\rangle = K(x, \mathbf{R}, t; x', \mathbf{R}', 0)$$

$$= \int\int \mathcal{D}x\, \mathcal{D}\mathbf{R}\, \exp\left\{\frac{i}{\hbar}\, S[x, \mathbf{R}]\right\} \tag{97a}$$

and

$$\left\langle y', \mathbf{Q}'|\exp\left(\frac{iHt}{\hbar}\right)|y, \mathbf{Q}\right\rangle = K^*(y, \mathbf{Q}, t; y', \mathbf{Q}', 0)$$

$$= \int\int \mathcal{D}y\, \mathcal{D}\mathbf{Q}\, \exp\left\{-\frac{i}{\hbar}\, S[y, \mathbf{Q}]\right\} \tag{97b}$$

The end-points of these path integrals are $x = x(t)$, $x' = x(0)$, $y = y(t)$, $y' = y(0)$, $\mathbf{R} = \mathbf{R}(t)$, $\mathbf{R}' = \mathbf{R}(0)$, $\mathbf{Q} = \mathbf{Q}(t)$, and $\mathbf{Q}' = \mathbf{Q}(0)$. In Eq. (97a) and (97b), the action S is defined by

$$S = S_A + S_I + S_R = \int_0^t (\mathcal{L}_A + \mathcal{L}_I + \mathcal{L}_R)dt' \tag{98}$$

where

$$\mathcal{L}_A = \frac{1}{2}\, M\dot{x}^2 - v(x) \tag{99a}$$

$$\mathcal{L}_R = \sum_i \frac{1}{2}\, m\dot{\mathbf{R}}^2 - \frac{1}{2}\sum_{i \neq j} v_R(R_i, R_j) \tag{99b}$$

$$\mathcal{L}_I = -\sum_i v_i(x, R_i) \tag{99c}$$

By use of Eq. (97a) and (97b), it is possible to write Eq. (96) as [12]:

$$\langle x, \mathbf{R} | \hat{\rho}(t) | y, \mathbf{Q} \rangle = \int dx' \, dy' \, d\mathbf{Q}' \, d\mathbf{R}' K(x, \mathbf{R}, t; x', \mathbf{R}', 0)$$

$$\times K^*(y, \mathbf{Q}, t; y', \mathbf{Q}', 0) \langle x', \mathbf{R}' | \hat{\rho}(0) | y', \mathbf{Q}' \rangle \tag{100}$$

where the density operator is taken to be the product of the density operator for system A and for the reservoir

$$\hat{\rho}(0) = \hat{\rho}_A(0) \hat{\rho}_R(0) \tag{101}$$

so that the density operator includes information on both system A and the reservoir. The reduced density matrix describing system A and the *influence* of the reservoir on system A is given by Eqs. (24) and (100):

$$\tilde{\rho}(x, y, t) = \int d\mathbf{R} \langle x, \mathbf{R} | \hat{\rho}(t) | y, \mathbf{R} \rangle$$

$$= \int dx' \, dy' \, d\mathbf{R}' \, d\mathbf{Q}' \, d\mathbf{R} K(x, \mathbf{R}, t; x', \mathbf{R}', 0)$$

$$\times K^*(y, \mathbf{R}, t; y', \mathbf{Q}', 0) \langle x', \mathbf{R}' | \hat{\rho}(0) | y', \mathbf{Q}' \rangle \tag{102}$$

With the relations defined by Eqs. (97–99), Eq. (102) may be written as [12]:

$$\tilde{\rho}(x, y, t) = \int dx' \, dy' \, J(x, y, t; x', y', 0) \rho_A(x', y', 0) \tag{103}$$

In this result, the propagator for the reduced density matrix is given by

$$J(x, y, t; x', y', 0) = \int \int \mathscr{D}x \, \mathscr{D}y \, \exp\left(\frac{iS_A[x]}{\hbar}\right) \exp\left(\frac{-iS_A[y]}{\hbar}\right) \mathscr{F}[x, y] \tag{104}$$

where the influence functional

$$\mathscr{F}[x, y] = \int d\mathbf{R}' \, d\mathbf{Q}' \, d\mathbf{R} \rho_R(\mathbf{R}', \mathbf{Q}', 0)$$

$$\times \int \int \mathscr{D}\mathbf{R} \, \mathscr{D}\mathbf{Q} \exp\left\{\frac{i[S_I[x, \mathbf{R}] - S_I[y, \mathbf{Q}] + S_R[\mathbf{R}] - S_R[\mathbf{Q}])}{\hbar}\right\} \tag{105}$$

describes the influence of the reservoir R on system A.

Equations (103–105) summarize key results of the Feynman–Vernon influence functional theory. The Feynman–Vernon influence functional theory is the basis for the Caldeira–Leggett model [12] of quantum Brownian motion. In this model

$$H_A = \frac{p^2}{2M} + v(x) \tag{106a}$$

$$H_R = \sum_k \frac{p_k^2}{2M} + \sum_k \frac{1}{2} m\omega_k^2 R_k^2 \tag{106b}$$

$$H_I = x \sum_k C_k R_k \tag{106c}$$

where ω_k is the harmonic frequency of the kth reservoir particle and C_k is the constant describing the coupling between the particle of mass M (system A) and the kth reservoir particle. The influence functional of Eq. (105) is known exactly [22, 23] for the Hamiltonian of Eq. (106a–c) when the reservoir is in equilibrium at temperature T [12]:

$$\mathscr{F}[x, y] = \exp\left\{ -\frac{i}{\hbar} \int_0^t \int_0^\tau [x(\tau) - y(\tau)][\alpha(\tau - s)x(s) - \alpha^*(\tau - s)y(s)]d\tau \, ds \right\} \tag{107}$$

where

$$\alpha(\tau - a) = \sum_k \frac{C_k^2}{2m\omega_k} \left[\exp(-i\omega_k(\tau - s)) + \frac{\exp(i\omega_k(\tau - s))}{\exp\left(\dfrac{\hbar\omega_k}{kT}\right) - 1} \right.$$
$$\left. + \frac{\exp(-i\omega_k(\tau - s))}{\exp\left(\dfrac{\hbar\omega_k}{kT}\right) - 1} \right] \tag{108}$$

this result assumes that the initial density matrix $\rho_R(0)$ pertains to a system in equilibrium [12, 23]:

$$\rho_R(\mathbf{R}', \mathbf{Q}', 0) = \prod_k \rho_R^{(k)}(\mathbf{R}_k', \mathbf{Q}_k', 0) \tag{109}$$

where

$$\rho_R^{(k)}(\mathbf{R}_{jk}', \mathbf{Q}_k', 0) = \frac{m\omega_k}{2\pi\hbar \sinh(\hbar\omega_k/kT)}$$

$$\times \exp\left[\frac{-m\omega_k}{2\pi \sinh^2(\hbar\omega_k/kT)} \left\{ (R_k'^2 + Q_k'^2) \cosh\left(\frac{\hbar\omega_k}{kT}\right) - 2R_k'Q_k' \right\} \right] \tag{110}$$

Following Zwanzig [31], Caldeira and Leggett introduced a phenomenological damping constant η as well as a continuum density $\rho_D(\omega)$ of reservoir oscillators through the relations

$$\rho_D(\omega)C^2(\omega) = \begin{cases} \dfrac{2m\eta\omega^2}{\pi}, & \omega < \Omega \qquad \text{(111a)} \\[2ex] 0, & \omega > \Omega \qquad \text{(111b)} \end{cases}$$

where Ω is the high-frequency cutoff for the oscillator distribution. Introducing a relaxation constant γ and a frequency shift $\Delta\omega$ through

$$\gamma = \frac{\eta}{2M} \tag{112}$$

$$(\Delta\omega)^2 = \frac{4\gamma\Omega}{\pi} \tag{113}$$

Caldeira and Leggett [12] demonstrate that

$$J(x, y, t; x', y', 0)$$

$$= \int\int \mathcal{D}x\, \mathcal{D}y \exp \frac{i}{\hbar} \left\{ S'_A[x] - S'_A[y] - M\gamma \int_0^t (x\dot{x} - y\dot{y} + x\dot{y} - y\dot{x})\, d\tau \right\}$$

$$\times \exp - \frac{1}{\hbar} \frac{2M\gamma}{\pi} \int_0^\Omega \omega \coth \frac{\hbar\omega}{kT}$$

$$\times \int_0^t\int_0^\tau [x(\tau) - y(\tau)] \cos \omega(\tau - s)[x(s) - y(s)]\, d\tau\, ds\, d\omega \tag{114}$$

where the renormalized action S'_A results from subtracting the term $\frac{1}{2}M(\Delta\omega)^2 x^2$ from the potential $v(x)$:

$$S'_A = \int_0^t [\tfrac{1}{2}M\dot{x}^2 - v(x)]\, d\tau + \int_0^t \tfrac{1}{2}M(\Delta\omega)^2 x^2\, d\tau$$

$$= \int_0^t [\tfrac{1}{2}M\dot{x}^2 - v'(x)]\, d\tau \tag{115}$$

In this result

$$v'(x) = v(x) - \tfrac{1}{2}M(\Delta\omega)^2 x^2 \tag{116}$$

where $v(x) = \frac{1}{2}M\omega^2 x^2$ the renormalized frequency is defined by

$$\omega' = \sqrt{\omega^2 - (\Delta\omega)^2}. \tag{117}$$

Equation (114) is the propagator for the reduced density matrix $\tilde{\rho}(x, y, t)$, mass M, interacting with the reservoir R, of harmonic dissipation centers. To derive a differential equation describing the evolution of $\tilde{\rho}(x, y, t)$, Caldeira and Leggett [12] considered

$$\tilde{\rho}(x, y, t + \varepsilon) = \int \int dx' \, dy' \, J(x, y, t + \varepsilon; x', y', t)\tilde{\rho}(x', y', t) \quad (118)$$

and expand to the high-temperature propagator

$$J(x, y, t; x', y', 0)$$

$$= \int \int \mathscr{D}x \, \mathscr{D}y \exp \frac{i}{\hbar} \left\{ S_A'[x] - S_A'[y] - M\gamma \int_0^t [x\dot{x} - y\dot{y} + x\dot{y} - y\dot{x}] \, d\tau \right\}$$

$$\times \exp - \frac{2M\gamma kT}{\hbar^2} \int_0^t [x(\tau) - y(\tau)]^2 \, d\tau \quad (119)$$

about t for infinitesimal values of ε to obtain the first-order term in ε:

$$\frac{\partial \tilde{\rho}}{\partial t} = -\frac{\hbar}{2Mi} \frac{\partial^2 \tilde{\rho}}{\partial x^2} + \frac{\hbar}{2Mi} \frac{\partial^2 \tilde{\rho}}{\partial y^2} - \gamma(x - y) \frac{\partial \tilde{\rho}}{\partial x} + \gamma(x - y) \frac{\partial \tilde{\rho}}{\partial y} + \frac{v'(x)}{i\hbar} \tilde{\rho}$$

$$- \frac{v'(y)}{i\hbar} \tilde{\rho} - \frac{2M\gamma kT}{\hbar} (x - y)^2 \tilde{\rho} \quad (120)$$

the high-temperature limit of Eq. (119) is defined by

$$2kT \geq \hbar\Omega \gg \hbar\omega' . \quad (121)$$

Equation (120) is the coordinate-space matrix element representation of the more general operator equation in $\hat{\tilde{\rho}}$ [12]:

$$\frac{\partial \hat{\tilde{\rho}}}{\partial t} = \frac{1}{i\hbar} [H', \hat{\tilde{\rho}}] + \frac{\gamma}{2i\hbar} [\{p, x\}, \hat{\tilde{\rho}}] - \frac{1}{\hbar^2} D[x[x, \hat{\tilde{\rho}}]]$$

$$+ \frac{\Delta}{\hbar^2} ([x, \hat{\tilde{\rho}}p] - [p, \hat{\tilde{\rho}}x]) \quad (122)$$

with

$$H' = \frac{p}{2M} + v'(x) \quad (123)$$

$$D = \eta kT \quad (124)$$

$$\Delta = -i\hbar\gamma \quad (125)$$

In Eq. (122), commutators are denoted by [,], while anticommutators are indicated by { , }.

By taking coordinate-space matrix elements of Eq. (122), Eq. (120) is recovered as the equation describing the evolution of the coordinate-space elements of the reduced density matrix $\tilde{\rho}$.

The Wigner transform of Eq. (58) may be expressed in terms of one-particle reduced quantities by making the replacements

$$P(x, p, t) \rightarrow w(x, p, t) \tag{126a}$$

$$\rho(x, y, t) \rightarrow \tilde{\rho}(x, y, t) \tag{126b}$$

With these definitions, the Wigner transform of Eq. (122) results in the phase-space Wigner distribution equation [12, 32]; in the limit $\hbar \rightarrow 0$

$$\frac{\partial w}{\partial t} + \frac{\partial(pw)}{\partial x} + \frac{\partial(Fw)}{\partial p} = 2\gamma \frac{\partial(pw)}{\partial p} + D \frac{\partial^2 w}{\partial p^2} \tag{127}$$

where

$$F = -\frac{\partial v'(x)}{\partial x} \tag{128}$$

is the classical force corresponding to the renormalized potential $v'(x)$. The right-hand side of Eq. (127) is the well-known Fokker–Planck collision operator; thus, the Caldeira–Leggett model reduces to the appropriate classical result in the limit $\hbar \rightarrow 0$.

From these results, it is clear that the Caldeira–Leggett model is successful in recovering the correct classical results from a fully quantum mechanical treatment of quantum dissipation; this treatment is, however, based on introducing phenomenological constants such as the damping constant η. In the continuing efforts to model quantum dissipation in ultrasmall electronic devices, models with features in common with the Caldeira–Leggett formalism are leading candidates for comprehensive device models.

Frensley [11] has performed numerical simulations relevant to the description of resonant tunneling in double-barrier quantum well diodes. As discussed previously [6, 7], the resonant tunneling current measured in double-barrier quantum well diodes exhibits NDR. To model these devices, Frensley [11] has not employed the full Caldeira–Leggett model; instead, dissipative interactions are modeled by the inclusion of a relaxation term with a relaxation time τ. The dynamical equation for the one-particle, reduced density matrix is (cf. Eq. (23)):

$$\frac{\partial \rho}{\partial t} = \frac{1}{i\hbar} [H, \rho] - \frac{1}{\tau} (\rho - \rho_0) \tag{129}$$

where H contains terms for the kinetic energy, the band energy potential of the quantum well heterostructure, and the electrostatic potential due to the externally applied voltages as well as due to the externally applied voltages as well as due to the charge of carrier and dopants. In Eq. (129), ρ_0 is the equilibrium density matrix. For a summary of the numerical results obtained by Frensley, the reader is referred to Ref. 11.

11.7.2 High-field Transport of Electrons in Polar Semiconductors

The problem of determining the velocity acquired by an electron in a finite electric field in a polar semiconductor [33] both illustrates the importance of quantum transport theory and demonstrates the inadequacy of conventional perturbation and Boltzmann transport techniques. In polar semiconductors, electrons under the influence of applied electric fields may be coupled strongly to the longitudinal optical (LO) phonon modes of the crystal lattice.

Since the energy of an LO phonon in polar semiconductors may be on the order of 0.1 eV, the mean free path for phonon emission is only a few angstroms and mean free emission time is on the order of a femtosecond. Under these conditions, quantum interference between emitted LO phonons must be included in transport calculations. Since the approximation of independent phonon emissions is invalid, accurate calculations must not be based on the assumptions of (1) localized Boltzman-like scattering events or (2) independent LO-phonon emissions, as in standard calculations based on the use of Fermi's Golden Rule.

To overcome these shortcomings, Thornber and Feynman [33] have used density matrix and path-integral techniques to estimate the velocity acquired by an electron in a finite applied electric field in a polar semiconductor. In particular, Ref. 33 demonstrates that the steady-state rate of LO-phonon emission minus LO-phonon absorption, at wave vector **k** is given by

$$\langle \hat{R}_k \rangle = \text{trace}(\rho \hat{R}_k) = \int \int \mathcal{D}\mathbf{x} \, \mathcal{D}\mathbf{x}' \exp i\Phi \qquad (130)$$

where $\Phi = S[x, x']/\hbar$ and the action $S[x, x']$ is derived from the Lagrangian for an electron in an electric field interacting with the polar semiconductor lattice through the Frohlich electron–LO-phonon interaction. In Eq. (130), R_k is defined in terms of the electron–LO-phonon coupling strength C_k and the LO-phonon creation operator $a_k^\dagger$ and annihilation operator a_k. Reference 33 provides a detailed description of each of these quantities and presents numerical results based on the evaluation of Eq. (130) by a quadratic, variational approximation.

11.7.3 Monte Carlo Simulations Based on Path-integral Techniques

Classical Monte Carlo simulation techniques are used widely to describe carrier transport in semiconductors and insulators [34]. These classical simulations assume carrier–scatterer interactions are sufficiently weak that first-order perturbation treatments are adequate. Another basic assumption of classical Monte Carlo simulations is that scattering events are localized in space and time; that is, it is assumed that the carrier mean free path is large compared with the carrier de Broglie wavelength. When these conditions are satisfied with some degree of accuracy, collision events may be treated as instantaneous, independent events and Boltzmann transport theory offers a potential solution to the carrier transport problem. In such cases, classical Monte Carlo techniques rely on the random generation of many classical trajectories to approximate the dynamics of an ensemble of charge carriers. In ultrasmall electronic structures, device features may be less than the carrier de Broglie wavelength, and electric fields may exceed 10^5 V/cm, with the consequence that electron–LO-phonon interactions in polar semiconductors may not be treated as independent, localized events. In short, the extremely successful techniques of classical Monte Carlo simulation are inadequate to model carrier transport in ultrasmall electronic structures.

Recently, Fischetti and DiMaria have presented [35] a quantum Monte Carlo scheme suitable for simulating high-field electron transport. In this scheme, the use of random trajectories in the classical Monte Carlo approach is replaced by a sampling technique permitting a stochastic evaluation of the quantum mechanical electron propagator. To treat the quantum mechanical phase and the phase interference between different trajectories properly, the quantum mechanical propagator is expressed in terms of the Feynman path integral over all possible paths from $(\mathbf{r}, t)$ to $(\mathbf{r}', t')$ [35]:

$$K(\mathbf{r}', t'; \mathbf{r}, t) = \int \int \mathscr{D}\mathbf{r}(\tau)\mathscr{D}\mathbf{k}(\tau)e^{iS(\mathbf{r}', t'; \mathbf{r}, t)/\hbar} \qquad (131)$$

where the effective action S is taken to be

$$S(\mathbf{r}', t', \mathbf{r}, t) = \int\limits_{(\mathbf{r}, t)}^{(\mathbf{r}', t')} d\tau[\hbar\mathbf{k}(\tau) \cdot \mathbf{r}(\tau) - \varepsilon_{\mathbf{k}}(\tau) - e\mathbf{E} \cdot \mathbf{r}(\tau)] \qquad (132)$$

In Eq. (132), $\mathbf{E}$ represents a uniform electric field and $\varepsilon_{\mathbf{k}}(\tau)$ is the renormalized energy of a quasi-particle of wave vector $\mathbf{k}$.

Fischetti and DiMaria have applied this quantum Monte Carlo technique to model high-field electron transport in silicon dioxide [35] and to model interference and phonon effects in experimental studies of ballistic transport in silicon dioxide [36]. In such applications, the renormalized quasi-particle energy $\varepsilon_{\mathbf{k}}(\tau)$ is the sum of the bare energy $W_{\mathbf{k}}$ of an electron of wave vector $\mathbf{k}$ and the proper electron self-energy $\Sigma(\mathbf{k}, \varepsilon_{\mathbf{k}})$ describing electron–phonon interactions:

$$\varepsilon_{\mathbf{k}} = W_{\mathbf{k}} + \sum (\mathbf{k}, \varepsilon_{\mathbf{k}}) \qquad (133)$$

In Eq. (133), the electron self-energy is estimated by use of the Dyson equation technique to sum all relevant electron–phonon interactions [35]; in this way, electron–phonon interactions are treated as interacting many-particle events.

The application of quantum Monte Carlo techniques to model carrier transport in ultrasmall electronic devices is in its infancy; few quantum Monte Carlo simulations have been attempted. It is clear, however, that quantum Monte Carlo techniques must be developed to model a wide variety of envisioned quantum-based electronic devices accurately.

11.7.4   Numerical Evaluation of the Wigner Function for Resonant Tunneling in Quantum Well Structures

The transient response of a quantum well resonant tunneling diode has recently been modeled [37] by numerically evaluating the Wigner distribution function. In these calculations, Frensley [37] successfully models the NDR of the quantum well resonant tunneling diode. For typical quantum well diode dimensions, the model predicts a peak-to-valley current switching time of about 200 fs.

The dynamical equation for the collisionless Wigner equation may be written as [8, 37] (cf. Eqs. (50a) and (50b)):

$$\frac{\partial P(x, k)}{\partial t} + \frac{\hbar k}{m} \frac{\partial P(x, k)}{\partial x} + \left(\frac{1}{2\pi}\right) \int dk' V(x, k - k') P(x, k') = 0 \quad (134)$$

where

$$V(x, k) = \frac{2}{\hbar} \int\limits_0^\infty dy \, \sin(ky)[v(x + \tfrac{1}{2}y) - v(x - \tfrac{1}{2}y)] \quad (135)$$

To approximate the open-system properties of the quantum well diode, Frensley [37] has treated the diode contacts as ideal particle reservoirs. Taking the diode boundaries shown in Fig. 11.1, along the x-direction, to be at $x = -a/2$ and at $x = +a/2$, the ideal-particle-reservoir boundary conditions require $P(-a/2, k)$, for $k > 0$, to be equal to distribution function of the left-hand reservoir; likewise, $P(a/2, k)$ for $k < 0$ must equal the distribution function of the right-hand reservoir, Frensley has applied these conditions under the assumption that both reservoirs are in equilibrium:

$$P\left(\frac{-a}{2}, k\right) = F(\mu_l, T_l), \qquad k > 0 \quad (136a)$$

$$P\left(\frac{a}{2}, k\right) = F(\mu_r, T_r), \qquad k < 0 \quad (136b)$$

where F is the Fermi distribution function integrated over transverse

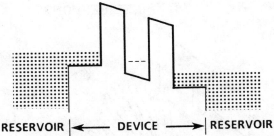

Figure 11.1 Energy band structure for the double-barrier quantum well diode of the type considered in Ref. 37. In the calculations this referrence, $Al_{0.3}Ga_{0.7}As$ barriers are 28 Å thick and the GaAs quantum well is 45 Å thick. The size-quantized state in the quantum well is indicated by the dashed line. The ideal particle reservoirs are shown to the left and right of the active region of the diode. (From Frensley [37]. Reprinted, with permission, from *Physical Review Letters*.)

momenta. In Eq. (136a), μ_l and T_1 are the Fermi level and equilibrium temperature, respectively, of the left-hand reservoir; similarly, in Eq. (136b), μ_r and T_r are the Fermi level and equilibrium temperature, respectively, of the right-hand reservoir. Figure 11.2 depicts the current–voltage curves for the quantum well diode potential of Fig. 11.3 for two cases: the Wigner function calculation and the standard scattering theory calculation based on the usual quantum mechanical treatment of transmission and

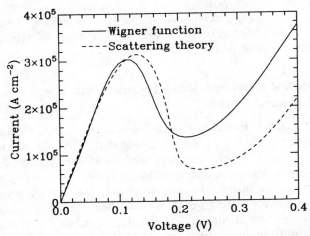

Figure 11.2 Current–voltage curves for the calculations described in the caption to Fig. 11.1. The Wigner function calculation is done for the case where the reservoir boundary conditions correspond to a temperature of 300 K. The scattering theory result is based on the usual quantum mechanical calculation where transmitted and reflected waves are matched at each boundary. (From Frensley [37]. Reprinted, with permission, from *Physical Review Letters*.)

TRANSIENT RESPONSE

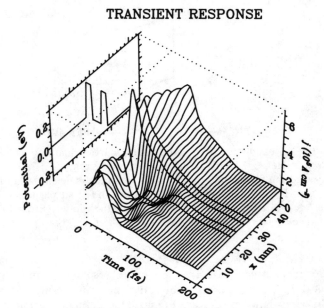

Figure 11.3 Transient response for the calculations described in the caption to Fig. 11.1. (From Frensley [37]. Reprinted, with permission, from *Physical Review Letters*.)

reflection from a given potential profile. Figure 11.3 displays Frensley's results for a dynamical calculation of the current–voltage characteristics; time scales of approximately 100 fs are predicted for switching the diode from maximum to minimum current.

11.7.5 Numerical Evaluation of the Feynman Path Integral for Dissipative Quantum Mechanical Systems

The Caldeira–Leggett model of dissipative quantum mechanical systems [12] provides a general framework for modeling dissipative interactions in electronic devices such as the quantum well diode. Recently, Schmid [38] formulated a quantum mechanical model for a particle coupled to a dissipative environment; in this model, the stochastic interaction associated with the environment is modeled so that its power spectrum is in accord with the fluctuation–dissipation theorem. The model of Schmid [38] bears a close similarity to the Caldeira–Leggett model [12].

Following the Feynman–Vernon theory [22], Schmid [38] takes the propagator for the density matrix (cf. Eqs. (103) and (118)) to be,

$$J(x, y, t; x', y', t') = \int \mathcal{D}x \, \mathcal{D}y \, \exp\left\{ \frac{i}{\hbar} \left(S[x] - S[y] \right) \right\} \mathcal{F}[x, y] \quad (137)$$

where

$$S[x, t] = \int_{t'}^{t} d\tau [\tfrac{1}{2} M\dot{x}^2 - V(x)] \tag{138}$$

and

$$\mathcal{F}[x, y] = \exp\left\{ \frac{-iM\gamma}{2\hbar} \int_{t'}^{t} d\tau [x(\tau) - y(\tau)][\dot{x}(\tau) + \dot{y}(\tau)] \right.$$

$$- \frac{M\gamma kT}{\hbar^2} \int_{t'}^{t} d\tau\, d\tau' [x(\tau) - y(\tau)] R(\tau - \tau')$$

$$\left. \times [x(\tau') - y(\tau')] \right\} \tag{139}$$

In these expressions, M, γ, k, and T have the same definitions as in Eq. (119). In Eq. (139), the correlation function $R(t)$ describes memory effects of the interaction with the bath and is given by [38]:

$$R(t) = \int \frac{d\nu}{2\pi} K(\nu) e^{i-\nu t} \tag{140a}$$

where

$$K(\nu) = \frac{\hbar\nu}{2kT} \coth\left(\frac{\hbar\nu}{2kT} \right) \tag{140b}$$

That is

$$R(t - t') = \frac{1}{2M\gamma kT} \langle \xi(t)\xi(t') \rangle \tag{141a}$$

where $\xi(t)$ describes a stochastic Gaussian process in the generalized Langevin equation:

$$M\ddot{x} = -M\gamma\dot{x} - \frac{\partial V(x)}{\partial x} + \xi(t). \tag{141b}$$

As described by Feynman and Vernon [22], the influence-functional treatment provides a means of modeling the interaction between a single electron and a system of phonons. For suitable values of the constants in Eqs. (137–141b), Eq. (139) describes the interaction of phonon modes of all orders on the electronic density matrix.

The path-integral formulation of Schmid's model facilitates the nonperturbative treatment of the multiphonon effects dominating the high-field

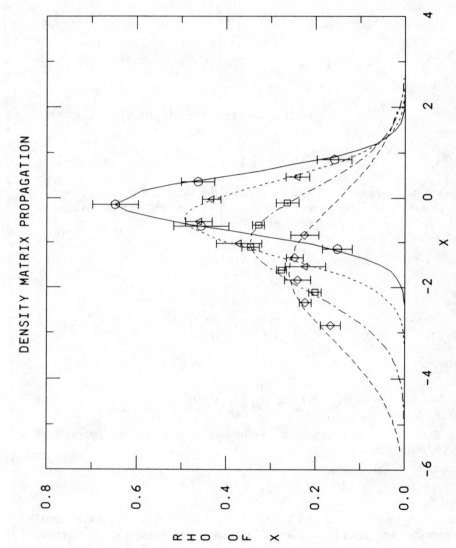

Figure 11.4 Density matrix results for the Monte Carlo path-integral calculations described in Ref. 39. (From Hess, et al. [39]. Reprinted, with permission, from *Superlattices and Microstructures*.)

energy loss in polar semiconductors. Indeed, Mason et al. [39] have recently adapted Schmid's model to formulate a numerical method of calculating the dynamical properties of the electronic density matrix for the coupled electron–phonon system; in this model, the semiconductor lattice is modeled as a many-phonon bath. This numerical model is based on Monte Carlo sampling of electronic trajectories and is suitable for treating the oscillatory nature of the propagator. Reference 39 describes the application of this numerical model to two physical systems: a classically damped harmonic oscillator and electron transport in an external field.

In the case of electron transport in an external field, the diagonal components of the electronic density matrix are computed for an electron subject to an electric field strong enough to induce multiple-phonon scattering effects. By selective sampling of trajectories, Mason et al. are able to compute the time-dependent diagonal components of the density matrix for this system at four times: $t = (1/4)\gamma$, $t = (1/2)\gamma$, $t = (3/4)\gamma$, and $t = 1/\gamma$ (Fig. 11.4). The results obtained are physically realistic and the successful application of quantum transport theory to the many-phonon problem represents a major advance in the theory of quantum-based electronic devices.[†]

REFERENCES

1. L. Esaki and R. Tsu, *IBM J. Res. Dev.* **14**, 61 (1970).
2. G. J. Iafrate, H. L. Grubin, and D. K. Ferry, *J. Phys. C.* **7**, 307 (1981).
3. M. A. Reed, R. T. Bate, K. Bradshaw, W. M. Duncan, W. R. Frensley, J. W. Lee, and H. D. Shin, *Proc. Int. Symp.* Electron, Ion, Proton Beams, 29th *J. Vac. Sci. Technol. B* [2] **4**, 358 (1986).
4. P. J. Stiles, *Surf. Sci.* **73**, 252 (1978).
5. A. Riesman, *Proc. IEEE* **71**, 550 (1983).
6. T. C. L. G. Sollner , W. D. Goodhue, P. E. Tannenwald, C. D. Parker, and D. D. Peck, *Appl. Phys. Lett.* **43**, 588 (1983).
7. T. C. L. G. Sollner, W. D. Goodhue, P . E. Tannenwald, C. D. Parker, D. D. Peck, and H. Q. Le, *IEEE Conf. Millimeter Waves Microwaves* **CH1917-4**, T51 (1983).
8. E. P. Wigner, *Phys. Rev.* **40**, 749 (1932); M. Hillney, R. F. O'Connell, M. O. Scully, and E. P. Wigner, *Phys. Rep.* **106**, 121 (1984).
9. H. L. Grubin, D. K. Ferry, G. J. Iafrate, and J. R. Barker, *in "VLSI Electronics"* (N. G. Einspruch, ed.), Vol. 3, Academic Press, New York, 1982.
10. M. Ploszajczak and M. J. Rhoades-Brown, *Phys. Rev. Lett.* **55**, 147 (1985).
11. W. R. Frensley, *J. Vac. Sci. Technol. B* [2] **3**, 1261 (1985); for a general treatment of the density matrix, see K. Blum, "Density Matrix Theory and Application." Plenum, New York, 1981.

[†](Editor's Note) For recent developments, see Ref. 40.

12. A. O. Caldeira and A. J. Leggett, *Physica A* (*Amsterdam*) **121A**, 587 (1983); *Ann. Phys.* (*N.Y.*) **149**, 374 (1983); **153**, 445(E) (1984).

13. M. A. Stroscio, *Superlattices Microstruct.* **2**, 45, 83 (1986).

14. S. Luryi, *Appl. Phys. Lett.* **47**, 490 (1985).

15. K. Meehan, N. Holonyak, Jr., J. M. Brown, M. A. Nixon, P. Gavrilovic, and R. D. Burnham, *Appl. Phys. Lett.* **45**, 549 (1984).

16. M. D. Camras, N. Holonyak, Jr., R. D. Burnham, W. Streifer, D. R. Scifres, T. L. Pooli, and C. Lindstrom, *J. Appl. Phys.* **54**, 5637 (1983).

17. K. Hirakawa, H. Sakaki, and J. Yoshino, *Phys. Rev. Lett.* **54**, 1279 (1985).

18. R. F. Kwasnick, M. A. Kastner, J. Melngailis, and P. A. Lee, *Phys. Rev. Lett.* **52**, 224 (1984); P. A. Lee, *ibid.* **53**, 2042 (1984); A. B. Fowler, A. Hartstein, and R. A. Webb, *ibid.* **48**, 196 (1982); R. G. Wheeler, K. K. Choi, A. Goel, R. Wisnieff, and D. E. Prober, *ibid* **49**, 1674 (1982); W. J. Skocpol, L. D. Jackel, E. L. Hu, R. E. Howard, and L. A. Fetter, *ibid.* p. 951.

19. S. W. Kirchoefer, R. Magno, and J. Comas, *Appl. Phys. Lett.* **44**, 1054 (1984).

20. R. A. Davies, M. J. Kelly, and T. M. Kerr, *Phys. Rev. Let.* **55**, 1114 (1985).

21. R. P. Feynman, *Rev. Mod. Phys.* **20**, 367 (1948).

22. R. P. Feynman and F. L. Vernon, Jr., *Ann. Phys.* (*N.Y.*) **24**, 118 (1963).

23. R. P. Feynman and A. R. Hibbs, "*Quantum Mechanics and Path Integrals.*" McGraw-Hill, New York, 1965; R. P. Feynman, R. B. Leighton and M. Sands, "*The Feynman Lectures on Physics.*" Addison-Wesley, Reading, MA, 1964.

24. J. R. Nix, in "*Progress in Particle and Nuclear Physics*" (D. Wilkinson, ed.), Vol. 2, Pergamon, New York, 1979.

25. M. P. A. Fisher and A. T. Dorsey, Phys. Rev. Lett. **54**, 1609 (1985); H. Grabert and U. Weiss, *ibid.* p. 1605.

26. P. M. Platzmann and P. A. Wolff, *Solid State Phys.*, *Suppl.* **13**,   (1973).

27. Equation (81) corresponds to Eq. (9) of Ref. 10: special thanks are due to M. J. Rhoades-Brown for confirming that Eq. (9) of Ref. 10 contains two typographical errors.

28. E. Moyal, *Proc. Cambridge Philos. Soc.* **45**, 99 (1949).

29. S. Washburn, R. A. Webb, R. V. Voss, and S. M. Faris, *Phys. Rev. Lett.* **54**, 2712 (1985).

30. T. C. L. G. Sollner, W. D. Goodhue, P. E. Tannenwald, C. D. Parker, and D. D. Peck, *Appl. Phys. Lett.* **43** 588 (1983); S. W. Kirchoefer, R. Magno, and J. Comas, *ibid.* **44**, 1054 (1984); R. A. Davies, M. J. Kelly, and T. M. Kerr, *Phys. Rev. Lett.* **55**, 1114 (1985).

31. R. Zwanzig, *J. Stat. Phys.* **9**, 215 (1973); R. Zwanzig, *J. Chem. Phys.* **33**, 1338 (1960).

32. H. Dekker, *Phys. Rev. A* **16**, 2116 (1977).

33. K. K. Thornber and R. P. Feynman, *Phys. Rev. B: Solid State* [3] **1**, 4099 (1970).

34. C. Jacoboni and L. Reggiani, *Rev. Mod. Phys.* **55**, 645 (1983).

35. M. V. Fischetti and D. J. DiMaria, *Phys. Rev. Lett.* **55**, 2475 (1985).

36. D. J. DiMaria, M. V. Fischetti, J. Batey, L. Dori, E. Tierney, and J. Stasiak, *Phys. Rev. Lett.* **57**, 3213 (1986).

37. W. R. Frensley, *Phys. Rev. Lett.* **57**, 2853 (1986); W. R. Frensley, *Phys. Rev. Lett.* **60**, 1589 (1988).

38. A. Schmid, *J. Low Temp. Phys.* **49**, 609 (1982).

39. B. A. Mason, K. Hess, R. E. Cline, Jr., and P. G. Wolynes, *Superlattices Microstruct.* **3**, 421 (1987).

40. "Hot carriers in semiconductors." *Solid-State Electronics*, vol. **31**, No. 314 (1988); N. C. Kluksdahl, A. M. Kriman, C. Ringhofer and D. K. Ferry, *Solid State Electron.* **31**, 743 (1988); R. K. Mains and G. I. Haddad, *J. Appl. Phys.* **64**, 3564 (1988); B. A. Mason and Karl Hess, *Physical Review* **B39**, 5051 (1989); L. F. Register, M. A. Stroscio and M. A. Littlejohn, *Superlattices and Microstructures*, **4**, 61–68 (1988).

INDEX